# FRACTURE SCALING

# FRACTURE SCALING

Edited by

ZDENĚK P. BAŽANT

*Department of Civil Engineering,*
*Northwestern University,*
*Evanston, Illinois, U.S.A.*

and

YAPA D.S. RAJAPAKSE

*Office of Naval Research,*
*Arlington, Virginia, U.S.A.*

Reprinted from *International Journal of Fracture*, Vol. 95, 1999

SPRINGER SCIENCE+BUSINESS MEDIA, B.V.

**A C.I.P. Catalogue record for this book is available from the Library of Congress.**

ISBN 978-94-010-5965-7 ISBN 978-94-011-4659-3 (eBook)
DOI 10.1007/978-94-011-4659-3

*Printed on acid-free paper*

## TABLE OF CONTENTS

*Part 4. Computational Aspects and Nonlocal or Gradient Models*



*Part 5. Size Effects in Concrete, Ice and Soils*

International Journal of Fracture **95:** vii–viii, 1999.


# Preface

The problem of scaling of structural failure caused by propagation of fracture or softening damage has recently acquired prominence in research. One reason for this has been the need to extrapolate from laboratory tests to much larger real structures, as well as the recognition that the strength of a brittle or quasibrittle structure may be significantly impaired when its size is increased beyond the usual range of dimensions. Another reason has been the need to bridge different scales in material modeling and to connect the models of physical phenomena occurring on different scales of the microstructure, which is a crucial task in the development of physically based material models.

The problem of size effect has a long history which can be traced back to the notebooks of Leonardo da Vinci in the Renaissance. Galileo speculated on the size effect in ropes and animal bones. The qualitative foundations of the statistical size effect have been laid down already in the 17th century, by Marriotte. However, little subsequent progress has been made until the first half of this century, in which the development of the weakest-link statistical model and introduction of extreme value statistics led to the completion of the theory of statistical size effect due to strength randomness by Weibull in 1939. After that, for almost half a century, the problem of size effect has not attracted any particular attention of solid mechanicians, who regarded it as a subject that should be relegated to statisticians.

The first hints that important size effects on structural strength that are not of statistical origin may exist appeared in the early 1970's in the studies of concrete. Since the mid 1980's, a large body of the literature has been accumulated on size effects due to stable propagation of fractures with a large fracture process zone. Furthermore, during the last decade, considerable advances have been made in the understanding of scale effects on the level of crystals in polycrystalline materials and down to atomistic dimensions. It has also been recognized that size effects must have played a significant role in a number of structural engineering disasters, although such a role was not perceived at the time of initial investigation (e.g., the sinking of Sleipner A oil platform, Norway, 1991, the failures of St. Francis Dam near Los Angeles in 1928 and Malpasset arch dam in the French Alps in 1954, the failures of Cypress Viaduct in Oakland and Han-Shin Viaduct in Kobe during the earthquakes of 1989 and 1995, the breakup of Liberty ships at the end of World War II, etc.).

The problems of size effect have been most acute in civil engineering where the design of very large structures cannot be verified by full scale laboratory testing and large size extrapolation is inevitable. It is therefore not surprising that the recent literature on concrete abounds with studies of the size effect, in which a part of the problem has already been clarified. During the last decade, development of large structures and structural members made of fiber composites has become a goal of major interest for, for example, load bearing fuselage panels, large ship hulls, ship decks, bulkheads and antenna shields.

In recognition of these developments, the Office of Naval Research has begun supporting a significant research effort in the problems of scaling of fracture in fiber composites, and generally in the problems of scaling in solid mechanics. These problems, which are of interest for many materials, especially all quasibrittle materials, share similar characteristics. Thus, progress in the understanding of scaling problems for one material may help progress for another material. This makes it clear that a dialogue between researchers in various fields of mechanics is highly desirable and should be promoted.

In view of the aforementioned needs and goals, it was decided to organize under the sponsorship of the Office of Naval Research, Arlington, Virginia, a Workshop on Fracture Scaling during June 10–12, 1999 at the University of Maryland, with Prof. Sung Lee as the local host. Since the problems of size effect and scaling have also become an object of principal attention to the recently formed RILEM (International Union of Testing and Research Laboratories for Materials and Structures) Technical Committee QFS, Quasibrittle Fracture Scaling, this committee has been invited to serve as the scientific co-sponsor of the Workshop.

The invited speakers of the Workshop were asked to submit the texts of their lectures well in advance of the Workshop to permit their publication in the present Special Issue intended for distribution at the Workshop. These papers, which have all been subjected to a regular anonymous peer review and subsequent revisions, are in the present special issue grouped under five themes: (1) Micromechanical Analysis, (2) Size Effects in Fiber Composites, (3) Scaling and Heterogeneity, (4) Computational Aspects and Nonlocal or Gradient Models, and (5) Size Effects in Concrete, Ice and Soils. The papers in the Special Issue will be subjected to extensive discussion at the workshop. The summaries of these discussions, to be written by invited leading experts, are planned for publication in a later issue of this Journal.

We would like to express our gratitude to Prof. Sung Lee of the University of Maryland for his willingness to provide the local organization of the Workshop at the University of Maryland. Thanks are also due to Deedra Holdhusen for her expert secretarial assistance in the handling of the peer reviews and preparation of the Special Issue.

*Evanston and Arlington*
*December 30, 1998*

ZDENĚK P. BAŽANT
*Northwestern University*
and
YAPA D.S. RAJAPAKSE
*Office of Naval Research*

*International Journal of Fracture* **95:** 1–17, 1999.


# Models of interface separation accompanied by plastic dissipation at multiple scales

YUEGUANG WEI and JOHN W. HUTCHINSON
*Division of Engineering and Applied Sciences, Harvard University,Cambridge, MA 02138*
*e-mail: hutchinson@husm.harvard.edu*

Received 19 November 1998; accepted in revised form 2 December 1998

**Abstract.** Two continuum mechanical models of interface fracture for interfaces joining materials where at least one undergoes plastic deformation are reviewed and examined critically. The embedded process zone model (EPZ model) has an adhesive zone, characterized by a work of separation and an interface strength, embedded within a continuum model of the adjoining materials. The SSV model imposes an elastic, plasticity-free layer of prescribed thickness between the interface and the surrounding elastic-plastic continuum. Crack advance requires the work of separation to be supplied by the local elastic crack tip field. The objective of each model is to provide the relation between the macroscopic interface toughness (the total work of fracture) and the work of separation. Under steady-state crack growth, the total work of fracture is the work of separation plus the work of plastic dissipation, the latter often far exceeding the former. It will be argued that each model has its own domain of validity, subject to the accuracy of conventional continuum plasticity at small scales, but neither is able to capture the dramatic trends which have been observed in macroscopic toughness measurements stemming from alterations in interface bonding conditions. A unified model is proposed which coincides with the two models in their respective domains of validity and provides a transition between them. Interface separation energy and interface strength (the peak separation stress) each play a central role in the unified model. Strain gradient plasticity is used to illustrate the effect of plastic deformation at the micron scale on the link between interface and macroscopic properties.

**Key words:** Adhesion, fracture, interface strength, interface toughness, plasticity.

## 1. Introduction

### *Linking the fracture process to macroscopic fracture behavior*

A zone of plasticity will generally surround the fracture process governing separation at a bi-material interface if at least one of the materials is a metal or polymer. Under quasi-static, steady-state propagation, the macroscopic toughness of the interface (the total work of fracture) is the sum of the work of separation and the plastic dissipation. Plastic dissipation contributes a large fraction of the macroscopic toughness for many interfaces. This paper examines mechanics models which relate the macroscopic work of fracture to the work of separation. Specifically, the embedded fracture process zone model (EPZ model) of Needleman (1984) and Tvergaard and Hutchinson (1992, 1993) and the plasticity-free strip model (SSV model) of Suo, Shih and Varias (1993) and Beltz et al. (1996) will be discussed with emphasis on delineating their domains of validity. A unified model will be proposed which coincides with the two models in their respective domains of validity and provides a transition between them.

In attempts to bridge from the microscopic to macroscopic scales in interface fracture, there is an important distinction that must be drawn between different classes of fracture processes (Evans et al., 1998). Where interface adhesion is controlled by atomic or molecular separation, the continuum plasticity description must give adequate estimates of stresses and plastic dissipation over the full range of scales for which it is used in the model. In particular, it must be capable of providing stresses at the micron to sub-micron scale, where conventional plasticity theory becomes suspect. There are important fracture processes (void nucleation, growth and coalescence in metals and crazing in polymers) which control separation on a scale measured in tens of microns or more. In these cases, conventional macroscopic plasticity theory is adequate to link all the way to the fracture process. The scale of the fracture process is consequently of essential importance in setting the requirements on the plasticity description used in fracture models.

Prior to detailing the three models, we begin by listing the parameters which appear in the models, indicating typical values for separation processes occurring at the at atomic or molecular scale and for those such as void growth which occur at larger scales. The thrust of the paper will be on adhesion models for separation processes occurring on the atomic or molecular scale.

*Interface parameters*

The primary parameters used to characterize separation of interfaces are the work of separation, $\Gamma_0$, and the peak separation stress $\hat{\sigma}$, which will also be referred to as the interface strength. The critical separation of the two materials on either side of the interface when the traction stress has dropped to zero $\delta_c$, is not an independent parameter because $\Gamma_0 \propto \hat{\sigma}\delta_c$, where the constant of proportionality depends of the shape of the traction-separation law. For *atomic separation* of a moderately strong metal/ceramic interface (Raynolds et al., 1996)

$$\Gamma_0 \approx 1\,\mathrm{Jm}^{-2}, \qquad \hat{\sigma} \approx 10\,\mathrm{GPa}, \qquad \delta_c \approx 10^{-10}\,\mathrm{m}. \tag{1.1}$$

The range of the separation parameters can be very large when the fracture process at a metal/ceramic interface is *void nucleation, growth and coalescence* within the metal. A representative set of values is

$$\Gamma_0 \approx 1\,\mathrm{kJm}^{-2}, \qquad \hat{\sigma} \approx 1\,\mathrm{GPa}, \qquad \delta_c \approx 1\,\mu\mathrm{m}. \tag{1.2}$$

The scale of the process is set by the spacing between voids or void nucleating particles in the metal at the interface. Most of work of the separation is plastic deformation consumed in the growth and coalescence of the voids on the fracture plane. Separation involving crazing at a strong interface between a polymer and a metal or ceramic can also lead to values such as (1.2), usually with somewhat smaller interface strengths and critical separations correspondingly larger.

The SSV model for atomic or molecular adhesive separation invokes an elastic layer of thickness $D$ from which plastic deformation is excluded lying between the plastic zone and the interface. This length quantity can be regarded as a material fitting parameter (Suo et al., 1993), or it can be estimated using dislocation concepts (Beltz et al., 1996; Lipkin et al., 1996). Typically, values are found to fall in the range

$$D \approx 10 \quad \text{to} \quad 100\,\mathrm{nm}. \tag{1.3}$$

*Continuum parameters*

The parameters employed in the most widely used continuum theories for elastic-plastic solids are Young's modulus $E$, Poisson's ratio $\nu$, yield stress in uniaxial tension $\sigma_Y$, and strain hardening exponent $N$. The $J_2$ flow theory of plasticity characterizing initially isotropic solids undergoing isotropic strain hardening has been used in nearly all the modeling thus far, and it will be used here as well for the studies based on 'conventional' theory. Conventional theories give an adequate description of plastic deformation as long as the scale of deformation field is somewhat above a micron. Experiments (hardness tests, wire torsion, film bending) conducted at the micron to sub-micron scale reveal much higher stresses at a given level of deformation than conventional plasticity would suggest. The conventional theories fail to account for significant elevations in strain hardening when small scale plastic deformation occurs in the presence of strain gradients. Deformations at the micron to sub-micron scale typically involve large numbers of dislocations such that a continuum approach retains its advantages over dislocation mechanics. The strong size effect requires the introduction of a material length scale $\ell$ into the constitutive law. In this paper, the strain gradient plasticity generalization of $J_2$ flow theory of Fleck and Hutchinson (1997) will be employed to illustrate the influence of the plasticity length parameter $\ell$ in fracture modeling.

*Plastic zone sizes*

There are two length quantities related to the size of the plastic zone for a bi-material interface crack (Tvergaard and Hutchinson, 1993)

$$R_0 = \frac{2}{3\pi(1-\nu^2)(1-\beta_D^2)}\left[1+\frac{(1-\nu_s^2)E}{(1-\nu^2)E_s}\right]^{-1}\frac{E\Gamma_0}{\sigma_Y^2}, \tag{1.4a}$$

$$R_{ss} = \frac{2}{3\pi(1-\nu^2)(1-\beta_D^2)}\left[1+\frac{(1-\nu_s^2)E}{(1-\nu^2)E_s}\right]^{-1}\frac{E\Gamma_{ss}}{\sigma_Y^2}, \tag{1.4b}$$

where $E$, $\nu$ and $\sigma_Y$ pertain to the plastically deforming material, $E_s$ and $\nu_s$ pertain to the elastic material, and $\beta_D$ is the second Dundurs elastic mismatch parameter which will be given later. Here, $\Gamma_{ss}$ denotes the total steady-state work of fracture (i.e. the remote energy release rate under small scale yielding conditions), and $R_{ss}$ is a rough estimate (to within a factor of about 2) of the size of the active plastic zone at the crack tip. Both $\Gamma_{ss}$ and $R_{ss}$ are computed quantities, unknown in advance. The length quantity $R_0$ is fundamental and appears prominently in all the models: it is specified in terms of the model parameters. By (1.4), $R_0$ can be thought of as the size of the plastic zone *if* the remote energy release rate were $\Gamma_0$. In other words, $R_0$ would be the size of the active plastic zone *if* the total work of fracture were not much larger than $\Gamma_0$. By (1.4), the two length quantities are related by $R_{ss}/R_0 = \Gamma_{ss}/\Gamma_0$. In the absence of elastic mismatch between the two materials, $R_0 = [3\pi(1-\nu^2)]^{-1}(E\Gamma_0/\sigma_Y^2)$.

For metal-ceramic interfaces undergoing atomic separation $\Gamma_0 \approx 1\,\mathrm{Jm}^{-2}$, and, typically

$$R_0 = 0.1 \quad \text{to} \quad 1\,\mu\mathrm{m}. \tag{1.5}$$

If separation occurs by the ductile void growth mechanism $\Gamma_0 \approx 1\,\mathrm{kJm}^{-2}$, and

$$R_0 = 0.1 \quad \text{to} \quad 1\,\mathrm{mm}. \tag{1.6}$$

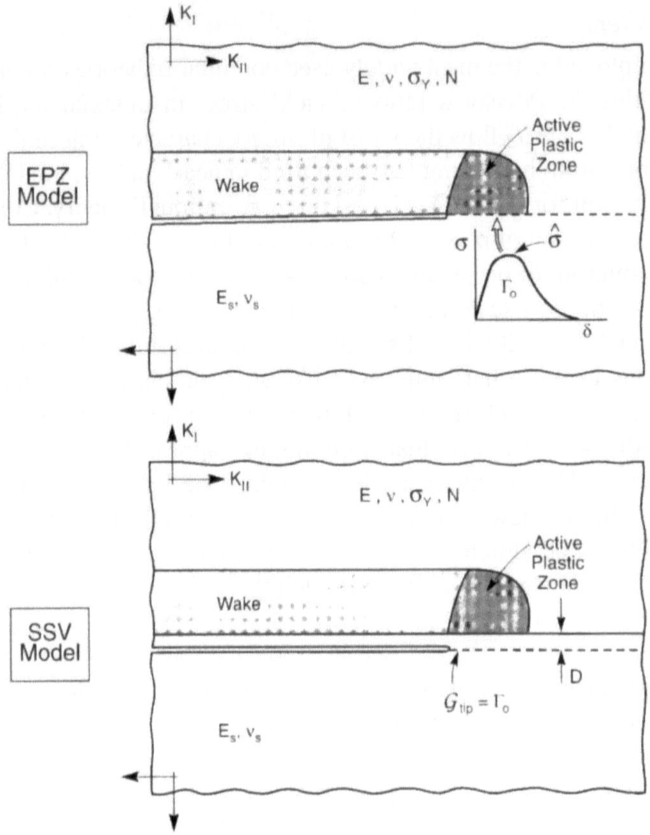


*Figure 1.* The embedded process zone model (EPZ model) and the plasticity-free strip model (SSV model) for a bi-material interface where the upper material is elastic-plastic and the lower material is elastic. The study is limited to steady-state crack growth in small scale yielding wherein the active plastic zone is small compared to the crack length. The finite length crack is replaced by a semi-infinite crack loaded remotely by the elastic $K$-field.

The plastic zone size $R_{ss}$ can be as small as $R_0$ if plastic dissipation is negligible, or it can be as much as 100 or more times $R_0$ when plastic dissipation is pronounced.

## 2. EPZ and SSV models

The embedded fracture process zone model (EPZ model) and the plasticity-free strip model (SSV model) share a number of common features. Both provide a link from interface separation to be the macroscopic scale using continuum descriptions of the elastic-plastic solids joined at the interface. Both aim to elucidate the role of plastic dissipation in amplifying the macroscopic toughness above the work of interface separation. The difference between the two models lies in the local separation criterion proposed for the interface.

The two models are shown in Figure 1 for the case where the material above the interface is elastic-plastic and that below is elastic. Plane strain conditions are assumed for both models. Moreover, in this paper, attention is restricted to small scale yielding wherein the plastic zone at the crack tip is sufficiently small compared to the crack length itself such that the asymptotic problem can be considered for two half spaces with a semi-infinite crack along the interface. The remote loading is prescribed using the stress intensity factors, $K_1$ and $K_2$, for the crack tip field of the elastic bi-material problem (Rice, 1988). That field has tractions on the interface specified by

$$\sigma_{22} + i\sigma_{12} = (K_1 + iK_2)(2\pi r)^{-1/2} r^{-i\varepsilon}, \tag{2.1}$$

where $r$ is the distance from the tip and $i = \sqrt{-1}$. The so-called oscillation index is given by

$$\varepsilon = \tfrac{1}{2} \ln \left( \frac{1 - \beta_D}{1 + \beta_D} \right), \tag{2.2}$$

where $\beta_D$ is the second Dundurs' mismatch parameter

$$\beta_D = \frac{1}{2} \frac{\mu(1 - 2\nu_s) - \mu_s(1 - 2\nu)}{\mu(1 - \nu_s) + \mu_s(1 - \nu)} \tag{2.3}$$

and $\mu$ and $\mu_s$ are the shear moduli of the upper and lower materials. The remote, or macroscopic, energy release is

$$G = \tfrac{1}{2}(1 - \beta_D^2) \left( \frac{1 - \nu^2}{E} + \frac{1 - \nu_s^2}{E_s} \right) (K_1^2 + K_2^2). \tag{2.4}$$

When $\varepsilon \neq 0$, the definition of mode mixity requires that a choice of distance $L$ ahead of the tip be made at which the relative amount of shear stress to normal stress acting on the interface is determined

$$\tan \psi = \frac{\sigma_{12}}{\sigma_{22}} = \frac{\mathrm{Im}[(K_1 + iK_2)L^{i\varepsilon}]}{\mathrm{Re}[(K_1 + iK_2)L^{i\varepsilon}]}. \tag{2.5}$$

This reduces to $\tan \psi = K_2/K_1$ if $\beta_D = 0$. The choice $L = R_{ss}$ is used the present study. The displacement components associated with the elastic bi-material singularity field are given by Rice et al. (1990) (see also, Tvergaard and Hutchinson, 1993).

In studies conducted until now, the conventional $J_2$ flow theory of plasticity has been used to characterize deformation in the upper half space. This isotropic hardening theory is based on the von Mises, or $J_2$, yield surface. The small strain version of the theory is employed, consistent with the fact that the strains at the tip of the steadily growing crack are indeed relatively small under conditions in which interface separation occurs. The tensile stress-strain relation used in the present study to represent the upper half space is

$$\begin{aligned} \varepsilon &= \sigma/E && \text{for } \sigma \leqslant \sigma_Y, \\ &= (\sigma_Y/E)(\sigma/\sigma_Y)^{1/N} && \text{for } \sigma > \sigma_Y, \end{aligned} \tag{2.6}$$

where $E$ is its Young's modulus and $\nu$ its Poisson's ratio. This information fully specifies $J_2$ flow theory. The lower half space is elastic with isotropic properties characterized by $E_s$ and $\nu_s$.

Conventional plasticity theories fail to account for important size-dependencies at small length scales (Fleck et al., 1994; Fleck and Hutchinson, 1997; Nix and Gao, 1998). Specifically, strain hardening is significantly elevated above the predictions of conventional plasticity when plastic deformation occurs at the micron to sub-micron scale in the presence of strain gradients. A crack tip induces strong strain gradients, and it is reasonable to expect that a zone

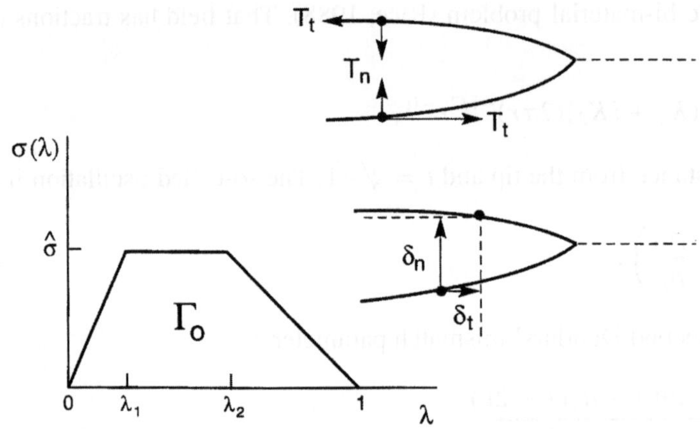


*Figure 2.* The traction-separation behavior characterizing the interface in the EPZ model and the unified model.

of gradient hardened material surrounds the crack tip. The associated traction elevation on the interface may significantly affect the link between macroscopic toughness and microscopic interface separation, particularly when the interface is strong. This connection will be explored in Section 4 where a strain gradient theory of plasticity will be substituted for $J_2$ flow theory to describe the upper half space.

The above specifications are common to all of models considered in this paper, including the unified model. Details specific to the individual models are now addressed.

### 2.1. THE EPZ MODEL

In the EPZ model a traction-separation law characterizing the interface fracture process is embedded as an internal boundary condition along the interface. In the present applications of the model, the fracture process zone along the interface lies between the plastic zone on one side of the interface and elastic material on the other. Once the parameters of the separation law are specified, the model can be used to compute the relation between crack advance and $G$. The primary quantity of interest here is the steady-state interface toughness $\Gamma_{ss}$, which is identified with the computed value of remote energy release rate $G$ in (2.4) needed to advance the crack in small scale yielding under steady-state conditions.

Following the notation for the law introduced in Tvergaard and Hutchinson (1993), let $\delta_n$ and $\delta_t$ be the normal and tangential components of the relative displacement of the crack faces across the interface, as indicated in Figure 2. Let $\delta_n^c$ and $\delta_t^c$ be critical values of these displacement components, and define a single dimensionless separation measure as

$$\lambda = \sqrt{(\delta_n/\delta_n^c)^2 + (\delta_t/\delta_t^c)^2}, \tag{2.7}$$

such that the tractions drop to zero when $\lambda = 1$. With $\sigma(\lambda)$ displayed in Figure 2, a potential from which the tractions are derived is defined as

$$\Phi(\delta_n, \delta_t) = \delta_n^c \int_0^\lambda \sigma(\lambda')\, d\lambda'. \tag{2.8}$$

The normal and tangential components of the traction acting on the interface in the fracture process zone are given by

$$T_n = \frac{\partial \Phi}{\partial \delta_n} = \frac{\sigma(\lambda)}{\lambda}\frac{\delta_n}{\delta_n^c}, \qquad T_t = \frac{\partial \Phi}{\partial \delta_t} = \frac{\sigma(\lambda)}{\lambda}\frac{\delta_t}{\delta_t^c}\frac{\delta_n^c}{\delta_t^c}. \tag{2.9}$$

The traction law under a purely normal separation ($\delta_t = 0$) is $T_n = \sigma(\lambda)$ where $\lambda = \delta_n/\delta_n^c$. Under a purely tangential displacement ($\delta = 0$), $T_t = (\delta_n^c/\delta_t^c)\sigma(\lambda)$ where $\lambda = \delta_t/\delta_t^c$. The peak normal traction under purely normal separation is $\hat{\sigma}$, and the peak shear traction is $(\delta_n^c/\delta_t^c)\hat{\sigma}$ in a purely tangential 'separation'. The work of separation per unit area of interface $\Gamma_0$ is given by (2.8) with $\lambda = 1$. For the separation function $\sigma(\lambda)$ specified in Figure 2,

$$\Gamma_0 = \tfrac{1}{2}\hat{\sigma}\delta_n^c[1 - \lambda_1 + \lambda_2]. \tag{2.10}$$

The parameters governing the separation law of the interface are the work of the fracture process $\Gamma_0$, the peak stress $\hat{\sigma}$, and the critical displacement ratio $\delta_n^c/\delta_t^c$, together with the factors $\lambda_1$ and $\lambda_2$ governing the shape of the separation function. Note that use of the potential ensures that the work of separation is $\Gamma_0$ regardless of the combination of normal and tangential displacements taking place in the process zone. Experience gained in the earlier studies suggests that the details of the shape of the separation law are relatively unimportant. The two most important parameters characterizing the fracture process in this model are $\Gamma_0$ and $\hat{\sigma}$. The parameter $\delta_n^c/\delta_t^c$ is the next most important, but the study of mixed mode interface toughness using this model (Tvergaard and Hutchinson, 1993) indicates that predictions are relatively insensitive to this parameter except when the loading is dominantly mode II. Then the peak in the shearing stress controls separation, and $\delta_n^c/\delta_t^c$ becomes important.

The condition for crack advance is attainment of $\lambda = 1$ at the current end of the traction-separation zone. In steady-state propagation, this condition must be imposed on the solution.

## 2.2. The SSV model

One limitation of the EPZ model as specified above is its failure to provide realistic predictions when the peak interface separation stress $\hat{\sigma}$ is prescribed to be at levels required for separation of strong interfaces, as will be evident from the numerical results described subsequently. In part, this limitation appears to be a consequence of the inadequacy of conventional plasticity to account for stress elevation in the region of high strain gradients at the tip of the crack (Wei and Hutchinson, 1997). Limitations of the EPZ model will be further discussed in Section 3 and 4. At this point, however, they serve to motivate the rationale for the SSV model. Under the assumption that dislocations emitted at the crack tip play a minimal role in crack propagation for the class of interfaces under study, Suo, Shih and Varias (1993) proposed a model capable of producing the high stresses at the crack tip necessary for atomic separation. They imposed an elastic, plasticity-free layer between the interface and the plastic zone. For the case in which the upper half space is elastic-plastic, a layer of thickness $D$ with the same elastic properties as the upper material is inserted above the interface, as shown in Figure 1. Thus, the interface crack tip lies fully within an elastic bi-material region and experiences stress intensity factors $K_{\mathrm{I}}^{\mathrm{tip}}$ and $K_{\mathrm{II}}^{\mathrm{tip}}$ and energy release rate

$$G_{\mathrm{tip}} = \tfrac{1}{2}(1 - \beta_D^2)\left(\frac{1-\nu^2}{E} + \frac{1-\nu_s^2}{E_s}\right)(K_{\mathrm{I}}^{\mathrm{tip}^2} + K_{\mathrm{II}}^{\mathrm{tip}^2}). \tag{2.11}$$

In this model, there is no explicit recognition of an interface separation law. Because the tractions on the interface are unbounded as the tip is approached, it is tacitly assumed that the length of any separation zone would be sufficiently small compared to $D$ such that the peak interface separation stress $\hat{\sigma}$ will always be attained. Thus, $\hat{\sigma}$ is not a parameter in this model. The criterion for crack propagation is simply $G_{\text{tip}} = \Gamma_0$, where $\Gamma_0$ is the local work of interface separation. Mode mixity effects could be introduced at this local level, but in this study $\Gamma_0$ will be taken to be mode-independent, as has been done in the case of the EPZ model.

The problem for the steady-state SSV model involves the computation of $G_{\text{tip}}$ as a function of the continuum properties of the two materials $D$, the remote energy release rate $G \equiv \Gamma_{ss}$, and $\psi$. The condition $G_{\text{tip}} = \Gamma_0$ is imposed to obtain $\Gamma_{ss}/\Gamma_0$. In passing, it can be mentioned that Tvergaard (1997) has shown for the mode I problem that replacing the strip by a circular elastic region with radius $D$ centered at the tip, results in little change in the predictions of the SSV model.

### 2.3. $\Gamma_{ss}/\Gamma_0$ FROM THE EPZ AND SSV MODELS

For the EPZ model, dimensional considerations dictate that $\Gamma_{ss}/\Gamma_0$ must have a functional dependence of the form

$$\frac{\Gamma_{ss}}{\Gamma_0} = F_{\text{EPZ}}\left\{\frac{\hat{\sigma}}{\sigma_Y}, N, \psi\right\}. \tag{2.12}$$

There is also some dependence on the moduli ratio $E/E_s$, the Poison ratios, and the dimensionless parameters $\lambda_1, \lambda_2$ and $\delta_n^c/\delta_t^c$, characterizing the details of the traction-separation law for the interface. However, the most important variables influencing $\Gamma_{ss}/\Gamma_0$ are those displayed explicitly in (2.12). It can be shown that (2.12) is otherwise independent of $\sigma_Y/E$. The dependence on the critical separation displacement $\delta_n^c$, is fully accounted for by the choice of dimensionless variables listed.

Apart from the elastic layer thickness $D$ in the SSV model, the fundamental length parameter in both models is $R_0$ defined in (1.4a). For the SSV model, the most important variables determining $\Gamma_{ss}/\Gamma_0$ are

$$\frac{\Gamma_{ss}}{\Gamma_0} = F_{\text{SSV}}\left\{\frac{R_0}{D}, N, \psi\right\}. \tag{2.13}$$

To set the stage for numerical results presented later in the paper, results are first presented for steady-state, mode I ($\psi = 0$) growth in a homogeneous elastic-plastic material (2.6). Curves of $\Gamma_{ss}/\Gamma_0$ as a function of $\hat{\sigma}/\sigma_Y$ for the EPZ model (Tvergaard and Hutchinson, 1992) and as a function of $R_0/D$ for the SSV model (Suo et al., 1993) are plotted in Figure 3 for three values of the strain hardening exponent $N$. In this case, the SSV model has a plasticity-free layer of total thickness $2D$ symmetrically located about the extended crack line.

From Figure 3 it can be noted that the ratio $R_0/D$, specifying the thickness of the plasticity-free layer in the SSV model plays a role similar to the normalized separation strength $\hat{\sigma}/\sigma_Y$, in the EPZ model. When based on conventional plasticity theory, the EPZ model is limited to values of normalized separation strength $\hat{\sigma}/\sigma_Y$ less than 4 or 5, depending on $N$ (c.f. Figure 3). At higher separation strengths, the stress levels achieved on the extended crack plane are not high enough to produce separation, and crack growth will not occur. By contrast, it will be seen that the SSV model becomes invalid at low separation strengths (at a given $\Gamma_0$) because

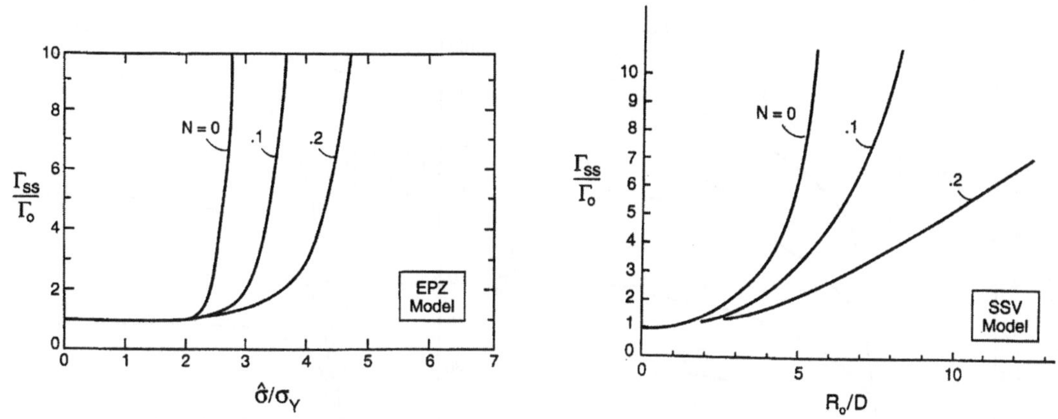


*Figure 3.* The ratio of steady-state macroscopic work of fracture to work of separation $\Gamma_{ss}/\Gamma_0$, for the EPZ and SSV models specialized to mode I growth in a homogeneous elastic-plastic solid. The EPZ results are from Tvergaard and Hutchinson (1992) with $\nu = 0.3$, $\lambda_1 = 0.15$ and $\lambda_2 = 0.5$. The SSV results are from Suo et al. (1993) with $\nu = 0.3$.

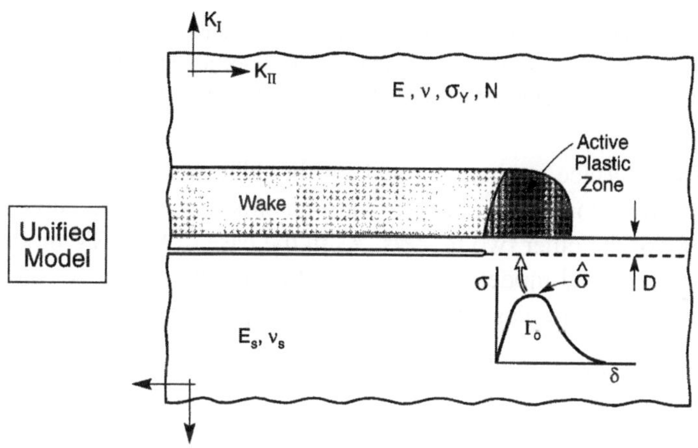


*Figure 4.* The unified model for crack growth along a bi-material interface under small scale yielding. The model incorporates the plasticity-free strip from the SSV model and the embedded traction-separation characterization of the interface from the EPZ model.

the length of the separation zone at the interface becomes comparable to or even larger than $D$. Then, the justification for tacitly assuming attainment of the peak separation stress in the SSV model is violated.

## 3. Unified model

Neither of the two models just discussed is capable of spanning the large observed variations in steady-state macroscopic toughness brought about by the effect of deleterious interface segregants on interface adhesion energy and strength (Evans et al., 1998). A unified model includes the respective limits of the two models and spans a much larger range of possible behaviors. The unified model (Figure 4) incorporates both the traction-separation description (2.9) of the interface used in the EPZ model and the plasticity-free layer of thickness $D$ of

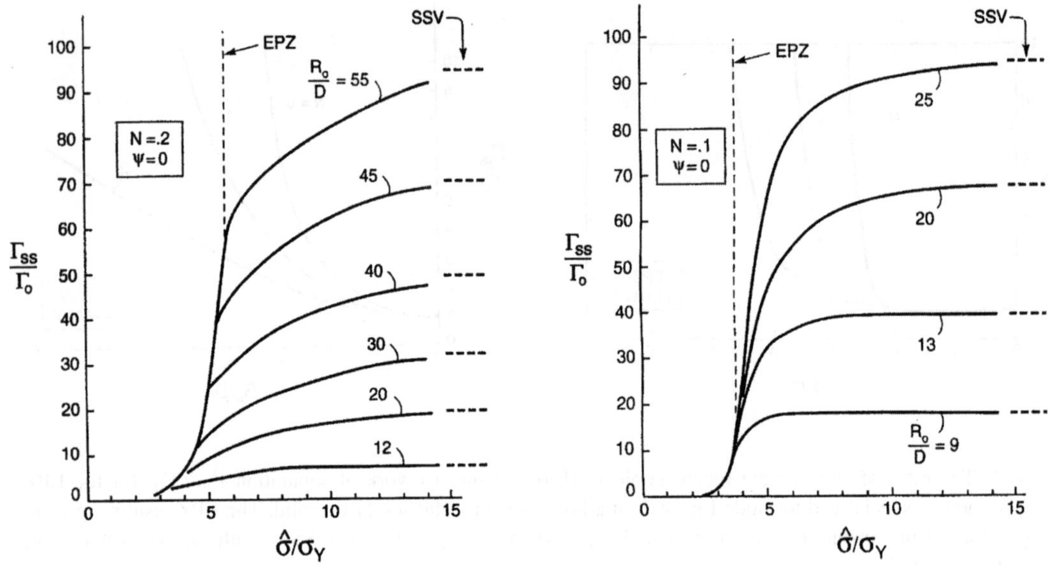


*Figure 5.* The ratio of the macroscopic work of fracture to work of separation, $\Gamma_{ss}/\Gamma_0$, for the unified model for mode I ($\psi = 0$) steady-state crack growth along a bi-material interface joining an elastic-plastic solid to a rigid solid. The parameters used in carrying out the calculations are $\nu = 0.3$, $\lambda_1 = 0.15$, $\lambda_2 = 0.5$ and $\delta_n^c/\delta_t^c = 1$.

the SSV model. The interface is characterized by the adhesion energy and strength $\Gamma_0$ and $\hat{\sigma}$, and the plasticity-free zone of width $D$. Consequently, the unified model has one more parameter than either of the other two models. As in the case of the SSV model, $D$ will again be regarded here as a material modeling parameter. For the unified model the set of the most important nondimensional variables on which $\Gamma_{ss}/\Gamma_0$ depends is

$$\frac{\Gamma_{ss}}{\Gamma_0} = F_{\text{UNIFIED}}\left\{\frac{\hat{\sigma}}{\sigma_Y}, \frac{R_0}{D}, N, \psi\right\}. \tag{3.1}$$

The additional parameters cited just beneath (2.12) are again relevant but continue to be less important than those identified in (3.1).

The numerical method used to generate solutions to the steady-state unified model is the same as that discussed in detail elsewhere for the EPZ model (Wei and Hutchinson, 1997). The same method has been used to solve the SSV model. Plastic deformation in crack growth problems is strongly history dependent. As depicted in Figure 4, an active plastic zone travels with the crack tip leaving behind a wake of residual plastic strains. The field equations are highly nonlinear. A finite element formulation is employed and iteration is used to directly achieve the steady-state solution.

Calculations have been performed to determine the functional dependencies in (3.1) for the case of mode I loading ($\psi = 0$) for a bi-material combination where the lower half-space is rigid ($E_s/E \to \infty$). For this case, the second Dundurs parameter reduces to

$$\beta_D = \frac{-(1-2\nu)}{2(1-\nu)} = -0.286 \quad (\text{for } \nu = 0.3). \tag{3.2}$$

The results are plotted in Figure 5 in the form of $\Gamma_{ss}/\Gamma_0$ as a function of $\hat{\sigma}/\sigma_Y$ at fixed values of $R_0/D$. Included in each of these figures as a dashed curve is the result for the EPZ

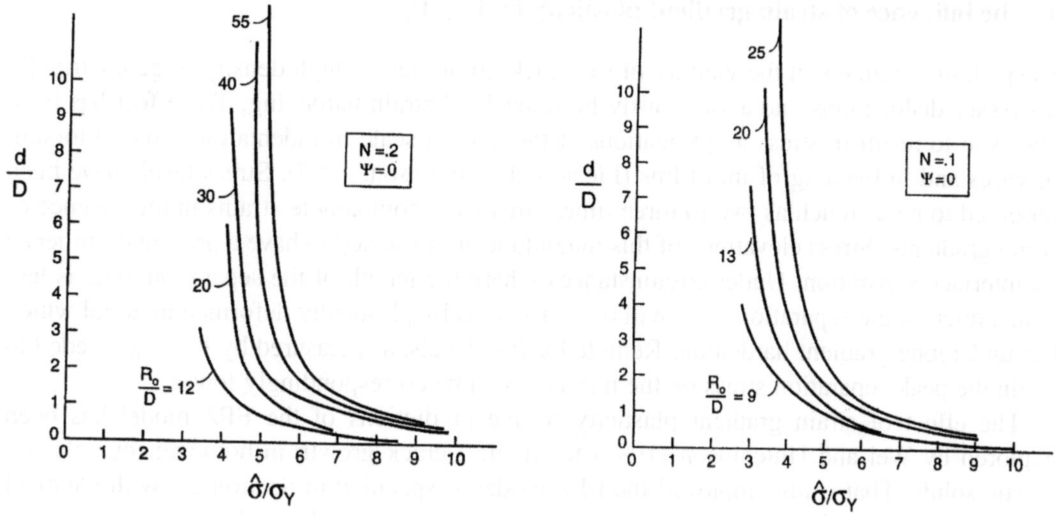


*Figure 6.* The ratio of the length of the separation zone at the interface to the thickness of the plasticity-free strip $d/D$, for the examples of Figure 5.

model. The limit corresponding to the SSV model is also indicated for each value of $R_0/D$. Recall that $D$ is not a parameter in the EPZ model, nor does $\hat{\sigma}$ appear in the SSV model. It is evident from the results in Figure 5 that the unified model reduces to the two earlier models in special limits and provides a transition between the two models in the intermediate range of parameters. Specifically, the unified model approaches the EPZ model in the limit of large $R_0/D$ as long as $\hat{\sigma}/\sigma_Y$ is within the range for which the EPZ model predicts crack advance. If $\hat{\sigma}/\sigma_Y$ lies outside this range, $\Gamma_{ss}/\Gamma_0$ is unbounded as $R_0/D$ becomes large. The unified model approaches the SSV model for large $\hat{\sigma}/\sigma_Y$ at all values of $R_0/D$.

The two limits to the unified model are readily understood in terms of the following considerations. First consider the approach to the EPZ model as $R_0/D$ becomes large. Suppose all parameters are fixed except $D$, which becomes small. Then the traction along the interface within the elastic strip will be approach the stresses imposed on it by the surrounding plastic zone. This limiting situation becomes identical to the EPZ model. Next, consider the approach to the SSV model. Now, suppose all parameters are fixed except $\hat{\sigma}$ and that it becomes large. At fixed $\Gamma_0$, the critical separation $\delta_n^c$ is inversely proportional to $\hat{\sigma}$. Thus, as $\hat{\sigma}$ becomes large the length of the separation zone decreases and the separation zone lies more deeply within the elastic strip. This is the situation envisioned in the formulation of the SSV model. Interface tractions can become arbitrarily large near the tip, and $\hat{\sigma}$ no longer plays a controlling role.

Further insight can be had by examining the ratio of the length of the separation zone at the interface $d$, to the thickness of the elastic strip $D$. The length of the separation zone is not known in advance and must be computed. For present purposes, it is defined as the distance $d$ along the interface from the point where the peak stress is first attained (i.e. where $\lambda$ in (2.7) attains $\lambda_1$) to the point of separation where $\lambda = 1$ (c.f. Figure 2). Figure 6 presents $d/D$ as a function of $\hat{\sigma}/\sigma_Y$ for the unified model for the same values of $R_0/D$ and $N$ taken above. It is evident that the tacit requirement of the SSV model, that $d/D$ be small, is only achieved for relatively large $\hat{\sigma}/\sigma_Y$. Conversely, at lower $\hat{\sigma}/\sigma_Y$, $d/D$ becomes large, and the presence of the elastic strip in the model become irrelevant such that the EPZ limit applies.

## 4. The influence of strain gradient plasticity on $\Gamma_{ss}/\Gamma_0$

Steep strain gradients in the vicinity of the crack tip produce a high density of geometrically necessary dislocations and anomalously high levels of strain hardening. The effect has been observed to result in stress amplifications at the micron scale in indentation tests, in torsion of wires and in bending of thin films (Fleck and Hutchinson, 1997). Stress levels have been observed to be as much as two to three times stresses at comparable strains in the absence of strain gradients. Stress elevations of this magnitude are expected to have a profound influence on interface separation. Under circumstances where the length of the separation zone is less than a micron, the separation zone will be surrounded by plastically deformed material which has undergone gradient hardening. Remote loading levels, as measured by $\Gamma_{ss}/\Gamma_0$, needed to attain the peak separation stress on the interface will be correspondingly lower.

The effect of strain gradient plasticity on the predictions of the EPZ model has been explored by Wei and Hutchinson (1997) for mode I crack growth in homogeneous elastic-plastic solids. That study employed the EPZ model as specified in Section 2.1 with identical elastic-plastic material properties prescribed for the upper and lower half-spaces. A strain gradient theory of plasticity was substituted for the conventional $J_2$ flow theory as the continuum description of the solid. Lipkin, Clarke and Beltz (1996) introduced strain gradient hardening into a version of the SSV model specialized to the onset of crack growth rather than steady-state growth. Their approach leads to the identification of $D$ as the distance from the tip where the stresses arising from strain gradient hardening merge continuously with those in the plasticity-free strip.

The strain gradient theory which will be used here is the same as that employed in the earlier EPZ study. It is a generalization of conventional plasticity theory that reduces to $J_2$ theory, both when strain gradients vanish or when the scale of the nonuniform deformation is larger than the constitutive length parameter $\ell$. In the generalized theory, the constitutive length parameter enters through strain gradient contributions to an effective strain quantity defined as

$$E_e^2 = \varepsilon_e^2 + \ell_{SG}^2 \eta_{ijk}^{(1)} \eta_{ijk}^{(1)} + \tfrac{2}{3} \ell_{RG}^2 \chi'_{ij} \chi'_{ij}, \tag{4.1}$$

where $\varepsilon_e$ is the standard Mises effective strain, $\eta_{ijk}^{(1)}$ is a particular combination of the strain gradients which includes both stretch and rotation gradients, and $\chi'_{ij}$ is the rotation gradient (Fleck and Hutchinson, 1997). Rotation gradients ($\ell_{RG}$) are found to have relatively little influence on the crack tip stresses. The most important strain gradient contribution for the crack growth problem is that involving stretch gradients ($\ell_{SG}$). In the present calculations, as well as in those reported by Wei and Hutchinson (1997), the choice $\ell_{SG} = \ell_{RG} \equiv \ell$ has been made, but the value assigned to $\ell_{RG}$ has relatively little effect on the results. Indentation is ideally suited for the measurement of $\ell_{SG}$ because it too is relatively insensitive to rotation gradients. Fitting indentation test data for a variety of metals (e.g. Ma and Clarke, 1995; Poole et al., 1996; Stelmashenko et al., 1993) with indentation predictions based on $J_2$ SGP theory suggests that $\ell_{SG}$ usually falls within the range from 0.25 to 1 $\mu$m (Begley and Hutchinson, 1998). (Wire torsion and film bending are dominated by rotation gradients and are insensitive to $\ell_{SG}$. From such tests, it is inferred that $\ell_{RG}$ is about 4 or 5 $\mu$m.) Full details of the $J_2$ SGP theory are given by Fleck and Hutchinson (1997) and by Wei and Hutchinson (1997), who provide formulas for the incremental version of the theory used in the crack growth models.

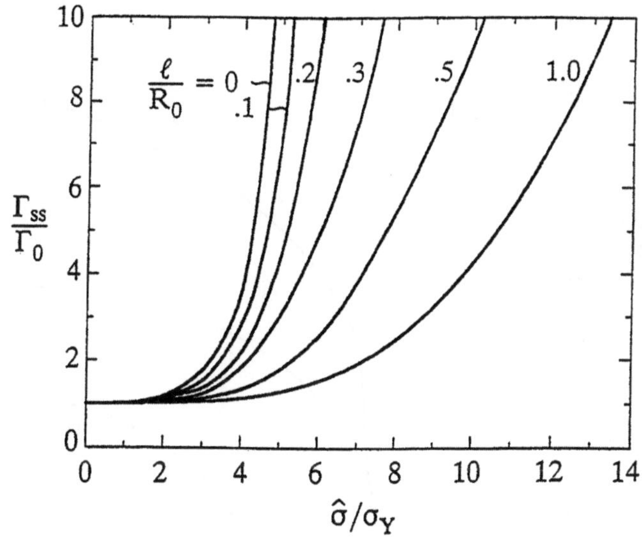


*Figure 7.* The influence of strain gradient hardening on $\Gamma_{ss}/\Gamma_0$ as predicted by the EPZ model for mode I growth in a homogeneous elastic-plastic solid characterized by $J_2$ SGP theory. The limit $\ell/R_0 = 0$ corresponds to conventional $J_2$ flow theory (Figure 3). The results are taken from Wei and Hutchinson (1997) and were computed with $N = 0.2$, $\nu = 0.3$, $\lambda_1 = 0.15$ and $\lambda_2 = 0.5$.

The influence of strain gradient hardening on $\Gamma_{ss}/\Gamma_0$ in the EPZ model appears through the dimensionless combination, $\ell/R_0$, such that for mode I crack growth, (2.12) becomes

$$\frac{\Gamma_{ss}}{\Gamma_0} = F_{\text{EPZ}}\left\{\frac{\hat{\sigma}}{\sigma_Y}, N, \frac{\ell}{R_0}\right\}. \tag{4.2}$$

Curves of $\Gamma_{ss}/\Gamma_0$ as a function of $\hat{\sigma}/\sigma_Y$ for various values of $\ell/R_0$ are displayed in Figure 7 for a homogeneous material with the tensile stress-strain curve (2.6) and a strain hardening index, $N = 0.2$. The limiting curve for $\ell/R_0 = 0$ is that for the EPZ model based on conventional $J_2$ plasticity theory. The influence of the dimensionless parameter $\ell/R_0$ on $\Gamma_{ss}/\Gamma_0$ is indeed large. At a given normalized peak separation stress $\hat{\sigma}/\sigma_Y$, $\Gamma_{ss}/\Gamma_0$ decreases sharply with increasing $\ell/R_0$. Whereas a crack in the conventional $J_2$ solid with $N = 0.2$ is unable to achieve crack-tip stresses large enough to undergo separation if $\hat{\sigma}/\sigma_Y = 5$, a crack in the $J_2$ SGP solid with $\ell/R_0 \geqslant 0.5$ achieves separation at normalized peak separation stresses twice that large.

The separation process determines whether strain gradient effects are likely to be important. Recall from (1.6) that $R_0$ is typically on the order of a fraction of a millimeter or more if the fracture process is void growth and coalescence. For such processes, $\ell/R_0$ is much too small for gradient effects to influence crack growth. The length of the fracture process zone will be large compared to $\ell$. By contrast, when the fracture process is atomic separation (c.f. (1.5)), $\ell$ and $R_0$ will often be comparable, and the strong gradient plasticity effects evident in Figure 7 should pertain.

The effect of $\ell/R_0$ on the *unified model* is shown in Figure 8, again for the case of a mode I crack in a homogeneous solid with the tensile stress-strain curve (2.6) with $N = 0.2$. Calculations for $\ell/R_0 = 0$ (conventional $J_2$ theory) are compared with those for a solid with $\ell/R_0 = 0.3$. As in the results just discussed, the only alteration to the unified model has been

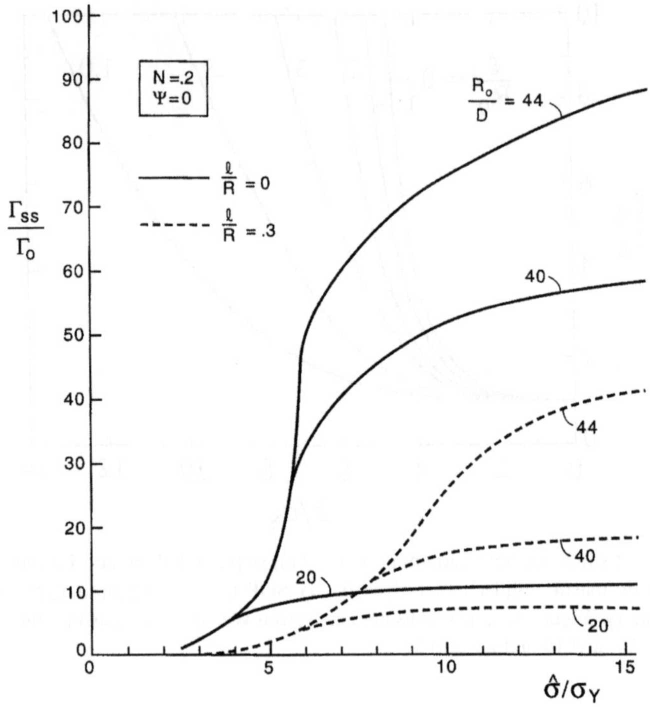


*Figure 8.* The influence of strain gradient hardening on $\Gamma_{ss}/\Gamma_0$ as predicted by the unified model for mode I growth in a homogeneous elastic-plastic solid characterized by $J_2$ SGP theory. The solid curves for $\ell/R_0 = 0$ coincide with the results for the conventional $J_2$ theory. The dashed curves are for the strain gradient theory with $\ell/R_0 = 0.3$. The calculations are made with $\nu = 0.3$, $\lambda_1 = 0.15$ and $\lambda_2 = 0.5$.

to substitute $J_2$ SGP theory for the conventional $J_2$ theory as the constitutive theory characterizing the upper and lower half spaces. The effect of $\ell/R_0$ on the macroscopic toughness prediction is similar to that noted for the EPZ model. Gradient hardening lowers the estimates of $\Gamma_{ss}/\Gamma_0$, significantly so at large $\hat{\sigma}/\sigma_Y$ and $R_0/D$. The effect occurs across the whole range of the unified model, from the EPZ limit to the SSV limit. It should not be surprising that strain gradient hardening at the micron scale also plays an important role in determining macroscopic toughness in the SSV model. For atomic separation (c.f. (1.3)), the thickness $D$ of the plasticity-free layer usually turns out to be much smaller than $\ell$. Thus, material just outside the strip experiences gradient hardening, resulting in an elevation of traction levels acting on the plasticity-free strip, thereby altering the energy release rate at the crack tip. This is an effect that has been modeled by Lipkin et al. (1996) for crack growth initiation.

## 5. Discussion

Magnification of the work of interface separation due to plastic deformation in the region surrounding the crack tip depends on a fairly large set of parameters. Earlier work by Tvergaard and Hutchinson (1993) based on the EPZ model has shown that there is a strong dependence of $\Gamma_{ss}/\Gamma_0$ on $\psi$. Mixed mode toughness can be much larger than mode I toughness. Similar trends are expected for the SSV and unified models. In the present work based on the unified model, there are four dimensionless parameters which are of primary importance in mode I

$$N, \frac{\hat{\sigma}}{\sigma_Y}, \frac{D}{R_0}, \frac{\ell}{R_0}. \tag{5.1}$$

While the interpretation of each of these dimensionless parameters is clear within the context of the unified model, a dependence on even three variables makes it difficult to connect the model to experiments (Bagchi and Evans, 1996; Evans et al., 1998). The situation for the fracture process of void nucleation, growth and coalescence simplifies considerably in this respect because the plasticity-free strip is irrelevant and $\ell/R_0 = 0$. The EPZ model, or equivalents (e.g. Rousselier et al., 1989; Xia and Shih (1995)), based on conventional plasticity theory is applicable. When the mechanism is atomic or molecular separation, the unified model helps to clarify of the limitations of the other models and provides guidance to the formulation of improved models. The following points summarize the most important implications of the unified model.

The interface strength, as measured by $\hat{\sigma}/\sigma_Y$, will generally be an important determinant of the total work of interface fracture when at least one of the solids deforms plastically. For atomistic separation processes, the work of separation $\Gamma_0$ and the interface strength $\hat{\sigma}$ are fundamental quantities which, in principle, can be computed using atomistic models (e.g. Raynolds et al., 1996). Both are affected by interface bonding conditions and interface segregants. Only when the length of the separation zone is small compared to the thickness of the plasticity-free strip will the interface strength cease to play a controlling role (c.f. Figures 5 and 6). This is the domain of validity of the SSV model. The length of the separation zone itself has a complicated dependence on $\hat{\sigma}/\sigma_Y$ and $D/R_0$.

The plasticity-free strip of the SSV and unified models remains the most nebulous aspect of the models. The self-consistent arguments of Beltz et al. (1996) and Lipkin et al. (1996) for the determination of $D$ provide insight into its validity as a region near the tip where continuum plasticity ceases to describe discrete dislocation interactions with each other and with the tip itself. The quantitative significant of the estimate of $D$ by Beltz et al. (1996) can be questioned on the grounds that these authors use conventional plasticity to bridge between the continuum scale and the scale where discrete dislocation interactions become important. The strain gradient effects cited for deformations at the micron scale are likely to be equally important in the determination of $D$. On the other hand, even with uncertainties surrounding the identification of the plasticity-free strip, the unified model emphasizes its importance. Whether behavior close to the crack tip where *any* continuum plasticity description breaks down can be characterized by elastic behavior is open to question and certainly requires further validation. However, it is clear that a near-tip zone must exist within which continuum plasticity will be inadequate, if separation occurs at the atomic scale. The sensitivity of the predictions of the SSV and unified models to $D$ provides evidence that this near-tip region, whether strictly elastic or not, has an important role to play in linking macroscopic toughness to separation at the atomic or molecular level.

The strong influence on macroscopic toughness of strain gradient hardening through the parameter $\ell/R_0$ also serves to illustrate the importance of behavior at intermediate scales between macroscopic and microscopic. Given the recent body of experimental evidence on stress elevation due to straining in the presence of strain gradients at the micron scale, it is hard to imagine this effect does not make its presence felt at crack tip. Quantitative predictions based on $J_2$ SGP theory reveal that the effects are indeed likely to be major if $\ell/R_0$ is of order unity, as will usually be the case when the separation process is an atomistic one. This conclusion applies to all the models considered in this paper.

The unified model provides a framework for further modeling developments. The model relates macroscopic crack-growth behavior to the fundamental parameters characterizing the interface, $\Gamma_0$ and $\hat{\sigma}$. If accurate discrete dislocation models existed for describing plastic deformation, they might be used to completely replace continuum plasticity in the unified model. However, given the long and difficult history of first principles modeling of even the basic aspects of strain hardening, it seems unlikely that continuum plasticity could be supplanted by more basic dislocation mechanics in quantitative fracture models in the near future. To put this into perspective, consider that the plastic zone associated with a relatively tough interface can be as large as several hundred microns, containing as many as $10^5$ dislocations. A more achievable goal might be to attempt further refinement of the 'plasticity-free' region using dislocation mechanics, continuing to rely on continuum plasticity outside this region. An approach along these lines might also enable the consideration of dislocation emission at the crack tip (Rice and Thomson, 1974), and its role in influencing the relation between macroscopic toughness and interface properties.

## Acknowledgments

This work was supported in part by the Office of Naval Research under grant N00014-96-0559, by the National Science Foundation under Grant CMS-9634632, and by the Division of Engineering and Applied Sciences, Harvard University.

## References

Bagchi, A. and Evans, A.G. (1996). The mechanics and physics of thin film decohesion and its measurement. *Interface Science* **3**, 169–193.

Begley, M.R. and Hutchinson, J.W. (1998). The mechanics of size dependent indentation. To be published in *Journal of Mechanics and Physics of Solids*.

Beltz, G.E., Rice, J.R., Shih, C.F. and Xia, L. (1996). A self-consistent model for cleavage in the presence of plastic flow. *Acta Materialia* **44**, 3943–3954.

Evans, A.G., Hutchinson, J.W. and Wei, Y. (1998). Interface adhesion: effects of plasticity and segregation. Harvard University Report MECH 337 (to be published in *Acta Materialia*).

Fleck, N.A. and Hutchinson, J.W. (1997). Strain gradient plasticity. *Advances in Applied Mechanics* **33**, 295–361.

Fleck, N.A., Muller, G.M., Ashby, M.F. and Hutchinson, J.W. (1994). Strain gradient plasticity: theory and experiment. *Acta Metallurgica et Materialia* **42**, 475–487.

Lipkin, D.M., Clarke, D.R. and Beltz, G.E. (1996). A strain-gradient model of cleavage fracture in plastically deforming materials. *Acta Materialia* **44**, 4051–4058.

Ma, Q. and Clarke, D.R. (1995). Size dependent hardness of silver single crystals. *Journal of Materials Research* **10**, 853–863.

Needleman, A. (1987). A continuum model for void nucleation by inclusion debonding. *Journal of Applied Mechanics* **54**, 525–531.

Nix, W.D. and Gao, H. (1998). Indentation size effects in crystalline materials: A law for strain gradient plasticity. *Journal of Mechanics and Physics of Solids* **46**, 411–425.

Poole, W.J., Ashby, M.F. and Fleck, N.A. (1996). Micro-hardness of annealed and work-hardened copper polycrystals. *Scripta Metallurgica et Materialia* **34**, 559–564.

Raynolds, J.E., Smith, J.E. Zhao, G.L. and Srolovitz, D.J. (1996). *Physical Reviews B* **53**, 13883–13890.

Rice, J.R. (1988). Elastic fracture concepts for interfacial cracks. *Journal of Applied Mechanics* **55**, 98–103.

Rice, J.R. and Thomson, R. (1974). Ductile vs. brittle behavior of crystals. *Philosophical Magazine* **29**, 73–97.

Rice, J.R., Suo, Z. and Wang, J.S. (1990). Mechanics and thermodynamics of brittle interfacial failure in bimaterial systems. In *Metal-Ceramic Interfaces*, (Edited by M. Ruhle, A.G. Evans, M.F. Ashby and J.P. Hirth), Pergamon Press, New York, 269–294.

Rousselier, G., Devaux, J.-C., Mottet, G. and Devesa, G. (1989). A methodology for ductile fractureanalysis based on damage mechanics: an illustration of a local approach of fracture. in *Nonlinear Fracture Mechanics: Vol. 11, Elastic-Plastic Fracture*, ASTM STP 995 (Edited by J.D. Landes, A. Saxena and J.G. Merkle) *American Society of Testing Materials*, Philadelphia, 332–354.

Stelmashenko, N.A., Walls, M.G., Brown, L.M. and Milman, Y.V. (1993). Microindentations on *W* and *Mo* oriented single crystals: An STM study. *Acta Metallurgica et Materialia* **41**, 2855–2865.

Suo, Z., Shih, C.F. and Varias, A.G. (1993). A theory for cleavage cracking in the presence of plastic flow. *Acta Metallurgica et Materialia* **41**, 1551–1557.

Tvergaard, V. (1997). Cleavage crack growth resistance due to plastic flow around a near-tip dislocation-free region. *Journal of Mechanics and Physics of Solids* **45**, 1007–1023.

Tvergaard, V. and Hutchinson, J.W. (1992). The relation between crack growth resistance and fracture process parameters in elastic-plastic solids. *Journal of Mechanics and Physics of Solids* **40**, 1377–1397.

Tvergaard, V. and Hutchinson, J.W. (1993). The influence of plasticity on mixed mode interface toughness. *Journal of Mechanics and Physics of Solids* **41**, 1119–1135.

Wei, Y. and Hutchinson, J.W. (1997). Steady-state crack growth and work of fracture for solids characterized by strain gradient plasticity, *Journal of Mechanics and Physics of Solids* **45**, 1253–1273.

Xia, L. and Shih, C.F. (1995). Ductile crack growth – I. A numerical study using computational cells with microstructurally-based length scales. *Journal of Mechanics and Physics of Solids* **43**, 223–259.

*International Journal of Fracture* **95:** 19–39, 1999.


# The effects of heterogeneity and anisotropy on the size effect in cracked polycrystalline films

R. BALLARINI[1]*, R.L. MULLEN[1] and A.H. HEUER[2]
[1]*Department of Civil Engineering, Case Western Reserve University, Cleveland, Ohio 44106-7201, U.S.A.*
[2] *Department of Materials Science and Engineering, Case Western Reserve University, Cleveland, Ohio 44106-7204*

Received 9 June 1998; accepted in revised form 2 December 1998

**Abstract.** A model is developed for quantifying the size effect due to heterogeneity and anisotropy in polycrystalline films. The Monte Carlo finite element calculations predict the average and standard deviation of the *microscopic* (local) stress intensity factors and energy release rate of a crack in a columnar aggregate of randomly orientated, perfectly bonded, orthotropic crystals (grains) under plane deformation. The boundary of the near-tip region is subjected to displacement boundary conditions associated with a *macroscopic* (far field or nominal) Mode-I stress intensity factor and average elastic constants calculated for the uncracked film with a large number of grains. The average and standard deviation of the microscopic stress intensity factors and energy release rate, normalized with respect to the macroscopic parameters, are presented as functions of the number of grains within the near-tip region, and the parameters that quantify the level of crystalline anisotropy. It is shown that for a given level of anisotropy, as long as the crack tip is surrounded by at least ten grains, then the expected value and standard deviation of the crack tip parameters are insensitive to the number of crystals. For selected values of crystalline anisotropy, the probability distributions of Mode-I stress intensity factor and stress ahead of the crack are also presented. The results suggest that the size effect due to heterogeneity and anisotropy is weak; crack initiation load and direction are governed only by the details of the grains in the immediate vicinity of the crack tip.

**Key words:** Anisotropy, heterogeneity, polycrystalline.

## 1. Introduction

In linear elastic fracture mechanics (LEFM) theory of homogeneous materials, the nominal stress $\sigma_n$ is expressed in terms of the stress intensity factor $K_{\mathrm{I}}$ as

$$\sigma_n = \frac{K_{\mathrm{I}}}{a^{1/2}F}, \tag{1}$$

where $a$ is the crack length and $F$ is a dimensionless function which depends on the loading and specimen geometry. In this one parameter system, the critical stress $\sigma_n^c$ is determined, for the desired loading, geometry, and crack length, by equating the stress intensity factor with the material's fracture toughness $K_{\mathrm{Ic}}$. The associated crack stress analysis for the homogeneous material produces a *macroscopic* (nominal or far-field) stress intensity factor, which will henceforth be denoted $K_{\mathrm{I}}^{\infty}$. The critical stress intensity factor criterion is equivalent to the condition that the energy release rate, defined for *self-similar* extension as $G_{\mathrm{I}}^{\infty} \equiv (K_{\mathrm{I}}^{\infty})^2/E'$, reaches a critical value $G_{\mathrm{Ic}} = K_{\mathrm{Ic}}^2/E' = 2\gamma$, where $E' = E$ for plane stress, $E' = E/(1-\nu^2)$

* Correspondence address: Civil Eng. Dept., Case Western Reserve University, Cleveland, OH 44106-7201, U.S.A. e-mail: rxb7@po.cwru.edu

for plane strain, $\gamma$ is the surface energy, $E$ is Young's modulus, and $\nu$ is Poisson's ratio. LEFM is based on the assumption that the 'process zone' in the immediate vicinity of the crack tip, within which irreversible deformation occurs and linear elasticity breaks down, is sufficiently small compared to other relevant length scales, so that there exists an annular region surrounding it within which the stresses are dominated by the elastic stress intensity factor. Moreover, the assumption of homogeneity implicitly assumes that the process zone extends from the crack tip to a distance much larger than the crystal (grain) size, and thus engulfs and in turn is engulfed by a sufficiently large number of grains that behave as a homogeneous medium. Abdel-Tawab and Rodin (1997) suggest that a material's tendency to behave homogeneously can be screened by the grain brittleness $GB$, defined as

$$GB \equiv \frac{d}{r_p}. \tag{2}$$

Here $d$ is the grain size, and $r_p$ is the size of the process zone, which can be estimated as

$$r_p = 0.1c\left(\frac{K_{\mathrm{I}c}}{\sigma_u}\right)^2, \tag{3}$$

where $\sigma_u$ represents a limiting stress appropriate for a given material system (yield stress for metals, tensile strength for brittle materials), and $c$ is a constant close to unity. The assumption of homogeneity is appropriate for $GB \ll 1$; if $GB$ is not sufficiently small, then the irreversible deformations at the crack tip do not overwhelm microstructure induced inhomogeneity, and the details of the microstructure surrounding the process zone may have a significant effect on the fracture process. Indeed, because of the randomness of the distribution and orientation of the grains, these effects will be of a statistical nature.

It has been established experimentally that the fracture toughness of most engineering materials depends on specimen dimensions. The amount of this so called 'size effect', which results from differences in the relative size of the process zone in geometrically similar but different sized specimens, varies from material to material. Much research has been conducted by the fracture mechanics community in the past twenty years aimed at describing and predicting the size effect. This effort has produced scaling laws through cohesive zone models (Cox and Marshall, 1994), dimensional analysis (Bazant, 1993) and fractal geometry concepts (Carpinteri, 1994). The main advantage of these formulations is that they can be easily implemented into analysis/design tools for predicting crack propagation in engineered structures. However, by smearing out the details of the deformation mechanisms involved in the fracture process, and by treating the phenomena deterministically and/or phenomenologically, they lose information that can be valuable in the interpretation of scatter in experimental data.

This paper focuses on characterization of the fracture mechanics parameters of a crack in a columnar aggregate of randomly orientated crystals under plane deformation. The following question is addressed: given the parameters that quantify the level of crystalline anisotropy, the number of grains in a unit volume, and a far field loading defined by $K_{\mathrm{I}}^{\infty}$, what are the average values and standard deviations of the *microscopic* (local) stress intensity factors $K_{\mathrm{I}}^{l}$ and $K_{\mathrm{II}}^{l}$ and energy release rate $G^{l}$? This information, which should shed some light on why the fracture toughness of brittle polycrystalline structures is prone to scatter and size effects, can be obtained only by using micromechanical models that treat the microstructural details and associated deformation mechanisms explicitly (Abdel-Tawab and Rodin, 1997; Ballarini et al., 1995; Jha et al., 1997; Jha, 1997). As stated in (Abdel-Tawab and Rodin, 1997), the ultimate goal of such models is to develop size effect maps analogous to those (Frost and Ashby, 1982) for deformation mechanisms, and that characterization of the individual size effects produced by inhomogeneity is the first step towards this end. The next logical step towards this end, which is being taken at the present time by the authors, is to include the effects of the microcracking that often develops in the vicinity of the crack tip. Abdel-Tawab and Rodin's analysis (Abdel-Tawab and Rodin, 1997), which addressed issues similar to those discussed in this paper, was developed for characterizing the fracture mechanics of crystalline columnar ice. Their approximate analytical and finite element plane strain models calculate the difference between $K_I^{\infty}$ and $K_I^l$ at the tip of a crack surrounded by a *large* number of brittle hexagonal crystals with perfectly bonded interfaces. This local parameter is needed because, for a heterogeneous material, $G_I^{\infty}$ and $\gamma$ in the fracture criterion should be replaced by $G^l$ and $\gamma^l$, where $\gamma^l$ is the local value of critical energy release rate. Abdal-Tawab and Rodin's contribution showed that, in an aggregate comprised of a very large number of grains, the grains remote from the crack tip do not induce significant differences between the macroscopic and microscopic stress intensity factors. For example, their calculations showed that even for highly anisotropic crystals, $K_I^l$ is only twenty percent less than $K_I^{\infty}$ ($K_{II}^l$ was negligible for the nominal mode-I loading). They concluded that for such materials the contribution of inhomogeneity to the size effect is relatively weak. Moreover, their results suggest that the crack tip parameters of materials with large values of $GB$ are controlled by only a few grains near the crack tip, and provide a rational explanation of why fracture toughness data exhibits significant statistical effects. As they pointed out, the results are expected to change dramatically if microcracking within grains or along grain boundaries is included in the models.

One of the issues that was not discussed in (Abdel-Tawab and Rodin, 1997) is fatigue. For brittle polycrystalline materials, crack growth under cyclic $K_I^{\infty}$, or under constant $K_I^{\infty}$ and corrosive environments, has been quantified with power law functions of the type $\mathrm{d}a/\mathrm{d}N = C(\Delta K_I^{\infty})^m$ and $\mathrm{d}a/\mathrm{d}t = D(K_I^{\infty})^p$, where $N$ and $t$ represent number of cycles and time, respectively, and $C, D, m$ and $p$ are experimentally measured constants. Because $m$ and $p$ are typically larger than, say 10, small heterogeneity induced fluctuations in local crack tip parameters may suddenly become important in lifetime predictions.

The Monte Carlo finite element model presented in this paper extends the calculations presented in (Abdel-Tawab and Rodin, 1997) to include *orthotropic* crystals and an *arbitrary* number of crystals within a unit volume. Furthermore, the model can be generalized to quantify the effects of microcracking in the process zone that often surrounds the crack tip in a brittle system. The model was developed to quantify the effects of heterogeneity and anisotropy on the *scatter* and potential size effects in the *nominal* fracture toughness (critical value of $K_I^{\infty}$) and fatigue data of microelectromechanical systems (MEMS), whose dimensions are of the same order as the microstructure. Using a first generation microfracture specimen, they reported (Ballarini et al., 1997) that the critical $J$ integral (which provided a nominal fracture toughness) of notched polycrystalline silicon, which is one of the principal materials currently used in the fabrication of MEMS devices, exhibited significant scatter, but was not sensitive to the number of crystals in the uncracked ligament. The number of crystals in the uncracked ligament was larger than ten. Figure 1(a) shows the second generation microfracture device that is currently being used by the authors to study fatigue and fracture of MEMS. The close-up view of the crack path through the originally uncracked ligament shown in Figure 1(b) clearly demonstrates that the characteristic lengths in MEMS devices are of the same order of magnitude as the grain size. These experiments are ideal for studying statistical effects

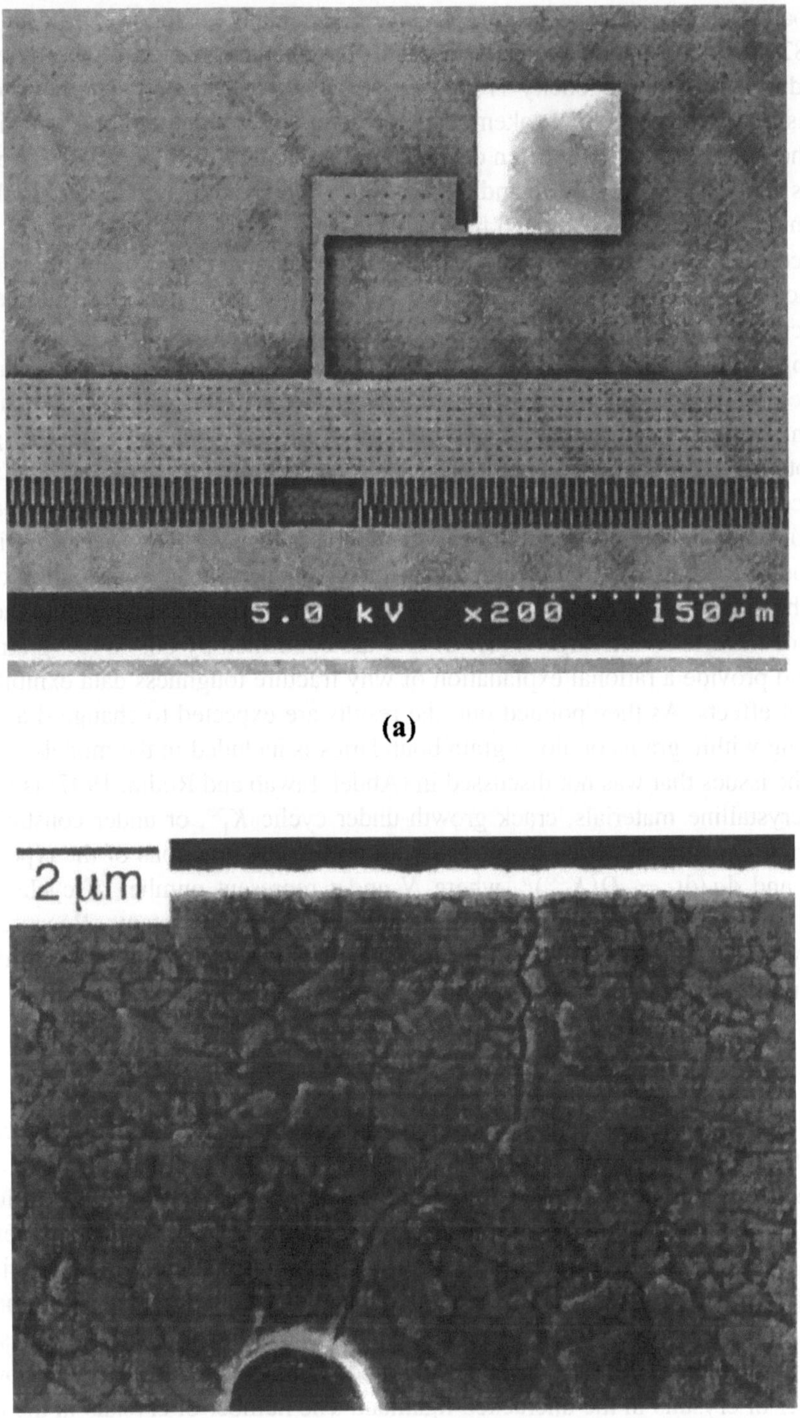


(b)

*Figure 1.* (a) Polycrystalline silicon microfracture specimen consisting of a cantilevered beam with a reduced section containing an initial smooth notch. The electrostatic load is provided by a comb actuator system, partially shown at the bottom of the figure; (b) Close-up of notch-tip showing the crack propagation path.

*Figure 2.* Cross-sectional transmission electron micrograph (TEM) of chemical vapor deposition film. This particular film was deposited using a two-step process, as indicated by the two layers of nearly columnar polysilicon.

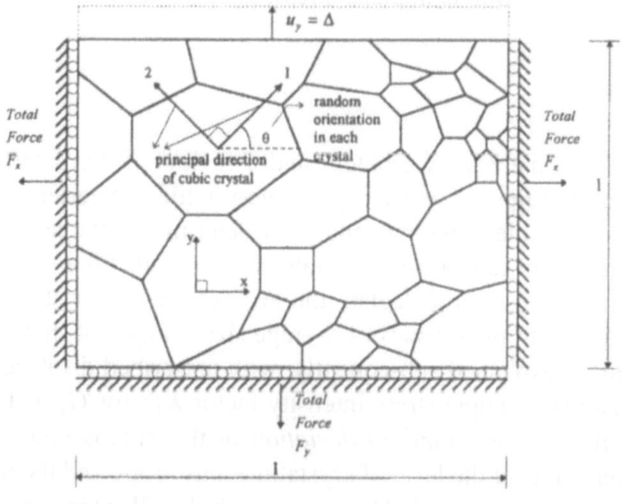


*Figure 3.* Poisson-Voronoi tessellation of microstructure for $n$-grains.

in the fracture mechanics of brittle materials; the integrated circuit processing technology used to fabricate the specimen involves batch fabrication, allowing production of a large number of identical specimens that are both dimensionally accurate and inexpensive. The details of the experimental program and results will be presented in a future communication.

Silicon possesses cubic symmetry, a $K_{Ic} \approx 0.8\,\text{MPa}\sqrt{\text{m}}$, $\sigma_{ult} = 1.2\,\text{GPa}$, and for the specimens shown in Figure 1, a crystal size $d \approx 0.5\mu\text{m}$. The resulting $GB \approx 10$ suggests that heterogeneity and anisotropy may play a significant role in the fracture behavior of

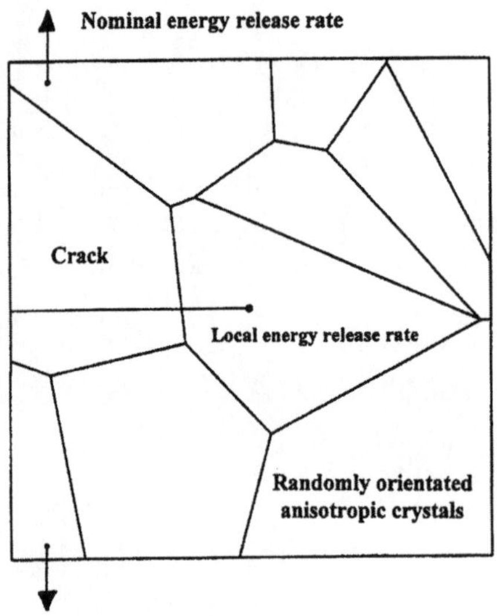


*Figure 4.* Poisson-Voronoi tessellation of a crack in an aggregate of crystals.

polycrystalline silicon devices that are comprised of a small number of grains (connecting beams are typically a few microns wide, and may be comprised of less than ten grains). As observed in the SEM micrograph in Figure 2, the microstructure of typical devices is (nearly) columnar. This supports the assumption of two dimensional behavior in subsequent finite element analyses. While the film is in plane stress, plane strain was assumed for convenience.

## 2. Near-tip mechanics

This section defines the mechanics problem and a review of the deterministic *dual length scale mechanics* near the tip of a crack in a polycrystalline material. Before proceeding to the fracture mechanics problem, the procedure for determining the nominal elastic constants of a columnar aggregate of anisotropic crystals is presented. Figure 3 shows the geometry and loading of the model used to calculate the average elastic constants of an aggregate of randomly orientated anisotropic crystals of average size $l$. Figure 4 shows the same aggregate with a crack, loaded, as explained subsequently, with prescribed displacements at the boundary consistent with a macroscopic stress intensity factor $K_I^{\infty}$ (or $G_I^{\infty}$). The problem consists of calculating the *average* and *standard deviation* of the microscopic crack tip parameters $K_I^l$, $K_{II}^l$ and $G^l$ as functions of the level of crystalline anisotropy and the number of grains in a unit volume. This information is needed because the crack will extend at a critical combination of the microscopic stress intensity factors $K_I^l$ and $K_{II}^l$, which are linearly related to $K_I^{\infty}$, but otherwise depend on the details of the microstructure in the immediate vicinity of the crack tip. Alternatively, the crack can be assumed to propagate at an energy release rate $G^l$ (which is defined subsequently) equal to a critical value $\gamma^l$. The propagation criterion in a single crystal, which is beyond the scope of this paper, is an important issue that is receiving attention (Azhdari and Nemat-Nasser, 1996; 1998). The problem of predicting crack propagation in the polycrystalline medium is complicated by the fact that function $F$ in (1) now depends on details such as the number of crystals in a unit area, the relative position of the crack tip within the crystal, and the orientations of the principal material directions within each crystal.

The crystalline structure in both models is approximated as a Poisson-Voronoi tessellation, consisting of $n$ grains, using the approach taken in (Kumar, 1992; Mullen et al., 1997). Each grain is assumed to possess either cubic or orthotropic symmetry.

With respect to the local coordinate axes 1–2 shown in Figure 3, which represent the principal material directions of a typical grain, the stress-strain relations are given, in matrix and component form, by Hooke's law in plane strain

$$\{\sigma\} \equiv \begin{Bmatrix} \sigma_{11} \\ \sigma_{22} \\ \sigma_{12} \end{Bmatrix} = [S]\{\varepsilon\} \equiv \begin{bmatrix} s_{11} & s_{12} & 0 \\ s_{12} & s_{22} & 0 \\ 0 & 0 & 2s_{44} \end{bmatrix} \begin{Bmatrix} \varepsilon_{11} \\ \varepsilon_{22} \\ \varepsilon_{12} \end{Bmatrix}, \tag{4}$$

where $\sigma_{ij}$ and $\varepsilon_{ij}$ $(i, j = 1, 2)$ are, respectively, components of the symmetric stress and strain tensors. System 1–2 is rotated through angle $\theta$ from the global $x$-axis. The stiffness coefficients in the global $x$-$y$ axes are calculated using standard transformation formulas. The level of anisotropy is quantified by the parameters $A, \overline{\nu}$ and $\lambda$, defined as

$$A \equiv \frac{2s_{44}}{s_{11} - s_{12}}, \tag{5a}$$

$$\overline{\nu} \equiv \frac{s_{12}}{s_{11} + s_{12}}, \tag{5b}$$

$$\lambda = \frac{s_{11}}{s_{22}}. \tag{5c}$$

For isotropic materials $A = 1, \overline{\nu} = \nu$, and $\lambda = 1$. The randomness of the orientation of the principal directions is introduced, in each of $m$ simulations ($m$ varied from 200–300), by assigning, to each cell, a uniformly distributed random angle $\theta$, where $0 < \theta < 2\pi$. The nominal Poisson's ratio and nominal plane strain Young's modulus are defined as $\nu^n \equiv \nu$ and $E^n/(1 - (\nu^n)^2)$, respectively. The normalized plane strain modulus $E$, is obtained by normalizing the nominal plane strain modulus with respect to the 'isotropic' plane strain modulus $E^{\text{isotropic}}$, which is obtained by equating the first entry in the stiffness matrix $S$, for $A = 1$ (for fixed $\overline{\nu}$), with that of an isotropic material, i.e. $E^{\text{isotropic}} \equiv c_{11}(1-2\overline{\nu})/(1-\overline{\nu}^2)$. For an aggregate comprised of a very large number of grains (an average of $n = 750$ grains within a unit area of film), Figures 5 show, as functions of $A$ and $\overline{\nu}$, the average values of the nominal Poisson's ratio and normalized plane strain Young's modulus. The most interesting result in these figures is the negative nominal Poisson's ratio that results from crystalline anisotropy, as first reported in (Mullen et al., 1997). As discussed in (Mullen et al., 1997), the nominal elastic constants are more sensitive to $\overline{\nu}$ than to $\lambda$.

Figures 6 present, for highly anisotropic cubic and orthotropic crystals, the fraction of occurrences of the average plane strain modulus. It is observed that these are nearly symmetric about the mean, and are associated with a well defined peak.

The near-tip mechanics is summarized through the local-global representation of the cracked polycrystalline plane region shown in Figure 7. Let $L$ represent the macroscopic length scale (crack length, uncracked ligament length, or any other relevant characteristic length), and recall that $l$ represents the average grain size. For a given (deterministic) aggregate, it is expected that for large values of $L/l$, a near-tip, *small scale heterogeneous* zone exists, and arguments analogous to those presented for small scale process zones can be used to study

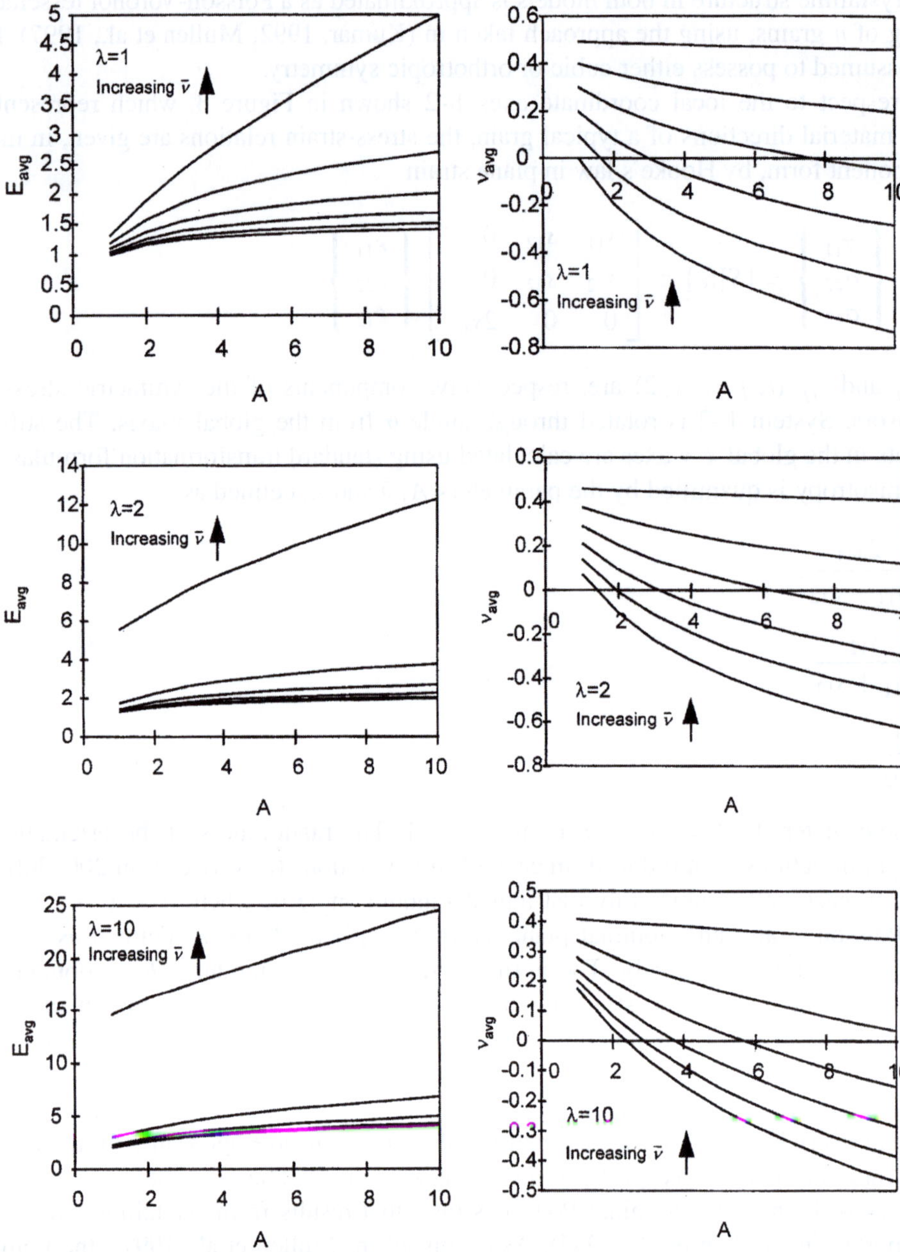


*Figure 5.* Average values of normalized plane strain Young's modulus and Poisson's ratio as functions of anisotropy parameters; $\bar{\nu} = 0, 0.1, 0.2, 0.3, 0.4, 0.5$.

the evolution of the near-tip stress fields (Ballarini et al., 1995; Jha et al., 1997; Jha, 1977). For example, in small scale yielding, the near-tip plastic zones that develop with increasing load disturb the singular elastic solution giving rise to other near-tip asymptotics such as the HRR fields in deformation plasticity and power law hardening materials (Rice and Rosengren, 1968; Hutchinson, 1968). Similarly, it is expected that for the polycrystalline material, the heterogeneity perturbs the singular fields predicted by the homogeneized aysmptotic solution giving rise to dual asymptotics consistent with the dual length scales in the near-tip region.

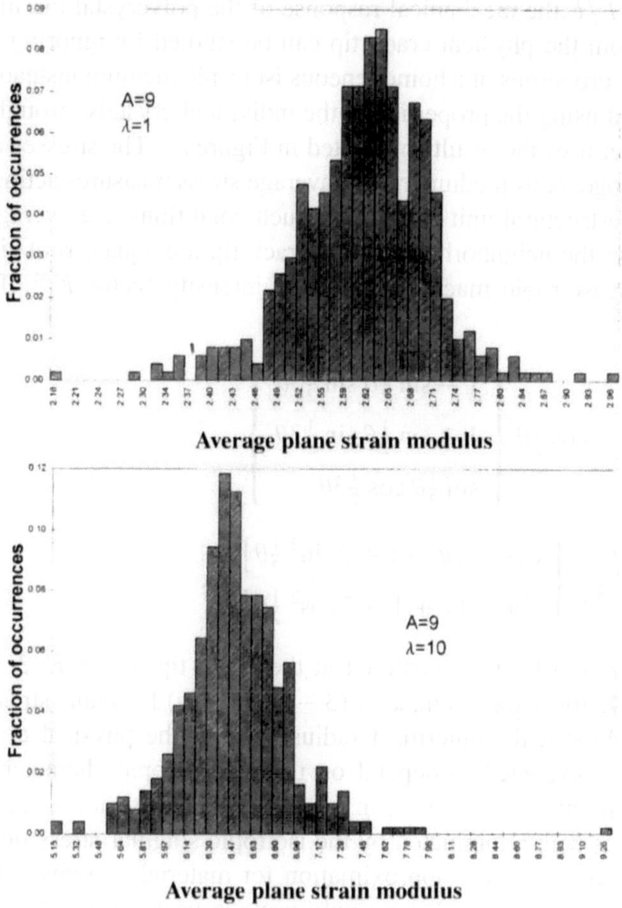


*Figure 6.* Distributions of average plane strain modulus.

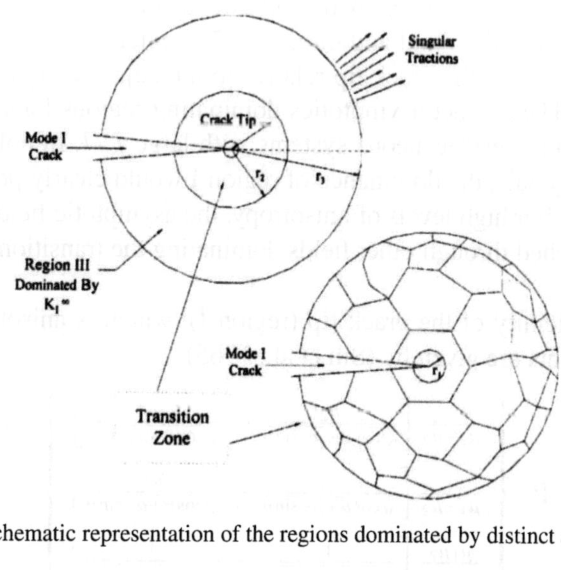


*Figure 7.* A schematic representation of the regions dominated by distinct asymptotics.

Moreover, for large $L/l$, the mechanical response of the polycrystalline material at distances sufficiently away from the physical crack tip can be studied by ignoring the microstructure while employing the properties of a homogeneous isotropic medium instead. The latter properties can be calculated using the properties of the individual crystals through homogeneization techniques; this paper uses the results presented in Figures 5. The stresses associated with the response of the homogeneous medium reflect average stress measures acting over the length of an appropriate microstructural unit cell. Under such conditions, the average unit cell stresses and displacements in the neighborhood of the crack tip are square root singular, and can be characterized via an isotropic macroscopic stress intensity factor $K_{\mathrm{I}}^{\infty}$. These are given by (Anderson, 1995)

$$\begin{Bmatrix} \sigma_{xx} \\ \sigma_{yy} \\ \sigma_{xy} \end{Bmatrix} = \frac{K_{\mathrm{I}}^{\infty}}{\sqrt{2\pi r}} \cos\tfrac{1}{2}\theta \begin{Bmatrix} 1 - \sin\frac{1}{2}\theta \sin\frac{1}{2}3\theta \\ 1 + \sin\frac{1}{2}\theta \sin\frac{1}{2}3\theta \\ \sin\frac{1}{2}\theta \cos\frac{1}{2}3\theta \end{Bmatrix}, \tag{6}$$

$$\begin{Bmatrix} u_x \\ u_y \end{Bmatrix} = \frac{K_{\mathrm{I}}^{\infty}}{2G}\sqrt{\frac{r}{2\pi}} \begin{Bmatrix} \cos\frac{1}{2}\theta \left[\kappa - 1 + 2\sin^2\frac{1}{2}\theta\right] \\ \sin\frac{1}{2}\theta \left[\kappa + 1 - 2\cos^2\frac{1}{2}\theta\right] \end{Bmatrix}, \tag{7}$$

where the $r$-$\theta$ polar coordinate is centered at the crack tip, $G = E/2(1+\nu)$ is the shear modulus, $\kappa = 3 - 4\nu$ for plane strain, $\kappa = (3-\nu)/(1+\nu)$ for plane stress.

As shown in Figure 7, the outermost radius $r_3$ from the physical crack tip of the zone dominated by $K_{\mathrm{I}}^{\infty}$ is expected to depend on the macroscopic characteristic length $L$, and in turn, $r_3$ is some fraction of $L$. Thus, in a homogeneized medium, region III as shown in Figure 7 is defined as being dominated by the isotropic singular stress field given by (6) and (7). While this may be a realistic approximation for material systems with large $L/l$ ratios, it is underscored that this approach may only provide useful stress field approximations at material points located within region III at distances from the physical crack tip which are much larger than the microstructural characteristic length $l$. As a result of the microstructure, it is clear that distinctly different asymptotics will dominate the mechanics at material points in region I (which contains the physical crack tip and extends to a distance $r_1$), and that these will determine the crack propagation load and direction. The extent $r_1$ of region I from the crack tip depends on the position of the crack tip relative to the adjacent crystal, and the degree of crystalline anisotropy. The distinct asymptotics dominating regions I and III discussed above are assumed to develop in heterogeneous systems with large $L/l$. In polycrystalline systems with nearly isotropic crystals, the dominance of region I would clearly prevail over region III, with $r_1$ extending to $r_3$. For high levels of anisotropy, the asymptotic fields in regions I and III are expected to be matched through other fields dominating the transition region II, extending from $r_1$ to $r_2$.

In the immediate vicinity of the crack tip (region I), which is anisotropic, the stress and displacement components are given by (Sih et al., 1965)

$$\begin{Bmatrix} \sigma_{xx} \\ \sigma_{yy} \\ \sigma_{xy} \end{Bmatrix} = \frac{K_{\mathrm{I}}^{l}}{\sqrt{2\pi r}} \mathrm{Re} \begin{Bmatrix} \frac{\mu_1\mu_2}{\mu_1-\mu_2}\left[\frac{\mu_2}{\sqrt{\cos\theta+\mu_2\sin\theta}} - \frac{\mu_1}{\sqrt{\cos\theta+\mu_1\sin\theta}}\right] \\ \frac{1}{\mu_1-\mu_2}\left[\frac{\mu_1}{\sqrt{\cos\theta+\mu_2\sin\theta}} - \frac{\mu_2}{\sqrt{\cos\theta+\mu_1\sin\theta}}\right] \\ \frac{\mu_1\mu_2}{\mu_1-\mu_2}\left[\frac{1}{\sqrt{\cos\theta+\mu_1\sin\theta}} - \frac{1}{\sqrt{\cos\theta+\mu_2\sin\theta}}\right] \end{Bmatrix}$$

$$+\frac{K_{\mathrm{II}}^{l}}{\sqrt{2\pi r}}\mathrm{Re}\left\{\begin{array}{l}\frac{1}{\mu_1-\mu_2}\left[\frac{\mu_2^2}{\sqrt{\cos\theta+\mu_2\sin\theta}}-\frac{\mu_1^2}{\sqrt{\cos\theta+\mu_1\sin\theta}}\right]\\ \frac{1}{\mu_1-\mu_2}\left[\frac{1}{\sqrt{\cos\theta+\mu_2\sin\theta}}-\frac{1}{\sqrt{\cos\theta+\mu_1\sin\theta}}\right]\\ \frac{1}{\mu_1-\mu_2}\left[\frac{\mu_1}{\sqrt{\cos\theta+\mu_1\sin\theta}}-\frac{\mu_2}{\sqrt{\cos\theta+\mu_2\sin\theta}}\right]\end{array}\right\}, \tag{8a}$$

$$\left\{\begin{array}{l}u_x\\ u_y\end{array}\right\} = K_{\mathrm{I}}^{l}\sqrt{\frac{r}{2\pi}}\mathrm{Re}\left\{\begin{array}{l}\frac{1}{\mu_1-\mu_2}[\mu_1 p_2(\cos\theta+\mu_2\sin\theta)^{1/2}-\mu_2 p_1(\cos\theta+\mu_1\sin\theta)^{1/2}]\\ \frac{1}{\mu_1-\mu_2}[\mu_1 q_2(\cos\theta+\mu_2\sin\theta)^{1/2}-\mu_2 q_1(\cos\theta+\mu_1\sin\theta)^{1/2}]\end{array}\right\}$$

$$+K_{\mathrm{II}}^{l}\sqrt{\frac{r}{2\pi}}\mathrm{Re}\left\{\begin{array}{l}\frac{1}{\mu_1-\mu_2}[p_2(\cos\theta+\mu_2\sin\theta)^{1/2}-p_1(\cos\theta+\mu_1\sin\theta)^{1/2}]\\ \frac{1}{\mu_1-\mu_2}[q_2(\cos\theta+\mu_2\sin\theta)^{1/2}-q_1(\cos\theta+\mu_1\sin\theta)^{1/2}]\end{array}\right\}, \tag{8b}$$

where Re(Im) denotes real (imaginary) part

$$p_j = c'_{11}\mu_j^2 + c'_{12} - c'_{16}\mu_j, \quad j = 1, 2, \tag{9a}$$

$$q_j = c'_{12}\mu_j + \frac{c'_{22}}{\mu_j} - c'_{26}, \tag{9b}$$

the $c'_{ij}$ are the flexibility coefficients (the components of the inverse of matrix C in (4)), transformed to the global axes, and $\mu_j$ are the roots (with positive imaginary parts) of the characteristic equation

$$c'_{11}\mu^4 - 2c'_{16}\mu^3 + (2c'_{12} + c'_{44})\mu^2 - 2c'_{26}\mu + c'_{22} = 0. \tag{10}$$

The energy release rate for self-similar extension is given as

$$G^l = -\tfrac{1}{2}K_{\mathrm{I}}^{l}c'_{22}\,\mathrm{Im}\left[\frac{K_{\mathrm{I}}^{l}(\mu_1+\mu_2)+K_{\mathrm{II}}^{l}}{\mu_1\mu_2}\right] + \tfrac{1}{2}K_{\mathrm{II}}^{l}c_{11}\,\mathrm{Im}[K_{\mathrm{II}}^{l}(\mu_1+\mu_2)+K_{\mathrm{I}}^{l}\mu_1\mu_2]. \tag{11}$$

Because the crystals in a real material are randomly orientated, the behavior of the stress intensity factors and of the stresses in the vicinity of the crack is more complicated; these physical parameters can only be discussed in terms of their probability distributions. Indeed, the scatter in the stresses ahead of the crack may be large enough to render the previous near-tip discussion irrelevant. In other words, distinct asymptotics may not be observed, as will be subsequently shown.

## 3. Near-tip finite element model

Consider the near-tip plane strain region shown in Figure 4. The physical dimensions of this zone are $RxR$ (with $R = 1$) such that $R \leqslant r_3$, where $r_3$ signifies the boundary of region III consistent with Figure 7. In this model, the Poisson-Voronoi tessellation approximation of the polycrystalline structure is retained. The region under consideration includes the crack tip, whose position within the crystal is a random variable. The crack surfaces are traction-free. Mode-I loading is applied by prescribing, along the boundary of the region, the displacements in (7). It should be understood that this model does not incorporate finite geometry

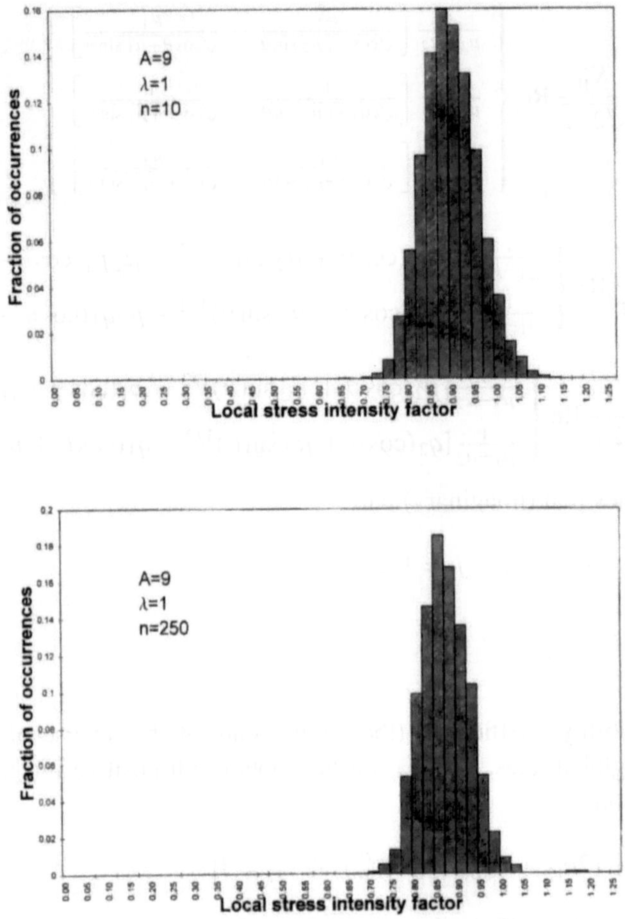


*Figure 8.* Fraction of occurrences of local stress intensity factor for highly anisotropic cubic crystals.

characteristics. In other words, $R$ does not directly represent a macroscopic length, rather it provides an indirect measure of a macroscopic characteristic length. The only requirement on $R$ is that stated earlier, i.e. $R \leqslant r_3$. Various morphologies corresponding to different $L/l$ are modeled by changing the number of crystals $n$ within the modeled region. It is expected that this approach will capture the various near-tip asymptotics and the transitional characteristics from region I to region III through region II. The relation between $R$ and $L$ could be quantified only by modeling specific finite geometries.

The microscopic stress intensity factors are calculated using the displacement correlation technique (Barsoum, 1977; Ingraffea and Manu, 1980), which involves equating the square root portion of the finite element displacement interpolation of the crack opening and crack sliding displacements, with the analytical values given by (8b). These in turn are used to calculate the local energy release rate.

## 4. Results

The average and standard deviation (denoted by subscripts avg and sd, respectively) of $K \equiv K_{\mathrm{I}}^{l}/K_{\mathrm{I}}^{\infty}$ and $G \equiv G^{l}/G_{\mathrm{I}}^{\infty}$ are summarized in Table 1. For $m$ simulations (the results presented

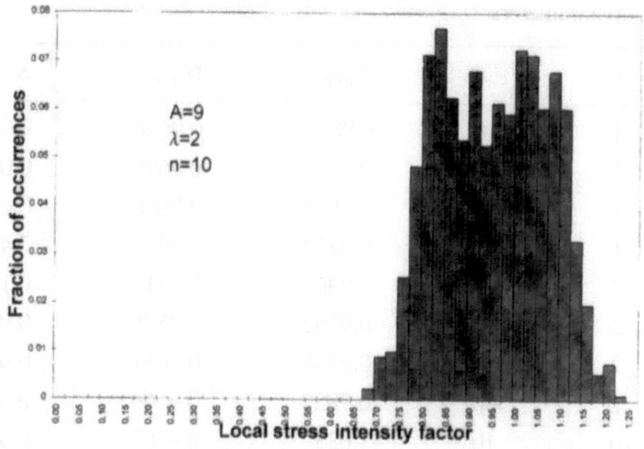


*Figure 9.* Fraction of occurrences of local stress intensity factor for highly anisotropic, slightly orthotropic crystals.

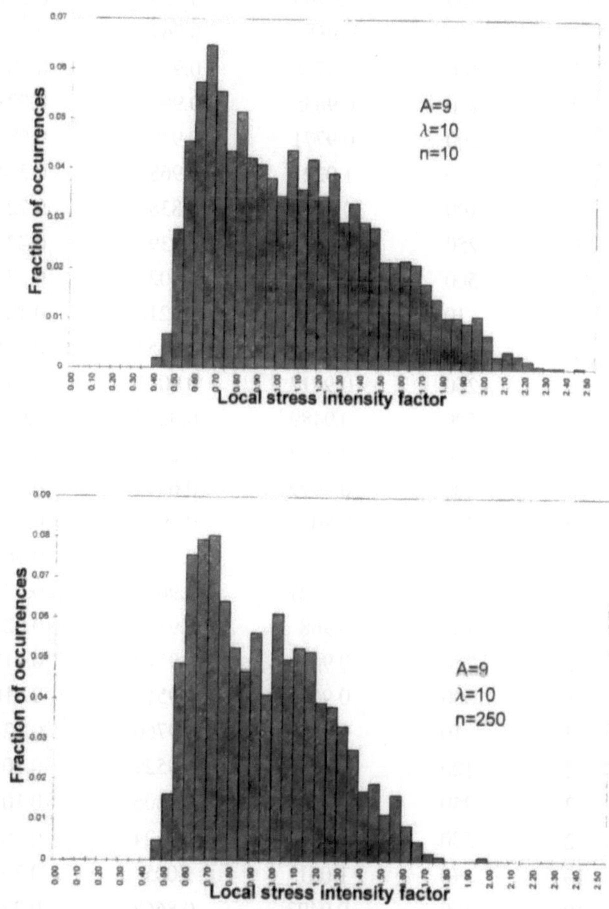


*Figure 10.* Fraction of occurrences of local stress intensity factor for highly anisotropic orthotropic crystals.

in Table 1 correspond to $m = 200$–$400$), the standard deviation of $x$ is defined as $\sqrt{(\Sigma_m x^2/(m-1)) - m x_{\text{avg}}^2/(m-1)}$. The results for the isotropic crystals ($A = \lambda = 1$) show that the errors in the calculated stress intensity factors and energy release rates are on the order of a few percent.

*Table 1.* Calculated crack tip parameters

| $A$ | $\overline{\nu}$ | $\lambda$ | $n$ | $K_{\text{avg}}$ | $G_{\text{avg}}$ | $K_{\text{sd}}$ | $G_{\text{sd}}$ |
|---|---|---|---|---|---|---|---|
| 1 | 0 | 1 | 10 | 0.9846 | 0.9696 | 0.00692 | 0.01398 |
| 1 | 0 | 1 | 100 | 0.9848 | 0.9699 | 0.00664 | 0.01377 |
| 1 | 0 | 1 | 250 | 0.9831 | 0.9666 | 0.00405 | 0.0077 |
| 1 | 0 | 1 | 500 | 0.9839 | 0.968 | 7.4E-05 | 0.00015 |
| 1 | 0 | 2 | 10 | 0.9969 | 0.9818 | 0.11 | 0.1203 |
| 1 | 0 | 2 | 100 | 0.9792 | 0.9575 | 0.08638 | 0.08356 |
| 1 | 0 | 2 | 250 | 0.9938 | 0.9651 | 0.083 | 0.07982 |
| 1 | 0 | 2 | 500 | 0.9852 | 0.9566 | 0.08823 | 0.08569 |
| 1 | 0 | 10 | 10 | 1.035 | 0.9767 | 0.3407 | 0.4097 |
| 1 | 0 | 10 | 100 | 0.992 | 0.8872 | 0.2745 | 0.2537 |
| 1 | 0 | 10 | 250 | 0.9601 | 0.8665 | 0.2644 | 0.2299 |
| 1 | 0 | 10 | 500 | 0.9303 | 0.8395 | 0.2191 | 0.1949 |
| 1 | 0.2 | 2 | 10 | 0.9853 | 0.9647 | 0.1117 | 0.1224 |
| 1 | 0.2 | 2 | 100 | 0.9778 | 0.9574 | 0.08969 | 0.09426 |
| 1 | 0.2 | 2 | 250 | 0.9835 | 0.96 | 0.09177 | 0.08819 |
| 1 | 0.2 | 2 | 500 | 0.9771 | 0.948 | 0.08347 | 0.08796 |
| 1 | 0.2 | 10 | 10 | 1.022 | 0.9687 | 0.3543 | 0.4279 |
| 1 | 0.2 | 10 | 100 | 0.9242 | 0.8389 | 0.2285 | 0.2121 |
| 1 | 0.2 | 10 | 250 | 0.9484 | 0.8393 | 0.2372 | 0.2289 |
| 1 | 0.2 | 10 | 500 | 0.9043 | 0.802 | 0.2329 | 0.1912 |
| 1 | 0.4 | 2 | 10 | 0.9558 | 0.9216 | 0.1005 | 0.1155 |
| 1 | 0.4 | 2 | 100 | 0.9485 | 0.9153 | 0.08509 | 0.09749 |
| 1 | 0.4 | 2 | 250 | 0.9417 | 0.9011 | 0.0798 | 0.08157 |
| 1 | 0.4 | 2 | 500 | 0.9489 | 0.9047 | 0.08087 | 0.07973 |
| 1 | 0.4 | 10 | 10 | 0.8912 | 0.8295 | 0.2974 | 0.3599 |
| 1 | 0.4 | 10 | 100 | 0.7999 | 0.6814 | 0.1737 | 0.1962 |
| 1 | 0.4 | 10 | 250 | 0.8087 | 0.6948 | 0.1591 | 0.1861 |
| 1 | 0.4 | 10 | 500 | 0.821 | 0.7024 | 0.16 | 0.1941 |
| 3 | 0 | 1 | 10 | 0.9727 | 0.9617 | 0.02514 | 0.04963 |
| 3 | 0 | 1 | 100 | 0.968 | 0.9521 | 0.0216 | 0.04259 |
| 3 | 0 | 1 | 250 | 0.968 | 0.9519 | 0.01872 | 0.03676 |
| 3 | 0 | 1 | 500 | 0.9675 | 0.951 | 0.01883 | 0.03715 |
| 3 | 0 | 2 | 10 | 0.9855 | 0.9766 | 0.1248 | 0.1485 |
| 3 | 0 | 2 | 100 | 0.978 | 0.9529 | 0.105 | 0.1255 |
| 3 | 0 | 2 | 250 | 0.9705 | 0.9406 | 0.1081 | 0.1212 |
| 3 | 0 | 2 | 500 | 0.966 | 0.9424 | 0.1034 | 0.1123 |
| 3 | 0 | 10 | 10 | 1.061 | 1.005 | 0.3984 | 0.4662 |
| 3 | 0 | 10 | 100 | 0.9493 | 0.8667 | 0.2769 | 0.2766 |
| 3 | 0 | 10 | 250 | 0.9744 | 0.8869 | 0.272 | 0.264 |
| 3 | 0 | 10 | 500 | 0.9866 | 0.8936 | 0.2886 | 0.2839 |
| 3 | 0.2 | 1 | 10 | 0.9732 | 0.9681 | 0.02889 | 0.05746 |
| 3 | 0.2 | 1 | 100 | 0.9729 | 0.9615 | 0.1134 | 0.1235 |
| 3 | 0.2 | 1 | 250 | 0.965 | 0.9515 | 0.02521 | 0.04959 |

*Table 1.* (Contd.). Calculated crack tip parameters

| $A$ | $\bar{\nu}$ | $\lambda$ | $n$ | $K_{avg}$ | $G_{avg}$ | $K_{sd}$ | $G_{sd}$ |
|---|---|---|---|---|---|---|---|
| 3 | 0.2 | 1 | 500 | 0.9631 | 0.9478 | 0.0253 | 0.04997 |
| 3 | 0.2 | 2 | 10 | 0.9922 | 0.9742 | 0.1181 | 0.1344 |
| 3 | 0.2 | 2 | 100 | 0.9708 | 0.9498 | 0.09824 | 0.1031 |
| 3 | 0.2 | 2 | 250 | 0.9827 | 0.9554 | 0.09156 | 0.09461 |
| 3 | 0.2 | 2 | 500 | 0.9822 | 0.9566 | 0.09097 | 0.09325 |
| 3 | 0.2 | 10 | 10 | 1.073 | 1.02 | 0.3866 | 0.4397 |
| 3 | 0.2 | 10 | 100 | 1.007 | 0.9227 | 0.2873 | 0.3059 |
| 3 | 0.2 | 10 | 250 | 0.9962 | 0.9105 | 0.2977 | 0.283 |
| 3 | 0.2 | 10 | 500 | 0.9834 | 0.9068 | 0.2634 | 0.2585 |
| 3 | 0.4 | 1 | 10 | 0.9597 | 0.9542 | 0.06316 | 0.118 |
| 3 | 0.4 | 1 | 100 | 0.9431 | 0.9221 | 0.05847 | 0.1119 |
| 3 | 0.4 | 1 | 250 | 0.9481 | 0.9313 | 0.05737 | 0.1074 |
| 3 | 0.4 | 1 | 500 | 0.9577 | 0.9498 | 0.0766 | 0.1456 |
| 3 | 0.4 | 2 | 10 | 0.9774 | 0.9557 | 0.1104 | 0.1238 |
| 3 | 0.4 | 2 | 100 | 0.9764 | 0.9425 | 0.09122 | 0.101 |
| 3 | 0.4 | 2 | 250 | 0.9766 | 0.9468 | 0.08848 | 0.09471 |
| 3 | 0.4 | 2 | 500 | 0.9688 | 0.9418 | 0.08373 | 0.08225 |
| 3 | 0.4 | 10 | 10 | 1.037 | 0.9866 | 0.3548 | 0.3915 |
| 3 | 0.4 | 10 | 100 | 0.9849 | 0.8779 | 0.2811 | 0.2693 |
| 3 | 0.4 | 10 | 250 | 0.9347 | 0.8322 | 0.245 | 0.211 |
| 3 | 0.4 | 10 | 500 | 0.9376 | 0.8399 | 0.2447 | 0.2211 |
| 5 | 0 | 1 | 10 | 0.9711 | 0.9648 | 0.04471 | 0.07122 |
| 5 | 0 | 1 | 100 | 0.9567 | 0.9364 | 0.03651 | 0.05684 |
| 5 | 0 | 1 | 250 | 0.9592 | 0.9412 | 0.03872 | 0.05649 |
| 5 | 0 | 1 | 500 | 0.9604 | 0.9398 | 0.04123 | 0.05033 |
| 5 | 0 | 2 | 10 | 0.9809 | 0.9708 | 0.122 | 0.1472 |
| 5 | 0 | 2 | 100 | 0.9621 | 0.9425 | 0.1037 | 0.119 |
| 5 | 0 | 2 | 250 | 0.9638 | 0.9488 | 0.1051 | 0.1296 |
| 5 | 0 | 2 | 500 | 0.95 | 0.932 | 0.09389 | 0.1007 |
| 5 | 0 | 10 | 10 | 1.055 | 1.022 | 0.4035 | 0.4877 |
| 5 | 0 | 10 | 100 | 0.9945 | 0.9171 | 0.3043 | 0.3055 |
| 5 | 0 | 10 | 250 | 0.9684 | 0.9034 | 0.2968 | 0.3025 |
| 5 | 0 | 10 | 500 | 0.9475 | 0.8725 | 0.2836 | 0.2766 |
| 5 | 0.2 | 1 | 10 | 0.9593 | 0.9577 | 0.03936 | 0.07897 |
| 5 | 0.2 | 1 | 100 | 0.9506 | 0.9396 | 0.0345 | 0.06848 |
| 5 | 0.2 | 1 | 250 | 0.9524 | 0.9431 | 0.03362 | 0.06683 |
| 5 | 0.2 | 1 | 500 | 0.952 | 0.9325 | 0.1455 | 0.1505 |
| 5 | 0.2 | 2 | 10 | 0.9759 | 0.9687 | 0.1191 | 0.1415 |
| 5 | 0.2 | 2 | 100 | 0.9581 | 0.9437 | 0.1003 | 0.1194 |
| 5 | 0.2 | 2 | 250 | 0.9723 | 0.9534 | 0.09271 | 0.1078 |
| 5 | 0.2 | 2 | 500 | 0.9712 | 0.9447 | 0.08757 | 0.101 |
| 5 | 0.2 | 10 | 10 | 1.077 | 1.063 | 0.4268 | 0.5295 |
| 5 | 0.2 | 10 | 100 | 0.9624 | 0.8901 | 0.2908 | 0.3069 |

*Table 1.* (Contd.). Calculated crack tip parameters

| $A$ | $\overline{\nu}$ | $\lambda$ | $n$ | $K_{avg}$ | $G_{avg}$ | $K_{sd}$ | $G_{sd}$ |
|---|---|---|---|---|---|---|---|
| 5 | 0.2 | 10 | 250 | 0.9841 | 0.892 | 0.286 | 0.27 |
| 5 | 0.2 | 10 | 500 | 0.9676 | 0.8927 | 0.2917 | 0.2902 |
| 5 | 0.4 | 1 | 10 | 0.9303 | 0.9299 | 0.04891 | 0.09878 |
| 5 | 0.4 | 1 | 100 | 0.9123 | 0.8944 | 0.05005 | 0.09767 |
| 5 | 0.4 | 1 | 250 | 0.9143 | 0.8976 | 0.04615 | 0.09006 |
| 5 | 0.4 | 1 | 500 | 0.9229 | 0.9153 | 0.05063 | 0.1015 |
| 5 | 0.4 | 2 | 10 | 0.9762 | 0.9643 | 0.1165 | 0.1416 |
| 5 | 0.4 | 2 | 100 | 0.9662 | 0.9458 | 0.09975 | 0.1183 |
| 5 | 0.4 | 2 | 250 | 0.9558 | 0.9297 | 0.09285 | 0.1054 |
| 5 | 0.4 | 2 | 500 | 0.9677 | 0.9406 | 0.09325 | 0.1083 |
| 5 | 0.4 | 10 | 10 | 1.105 | 1.054 | 0.3949 | 0.466 |
| 5 | 0.4 | 10 | 100 | 0.9975 | 0.8837 | 0.2743 | 0.2613 |
| 5 | 0.4 | 10 | 250 | 0.9845 | 0.8991 | 0.2702 | 0.2659 |
| 5 | 0.4 | 10 | 500 | 0.9684 | 0.8628 | 0.2486 | 0.2308 |
| 9 | 0 | 1 | 10 | 0.9579 | 0.956 | 0.04177 | 0.08366 |
| 9 | 0 | 1 | 100 | 0.9386 | 0.9162 | 0.02891 | 0.0563 |
| 9 | 0 | 1 | 250 | 0.9372 | 0.9134 | 0.02816 | 0.05491 |
| 9 | 0 | 1 | 500 | 0.9444 | 0.9254 | 0.04802 | 0.07167 |
| 9 | 0 | 2 | 10 | 0.9678 | 0.9596 | 0.1291 | 0.1637 |
| 9 | 0 | 2 | 100 | 0.9439 | 0.922 | 0.1051 | 0.1243 |
| 9 | 0 | 2 | 250 | 0.9269 | 0.9031 | 0.09488 | 0.1116 |
| 9 | 0 | 2 | 500 | 0.9367 | 0.9143 | 0.09702 | 0.1162 |
| 9 | 0 | 10 | 10 | 1.038 | 1.041 | 0.4261 | 0.5412 |
| 9 | 0 | 10 | 100 | 0.9199 | 0.8478 | 0.3257 | 0.3755 |
| 9 | 0 | 10 | 250 | 0.9021 | 0.8214 | 0.2705 | 0.2901 |
| 9 | 0 | 10 | 500 | 0.9342 | 0.8535 | 0.3035 | 0.3358 |
| 9 | 0.2 | 1 | 10 | 0.94 | 0.948 | 0.04818 | 0.09767 |
| 9 | 0.2 | 1 | 100 | 0.9255 | 0.9172 | 0.04194 | 0.08291 |
| 9 | 0.2 | 1 | 250 | 0.9188 | 0.9035 | 0.03489 | 0.06888 |
| 9 | 0.2 | 1 | 500 | 0.9227 | 0.9111 | 0.03523 | 0.06972 |
| 9 | 0.2 | 2 | 10 | 0.9676 | 0.9695 | 0.1692 | 0.2056 |
| 9 | 0.2 | 2 | 100 | 0.9378 | 0.9171 | 0.0987 | 0.123 |
| 9 | 0.2 | 2 | 250 | 0.935 | 0.9131 | 0.1016 | 0.1204 |
| 9 | 0.2 | 2 | 500 | 0.9463 | 0.9269 | 0.09811 | 0.1199 |
| 9 | 0.2 | 10 | 10 | 1.073 | 1.073 | 0.4394 | 0.5444 |
| 9 | 0.2 | 10 | 100 | 0.9414 | 0.866 | 0.3021 | 0.31 |
| 9 | 0.2 | 10 | 250 | 0.9385 | 0.8598 | 0.2882 | 0.3136 |
| 9 | 0.2 | 10 | 500 | 0.921 | 0.8624 | 0.305 | 0.3383 |
| 9 | 0.4 | 1 | 10 | 0.9006 | 0.9132 | 0.05919 | 0.1214 |
| 9 | 0.4 | 1 | 100 | 0.8835 | 0.8786 | 0.06104 | 0.1216 |
| 9 | 0.4 | 1 | 250 | 0.875 | 0.8609 | 0.054 | 0.1073 |
| 9 | 0.4 | 1 | 500 | 0.869 | 0.8504 | 0.06049 | 0.1189 |
| 9 | 0.4 | 2 | 10 | 0.963 | 0.9636 | 0.127 | 0.1614 |

*Table 1.* (Contd.). Calculated crack tip parameters

| $A$ | $\overline{\nu}$ | $\lambda$ | $n$ | $K_{avg}$ | $G_{avg}$ | $K_{sd}$ | $G_{sd}$ |
|---|---|---|---|---|---|---|---|
| 9 | 0.4 | 2 | 100 | 0.9376 | 0.9131 | 0.1017 | 0.1272 |
| 9 | 0.4 | 2 | 250 | 0.9387 | 0.9093 | 0.09749 | 0.1212 |
| 9 | 0.4 | 2 | 500 | 0.9396 | 0.9168 | 0.09734 | 0.1183 |
| 9 | 0.4 | 10 | 10 | 1.089 | 1.064 | 0.3915 | 0.4936 |
| 9 | 0.4 | 10 | 100 | 0.9735 | 0.9041 | 0.3105 | 0.318 |
| 9 | 0.4 | 10 | 250 | 1.009 | 0.9125 | 0.2921 | 0.304 |
| 9 | 0.4 | 10 | 500 | 0.9823 | 0.8908 | 0.2944 | 0.3047 |

For cubic crystals ($\lambda = 1$), the effects of anisotropy and heterogeneity on the crack tip parameters are relatively small. Consider the results for $n = 500$; for the highest levels of anisotropy considered ($A = 9, \overline{\nu} = 0.4$), the average stress intensity factor is slightly reduced to $\approx 0.87 \pm 0.06$ ($\pm x$ represents $\pm$ one standard deviation), and the average energy release rate is reduced to $\approx 0.85 \pm 0.12$. Moreover, the results are not sensitive to $n$. Polycrystalline silicon is defined by $A = 1.65, \overline{\nu} = 0.282$; for all values of $n$, the average stress intensity factor is $\approx 0.93 \pm 0.02$, and the average energy release rate is $\approx 0.91 \pm 0.03$.

The results for orthotropic crystals show that anisotropy and heterogeneity produce significant effects for highly anisotropic crystals. For example, if $n = 500$, $A = 9$, $\lambda = 10$, and $\overline{\nu} = 0.4$, the stress intensity factor is $\approx 0.98 \pm 0.29$, and the energy release rate is $\approx 0.89 \pm 0.30$. Moreover, for highly anisotropic crystals, the results are significantly influenced by the number of grains; for $n = 10$, the standard deviations of the stress intensity factor and energy release rate, for the previous example, increase to 0.39 and 0.49, respectively.

Figures 8–10 show, respectively, the fraction of occurrences of the local Mode-I stress intensity factor for highly anisotropic cubic crystals, highly anisotropic 'slightly' orthotropic crystals, and high anisotropic orthotropic crystals. It is observed that for cubic crystals, the distribution is relatively narrow and nearly symmetric about the mean, while for orthotropic crystals the distribution is wider and not symmetric. Moreover, for orthotropic crystals, a small population of realizations is associated with stress intensity factors that are more than double of the nominal. It is interesting to note that unsymmetric distributions of stress intensity factors were also predicted by the model developed by Scavia and Carpinteri (1991) in their study of the interaction between a crack with a random distribution of surrounding microcracks. However, while Figure 10 indicates peaks in the shielding range and a long tail in the amplification range, similar plots in (Scavia and Carpinteri, 1991) show that distributed microcracking leads to peaks in the amplification range and a long tail in the shielding range.

The scatter of the normalized stress $\sigma \equiv \sqrt{2\pi R}\sigma_{yy}/K_{\mathrm{I}}^{\infty}$ ahead of the crack for highly anisotropic cubic crystals, and highly anisotropic orthotropic crystals, are shown in Figures 11 and 12, respectively. In these figures the distance from the crack tip $r$, actually represents the nondimensional distance $r/R$, with $R = 1$. It is observed that for cubic and orthotropic crystals, even at high levels of anisotropy, the dual length scale asymptotics disappear within the scatter; the stress is square root singular, with upper and lower bounds modulated by the deviation of the local stress intensity factor from the mean. Furthermore, the distribution of stress is nearly independent of the number of crystals $n$.

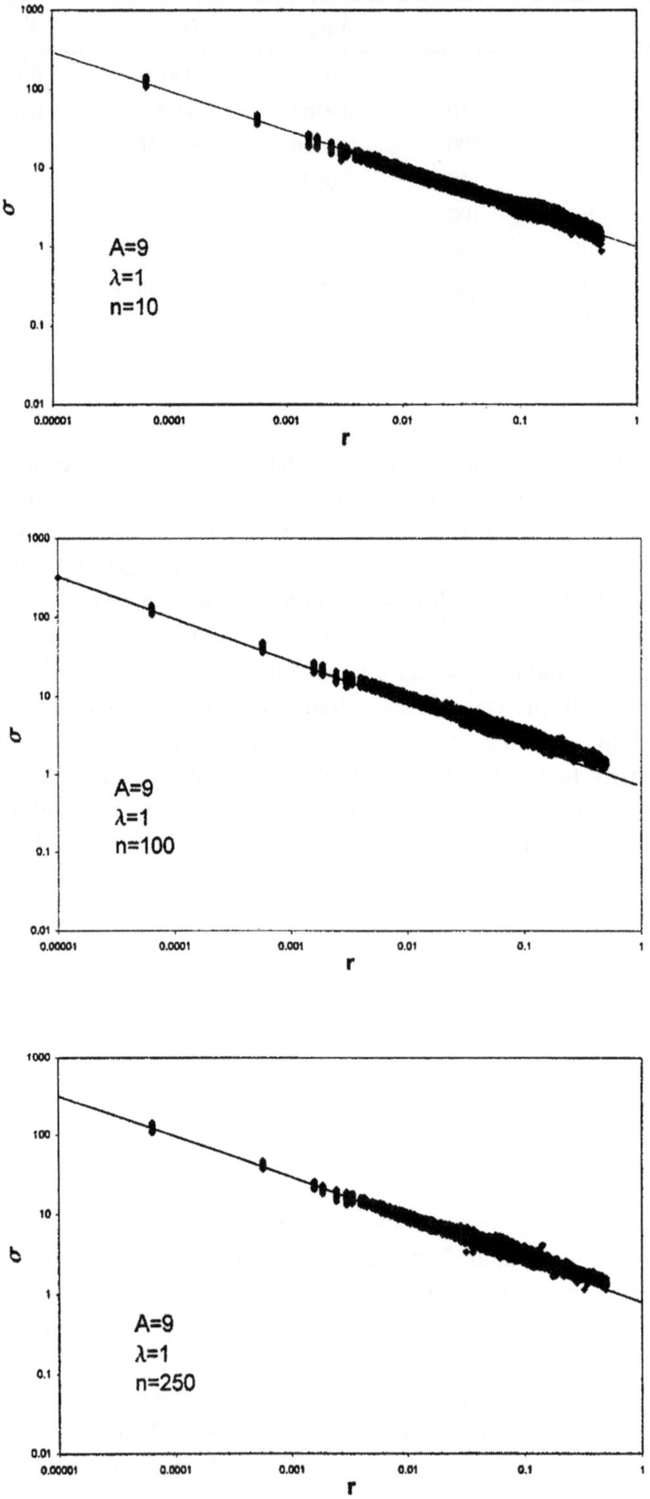


*Figure 11.* Scatter of normalized stress ahead of crack for several values of $n(\lambda = 1, A = 9, \overline{\nu} = 0.4)$.

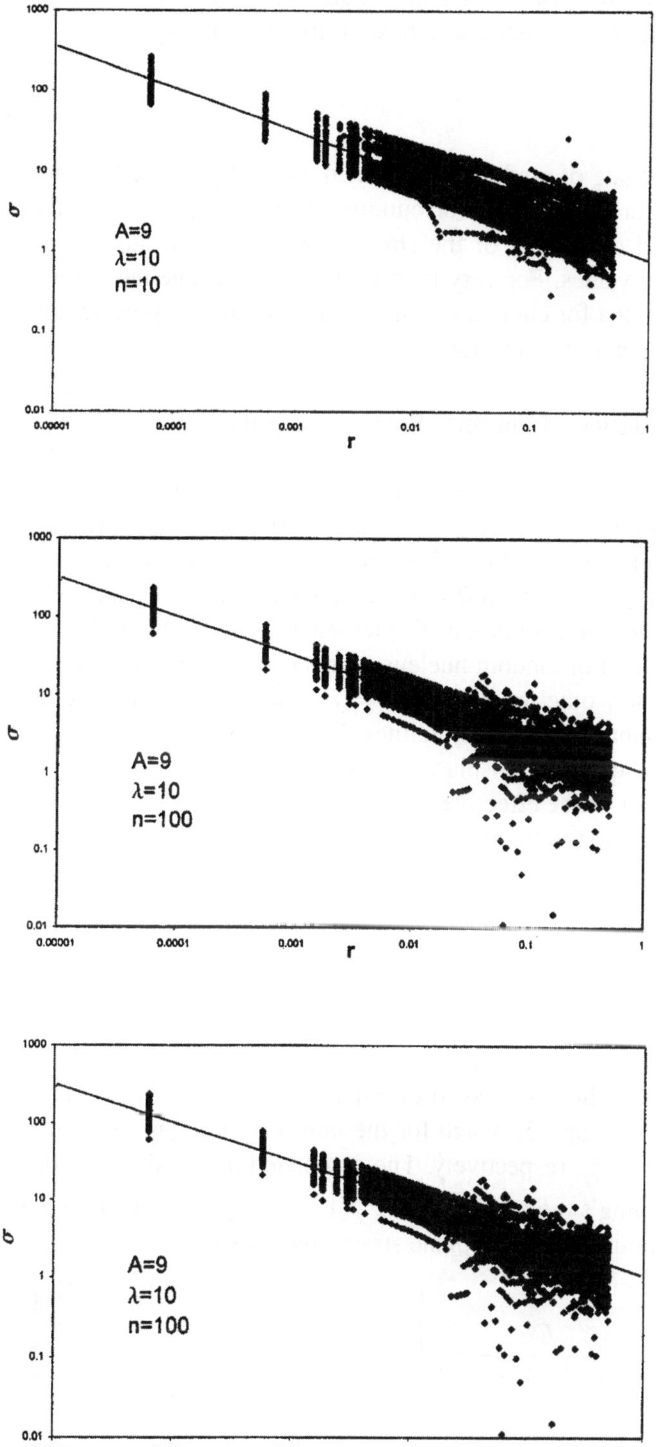


*Figure 12.* Scatter of normalized stress ahead of crack for several values of $n(\lambda = 10, A = 9, \overline{\nu} = 0.4)$.

It is important to note that the effects of anisotropy are expected to increase drastically if cracking within crystals or along grain boundaries is taken into account. The authors are currently calculating these effects, and those of finite geometry.

## 5. Conclusions

In a columnar aggregate of randomly orientated, perfectly bonded, orthotropic crystals under plane deformation, only the grains surrounding the crack tip significantly perturb the average values and standard deviations of the stress intensity factors and energy release rate from the nominal applied values. For very high levels of crystalline anisotropy, the average values and standard deviations for cubic and orthotropic crystals are perturbed by approximately 10 percent and 30–40 percent, respectively.

## Appendix. Calculation of nominal elastic constants

With respect to the global coordinate system $x$-$y$-$z$ shown in Figure 3, the film is analyzed under plane strain conditions, $u_z = \partial(\,)/\partial z = 0$, where $u_z$ is the displacement in the $z$ direction, and ( ) denotes any physical quantity. For a given number of grains $n$, the topology of the microstructure within the *unit* volume shown in Figure 3 is approximated, for *each* of $m$ simulations, as a Voronoi tessellation. The tessellation is constructed by first generating, using a Poisson process, a set of random nucleus points $\boldsymbol{r}_i$ $(i = 1, n)$. The set of all points which are closer to $\boldsymbol{r}_i$ than to any other nucleus $\boldsymbol{r}_j$ $(i \neq j)$ comprise the $i$th Voronoi cell. The assembly of the $n$ convex, planar edged cells, defines the tessellation. Each grain is discretized with a sufficient number of quadratic displacement finite elements.

For determination of the effective elastic constants, the film *structure* is loaded, as shown in Figure 3, through the following boundary conditions

$$u_y = \Delta, \quad y = 1, \tag{A1}$$

$$u_x = \sigma_{xy} = 0, \quad x = 0, \quad x = 1, \tag{A2}$$

$$u_y = \sigma_{xy} = 0, \quad y = 0. \tag{A3}$$

These are associated with nominal (average) strain components $\varepsilon^n_{yy} = \Delta/1$, $\varepsilon^n_{xx} = 0$, $\varepsilon^n_{xy} = 0$. The finite element method is used to calculate the total reaction forces (per unit thickness) $F_x$ and $F_y$ shown in Figure 3, which for the unit length edges, correspond to nominal stress components $\sigma^n_{xx}$ and $\sigma^n_{yy}$, respectively. The associated nominal values of the Poisson ratio $\nu^n$ and plane strain Young's modulus $\overline{E}^n \equiv E^n/[1 - (\nu^n)^2]$ are defined by invoking Hooke's law for the *isotropic* continuum under plane strain conditions

$$\begin{Bmatrix} \sigma^n_{xx} = F_x \\ \sigma^n_{yy} = F_y \\ \sigma^n_{xy} = 0 \end{Bmatrix} = \frac{E^n}{(1+\nu^n)(1-2\nu^n)} \begin{bmatrix} 1-\nu^n & \nu^n & 0 \\ \nu^n & 1-\nu^n & 0 \\ 0 & 0 & (1-2\nu^n) \end{bmatrix} \begin{Bmatrix} \varepsilon^n_{xx} = 0 \\ \varepsilon^n_{yy} = \Delta \\ \varepsilon^n_{xy} = 0 \end{Bmatrix}. \tag{A4}$$

Equations (A4) provide

$$\nu^n = \frac{1}{1 + \frac{F_y}{F_x}}, \tag{A5}$$

$$\overline{E}^n = \frac{E^n}{[1-(\nu^n)^2]} = \frac{(1+\nu^n)(1-2\nu^n)}{\nu^n[1-(\nu^n)^2]}\frac{F_x}{\Delta}. \tag{A6}$$

The steps described above are repeated $m$ times for each value of $n$.

$\overline{E}^n$ is normalized with the 'isotropic' plane strain Young's modulus $\overline{E}^{\text{isotropic}}$ obtained by equating the first entry in the stiffness matrix for cubic symmetry, for $A = 1$ (for fixed $\overline{\nu}$) with that of an isotropic material, i.e.

$$\overline{E}^{\text{isotropic}} \equiv \frac{c_{11}(1-2\overline{\nu})}{(1-\overline{\nu})^2}. \tag{A6}$$

## Acknowledgements

This research was funded by the National Science Foundation under Grants MSS94-16752 and CMS97-13997, and by ARPA under Grant DABT 63-92-C-0032. The first author acknowledges fruitful discussions with Professor Greg Rodin.

## References

Abdel-Tawab, K. and Rodin, G. (1997). *MSS&M Report 97/12*, Research Center for Mechanics of Solids Structures & Materials, The University of Texas at Austin.

Anderson, T.L. (1995). *Fracture Mechanics Fundamentals and Applications*, 2nd Edition, CRC Press.

Azhdari, A. and Nemat-Nasser, S. (1996). *Journal of Mechanics and Physics of Solids* **44**, 929–951.

Azhdari, A. and Nemat-Nasser, S. (1998). To appear in *Mechanics of Materials*.

Ballarini, R., Charalambides, P.G. and Islam, S. (1995). *International Journal of Fracture* **70**, 275–304.

Ballarini, R., Mullen, R.L., Yin, Y., Kahn, H., Stemmer, S. and Heuer, A.H. (1997). *Journal of Materials Research* **12**, 915–922.

Barsoum, R.S. (1977). *International Journal of Numerical Methods in Engineering* **11**, 85.

Bazant, Z.P. (1993). *ASCE Journal of Engineering Mechanics* **119**, 1828–1844.

Carpinteri, A. (1994). *Mechanics of Materials* **18**, 89–101.

Carpinteri, A. and Scavia, C. (1991). In *Fracture Processes in Concrete, Rock and Ceramics*, (Edited by J.G.M. van Mier, J.G. Rots and A. Bakker), E.& F.N. Spon, London, 173–182.

Cox, B.N. and Marshall, D.B. (1994). *Acta Metallurgica et Materialia* **42**, 341–363.

Frost, H.J. and Ashby, M.F. (1982). *Deformation Mechanism Maps*, Pergamon Press, Oxford.

Hutchinson, J.W. (1968). *Journal of the Mechanics and Physics of Solids* **16**, 13–31.

Ingraffea, A.R. and Manu, C. (1980). *International Journal of Numerical Methods in Engineering* **15**, 1427–1445.

Jha, M. (1997). *Ph.D. Dissertation*, Department of Mechanical Engineering, University of Maryland at Baltimore County.

Jha, M., Charalambides, P.G. and Ballarini, R. (1997). *International Journal of Solids and Structures* **34**, 1849–1871.

Kumar, S. (1992). *Ph.D. Dissertation*, Department of Engineering Science and Mechanics, Pennsylvania State University.

Mullen, R.L., Ballarini, R., Yin, Y. and Heuer, A.H. (1997). *Acta Materialia* **45**, 2247–2255.

Rice, J.R. and Rosengren, G.F. (1968). *Journal of the Mechanics and Physics of Solids* **16**, 1–12.

Sih, G.C., Paris, P.C. and Irwin, G.R. (1965). *International Journal of Fracture Mechanics* **1**, 189–203.

*International Journal of Fracture* **95:** 41–50, 1999.


# Size effect relations associated with cohesive zone type fracture at a blunt stress concentration

E. SMITH
*Manchester University – UMIST Materials Science Centre, Grosvenor Street, Manchester M1 7HS, United Kingdom*

Received 24 June 1998; accepted in revised form 12 November 1998

**Abstract.** The paper is concerned with the formulation of size effect relations associated with the effect of a blunt stress concentration on the failure of a quasi-brittle material, where the failure can be modelled in terms of a two-dimensional embedded blunt flaw within an infinite solid, or a blunt flaw in the surface of a semi-infinite solid. In order to obtain analytical results so that the role of material and geometrical parameters in the size effect relations can be clearly seen, the paper is primarily based on an idealised cohesive zone description of material failure, and on Mode III analyses, though some consideration is given to the practically more appropriate Mode I situation. The size effect relations are expressed in terms of relations between the maximum sustainable effective peak stress at the flaw root or an effective stress intensity, and material and geometrical parameters of the flaw. If the flaw is sufficiently large, the geometry effect simplifies to the effect of the flaw root radius.

**Key words:** Quasi-brittle materials, blunt stress concentration, cohesive zone model, size effect relations, scaling effects, material and geometrical parameters.

## 1. Introduction

This paper is concerned with the behaviour of brittle and quasi-brittle materials such as concrete, sea ice, rocks, tough ceramics and composites. With such materials, it is a well known experimental fact that the effective failure stress of an uncracked structure scales with its dimensions. This is best, and very simply, illustrated by considering the case of an uncracked bend beam specimen for which the maximum tensile stress occurs at the tensile surface. Assuming elastic behaviour and a beam depth $d$ and beam thickness $B$, the bending moment $M$ is equal to $Bd^2\sigma_s/6$ where $\sigma_s$ is the tensile stress at the surface. Experiments, for example with unreinforced concrete beams (Ozbolt and Eligehausen, 1996; Elices et al., 1996), have shown that the maximum moment $M_m$ that a beam is able to sustain is equal to $Bd^2\sigma_s/6$ with $\sigma_s = \sigma_{\mathrm{EFF}}$, and where $\sigma_{\mathrm{EFF}}$ (the effective tensile failure stress) decreases as the beam depth $d$ increases. It is the geometry effect, for which this $\sigma_{\mathrm{EFF}}$-$d$ dependency is an example, that is the concern of this paper. It has been emphasized (Li and Bazant, 1996) that the underlying cause of this type of geometry dependence is linked to the experimental observation that quasi-brittle materials exhibit damage zones that are able to grow in a stable manner up to the attainment of maximum load. That being the case, the dominant source of the geometry dependence is not randomness of strength, but rather it is deterministic, and is related to the way in which such a damage zone develops.

The simplest way of representing a damage zone is to use the so-called cohesive zone description, whereby a single infinitesimally thin two-dimensional cohesive zone starts to form at a planar surface, or at the root of a stress concentration, when the tensile stress at the surface or root attains some critical value $p_c$ (assuming plane strain deformation conditions). As the applied loading increases, the zone spreads into the interior of the structure. The zone can be characterised by a material specific relation between the tensile stress ($p$) and the relative displacement ($v$) between the zone faces, with $p$ being a maximum (with value $p_c$) at the leading edge of the cohesive zone. The zone is said to be fully developed when the stress falls to zero at the trailing edge of the zone, a situation that is assumed to be attained when the displacement $v$ attains a critical value $v_c$. With a general $p$-$v$ cohesive zone behaviour, the maximum stress, and thereby failure, is attained prior to the cohesive zone's full development. However, so as to simplify the considerations, it will, in this paper, be assumed that the stress $p$ within the cohesive zone remains constant at the value $p_c$ until the displacement $v$ attains the critical value $v_c$ when the stress $p$ is assumed to fall abruptly from $p_c$ to zero. This is the classic DBCS (Dugdale, 1960; Bilby et al., 1963) representation that is frequently used to model stress relaxation phenomena. With this specific cohesive zone behaviour, the attainment of maximum stress, and thereby failure, is associated with the full development of the cohesive zone.

The cohesive zone description, particularly when coupled with the DBCS representation, has been used extensively to quantify the maximum stress-geometry dependence when a structure contains a crack (see for example Li and Bazant, 1994; Elices et al., 1996; Smith 1996). Furthermore the author (Smith, 1997) has used the DBCS representation to model the uncracked bend beam configuration, and has shown that the effective tensile failure stress $\sigma_{\mathrm{EFF}}$ is critically dependent on the applied loading induced stress gradient beneath the surface. $\sigma_{\mathrm{EFF}}$ increases as the steepness of this gradient increases, and we therefore have a simple physical explanation as to why the effective tensile failure stress for an uncracked beam increases as the beam depth decreases.

This paper extends these considerations by considering the formation of a cohesive zone at the surface of a blunt stress concentration, instead of a planar surface or a sharp crack. The objective is to formulate size effect relations, and in order to obtain analytical results so that the interplay between material and geometrical parameters can be clearly seen, the paper is based on Mode III analyses, though some consideration is given to the practically more appropriate Mode I situation. A preliminary consideration of the problem has already been presented (Smith,1998), and the present paper extends the earlier paper's work.

## 2. Cohesive zone formation at a cylindrical hole: Use of mode III simulation model results

Consider the Mode I situation where there is a two-dimensional elliptic cylindrical hole in an infinite solid (Figure 1) which is subjected to the applied tensile stress $\sigma$. The semi-major axis is $a$ and the semi-minor axis is $b$, whereupon the radius of curvature of the hole at the ends of the major axis is $\rho = b^2/a$. Cohesive zones of the DBCS type described in the preceding section emanate from the root of the hole as shown in Figure 1. This model, if cut down the middle, provides an approximate representation of the situation where a cohesive zone emanates from the root of a cylindrical flaw in the surface of a semi-infinite solid. [The representation is, of course, exact for the corresponding Mode III situation.]

The anti-plane strain (Mode III) analogue of the plane strain (Mode I) model in Figure 1 can be analysed to give closed form solutions, thereby providing a clear picture of the way in which the various geometrical and material parameters interact. Thus the Mode III results

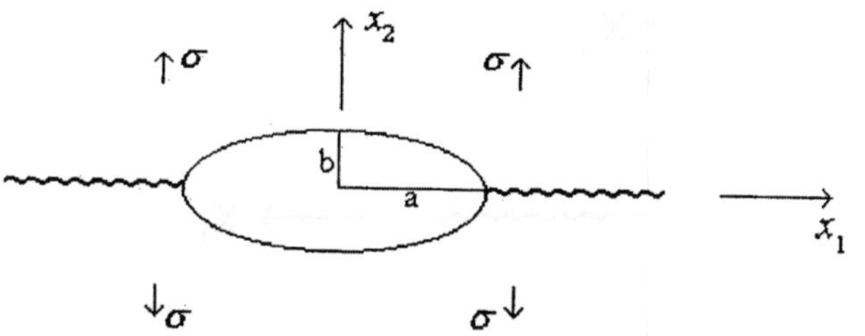


*Figure 1.* The model of an elliptic cylindrical hole in an infinite solid that is subjected to an applied tensile stress $\sigma$; there are cohesive zones at the roots of the hole.

(Smith, 1967) will be applied to the Mode I situation with the shear modulus being replaced by $E_0/2$ where $E_0$ is the reduced Young's modulus of the material, i.e., $E_0 = E/(1-\nu^2)$, $E$ being Young's modulus and $\nu$ being Poisson's ratio. The results give the maximum applied stress $\sigma$ required for failure, i.e., for the relative displacement at a flaw root to attain a critical value $v_c$, as

$$\frac{\sigma}{p_c} = \frac{(b/a)}{\{1+(b/a)\}} + \frac{2}{\pi\{1+(b/a)\}}\sec^{-1}\exp\left\{\frac{\pi E_0 v_c}{8p_c a}\right\}. \tag{1}$$

If the effective failure stress, i.e. the elastically calculated stress at a flaw root, is $\sigma_{EFF}$, i.e.

$$\frac{\sigma_{EFF}}{\sigma} = 1 + \frac{a}{b} = 1 + \left(\frac{a}{\rho}\right)^{1/2}, \tag{2}$$

it follows from relations (1) and (2) that

$$\frac{\sigma_{EFF}}{p_c} = 1 + \frac{2}{\pi}\left(\frac{a}{\rho}\right)^{1/2}\sec^{-1}\exp\left\{\frac{\pi E_0 v_c}{8p_c a}\right\}^{1/2}. \tag{3}$$

Furthermore, the Mode III analytical results give the cohesive zone size ($s$) at failure (the maximum stress situation) as

$$\frac{s}{a} = \exp\chi - 1 + \left[\frac{\rho}{a}\{(\exp\chi)^2 - 1\}\right]^{1/2} \tag{4}$$

with

$$\chi = \frac{\pi E_0 v_c}{8p_c a}. \tag{5}$$

Now with the DBCS representation of a cohesive zone, the fracture toughness $K_{IC}$ associated with the extension of a crack under LEFM conditions, i.e., where the cohesive zone size is small compared with all the geometrical parameters associated with a configuration, is given by the relation

$$K_{IC} = \{E_0 p_c v_c\}^{1/2}. \tag{6}$$

Then relations (3) and (5) can be written respectively as

$$\frac{\sigma_{EFF}}{p_c} = 1 + \frac{2}{\pi}\left(\frac{a}{\rho}\right)^{1/2}\sec^{-1}\exp\chi \tag{7}$$

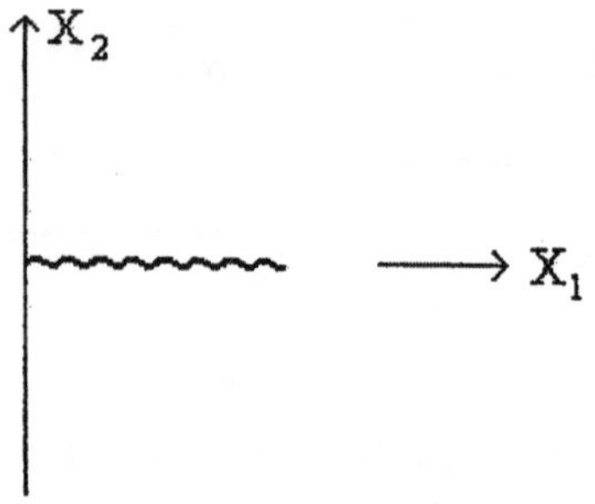


*Figure 2.* The planar surface model.

and

$$\chi = \frac{\pi K_{IC}^2}{8 p_c^2 a}. \tag{8}$$

Now if a hole is sufficiently large, i.e., $a$ is sufficiently large, that $\chi$ is less than 0.6, then

$$\sec^{-1} \exp \chi \sim \{2\chi\}^{1/2} \tag{9}$$

to within an accuracy of 10 percent. It then follows from relations (7)–(9) that a reasonably accurate expression for $\sigma_{EFF}$ is

$$\frac{\sigma_{EFF}}{p_c} = 1 + \frac{K_{IC}}{p_c (\pi \rho)^{1/2}}. \tag{10}$$

This size effect relation is such that the second term on the right-hand side, which can be viewed as the 'flaw strengthening contribution' to $\sigma_{EFF}$, is accurate to within 10 percent (it overpredicts $\sigma_{EFF}$), irrespective of the magnitude of $\rho$, provided that $\chi = \pi K_{IC}^2 / 8 p_c^2 a$ is less than 0.6. $K_{IC}^2 / p_C^2 \rho$ can be viewed as a scaling parameter associated with the size effect relation between $\sigma_{EFF}$ and the flaw dimensions.

Relation (4) gives the cohesive zone size at failure, and this relation can be written in the alternative but equivalent form

$$\frac{s}{\rho} = \frac{a}{\rho}\{\exp \chi - 1\} + \left\{\frac{a}{\rho}\{(\exp \chi)^2 - 1\}\right\}^{1/2}. \tag{11}$$

If $\chi$ is sufficiently small, then $s/a$ and $s/\rho$ are both small, and relation (10) for $\sigma_{EFF}$ is appropriate. Vice-versa, if $s/a$ and $s/\rho$ are both small, then $\chi$ must be small and again relation (10) for $\sigma_{EFF}$ is appropriate. Intuitively, it is expected that if failure occurs in a situation where the cohesive zone size is small compared with both the hole size and hole root radius, i.e. $s/a$ and $s/\rho$ are both small, then this situation can be simulated in terms of a cohesive zone emanating from the planar surface of a semi-infinite solid with the 'applied stress' distribution being the linear stress distribution immediately ahead of the actual hole root. The next section shows that this is indeed the case, and we therefore obtain relation (10) via an alternative procedure, albeit for a very limited range of conditions.

## 3. The planar surface simulation procedure: Use of mode III results

Consider the situation where a cohesive zone emanates from the planar surface of a semi-infinite solid (Figure 2). With Mode I loading and with $x$ being measured from $X_1 = 0$ along the $X_1$ axis, the tensile stress along the plane $X_2 = 0$ in the absence of the cohesive zone is

$$\sigma(x) = \sigma_L\left(1 - \frac{x}{h}\right) = K_t\sigma\left(1 - \frac{x}{h}\right) \tag{12}$$

when $x$ is small. This stress simulates the tensile stress in the immediate vicinity of a root of a hole with $\sigma$ being the applied stress, $K_t$ is the elastic stress concentrations factor, $\sigma_L$ is the stress at the hole surface and $h$ is a length parameter. Following the pattern described in the preceding section, and with $p_c$ being the uniform tensile stress within the cohesive zone, the maximum stress situation, i.e. failure, is assumed to be attained when the relative displacement at the trailing edge of the cohesive zone attains a critical value $v_c$. We will use the solutions for the corresponding Mode III situation, which means that we can then use the isolated crack-infinite body solutions, by making a cut along the plane $X_1 = 0$. Applying these solutions (Tada et al., 1973) to the Mode I loading state, the conditions for stress finiteness ($= p_c$) at the leading edge of the cohesive zone is

$$\tfrac{1}{2}\pi(\sigma_L - p_c) - \frac{\sigma_L s}{h} = 0, \tag{13}$$

where $s$ is the cohesive zone length, while the relative displacement condition at $x = 0$ is

$$v_c = \frac{4s(\sigma_L - p_c)}{E_0} - \frac{4s^2\sigma_L}{\pi E_0 h}. \tag{14}$$

Now $\sigma_L = \sigma_{EFF}$, the effective failure stress appropriate for use in a traditional elastic stress analysis, whereupon elimination of $s$ between relations (13) and (14) gives

$$\frac{\sigma_{EFF}}{p_c} = 1 + \left\{\frac{8\psi}{\pi^2}\left(1 + \frac{2\psi}{\pi^2}\right)\right\}^{1/2} + \frac{4\psi}{\pi^2} \tag{15}$$

with

$$\psi = \frac{\pi K_{IC}^2}{8p_c^2 h}, \tag{16}$$

assuming, as before, that $K_{IC} = \{E_0 p_c v_c\}^{1/2}$. Relations (13) and (15) give the cohesive zone size at failure as

$$\frac{s}{h} = \left\{\frac{8\psi}{\pi^2}\left(1 + \frac{2\psi}{\pi^2}\right)\right\}^{1/2} - \frac{4\psi}{\pi^2}. \tag{17}$$

It is expected that the planar surface simulation procedure only gives accurate predictions for the actual hole situation when $s/h$ is small, i.e. $\psi$ is small (see relation (17), and then relation (15)) can be expanded in terms of increasing powers of $\psi$ when, to the second term, we arrive at the size effect relation

$$\frac{\sigma_{EFF}}{p_c} = 1 + \frac{K_{IC}}{p_c(\pi h)^{1/2}}. \tag{18}$$

We now require the value of the geometrical parameter $h$ that characterises, for Mode III loading, the stress distribution ahead of the surface of an actual elliptic hole. For the model in Figure 1 and with $x$ being the distance along the plane $X_2 = 0$ ahead of the hole root at $X_1 = a$, the appropriate stress, i.e. the $p_{23}$ component, is given from the results of an earlier paper (Smith, 1968) by the expression

$$\sigma(x) = \frac{\sigma}{(a-b)}\left\{\frac{a(a+x)}{\{b^2+2ax+x^2\}^{1/2}} - b\right\}. \tag{19}$$

Noting that the radius of curvature at the hole root is $\rho = b^2/a$, relation (19) shows that, provided $x/\rho$ and $x/a$ are both small, then the stress distribution in the immediate vicinity of the hole root, i.e. small $x$, is of the form given by relation (12) with $K_t = 1 + (a/\rho)^{1/2}$ and, most importantly, with $h = \rho$. That being the case, it follows from relation (18) that we arrive at the simple $\sigma_{\mathrm{EFF}}$-$\rho$ relation as given by relation (10). In other words we have derived relation (10) by using the planar surface simulation procedure coupled with a knowledge of the stress distribution ahead of the actual flaw, rather than by performing a cohesive zone analysis for the actual flaw. In making this comment, however, we should note that $s/\rho$ and $s/a$ must both be small (see the comment immediately after relation (19)). This means (see relation (17)) that $\psi$ must be small, and $\chi$ must be small (see relation (4)). Consequently it is important to appreciate that relation (10) has been derived, via the planar surface simulation procedure, only for the restricted range of conditions where $\psi = \pi K_{\mathrm{IC}}^2/8p_c^2\rho$ and $\chi = \pi K_{\mathrm{IC}}^2/8p_c^2$ a are both small, whereas the actual model analysis (Section 2) leading to relation (10) has been derived subject to only the small $\chi$ restriction.

The equivalence demonstrated in this section highlights the potential for using the planar surface simulation procedure coupled with a linear 'applied' stress distribution, in lieu of a detailed cohesive zone analysis for an actual hole, expecting this approach to be reasonably accurate when the cohesive zone size is small compared with both the hole size and root radius. The equivalence has been demonstrated in this section for the Mode III situation, but we have no reason to believe that the planar surface simulation procedure cannot be equally well applied to the Mode I situation. Thus in the next section we derive an expression for the effective failure stress $\sigma_{\mathrm{EFF}}$, from a planar surface analysis, on the basis of actual Mode I results.

## 4. The planar surface simulation procedure: Mode I results

Consider the Mode I situation where there is a cohesive zone emanating from the planar surface of a semi-infinite solid (Figure 2). It is assumed that the tensile stress along the plane $X_2 = 0$ in the absence of the cohesive zone is given by relation (12) where $x$ is measured from $X_1 = 0$ along the $X_1$ axis. In this section, the objective is to obtain a relation of the form (18), using actual mode I results rather than by making use of Mode III results, as was the case in the preceding section. The quoted results of Tada et al. (1973) show that the condition for stress finiteness at the leading edge of the process zone (see Figure 2) is

$$0.52\pi(\sigma_{\mathrm{L}} - p_c) - \frac{\sigma_{\mathrm{L}} s}{h} = 0, \tag{20}$$

where $s$ is the cohesive zone length, while the relative displacement condition at $x = 0$ is

$$v_c = \frac{5.83s(\sigma_{\mathrm{L}} - p_c)}{E_0} - \frac{1.77s^2\sigma_{\mathrm{L}}}{E_0 h}, \tag{21}$$

these relations being analogous to respectively relations (13) and (14) for the Mode III analysis. With $\sigma_L = \sigma_{EFF}$, the effective failure stress, elimination of $s$ between relations (20) and (21) gives, to the second term in increasing powers of $K_{IC}/p_c\sqrt{h}$, and again with $K_{IC} = \{E_0 p_c v_c\}^{1/2}$,

$$\frac{\sigma_{EFF}}{p_c} = 1 + \frac{0.81 K_{IC}}{p_c(\pi h)^{1/2}}, \tag{22}$$

which is a size effect relation that is analogous to relation (18) from a Mode III analysis. We immediately see that there is a close correspondence between the two sets of results.

The parameter $h$ describing the stress distribution (see relation (12)) ahead of the hole (Mode I loading) is given by the expression (Smith, 1998)

$$h = \frac{\rho}{2}\left\{1 + \frac{1}{2}\left(\frac{\rho}{a}\right)^{1/2}\right\} \Big/ \left\{1 + \frac{3}{4}\left(\frac{\rho}{a}\right)^{1/2}\right\}, \tag{23}$$

an expression which checks with the well known solution (Knott, 1973) for a circular hole; i.e., $\rho = a$, when $h = 3a/7$. Expressions (22) and (23) allow the effective failure stress to be given by the size effect relation

$$\frac{\sigma_{EFF}}{p_c} = 1 + \frac{1.20 K_{IC}}{p_c(\pi \rho)^{1/2}}, \tag{24}$$

where the second term on the right-hand side of this relation is accurate to within 4 percent for holes with sharp roots, i.e. $\rho/a$ lying within the range $0 \to 1$. For holes with blunt roots, i.e. $\rho/a$ lying within the range $1 \to \infty$, an approximate expression for $\sigma_{EFF}$ is

$$\frac{\sigma_{EFF}}{p_c} = 1 + \frac{1.32 K_{IC}}{p_c(\pi \rho)^{1/2}}, \tag{25}$$

where the second term on the right-hand side of this relation is accurate to within 6 percent. For a circular hole, i.e. $\rho/a = 1$, expressions (22) and (23) give the effective failure stress as

$$\frac{\sigma_{EFF}}{p_c} = 1 + \frac{1.24 K_{IC}}{p_c(\pi \rho)^{1/2}}. \tag{26}$$

In all three size effect relations: (24), (25) and (26), $K_{IC}^2/p_c^2\rho$ is the appropriate scaling parameter relating $\sigma_{EFF}$ with the flaw dimensions.

Now Vitek (1976) has undertaken a numerical analysis of the Mode I model in Figure 1, based on a representation of the displacement discontinuity within the cohesive zones in terms of discrete dislocations, and a reduction of the problem to the solution of a system of linear equations. When $K_{IC}/p_c a^{1/2}$ is small, Vitek's curve-fitted results for $0 < \rho/a < 1$ (holes with sharp roots) can be reframed so as to give the following expression for the effective failure stress

$$\frac{\sigma_{EFF}}{p_c} = 1 + \frac{K_{IC}}{p_c(\pi \rho)^{1/2}}\left\{\frac{1}{2} + \frac{(\rho/a)^{1/2}}{\left\{1 + \frac{1}{2}\left(\frac{\rho}{a}\right)^{1/2}\right\}^3}\right\}^{-1}, \tag{27}$$

reducing to the simplified size effect relation

$$\frac{\sigma_{EFF}}{p_c} = 1 + \frac{1.26 K_{IC}}{p_c(\pi\rho)^{1/2}} \tag{28}$$

where the second term is accurate to within 12 percent for $0.1 < \rho/a < 1$.

There is good agreement between relations (24) and (28) for holes with sharp roots, though it should be noted that assuming the Mode III arguments carry over to the Mode I situation, then the derivation of relation (24) is based on $K_{IC}/p_c a^{1/2}$ and $K_{IC}/p_c \rho^{1/2}$ both being small.

## 5. Expression of the failure criterion in terms of an effective fracture toughness $K_{EFF}$

So far in this paper, the failure criterion has been expressed in terms of an effective failure stress $\sigma_{EFF}$, as measured at the flaw root surface. Alternatively, but equivalently, the criterion can be expressed in $K$ terms by writing the effective $K$ value, $K_{EFF}$, for failure as $K_{EFF} = \sigma(\pi a)^{1/2}$; this approach might be particularly attractive when considering fracture arising as a consequence of the presence of sharp holes. Then, using the Mode III simulation results (see Section 2), it follows from relations (1), (6) and (8) that

$$\begin{aligned}\frac{K_{EFF}}{K_{IC}} &= \frac{\sigma}{p_c} \cdot \frac{\pi}{(8\chi)^{1/2}} \\ &= \frac{\pi}{(8\chi)^{1/2}} \left\{ \frac{(\rho/a)^{1/2}}{(1+(\rho/a)^{1/2})} + \frac{2}{\pi(1+(\rho/a)^{1/2})} \sec^{-1} \exp \chi \right\}.\end{aligned} \tag{29}$$

Following the arguments in Section 2, with a hole that is sufficiently large that $\chi \equiv \pi K_{IC}^2/8p_c^2 a$ is less than 0.6, it follows that relation (29) can be expressed to a high degree of approximation in the simplified form

$$\frac{K_{EFF}}{K_{IC}} = \frac{1}{\{1+(\rho/a)^{1/2}\}} \left\{ 1 + \frac{p_c(\pi\rho)^{1/2}}{K_{IC}} \right\} \tag{30}$$

and, in the limit where the hole is very large, this size effect relation simplifies to

$$\frac{K_{EFF}}{K_{IC}} = 1 + \frac{p_c(\pi\rho)^{1/2}}{K_{IC}} \tag{31}$$

and again we see the importance of the scaling parameter $K_{IC}^2/p_c^2\rho$.

The large hole expression (30) can, of course, be obtained for large $p_c\rho^{1/2}/K_{IC}$ and large $p_c a^{1/2}/K_{IC}$, using the arguments in Section 3. Furthermore, using the arguments in Section 4, and noting that the elastic stress concentration factor is $K_t = 1 + 2(a/\rho)^{1/2}$ for Mode I loading, it follows that the relation, for Mode I loading, corresponding to relation (30), for Mode III loading, as obtained by using the planar surface simulation procedure appropriate for large $p_c\rho^{1/2}/K_{IC}$ and large $p_c a^{1/2}/K_{IC}$, is (for sharp holes – see relation (24))

$$\frac{K_{\mathrm{EFF}}}{K_{\mathrm{IC}}} = \frac{1}{\{2 + (\rho/a)^{1/2}\}}\left\{1.20 + \frac{p_c(\pi\rho)^{1/2}}{K_{\mathrm{IC}}}\right\} \tag{32}$$

which, in the limit where the hole is very large, i.e. $a$ is very large, simplifies to

$$\frac{K_{\mathrm{EFF}}}{K_{\mathrm{IC}}} = \frac{1}{2}\left\{1.20 + \frac{p_c(\pi\rho)^{1/2}}{K_{\mathrm{IC}}}\right\}. \tag{33}$$

## 6. Discussion

This paper has been concerned with the formulation of size effect relations associated with the effect of a blunt stress concentration on the failure of a quasi-brittle material where the failure can be modelled in terms of a (two-dimensional) embedded elliptic flaw within an infinite solid, a model which can be used to provide an approximate representation of the situation where there is a blunt elliptic flaw in the surface of a semi-infinite solid. The model has been coupled with a cohesive zone description of material failure, and the DBCS uniform stress cohesive zone stress-displacement law. So as to obtain analytical results so that the role of material and geometrical parameters in the size effect relations can be clearly seen, the paper has also been primarily based on Mode III analyses (Sections 2 and 3), though consideration has also been given (Section 4) to the practically more appropriate Mode I analysis. Failure has been expressed in terms of an effective fracture toughness $K_{\mathrm{EFF}}$ (Section 5) but, probably more importantly, also in terms of the maximum sustainable effective peak stress $\sigma_{\mathrm{EFF}}$ at the flaw root, this stress being related to material and geometrical parameters (Sections 2–4).

With a flaw length $2a$ and flaw root radius $\rho$, a key result is that with a sufficiently large flaw, i.e. $a$ is sufficiently large, the size effect dependence of $\sigma_{\mathrm{EFF}}$ reduces to a dependency on the root radius $\rho$ (see relation (10)), and $K_{\mathrm{IC}}^2/p_c^2\rho$ can be viewed as an appropriate scaling parameter. $\sigma_{\mathrm{EFF}}$ increases as the root radius decreases, this being due to the fact that the $\sigma_{\mathrm{EFF}}$ increase stems from the gradient in the stress ahead of the hole root, this gradient being steeper the smaller is the root radius. As indicated in the Introduction, a similar effect arises with regard to the behaviour of the uncracked bend beam configuration where $\sigma_{\mathrm{EFF}}$ is critically dependent on the applied loading induced stress gradient beneath the surface, which increases as the beam depth decreases. As an illustration of the magnitude of the 'flaw strengthening' effect for a quasi-brittle material, if we use input values of specific fracture energy $G_F = K_{\mathrm{IC}}^2/E_0 = 100\,\mathrm{N/m}$ ($G_F$ is equal to $p_c v_c$ with the constant stress cohesive zone representation), $p_c = 3\,\mathrm{MPa}$, $E_0 = 30\,\mathrm{GPa}$, which might be appropriate for ordinary concrete, then with $a = 1$ metre, we see (relation (8)) that $\chi$ is equal to 0.13. If we then use the Mode I relationship (26), we see that with a circular hole, i.e. $\rho = a = 1$ metre, then $\sigma_{\mathrm{EFF}}/p_c = 1.40$, indicating a flaw strengthening of 40 percent.

Before closing this discussion, it is worth mentioning that although this paper has been concerned with the modelling of the behaviour of a quasi-brittle material in a situation where there is a 'tensile' cohesive zone, similar considerations should also be relevant to situations where there is a 'compressive' cohesive zone, e.g. the compressive failure of a notched composite. This situation has already been quantified (Sutcliffe and Fleck, 1993; Smith, 1994) in relation to the formation of a 'compressive' cohesive zone ahead of a very sharp notch.

## References

Bilby, B.A., Cottrell, A.H. and Swinden, K.H. (1963). The spread of plastic deformation from a notch. *Proceedings of the Royal Society, A*, **272**, 304–314.

Dugdale, D.S. (1960). Plastic yielding from slits in steel sheets. *Journal of the Mechanics and Physics of Solids* **8**, 100–104.

Elices, M., Guinea, G.V. and Planas, J. (1996). Prediction of size effect in concrete and reinforced concrete structures. *Size-Scale Effects in the Failure Mechanisms of Materials and Structures* (Edited by A. Carpinteri), E and FN SPON, 309–324.

Knott, J.F. (1973). *Fundamentals of Fracture Mechanics*, Butterworths, London, U.K.

Li, Y.N. and Bazant, Z.P. (1994). Eigenvalue analysis of size effect for cohesive crack model. *International Journal of Fracture* **66**, 213–226.

Li, Y.N. and Bazant, Z.P. (1996). Scaling of cohesive fracture (with ramification to fracture cracks). *Size-Scale Effects in the Failure Mechanisms of Materials and Structures* (Edited by A. Carpinteri), E and FN SPON, 274–289.

Ozbolt, J. and Eligehausen, R. (1996). Size effect in concrete and reinforced concrete structures. *Size-Scale Effects in the Failure Mechanisms of Materials and Structures* (Edited by A. Carpinteri), E and FN SPON, 290–306.

Smith, E. (1967). Fracture initiation at a sharp notch. *Proceedings of the Royal Society, A* **299**, 455–463.

Smith, E. (1968). The interaction between dislocations and inhomogeneities. *International Journal of Engineering Science* **6**, 129–143.

Smith, E. (1994). The compressive failure of notched composites: The effects of geometry and microbuckling softening law. *International Journal of Fracture* **66**, 33–44.

Smith, E. (1996). Critical dimensions for scaling effects in the fracture of quasi-brittle materials. *Size-Scale Effects in the Failure Mechanisms of Materials and Structures* (Edited by A. Carpinteri), E and FN SPON, 340–356.

Smith, E. (1997). The maximum load size effect for uncracked brittle structures. *Journal of Materials Science* **32**, 3939–3942.

Smith, E. (1998). The effective tensile failure stress of an uncracked brittle structure: failure at a blunt stress concentration. *Journal of Materials Science* **33**, 29–33.

Sutcliffe, M.P.F. and Fleck, N.A. (1993). Effect of geometry on compressive failure of notched composites. *International Journal of Fracture* **59**, 115–132.

Tada, H., Paris, P.C. and Irwin, G.R. (1973). *The Stress Analyses of Cracks Handbook*, Del Research Corporation, Hellertown, Pa, U.S.A.

Vitek, V. (1976). Yielding from a crack with finite root-radius loaded in uniform tension. *Journal of the Mechanics and Physics of Solids* **24**, 67–76.

*International Journal of Fracture* **95:** 51–65, 1999.


# Scaling laws for elastoplastic fracture

A.G. ATKINS
*Department of Engineering, University of Reading, Reading RG6 6AY UK, e-mail: A.G.Atkins@reading.ac.uk*

Received 28 December 1998; accepted in revised form 31 December 1998

**Abstract.** Scale or size effects in fracture result from the interaction of some energies dependent upon volume and other energies dependent on area (cube-square scaling). The known scaling laws within the lefm and nlefm ranges and within the rigid-plastic range are highlighted along with applications. This paper derives for the first time the scaling laws for elastoplastic fracture based on linear power-law behaviour which span the different regimes from lefm at one end of the spectrum to extensive ductile fracture at the other. The existence of a master-curve of normalised load $\bar{X}$ vs normalised displacement $\bar{u}$ is demonstrated on which fall all results from different size geometrically-similar testpieces up to first cracking. Crack propagation in larger bodies begins at smaller normalised loads and displacements than geometrically-similar small bodies. Large bodies behave as if their fracture toughness were given by $(R/\lambda)$, where $\lambda(>1)$ is the scaling factor, rather than by the material value $R$. Propagation behaviour is path-dependent and each size cracked body has its own $\bar{X}-\bar{u}$ propagation plot. This explains departures from 'geometric' ($\lambda^3$) scaling well-known in the literature. Comparison is made with old and recent experimental results.

*List of Symbols*

$A$ crack area
$a$ crack length
$a_o$ starting crack length
$B$ thickness
$b$ remaining ligament given by $(W-a)$
$D$ parameter give by $\sigma_0(B/S)W^2$
$J_c$ critical value of the $J$-integral
$J_e$ elastic component of $J$
$J_P$ plastic component of $J$
$J_R$ $J$-resistance curve
$R$ fracture toughness
$S$ span (in bending)
$u$ load-point displacement
$\bar{u}$ scaled load-point displacement given by $u/\lambda$
$V$ volume
$W$ width of cracked body
$w$ energy density
$X$ load
$\bar{X}$ scaled load given by $X/\lambda^2$
$\Gamma$ plastic work
$\Lambda$ elastic strain energy
$\lambda$ scaling factor, i.e. ratio of characteristic lengths in different size bodies
$\sigma$ stress
$\Omega$ complementary strain energy

*Superscripts and subscripts*

$e$ elastic
extra extra
$m$ model
$nl$ non-linear
$P$ plastic
$p$ prototype
tot total

## 1. Introduction

Experiments over a wide size range, and extending back many years, show that there is a 'scale effect' or 'size effect' in the mechanics of fracture (Stanton and Batson, 1921; Docherty, 1932, 1935; Brown et al., 1947; Shearin et al., 1948; Irwin, 1954; Lubahn, 1955; Lubahn and Yukawa, 1958; Booth et al., 1983 and Hagiwara et al., 1983). By 'scale effect' is meant that components and structures, made of materials which are appreciably ductile in laboratory-size testpieces, behave in a progressively less ductile fashion the larger they get. Eventually, above some critical size, their behaviour is globally elastic, and fractures are brittle. Conversely, when deformation zones are kept very small in normally-brittle materials, plastic deformation is possible, as shown by limiting sizes in the comminution of powders and by the ability to micro-machine glass. (Of course, other effects enter the picture for thin film or micron-size structures but these are beyond the scope of the present paper). Such size effects are over and above any questions of material inhomogeneity, non-uniform heat treatment, thickness-dependence of fracture toughness and so on in the tested material. In many investigations, considerable care was taken to ensure, as far as practicable, uniform microstructure in small and large testpieces. Notched bending is a convenient method of testing since the large loads required in scaled large testpieces are readily provided by compression machines, and in such cases smaller samples have often been manufactured from undeformed broken parts of larger pieces.

Irwin in 1954 recognised that size-effects were inherent in the mechanics of fracture: he remarked that (Irwin, 1954) '... disregarding the size effects due to changes in materials properties with size caused by problems of metallurgical processing, previous studies have shown that fracture size effects of significant magnitude exist in solids whether they are brittle or ductile and whether the material is a metal or a plastic. The most elaborate precautions to eliminate differences due to material quality, lack of precise similarity of dimensional ratios, specimen surface preparation, and the experimental loading devices have failed to remove fracture size effects. It is apparent that these are fundamental to the nature of the severing process...'. The origins of size effects in fracture arise because recoverable elastic energy, and irreversible remote plasticity, are volume-dependent quantities, but fracture work depends only on area. Thus cube/square energy scaling principles are inherent in all mechanics of fracture.

In this paper, we use the terms elastic-, elastoplastic-, and plastic-fracture, defined as follows. When a body is loaded, elastic deformations are first experienced and if fracture occurs before yielding, we have *elastic fracture*, the body behaving in a globally-reversible fashion, with all the fracture irreversibilities being confined to boundary layers alongside the crack. The original body may be regained by fitting back together the broken pieces. In such cases, when the load ($X$)-displacement ($u$) behaviour is linear, calculations are performed using linear elastic fracture mechanics (LEFM); when the $X$-$u$ behaviour is non-linear (but still globally reversible, of course) calculations are performed using non-linear elastic fracture mechanics (NLEFM). When yielding precedes fracture, we have *elastoplastic fracture* (ELPL fracture) when cracking occurs at deformations not much into the plastic range or *plastic fracture* (PL fracture) when the plastic deformation is extensive. In these latter cases, the behaviour is path-dependent and the broken pieces are permanently distorted. Other things being equal, elastic ('brittle') fracture occurs in large bodies, plastic ('ductile') fracture in small, with elastoplastic fracture in between. The property that is invariant with size in the analysis is the fracture toughness $R$.

The scaling laws for loads, displacements, stresses, energy consumed etc have been determined for fracture in geometrically-similar bodies *within* the linear (Gurney and Hunt, 1967; Atkins and Caddell, 1974) and non-linear elastic range (Mai and Atkins, 1975); and for non-proportionally-scaled elastic bodies (Mai and Atkins, 1978; Atkins, 1995); likewise for fracture *within* the plastic range (Atkins, 1988). Different size geometrically-similar bodies are those where all dimensions change by the same 'scaling factor' $\lambda$ (ratio of characteristic lengths); non-proportional scaling concerns four separate scaling factors for 'length', 'height', 'width' and 'crack length' in different size bodies. Some important results from these previous analyses are

(i) in LEFM, $\sigma_p = \sigma_m/\sqrt{\lambda}$,
where $\sigma_p$ is the cracking stress for the 'prototype' ($p$) body and $\sigma_m$ that for the reference 'model' body ($m$);
(ii) in power-law NLEFM where $X = (\text{constant})u^n$, $\sigma_p = \sigma_m/\lambda^{n/(n+1)}$;
(iii) in lefm and true nlefm, energies scale as $\lambda^2$, the only dissipated work being fracture work;
(iv) in plastic fracture, energies scale as $\lambda^x$ where $2 < x < 3$. (In plasticity with no fracture, energies scale as $\lambda^3$ – sometimes called 'geometric scaling', but not to be confused with geometrically-similar scaled bodies).

These results have been successfully applied to practical scaling problems, e.g. burst tests on non-proportionally-scaled mild steel pipes (Mai and Atkins, 1978); ice-covered towing tanks (Atkins and Caddell, 1974; Atkins, 1975; 1986); and various ductile fracture scaling experiments (Atkins, 1988).

Save for one special cracked geometry (the double cantilever beam testpiece) (Atkins and Chen Zhong, 1996), the scaling laws for elastoplastic fracture spanning all three different regimes (elastic, elastoplastic and plastic) have not been determined. This paper now derives the generalised scaling laws for elastoplastic fracture for arbitrary cracked bodies. Before deriving the scaling laws for elastoplastic fracture in Section 3, we must review the conventional way of dealing with elastoplastic fracture.

## 2. Nonlinear elastic fracture, real elastoplastic fracture, $J_R$ curves etc

Crack *initiation* is successfully predicted for elastoplastic fracture using nonlinear elastic fracture mechanics (NLEFM) *providing* that there is no unloading before crack initiation

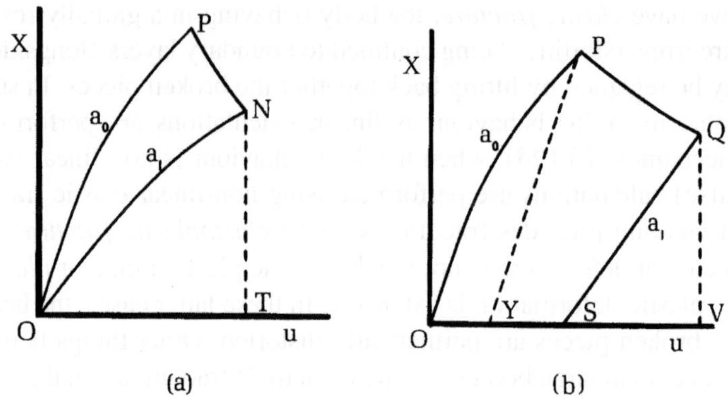


*Figure 1.* Differences between load ($X$) – displacement ($u$) – crack length (a) behaviour in (a) true nlefm and (b) real elpl fracture. In (a) loading along OP, crack propagation along PN and unloading along NO is globally reversible and path-independent. In (b) the corresponding path OPQS is partly irreversible and path-dependent.

(and that the loading is proportional). This is permissible because the body cannot distinguish between reversible and irreversible energies while the load increases. To quantify, for a crack of length $a_0$, the fracture initiation load at point P (Figure 1) in terms of geometry and material properties, it does not matter that after some crack propagation to crack length $a_1$ under decreasing load, a true non-linear elastic body would unload down NO (Figure 1(a)), whereas a real elastoplastic body would unload down QS (Figure 1(b)). In NLEFM, work area OPNO $= R\Delta A = RB(a_1 - a_0) = J_c B(a_1 - a_0)$ where $\Delta A$ is the increase of crack area, $B$ is plate thickness, $R$ is the specific essential work of fracture of the material (the fracture toughness) and $J_c$ is the critical value of the $J$-integral at crack initiation, which is the same as $R$. Note that $R$ need not be constant, but it is a *crack tip* toughness and should be distinguished from $J_R$ (see later). In real elastoplastic fracture, area OPQS comprises the same fracture work but additionally includes plastic work remote from the crack tip and residual elastic strain energy (the body having been unloaded from a non-uniform plastic flow field at Q in Figure 1(b)).

The locations of N and Q in Figures 1(a) and (b) are different; in particular, the displacement $u_N$ in the true NLEFM case (Figure 1(a)) is smaller than $u_Q$ in real elastoplastic fracture (Figure 1(b)), because all of the stored elastic strain energy ONT is available for fracture in Figure 1(a), whereas only SQV is available in Figure 1(b). In other words, in real ELPL crack propagation, extra external work has to be performed to achieve the same crack advance. The great problem in elastoplastic fracture mechanics is how to partition work areas such as OPQS into fracture work, plastic work and residual elastic strain energy; and how to predict the path-dependent locus PQ in elastoplastic fracture - which does not coincide with the path-independent locus PN of true nlefm which is predictable from theory.

Conventional wisdom uses the so-called $J_R$ curve to describe ductile crack propagation, where no attempt is made to distinguish between crack-tip fracture work and remote plastic flow. $J_R$ (or increments of $J_R$) are determined from work areas OPQV (or their increments) normalised by the increase in crack area, and is customarily broken down into components $J_e$ (related to SQV) and $J_p$ (related to YPQS). At one time, it was believed by some workers that $J_R$ was a material property, but many experiments (including those under '$J$-control'

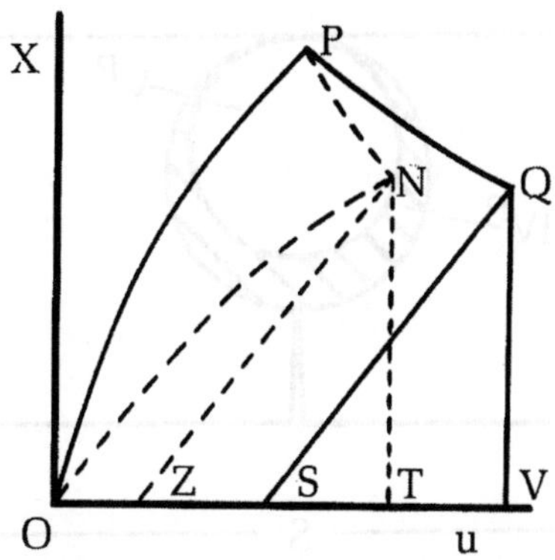


*Figure 2.* For the same increase in crack length, extra work given by TNPQV is required in elastoplastic fracture compared with nonlinear elastic fracture. This comes about because only ZNT is recoverable and not all of ONT as in true nlefm. Displacements in real elpl fracture are thus greater at the same load than in nle fracture.

(Kanninen and Popelar, 1985; Xia and Shih, 1995)) show that $J_R$ depends on cracked geometry and crack tip constraint.

The energy rate balance for linear and non-linear elastic cracking is

$$X\,\mathrm{d}u = \mathrm{d}\Lambda + R\,\mathrm{d}A, \tag{1}$$

where $X$ is load, $u$ load point displacement, $\Lambda$ is elastic strain energy, $R$ is crack tip fracture toughness and $A$ crack area. Whence

$$R = -\left(\frac{\partial \Lambda}{\partial A}\right)_u = +\left(\frac{\partial \Omega}{\partial A}\right)_X, \tag{2}$$

with the complementary elastic strain energy $\Omega = (Xu - \Lambda)$. Equation (2) governs all globally-elastic fracture; in linear elastic fracture mechanics (LEFM) $\Lambda = \Omega = \frac{1}{2}Xu$, but in NLEFM $\Lambda \neq \Omega$. Elastoplastic fracture is governed by

$$X\,\mathrm{d}u = \mathrm{d}\Lambda + \mathrm{d}\Gamma + R\,\mathrm{d}A, \tag{3}$$

where $\Gamma$ is plastic work remote from the crack tip and is separated from the work of fracture in (3). Up to crack initiation, even though $\Gamma$ is irreversible, $(\Lambda + \Gamma)$ may be lumped together, given the equivalence of non-linear elasticity and Hencky total strain plasticity, so that in both true NLEFM and real elpl, the *initiation* at points P in Figure 1 are given by

$$R = J_c = -\frac{\partial}{\partial A}(\Lambda + \Gamma)_u. \tag{4}$$

When (4) is applied to *propagation* in real elpl bodies, however, erroneous predictions arise for it is the PN locus that is being predicted (as if $\Gamma$ were all recoverable during unloading) rather than the PQ locus in Figure 1.

A recent closed-form, path-dependent elastoplastic crack propagation solution for the double cantilever beam (DCB) geometry (Atkins et al., 1998) gives a connexion between the NLEFM PN locus and the real elpl PQ locus which is algebraically exact for this geometry. It turns out that, on an incremental basis, the extra work that has to be done is equal to the energy that would have been recovered had the body been fully reversible. In Figure 2 the integrated

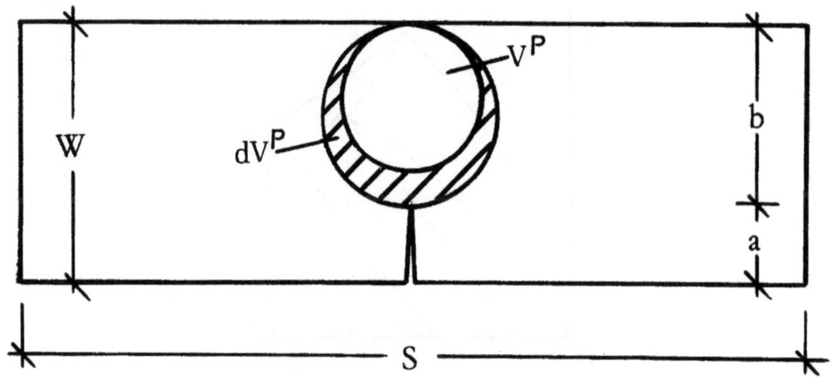


*Figure 3.* The volume of the plastically-deforming region *decreases* during crack propagation in geometries with finite-size ligaments. The increment of volume $\mathrm{d}V$ is thus negative and the increment of $w\,\mathrm{d}V$ (where $w$ is energy density) is also negative. In true nlefm, negative $w\,\mathrm{d}V$ means energy recovery and is permissible; in real elpl fracture, only the linear elastic component of $w\,\mathrm{d}V$ is recovered and, for the plastic work component, $w\,\mathrm{d}V$ must be set at zero. This leads to the requirement of extra work for the same amount of crack propagation in real elastoplastic bodies compared with nonlinear elastic bodies.

extra work area TNPQV is necessary because only ZNT is recoverable and not all of ONT as in true NLEFM. In many cases of elastoplastic crack propagation, the loads at N and Q are approximately the same (exactly the same for the DCB geometry) for the same crack advance (same remaining ligament) since they are controlled by the relevant (work hardening) plastic collapse expression (Hu and Albrecht, 1991; Miller, 1988). While the algebraically-exact relationship for the DCB geometry is unlikely to be a general truth for other cracked bodies because the DCB testpiece is peculiar in that cracking always proceeds into virgin material, it may, even so, be a very good approximation for elastoplastic crack propagation in bodies with finite size ligaments,

Since both $\Lambda$ and $\Gamma$ may be written as a (mean work/volume)(volume)= $wV$, we may rewrite (3) as

$$X\,\mathrm{d}u = w^e\,\mathrm{d}V^e + V^e\,\mathrm{d}w^e + w^P\,\mathrm{d}V^P + V^P\,\mathrm{d}w^P + R\,\mathrm{d}A, \tag{5}$$

with superscripts *e* for elastic and *p* for plastic (Atkins, 1990). Up to initiation, (5) applies both to nlefm and real elpl. During propagation over finite size ligaments $\mathrm{d}V^P$ is *negative* and $W^P\,\mathrm{d}V^P$ is negative; for example, in a notched beam where plasticity is confined to an approximately circular patch with ligament as diameter, Figure 3, the hatched region delineates the negative $\mathrm{d}V^P$. In NLEFM, negative $w^P\,\mathrm{d}V^P$ is permissible and it indicates an increment of *recovered* work; for real elastoplastic crack propagation however $w^P\,\mathrm{d}V^P$ must be set at zero since plastic work cannot be recovered. It follows, from the two versions of (5), that

$$X_{elpl}\,\mathrm{d}u_{elpl} - X_{nle}\,\mathrm{d}u_{nle} = -w^P\,\mathrm{d}V^P, \tag{6}$$

which if $X_{elpl} \approx X_{nle}$ from collapse formulae (Hu and Albrecht, 1991; Miller, 1988) becomes

$$X\,\mathrm{d}u_{\mathrm{extra}} = -w^P\,\mathrm{d}V^P, \tag{7}$$

where $u_{\mathrm{extra}} = (u_{elpl} - u_{nle})$ and where the mean plastic work density $w^P$ is given by

$$w^P = \left(\frac{\Gamma}{V^P}\right), \tag{8}$$

which is work area ONZ $\div V^P$ (Figures 2 and 3). $X\,du_{\text{extra}}$ is positive in (7) because $dV^P$ is negative. Equation (7) may be used to determine the path-dependent elpl PQ locus in Figure 2 and its shift to greater $u$ beyond the NLEFM displacements along PN. Equation (5) with negative $w^P\,dV^P$ (equivalent to (4)), when applied to propagation, predicts locus PN. Knowledge of the PQ locus is essential for proper determination of scaling laws through to complete fracture – otherwise we have merely NLEFM scaling laws.

## 3. Analysis

### 3.1. NLEFM SCALING LAWS FOR LINEAR-POWER LAW MATERIAL

Although the scaling laws for bodies having either linear elastic or power-law elastic compliances separately have been determined (Gurney and Hunt, 1967; Atkins and Caddell, 1974; Mai and Atkins, 1975), the laws for *combined* linear-power law elastic behaviour (analogous to fully-reversible Ramberg–Osgood behaviour) have not been determined. As they are the precursor of the full elastoplastic scaling laws, we derive them now.

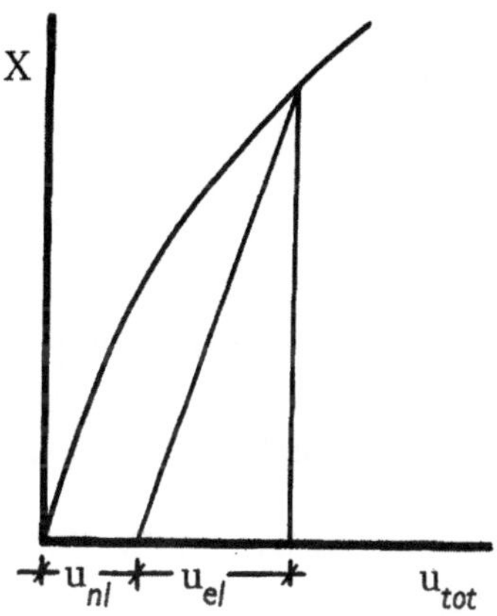


*Figure 4.* The total load point displacement $u_{\text{tot}}$ is comprised of a linear component $u_{el}$ and a non-linear component $u_{nl}$. In true non-linear behaviour, both $u_{el}$ and $u_{nl}$ are reversible, but in elastoplastic behaviour $u_{nl}$ is irreversible.

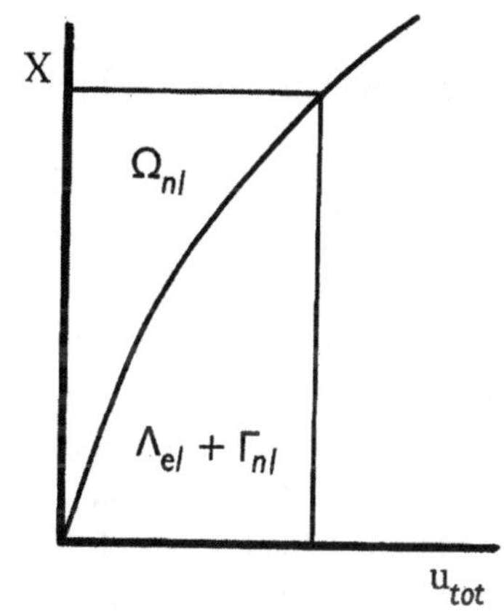


*Figure 5.* Meanings of linear elastic strain energy ($\Lambda_{el}$), non-linear strain energy ($\Gamma_{nl}$) and complementary strain energy ($\Omega_{nl}$) in relation to non-linear load-displacement plots.

The total displacement $u_{\text{tot}}$ is given by (Figure 4)

$$u_{\text{tot}} = u_{el} + u_{nl}. \tag{9}$$

Now $u_{el} = XC$ with $C$ the linear elastic compliance of the cracked body. The non-linear load-displacement plot is presumed to be given by

$$X = X(a, u_{nl}), \tag{10}$$

with a the crack length. For bending testpieces for example we have

$$X = \sigma_0 \left(\frac{B}{S}\right) W^2 \left(\frac{b}{W}\right)^Z \left(\frac{u_{nl}}{W}\right)^n = D \left(\frac{b}{W}\right)^Z \left(\frac{u_{nl}}{W}\right)^n, \tag{11}$$

where the variables have been separated (experiments show that this is acceptable) and

$$D = \sigma_0 \left(\frac{B}{S}\right) W^2.$$

As shown in Figure 3, $W$ is the specimen width, $B$ the thickness, $S$ the span and $b = (W - a)$ the ligament; $z$ and $n$ are empirical constants ($z \approx 2$ from plastic collapse algebra and $n$ relates to the power law stress-strain relation for the material). Hence

$$u_{\text{tot}} = XC + W \left[ \frac{X}{D(\frac{b}{W})^Z} \right]^{1/n}. \tag{12}$$

The complementary strain energy $\Omega_{nl} = \int u_{\text{tot}} \, dX$ (Figure 5) and is

$$\Omega_{nl} = \frac{X^2 C}{2} + \frac{W}{[D[\frac{b}{W}]^Z]^{1/n}} \frac{X^{(1/n)+1}}{(\frac{1}{n}+1)} \tag{13}$$

Whence, from basic NLEFM relations (Gurney and Ngan, 1971)

$$R = \left.\frac{\partial \Omega_{nl}}{B \partial a}\right|_X = \frac{X^2}{2B} \frac{dC}{da} + \frac{W^{1+(Z/n)} X^{(1/n)+1}}{D^{1/n}(\frac{1}{n}+1)B} \left(\frac{Z}{n}\right) \frac{1}{b(\frac{Z}{n}+1)}. \tag{14}$$

Equation (14) is the nlefm solution for $R$ loci in a fully-reversible cracked body displaying 'linear-power law' $Xu$ relations. Notice that (14) has the form of $R = J_e + J_{nl}$, as employed in $J_c$ and $J_R$ procedures. Notice also the connexion between (9) and (14) with the EPRI/GE estimation procedures (Kumar et al., 1981).

The fracture load scaling laws for an elastic linear-power law material follow from solution of (14) for $X$; those for $u_{\text{tot}}$ from (12) using $X$ from (14) and so on (see later).

We note, for later use, that

$$\begin{aligned} \Gamma_{nl} &= Xu_{\text{tot}} - \Omega_{nl} - \tfrac{1}{2}X^2 C \\ &= \frac{1}{(n+1)} \frac{W X^{(1/n)+1}}{[D(\frac{b}{W})^Z]^{1/n}} \end{aligned} \tag{15}$$

using (13).

### 3.2. Normalised 'master-curve' of scaled loads and scaled displacements

Much work on fracture scaling has been concerned with trying to explain departures from 'geometric' scaling where loads carried at every stage of the test would be proportional to the squares of linear dimensions and the deflexions proportional directly to the linear dimensions. It was noticed (Docherty, 1935; Shearin et al., 1948; Hagiwara et al., 1983) that *up to* fracture, plots of (load/area) vs (displacement/reference length) from different size testpieces did follow the same curve but that crack initiation took place at different points along the curve (Figures 6 and 7). Small bodies of the same material underwent much greater normalised displacements before cracking; larger pieces broke earlier, and very large bodies could break in the initial

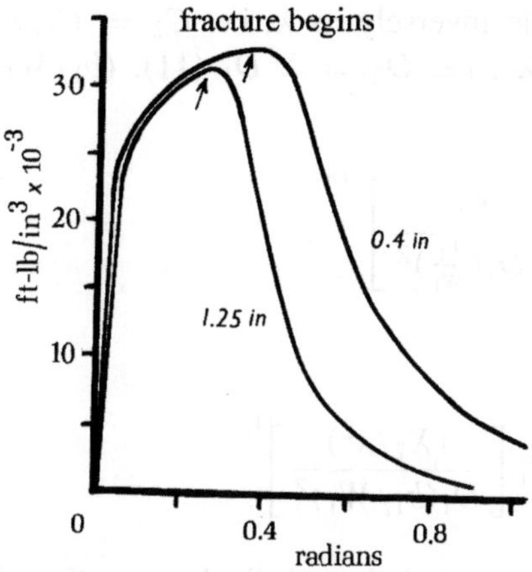


*Figure 6.* Reduced moment vs angle of bend of scaled 3-point bend testpieces (Shearin et al. 1948), showing existence of 'master-curve' of normalised load (or moment) vs normalised displacement (or rotation) prior to fracture of geometrically similar testpieces. Crack initiation occurs at bigger reduced moments and bigger angles of bend in smaller testpieces.

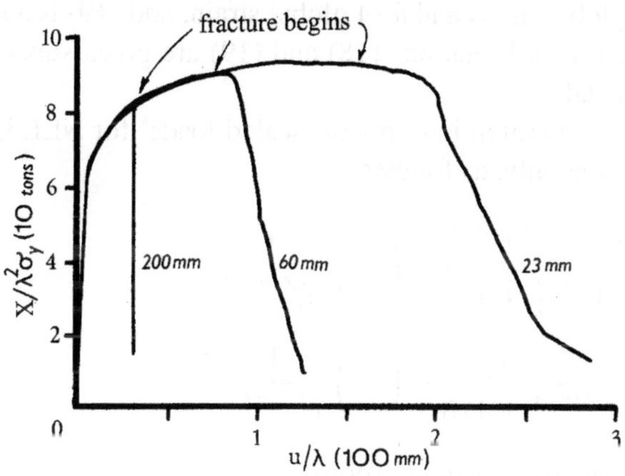


*Figure 7.* Normalised load vs normalised deflexion for geometrically-similar I-beams, notched inside the flange in tension with thumbnail geometrically-similar cracks of depth-to-surface length ratio about 0.06 (Hagiwara et al., 1983). The I-beams were fabricated from plate of different thicknesses. Since the mechanical properties of the three plate thicknesses used were not identical (the yield strengths being 29.6 kgf/mm$^2$ (20 mm thick); 28.3 (6 mm) and 15.9 (2.3 mm), the load is normalised not only by $\lambda^2$ but also by the yield strength (note: the ordinate of the graph in the original paper, as here, is labelled 'tons' even so). Plots demonstrate a 'master-curve' before fracture, and that cracking starts at greater scaled loads and scaled displacements the smaller the test beam.

linear elastic regime. We shall now show that the existence of the master curve follows from (12).

For two geometrically similar bodies with scaling factor $\lambda$

$$
\begin{aligned}
u_{\mathrm{tot}\,1} &= X_1 C_1 + W_1 \left[ \frac{X_1}{D_1 (\frac{b_1}{W_1})^Z} \right]^{1/n}, \\
u_{\mathrm{tot}\,2} &= X_2 C_2 + W_2 \left[ \frac{X_2}{D_2 (\frac{b_2}{W_2})^Z} \right]^{1/n}.
\end{aligned}
\tag{16}
$$

Now the compliances $C$ scale inversely as $\lambda$, i.e. $C_2 = C_1/\lambda$ (Atkins and Caddell, 1974; Atkins, 1990). $D$ scales as $\lambda^2$, i.e. $D_2 = \lambda^2 D_1$ (11); $(b_1/W_1) = (b_2/W_2)$; $W_2 = \lambda W_1$. Therefore

$$u_{\mathrm{tot}2} = \frac{X_2 C_1}{\lambda} + \lambda W_1 \left[ \frac{X_2}{\lambda^2 D_1 (\frac{b_1}{W_1})^Z} \right]^{1/n}. \tag{17}$$

Dividing by $\lambda$,

$$\left(\frac{u_{\mathrm{tot2}}}{\lambda}\right) = \left(\frac{X_2}{\lambda^2}\right) C_1 + W_1 \left[ \frac{(X_2/\lambda^2)}{D_1 (b_1/W_1)^Z} \right]^{1/n} \tag{18}$$

or, defining $(u_{\mathrm{tot}}/\lambda = \bar{u})$ as the 'scaled total displacement' and $(X/\lambda^2 = \bar{X})$ as the 'scaled load', it follows that

$$(\bar{u}) = \bar{u}_{el} + \bar{u}_{nl} = (\mathrm{const}_1)(\bar{X}) + (\mathrm{const}_2)(\bar{X})^{1/n}. \tag{19}$$

$\bar{X}$ is some measure of global stress and $\bar{u}$ of global strain, and (19) is a master curve of scaled displacement and scaled load. Equations (18) and (19) are given support by the results prior to fracture in Figures 6 and 7.

Equation (14) may be rewritten in terms of 'scaled loads' for NLE initiation and propagation, or for ELPL initiation only, as follows:

$$\begin{aligned} R &= \frac{X_1^2}{2B_1}\frac{\mathrm{d}C_1}{\mathrm{d}a_1} + \frac{W_1^{1+(Z/n)} X_1^{(1/n)+1}}{D_1^{1/n}(\frac{1}{n}+1)B_1}\left(\frac{Z}{n}\right)\frac{1}{b_1^{(Z/n)+1}}, \\ R &= \frac{X_2^2}{2B_2}\frac{\mathrm{d}C_2}{\mathrm{d}a_2} + \frac{W_2^{1+(Z/n)} X_2^{(1/n)+1}}{D_2^{1/n}(\frac{1}{n}+1)B_2}\left(\frac{Z}{n}\right)\frac{1}{b_1^{(z/n)+1}}. \end{aligned} \tag{20}$$

Now (Atkins and Caddell, 1974; Atkins, 1990)

$$\frac{\mathrm{d}C_2}{\mathrm{d}A_2} = \left(\frac{1}{\lambda^3}\right)\frac{\mathrm{d}C_1}{\mathrm{d}A_1}; \qquad \text{i.e.} \left(\frac{1}{B_2}\frac{\mathrm{d}C_2}{\mathrm{d}a_2}\right) = \left(\frac{1}{\lambda^3}\right)\left(\frac{1}{B_1}\frac{\mathrm{d}C_1}{\mathrm{d}a_1}\right),$$

whence

$$\begin{aligned} R &= \frac{X_2^2}{\lambda^3}\left(\frac{1}{2B_1}\frac{\mathrm{d}C_1}{\mathrm{d}a_1}\right) + \frac{\lambda^{1+(Z/n)} W_1^{1+(Z/n)} X_2^{(1/n)+1}}{\lambda^{2/n} D_1^{1/n}(1+\frac{1}{n})\lambda B_1}\left(\frac{Z}{n}\right)\frac{1}{\lambda^{(Z/n)+1} b_1^{(Z/n)+1}} \\ &= \frac{X_2^2}{\lambda^3}\left(\frac{1}{2B_1}\frac{\mathrm{d}C_1}{\mathrm{d}a_1}\right) + \frac{X_2^{(1/n)+1}}{\lambda^{(2/n)+1}}\left[\frac{W_1^{1+(Z/n)}}{D_1^{1/n}(1+\frac{1}{n})B_1}\left(\frac{Z}{n}\right)\frac{1}{b_1^{(Z/n)+1}}\right], \end{aligned} \tag{21}$$

that is, for geometrically-similar bodies,

$$R = \lambda(\bar{X})^2(\mathrm{const}_3) + \lambda(\bar{X})^{(1/n)+1}(\mathrm{const}_4). \tag{22}$$

Notice that if the components of (22) are taken separately, they agree with the known results for (a) just LEFM, ie first term alone, $X \propto \lambda^{3/2}$ (Atkins and Caddell, 1974; Atkins, 1990) and (b) just power law NLEFM i.e. second term alone, $X \propto \lambda^{n+2/n+1}$ (Mai and Atkins, 1975; Atkins, 1990) (also, in the latter case, if $n = 1$, $X \propto \lambda^{3/2}$ again). Notice also (and this is very interesting) that (22) may be written

$$\left(\frac{R}{\lambda}\right) = (\text{const}_3)(\bar{X})^2 + (\text{const}_4)(\bar{X})^{(1/n)+1}, \tag{23}$$

i.e. bigger geometrically-similar bodies behave as if they have a 'reduced toughness' given by $(R/\lambda)$, and bigger bodies 'fall off' the master curve at crack initiation earlier than small bodies. This is confirmed by the initiation results in Figures 6 and 7.

### 3.3. Scaling laws for real elastoplastic behaviour

While crack initiation in Figures 6 and 7 may be predicted from NLEFM, the propagation loci in Figures 6 and 7 for real elastoplastic bodies depend on the requirement for extra work, because the body is not fully reversible, as described in Section 2. For propagation in real elastoplastic bodies

$$u = u_{el} + u_{nl} + u_{\text{extra}}. \tag{24}$$

Over finite ligaments

$$X\,\mathrm{d}u_{\text{extra}} = -w^p\,\mathrm{d}V^P. \tag{7 bis}$$

When $V^P \propto b^2$, as in bending (Figure 3), $\mathrm{d}V/V = 2\,\mathrm{d}b/b$. Again $w^p = (\Gamma_{nl}/V^P)$ (8), with $\Gamma_{nl}$ given by (15), so

$$\begin{aligned} \mathrm{d}u_{\text{extra}} &= -\frac{\Gamma_{nl}}{V}\frac{\mathrm{d}V}{X} \\ &= -\frac{-WX^{1/n}2\,\mathrm{d}b}{(D\lfloor b/W\rfloor^Z)^{1/n}(n+1)b} \\ &= +\frac{2u_{nl}\,\mathrm{d}a}{(n+1)b}, \end{aligned} \tag{25}$$

using (11). To obtain $u_{\text{extra}}$, we need $u_{nl}$ as a function of $b$ in terms of $R$ to permit integration. That is, (14) has to be solved for $X$ and then $u_{nl}$ evaluated from (10) for use in (25). Equation (25) may also be written in terms of normalised extra displacement $\bar{u}_{\text{extra}} = u_{\text{extra}}/\lambda$.
In addition to extra displacements, the extra work consumed may be determined. Equation (7) becomes, using (25),

$$X\,\mathrm{d}u_{\text{extra}} = +\left[\frac{XW}{(n+1)}\left[\frac{X}{D(b/W)^Z}\right]^{1/n}\right]\frac{2\,\mathrm{d}a}{b}. \tag{26}$$

Once again, this requires solution of (14) for $X$ in terms of b and R for integration. When $X$ is given approximately by plastic collapse algebra, i.e. by (11), the extra work scales as $\lambda^3$ as shown by writing out (26) for two geometrically-similar bodies. That is

$$\frac{\text{extra work}_2}{\text{extra work}_1} = \frac{\left[\dfrac{X_2^{1+(1/n)}W_2}{(n+1(D_2(b/W)_2^Z)^{1/n}}\right]\dfrac{2\,\mathrm{d}a_2}{b_2}}{\left[\dfrac{X_1^{1+(1/n)}W_1}{(n+1)(D_1(b/W)_1^Z)^{1/n}}\right]\dfrac{2\,\mathrm{d}a_1}{b_1}}. \tag{27}$$

From (11), $X$ scales as $\lambda^2$; $D$ scales as $\lambda^2$; $W$ scales as $\lambda$; and the $(b/W)$ and $(\mathrm{d}a/b)$ ratios will be the same in geometrically-similar bodies. Hence

$$\frac{\text{extra work}_2}{\text{extra work}_1} = \lambda^3. \tag{28}$$

Since the (fully-reversible) NLEFM work scales as $\lambda^2$ (Mai and Atkins, 1975), the total work in elastoplastic fracture (when the fracture loads are given approximately by plastic collapse algebra) scales as the sum of terms in $\lambda^2$ and $\lambda^3$. Hence the experimental result in predominantly ductile fracture that energy scales as $\lambda^x$ where $2 < x < 3$ (Atkins, 1988).

## 4. Comparison with experiment

The theory given in this paper predicts both the load-displacement behaviour through to complete fracture as well as energies absorbed, for a linear power-law material. Many older papers in the literature report energies (sometimes partitioned into energies up to fracture (or at least to maximum load which is not necessarily the same thing), and the remaining work to fracture), but few give the raw data of load-displacement-crack length that would enable a full comparison to be made with theory. In Shearin et al. (1948), for example, 89 different combinations of materials and specimen sizes are discussed most of which had three repeat tests, but only *one* scaled diagram (reproduced as Figure 6 here) is given!

For energies up to crack initiation, Shearin et al. (1948) give normalised $\Gamma_{nl}$ values (their $W_1$) for different size testpieces which we shall call $\bar{\Gamma}_{nl}$. From (15) and (19), it may be shown that

$$\bar{\Gamma}_{nl} = \frac{(\text{const}_2)\bar{X}^{(1/n)+1}}{(n+1)}. \tag{29}$$

It follows that for toughness $R$ invariant with size,

$$(\text{ratio of Shearin } W_1 \text{ values}) = (\text{ratio of } \bar{X} \text{ at fracture})^{(1/n)+1}. \tag{30}$$

Equivalently, using (19)

$$(\text{ratio of Shearin } W_1 \text{ values}) = (\text{ratio of } \bar{u}_{nl} \text{ at fracture})^{n+1}. \tag{31}$$

Log plotting of the normalised $\bar{X}$ $\bar{u}_{nl}$ master-curve data in Figure 6 up to fracture gave $n = 0.09$, which is reasonable for steel. The average $W_1$ values given in Table II of (Shearin et al., 1948), are 707 ft lbs/cu in for the 0.4 in size beams and 499 ft lbs/cu in for the 1.25 in beams. From (30) the theoretical ratio of non-dimensional moments at fracture for these two sizes is $(499/707)^{n/(1+n)} = (499/707)^{0.09/1.09} = 0.97$; and the theoretical ratio of plastic bend angles at fracture from (31) is $(499/707)^{1/1.09} = 0.73$. The experimental peak moment ratio in Figure 6 is $31.4/33.1 = 0.95$ and the experimental ratio of plastic bend angles at peak moment is about $0.21/0.30 = 0.7$. The agreement between 0.97 and 0.95, and between 0.73 and 0.7, is not too bad. The reduced moment-angle of bend plot was not given for the 1.875 in size testpiece of the same material. The average $W_1$ value is 343 ft lbs/cu in, whence the reduced moment at fracture is predicted to be $(343/707)^{0.09/1.09} = 0.94$ of that for the 0.4 in testpiece, ie $0.94\ (33.1) = 31.1$ ft lbs/cu in; and the plastic bend angle is predicted to be

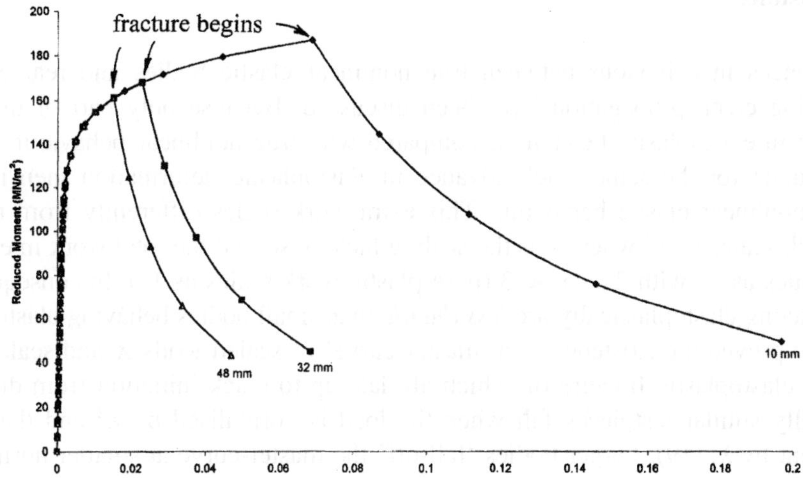


*Figure 8.* Reduced moment vs normalised displacement predictions for scaled 3-point bend bars with $(a_0/W) = 0.2$, $B = W$, $S = 4W$ and $W = 10$ mm, 32 mm and 48 mm. $E = 210$ GPa, $\sigma_y = 954$ MPa, $n = 0.1$, $R = 200$ kJ/m$^2$. Larger bodies crack earlier.

$(343/707)^{1/1.09} = 0.52$ of the 0.4 in testpiece, ie $0.52(0.30) = 0.16$ radian. These values fit the trends of Figure 6.

The Hagiwara et al. (1983) data shown in Figure 7 for notched fabricated I-beams have an n value also of 0.09. From (19) the ratio of $\bar{u}_{nl}$ in different size testpieces varies as the (ratio of corresponding $\bar{X})^{1/n}$. We have for the $(\bar{u}_{nl}, \bar{X})$ pairs at fracture: (24 mm, 81 tons) for 200 mm deep notched I-beams; (74 mm, 90 tons) for 60 mm deep beams; and (130 mm, 92 tons) for 23 mm deep beams. Taking first the 200 mm and 60 mm deep beam results, we find

$$(\tfrac{24}{74}) = 0.32 \quad \text{and} \quad (\tfrac{81}{90})^{1/0.09} = 0.31,$$

which agrees well. Taking the $\bar{u}_{nl}$ values for the 200 mm and 23 mm beams, the fracture load of the 23 mm beam is predicted to be 1.16 that of the 200 mm beam, i.e. $1.16 \times 81 = 94$ tons; the 60 mm and 23 mm $\bar{u}_{nl}$ values predict $\bar{X} = 95$ tons for the 23 mm beam. The experimental result is some 93 tons. We recall (caption to Figure 7), that the 2.3 mm plates from which the 23 mm deep beams were fabricated had a yield strength of some 16 kgf/mm$^2$ compared with about 29 kgf/mm$^2$ for the larger plates. No measurements of fracture toughness were made; it could very well be that $R$ of the thinner plate was different. Even so, the experimental theoretical comparisons for fracture load are good. We note also that the $\bar{X} - \bar{u}$ plot is very flat at the fracture initiation point and big changes in $\bar{u}$ correspond with minor changes in $\bar{X}$.

Figure 8 shows theoretical predictions of reduced moment vs normalised displacement for 3-point notched bend bars of a low carbon steel with $(a_0/W) = 0.2$, $B = W$, $S = 4W$ and $W = 10$, 32 and 48 mm; $E = 210$ GPa, $\sigma_y = 954$ MPa, $n = 0.1$ and $R = 200$ kJ/m$^2$. The experimental behaviour shown in Figures 6 and 7 is predicted. The discontinuities in slope at fracture in the theoretical plots would be removed were the practical *crack-tip* $R(\Delta a)$ incorporated.

## 5. Conclusions

The differences in behaviour between true nonlinear elastic bodies and real elastoplastic bodies during crack propagation have been discussed. Because only part of the energy is recoverable in elastoplastic behaviour, compared with true nonlinear behaviour, more work has to be done for the same crack advance in elastoplastic deformation then in globally-reversible nonlinear elastic behaviour. This extra work scales differently from the fracture work (which scales as $\lambda^2$ where $\lambda$ is the scaling factor), so that the total work in elastoplastic fracture scales as $\lambda^x$ with $2 < x < 3$ (pure plastic work scales as $\lambda^3$). In consequence large bodies behaving elastoplastically are *less ductile* than small bodies behaving elastoplastically.

We have proved the existence of a 'master-curve' of scaled loads $\bar{X}$ and scaled displacements $\bar{u}$ in elastoplastic fracture on which all data up to crack initiation from different size geometrically similar testpieces fall when the load is normalised by $\lambda^2$ and the load-point displacement by $\lambda$, (19). Larger bodies 'fall-off' the master-curve at smaller normalised displacements. Furthermore we have shown that the larger bodies (made of the same material as the smaller with the same $E$, $\sigma_y$, $R$ etc) behave as if their fracture toughness were $(R/\lambda)$ rather than $R$, (23).

During propagation the behaviour is path-dependent and each size testpiece has its own $\bar{X} - \bar{u}$ locus. This affects the total normalised energy absorbed to give complete fracture and explains the departures from 'geometric' ($\lambda^3$) scaling reported in the old literature. The actual scaling laws for loads, displacements and energies in elastoplastic materials following a linear-power law relationship are those given in this paper and encompass lefm at one end of the spectrum across to extremely ductile behaviour at the other. These changes in behaviour of the same material are produced by changes of size and replicate experimental results.

## Acknowledgements

I am very grateful to Chris Paul for performing numerical calculations, and to Chen Zhong and Brian Cotterell for continuing discussions.

## References

Atkins, A.G. Chen Zhong and Cotterell, B. (1998). The essential work of fracture and $J_R$ curves for the double cantilever beam specimen: an examination of elastoplastic crack propagation. *Proceedings of the Royal Society of London* **A454**, 815–833.

Atkins, A.G. (1990). Scaling of $J_R$ curves. *International Journal of Fracture* **43**, R43–R46.

Atkins, A.G. and Caddell, R.M. (1974). The laws of similitude and crack propagation. *International Journal of Mechanical Sciences* **16**, 541–548.

Atkins, A.G. (1995). Non-proportionally scaled lefm components and the scaling of elastic eta factors. *International Journal of Fracture* **70**, R23–R27.

Atkins, A.G. (1988). Scaling in combined flow and fracture. *International Journal of Mechanical Sciences* **30**, 173–191.

Atkins, A.G. (1975). Ice breaking modelling. *Journal of Ship Research* **19**, 40–43.

Atkins, A.G. (1968). Scaling in ice fracture mechanics: non-proporationately scaled and non-linear behaviour in *Ice Technology*. (Edited by C.Murthy, J. Connor, A. Brebbia), Berlin, Springer Verlag, 67–75.

Atkins, A.G. and Chen Zhong, (1996). Scaling in elastoplastic fracture mechanics in *Size-Scale Effects in the Failure Mechanisms of Materials and Structures*. (Edited by A.Carpinteri), (Proc. IUTAM Conference on Scaling, Turin) London, E&FN Spon, 473–482.

Booth, E., Collier, D. and Miles, J. (1983). Impact scalability of plated steel structures in *Structural Crashworthiness*. (Edited by N. Jones and T. Wierzbicki), Butterworths, London, 136–174.

Brown, W.F., Lubahn, J.D. and Ebert, L.J. (1947). Effects of section size on the static notch bar tensile properties of mild steel plate. *Welding Journal* **26**, 554s–559s.

Docherty, J.G. (1932). Bending tests on geometrically similar notched bar specimens. *Engineering* **133**, 645–647.

Docherty, J.G. (1935). Slow bending tests on large notched bars. *Engineering* **139**, 211–213.

Gurney, C. and Ngan, K.M. (1971). Quasistatic crack propagation in non-linear structures. *Proceedings of Royal Society of London* **A325**, 207–222.

Gurney, C. and Hunt, J. (1967). Quasi-static crack propagation. *Proceedings of the Royal Society* **A299**, 508–524.

Hagiwara, K., Takanabe, H. and Kawano, H. (1983). A proposed method of predicting ship collision damage. *International Journal of Impact Engineering* **1**, 257–279.

Hu, J.M. and Albrecht, P. (1991). Limit load solution and loading behaviour of $C(T)$ fracture specimen. *International Journal of Fracture* **52**, 19–45.

Irwin, G.R. (1954). The effect of size upon fracturing in *Symposium on Effects of Temperature on Brittle Behaviour*, ASTM, Philadelphia, 176–194.

Kanninen, M.F. and Popelar, C.H. (1985). *Advanced Fracture Mechanics*. Clarendon Press, Oxford.

Kumar, V., German, M.D. and Shih, C.F. (1981). An engineering approach for elastic-plastic fracture analysis (NP-1931/1237-1) Palo Alto, Electric Power Research Institute.

Lubahn, J.D. (1955). Room temperature crack propagation and size effect on mild steel. *Welding Journal* **34**, 518s–528s.

Lubahn, J.D. and Yukawa, S. (1958). Size effects in slow notch-bend tests of a nickel-molybdenum-vanadium steel. *Proceedings of American Society for Testing and Materials* **58**, 661–677.

Mai, Y-W. and Atkins, A.G. (1975). Scale effects and crack propagation in non-linear elastic structures. *International Journal of Mechanical Sciences* **17**, 673–675.

Mai, Y-W. and Atkins, A.G. (1978). Crack propagation in non-proportionally scaled elastic structures. *International Journal of Mechanical Sciences* **20**, 437–449.

Miller, A.G. (1988). Review of limit loads of structures containing defects. *International Journal of Pressure Vessels and Piping* **32**, 1–50.

Shearin, P.E., Ruark, A.E. and Trimble, R.M. (1948). Size effects in steels and other metals from slow notch bend tests in *Fracturing of Metals*, ASM, Cleveland, 167–188.

Stanton, T.E. and Batson, R.G.C. (1921). On the characteristics of notched-bar impact tests. *Proceedings of the Institution of Civil Engineers* **211**, 67–87.

Tada, H., Paris, P.C. and Irwin, G.R. (1973). *The Stress Analysis of Cracks Handbook*, Del Research Corporation Providence, R.I.

Xia, L. and Shih, C.F. (1995). Ductile crack growth – I. A numerical study using computational cells with microstructurally-based length scales. *Journal of the Mechanics and Physics of Solids* **43**, 233–259.

*International Journal of Fracture* **95:** 67–88, 1999.


# Scale effects in the initiation of cracking of a scarf joint

D. LIU and N.A. FLECK
*Cambridge University, Engineering Department, Trumpington Street, Cambridge, CB2 1PZ, U.K. e-mail: nafl@eng.cam.ac.uk*

Received 23 July 1998; accepted in revised form 31 December 1998

**Abstract.** The singular stress field at the interface-corner of a bi-material scarf joint is analysed for a strip of finite width, $w$, under remote tension and bending. The two substrates are taken as linear elastic and isotopic. The intensity of the singular stress field is calculated using a domain integral method, and is plotted as a function of joint geometry and material mismatch parameters. It is envisaged that the intensity of singularity can serve as a valid fracture criterion provided the zone of nonlinearity is fully embedded within the singular elastic field. It is assumed that fracture initiates when the magnitude of the corner singularity attains a critical value; consequently, the fracture strength of the joint depends upon the size of the structure. In addition, the interfacial stress intensity factor and the associated $T$-stress are determined for an edge interfacial crack. When the crack is short with respect to the width of the strip, the stress intensity factor is dominated by the presence of the corner singularity; a boundary layer formulation is used to determine the coupling between the crack tip field and the interface-corner field. The solution suggests that an interfacial crack grows unstably with a rapidly increasing energy release rate, but with only a small change in mode mix.

**Key words:** Scale effects, size effects, interfacial fracture, initiation of failure, scarf joint, domain integral, corner singularity.

## 1. Introduction

Adhesive bonding of dissimilar parts is increasingly used in the aerospace, automotive and electronic industries as this technology offers a number of attractive features, including high strength to weight ratio, improved appearance, improved corrosion resistance and cost effectiveness (Kinloch, 1987). Scarf joints are preferable to butt joints as the adhesive is loaded by a combination of shear and tension rather than in a tensile peel mode, giving higher joint strengths. The failure of bonded joints often initiates at the corner singularity where the interface intersects a traction-free edge. Therefore, the analysis of the singular stress field at an interface-corner is fundamental to an understanding of the initiation of failure from the free-edge of a joint.

Consider the problem of two wedges of apex angles $\theta_1$ and $\theta_2$, made from linear elastic and isotropic solids, 1 and 2, respectively, as sketched in Figure 1. A stress singularity of algebraic type $Hr^{\lambda-1}$ exists at the interface-corner for $\lambda$ less than unity. Here, $r$ is the radial distance from the corner, $H$ is the intensity of the singularity and the order of the stress singularity, $\lambda - 1$, may be real or complex. The $H$-singularity occurs only within a local region near the interface-corner and is a free-edge effect (Akisanya and Fleck, 1997). An asymptotic analysis has been performed for an extensive range of material combinations and edge geometries, $\theta_1$ and $\theta_2$ (Bogy, 1971; Hein and Erdogan, 1971); the magnitude of $\lambda$ depends upon the local wedge geometry and the mismatch in elastic parameters, but not upon the details of the overall geometry of the body. The intensity $H$ characterises the amplitude of the singular stress field;

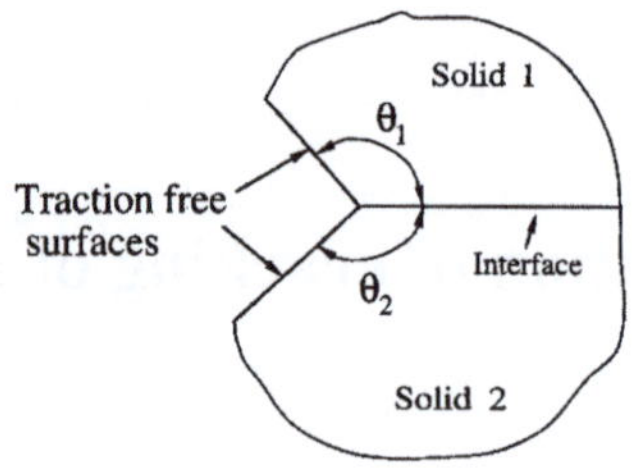


*Figure 1.* General configuration at the free-edge of two bonded dissimilar materials.

$H$ scales linearly with the applied load on the body, and depends upon the degree of elastic mismatch, the overall geometry of the body, and upon the details of the loading on the body. A number of researchers (Gradin, 1982; Groth, 1988; Hattori et al., 1989) have proposed that $H$ can be used as a crack initiation parameter, analogous to the crack tip stress intensity factor $K$ for the onset of crack growth from a pre-existing crack. Failure is assumed to initiate at an interface-corner when the magnitude of $H$ attains a critical value, $H_c$. The critical value $H_c$ must be determined by experiments, or from a micromechanical model of the initiation of fracture. For example, we might envisage that the critical value $H_c$ is associated with a normal stress component on the interface exceeding a critical value over a critical distance from the interface-corner.

Several experimental studies have validated the use of an $H$-based failure criterion for bi-material joints (Gradin, 1982; Groth, 1988; Hattori et al., 1989; Reedy and Guess, 1993; Qian and Akisanya, 1998). It is found that the critical value of the intensity $H$ at crack initiation is insensitive to the global details of joint geometry, for a given wedge geometry and material combination. Finite element analysis has shown that the zone of material nonlinearity at failure is embedded within a region dominated by the $H$-field, and so the elastic parameter $H_c$ is valid for the experimental systems considered (Reedy, 1993; Groth and Brottare, 1988). Since $H$ is of dimension stress times length to the power $(1 - \lambda)$ the failure criterion $H = H_c$ implies that the remote failure stress scales with the leading dimension of the specimen to the power $(\lambda - 1)$. A specimen size effect is predicted such that the strength decreases with increasing specimen size. There is a growing body of evidence to support such a size effect (see for example, Reedy and Guess, 1993; Qian and Akisanya, 1998). Alternative sources of a size effect are the fractal nature of microcracking (Carpinteri, 1996), the statistics of pre-existing flaws (Weibull, 1939) and the dependence of the fracture process zone upon the structural size (Bazant, 1997).

Application of an $H$-based failure criterion requires detailed calibrations for the intensity $H$ as a function of the joint overall geometry, material combinations and remote loading. Two methods have been used to extract the intensity $H$ from finite element analysis:

(i) a comparison of the finite element solution for stress and displacement with the asymptotic field; and

(ii) a more accurate procedure, based on a contour integral method (Stern *et al.*, 1976; Sinclair et al., 1985; Carpenter and Byers, 1987).

The first method was used to determine $H_c$ in most of the experimental studies mentioned above. This method has also been used in the theoretical analysis of free-edge singularities for mechanical and thermal loading; for example, Reedy (1990, 1993) considered a thin elastic layer sandwiched between two rigid substrates, and Munz and Yang (1992, 1993) examined elastic bi-material wedges of various wedge angles. Akisanya and Fleck (1997) extended the second method to extract $H$ by developing a domain integration method. They analysed the case of a butt joint between two long strips of dissimilar elastic solids. Using the same methodology, Qian and Akisanya (1997a, b) calculated the magnitude of the intensity $H$ for a scarf joint between two long dissimilar strips, subjected to remote tension and thermal loading; they also considered a scarf joint comprising a thin sandwich layer between identical substrates. They found that a much denser mesh was needed to extract accurate values for $H$ from the singular stress and displacement field (method 1) compared with the domain integral method (method 2).

### 1.1. Level of singularity in a scarf joint

In this study we shall address the scarf joint shown in the insert of Figure 2. The joint comprises two elastic strips made from dissimilar isotropic, elastic solids, and the bond line is inclined at an arbitrary inclination $\omega$ to the free surface. For the purposes of an asymptotic analysis of the stress state at the interface-corner, it may be assumed that the joint is subjected to arbitrary remote loading in equilibrium. For such a problem, Dundurs (1969) has shown that the stress distribution depends on only two combinations of the elastic constants. The two elastic mismatch parameters are defined by

$$\alpha = \frac{\mu_1(\kappa_2+1)-(\kappa_1+1)\mu_2}{\mu_1(\kappa_2+1)+(\kappa_1+1)\mu_2}, \tag{1a}$$

and

$$\beta = \frac{\mu_1(\kappa_2-1)-(\kappa_1-1)\mu_2}{\mu_1(\kappa_2+1)+(\kappa_1+1)\mu_2}, \tag{1b}$$

where the subscripts refer to materials 1 or 2. $E_i$ and $\nu_i$ denote Young's modulus and Poisson's ratio for material $i$, respectively, the shear modulus is $\mu_i$ ($\equiv E_i/2(1+\nu_i)$) and $\kappa_i \equiv 3-4\nu_i$ for plane strain conditions. The $(\alpha, \beta)$ values for typical material combinations are concentrated between the $\beta = 0$ and $\beta = \alpha/4$ lines in $\alpha - \beta$ space (Suga et al., 1988). In the current paper, we restrict our attention to material combinations with $\beta = 0$ and $\beta = \alpha/4$, and we shall consider the bi-material body to be sufficiently thick for plane strain conditions to prevail.

Recall that the displacement field $u_i$ at a distance $r$ from the interface-corner of the scarf joint scales as $Hr^{\lambda}$, and the elastic stress and strain distributions scale as $Hr^{\lambda-1}$. The eigenvalue problem for $\lambda$ has already been addressed by Qian and Akisanya (1997a), and the details are omitted here; the analysis follows that laid down by Stern et al. (1976) and by Carpenter and Byers (1987). Contours of $\lambda$ for the interface-corner of a scarf joint are shown in Figure 2 as a function of the material mismatch parameter $\alpha$ and the interface inclination $\omega$. Without loss of generality, we consider the range $0° \leqslant \omega \leqslant 90°$ and $|\alpha| \leqslant 1$. Results for $\lambda$ are displayed for $\beta = 0$, and for $\beta = \alpha/4$. Recall that for $\lambda < 1$ the stresses are unbounded (singular) as the interface-corner is approached. For $\lambda \geqslant 1$ the stresses remain finite (nonsingular) at the interface-corner.

Consider first the case $\beta = 0$. The eigenvalue $\lambda$ is real and the stresses are singular ($\lambda < 1$) or nonsingular ($\lambda \geqslant 1$) depending upon the particular choice of values for $\alpha$ and $\omega$. For the special case $\alpha = 0$ the material mismatch vanishes and the asymptotic stress state consists of uniform tension parallel to the free surface, giving $\lambda = 1$. A solution comprising uniform

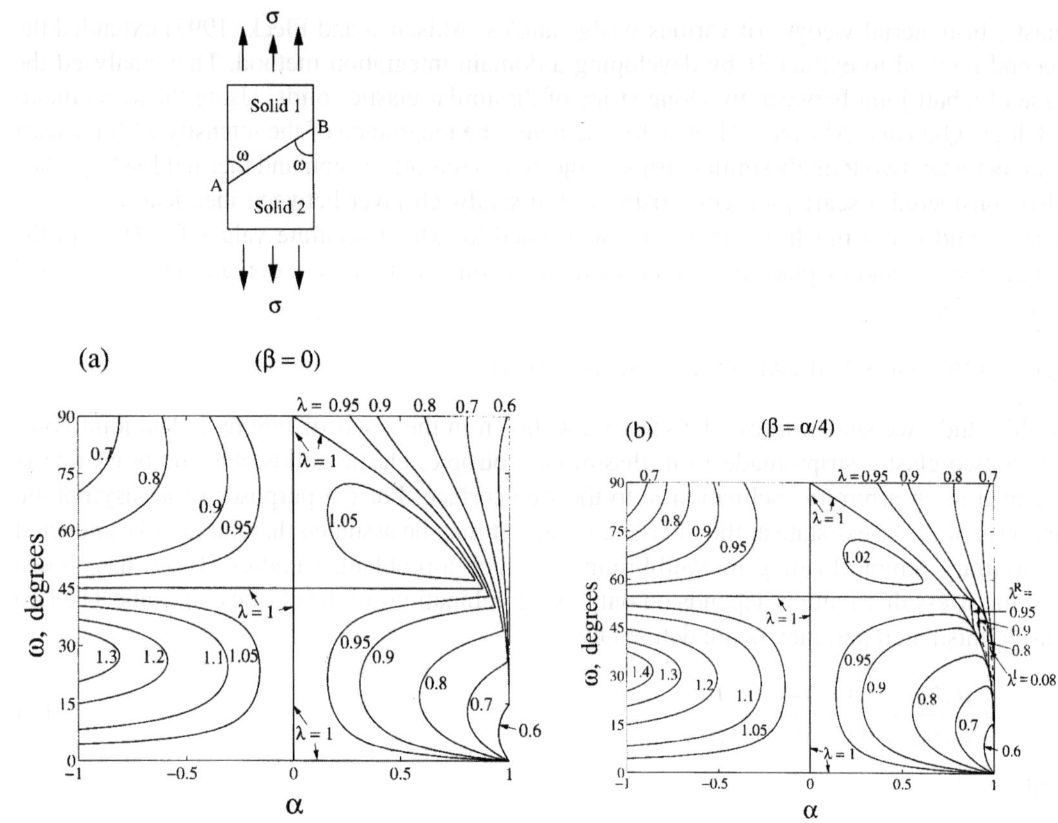


*Figure 2.* The level of displacement singularity λ at the interface-corner of a scarf joint, as a function of the material mismatch parameter $\alpha$ and the inclination $\omega$ of the interface. (a) $\beta = 0$, and (b) $\beta = \alpha/4$.

tension parallel to the free surface also occurs for the bi-material case when $\omega = 45°$ and $\alpha = 0$. For an inclination $\omega$ close to 45° the level of singularity λ is modest, and the stress state at the interface-corner is predominantly shear parallel to the interface. The level of singularity λ also approaches unity for the case of a long and slender scarf joint, with $\omega$ close to 0°. We further note from Figure 2a that the stress state at the interface-corner is always less singular than the crack case ($\lambda = 0.5$) for the full range of $\omega$ and $\alpha$. The solution is most singular for $\alpha$ close to unity for all values of $\omega$, and for $\alpha \to -1$ with $45° < \omega \leqslant 90°$.

At the two opposing interface-corners of a scarf joint, the inclination $\omega$ is equal in magnitude but the mismatch parameter $\alpha$ jumps in sign. Consequently, the level of singularity is different at the two opposing interface-corners of the joint. For the purposes of the following discussion, we shall consider the lower material 2 to be stiffer than the upper material 1. Then, $\alpha$ is negative for the left-hand interface-corner and $\alpha$ is positive for the right-hand interface-corner of the joint, labelled A and B, respectively, in Figure 2(a). For $\omega$ less than 45°, the stress state is nonsingular at the left-hand corner and is singular at the right-hand interface-corner. For $\omega$ greater than 45° the left-hand corner is always singular; also, the right-hand corner is singular for a large value of material mismatch, but is nonsingular if $\alpha$ is sufficiently close to zero. For the limiting case of a butt joint ($\omega = 90°$), the stress states at both interface-corners of the strip are identical, and are singular for $\alpha \neq 0$.

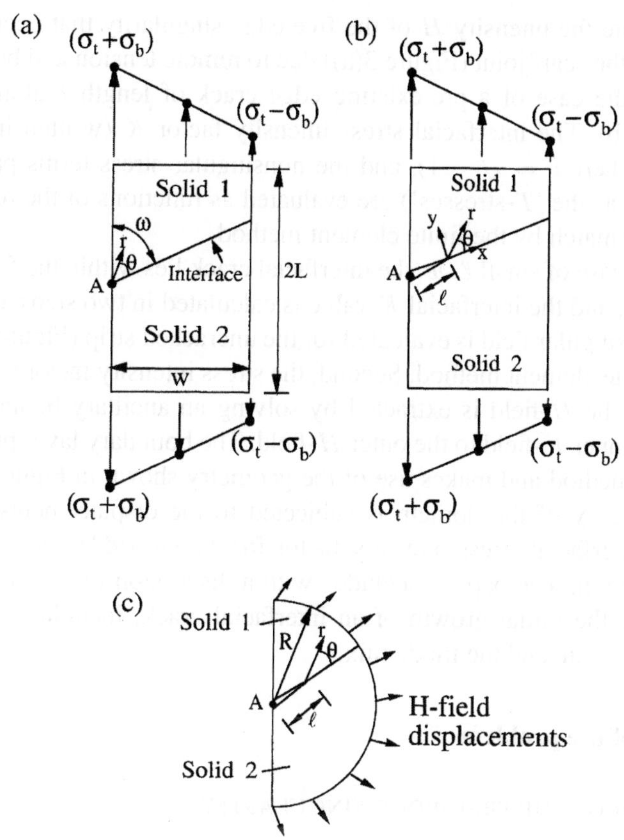


*Figure 3.* The bi-material geometries. (a) An uncracked bi-material strip with perfect adhesion between the two materials. (b) A bi-material strip with an interfacial edge crack of length $\ell$. (c) A boundary layer geometry comprising a short interfacial edge crack embedded within the free-edge singular field.

Now consider the case $\beta = \alpha/4$, see Figure 2(b). In general, the contours of $\lambda$ shift only slightly compared with the case $\beta = 0$. The horizontal line of $\lambda = 1$ shifts a little up the $\omega$-axis to the location $\omega = 54.7°$. But there is an added complication. A small region now exists in the $\alpha, \omega$ plane for which the stresses are singular and $\lambda$ is complex. However, we note from Figure 2(b) that the imaginary part of $\lambda$ is small. This region of complex singularity is thought to be of limited practical relevance, and is not considered further in our analysis.

### 1.2. Scope of the present study

We consider the scarf joint geometry shown in Figure 3(a). Two infinitely long strips, made from dissimilar isotropic elastic solids, are perfectly bonded along an interface oriented at an angle $\omega$ in the range 0–90° to the free surface. The strip, of width $w$ is subjected to a remote uniform tensile stress $\sigma_t$ giving rise to an axial tension $P$, or to a linear stress field of magnitude $\sigma_b$ resulting in a bending moment $M$. The length $2L$ of the strip is assumed to be much greater than the width $w$, and so only the single length scale $w$ enters the singularity analysis.

We shall calculate the intensity $H$ of the free edge singularity that occurs at the left-hand interface-corner of the scarf joint (Figure 3(a)) due to remote tension and bending. In addition, we shall consider the case of a pre-existing edge crack of length $\ell$ along the interface, as shown in Figure 3(b). The interfacial stress intensity factor $K$ (written in complex form as $K = K_1 + iK_2$, where $i = \sqrt{-1}$), and the nonsingular stress terms parallel to the crack surface (referred to as the '$T$-stresses') are evaluated as functions of the relative crack length $\ell/w$ and elastic mismatch by the finite element method.

For the limiting case of small $\ell/w$, the interfacial crack lies within the free-edge singularity characterised by $H$, and the interfacial $K$ value is calculated in two steps as follows. First, the intensity $H$ for the singular field is evaluated for the uncracked strip (Figure 3(a)) via a contour integral and the finite element method. Second, the stress intensity factor for a short interfacial crack lying within the $H$-field is extracted by solving an ancillary boundary layer problem which couples the inner $K$-field to the outer $H$-field. The boundary layer problem is solved by the finite element method and makes use of the geometry shown in Figure 3(c). The external semi-circular boundary of the domain is subjected to the displacements of the asymptotic $H$-field, and the interfacial stress intensity factor for the embedded crack is extracted using a $J$-integral evaluation. The paper concludes with a discussion of the effect of the interface orientation $\omega$ upon the initial growth of an interfacial crack, including a calculation of the strain energy release rate and the mode mix.

## 2. Formulation of the problem

### 2.1. Intensity $H$ of the free-edge singularity

Let $(r, \theta)$ be cylindrical polar co-ordinates centred at the interface-corner $A$ of the bi-material geometry shown in Figure 3(a). The singular field near the interface-corner is of the form

$$\sigma_{ij}^m = Hr^{\lambda-1} f_{ij}^m(\alpha, \beta, \omega, \theta),$$

$$u_i^m = \frac{1}{2\mu_m} Hr^{\lambda} g_i^m(\alpha, \beta, \omega, \theta) \tag{2.1}$$

where $(i, j) \equiv (r, \theta), m (= 1, 2)$ is the material index, and $f_{ij}^m$ and $g_i^m$ are known functions of $(\omega, \theta)$ and of the material mismatch parameters $(\alpha, \beta)$; explicit expressions for $f_{ij}^m$ and $g_i^m$ are given by Qian and Akisanya (1997a) and are not repeated here. We wish to evaluate $H$ as a function of material elastic parameters $(\alpha, \beta)$ for remote tension of magnitude $\sigma_t$ and for a bending stress field of magnitude $\sigma_b$, as shown in Figure 3(a).

The intensity $H$ of the free-edge singularity will hereafter be referred to as the *free-edge intensity factor*. $H$ is defined such that at a distance $r$ from the interface-corner and along the interface ($\theta = 0$), the stress component normal to the interface, $\sigma_{\theta\theta}$, in the region dominated by the singularity is given by

$$\sigma_{\theta\theta}^1 = \sigma_{\theta\theta}^2 = Hr^{\lambda-1}, \tag{2.2}$$

where the superscripts 1 and 2 denote the two materials. The free-edge intensity factor $H$ depends, in general, upon the free-edge geometry, elastic mismatch parameters and upon the

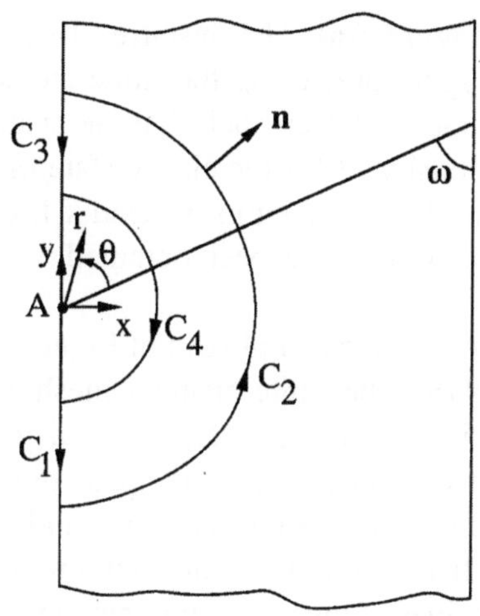


*Figure 4.* A closed integration path around the interface-corner A.

remote loading. Dimensional considerations dictate that $H$ is related to the geometry and material elastic properties by

$$H = \sigma w^{1-\lambda} \mathbf{a}(\alpha, \beta, \omega) \tag{2.3}$$

where $\sigma$ equals the remote tensile stress $\sigma_t$ for remote tension and $\sigma = \sigma_b$ for remote bending, $w$ is the width of the strip and $\mathbf{a}$ is a real dimensionless function of the elastic parameters $\alpha$ and $\beta$. Values for $\mathbf{a}$ are calculated using a contour integral method, based on Betti's reciprocal theorem; it involves a convolution of the asymptotic field of the corner singularity with a finite element solution. This contour integral method has been used by various authors to obtain the stress intensities for various crack and notch geometries, see for example, Stern et al. (1976), Sinclair et al. (1985), Carpenter and Byers (1987), and Akisanya and Fleck (1997). The method is based on Betti's reciprocal law (Sokolnikoff, 1956), and is outlined below.

Consider a closed contour $C(= C_1 + C_2 + C_3 + C_4)$ around the interface-corner A, as shown in Figure 4. Betti's reciprocal law can be stated as

$$\oint_C (\sigma_{ij} u_i^* - \sigma_{ij}^* u_i) n_j \, \mathrm{d}s = 0, \tag{2.4}$$

where $(i, j) \equiv (r, \theta)$ represent polar co-ordinates centred at the interface-corner, $(\sigma_{ij}, u_i)$ are the free-edge singular stress and displacement fields given by (2.1), $(\sigma_{ij}^*, u_i^*)$ are auxiliary fields satisfying the same boundary conditions as $(\sigma_{ij}, u_i)$, $n_j$ is the outward unit normal to $C$, and $\mathrm{d}s$ is an infinitesimal line segment of $C$. The integration in (2.4) is performed in an anticlockwise sense around $C$. By appropriate choice of the auxiliary field $(\sigma_{ij}^*, u_i^*)$, the integral (2.4) can be used to determine the free-edge intensity factor $H$.

In the evaluation of $H$ it is convenient to choose the auxiliary fields $(\sigma_{ij}^*, u_i^*)$ as the free-edge singular stress and displacement fields given by (2.1) with intensity $H^*$ and with $\lambda$ replaced by $\lambda^* = -\lambda$. Then, the starred fields $(\sigma_{ij}^*, u_i^*)$ satisfy the same boundary conditions as those for the unstarred fields $(\sigma_{ij}, u_i)$. These boundary conditions are: (i) traction-free boundaries along $\theta = \omega - \pi$ and $\theta = \omega$, and (ii) continuity of displacements and stress components $(\sigma_{\theta\theta}, \sigma_{r\theta})$ along the interface, where $(r, \theta)$ are polar co-ordinates centred at the interface-corner, as defined in Figure 3(a). The unstarred fields $(\sigma_{ij}, u_i)$ are obtained for the bi-material strip geometry of Figure 3(a) using the finite element method. The value of the intensity $H^*$ for the auxiliary field is chosen such that the evaluation of (2.4) by the contour integration method gives the intensity $H$ for the elastic state of interest. A full description of the procedure for the evaluation of $H$ is given in Appendix B of Akisanya and Fleck (1997), and is omitted here. Once the value of $H$ has been obtained, we calculate the nondimensional constant $\mathbf{a}(\alpha, \beta, \omega)$ via (2.3).

Elastic analysis of the uncracked bi-material strip of Figure 3(a) has been carried out using the finite element code ABAQUS[1] The finite element mesh consists of 928 to 1564 eight-noded plane strain isoparametric, quadrilateral elements of the bi-material strip. The mesh near the interface-corner is refined in order to give adequate numerical accuracy in the vicinity of the singularity. The width of the strip $w$ is taken as unity and the strip length $2L = 20w$. The results for the coefficient **a** are listed in Table 1 for both remote tension and remote bending, over the full range of practical values of the material parameters $\alpha$ and $\beta$, and of the inclination $\omega$.

*Table 1(a).* Magnitude of $\mathbf{a}(\alpha, \beta = 0, \omega)$ for tension.

| $\alpha$ | $\omega = 15°$ | $\omega = 30°$ | $\omega = 60°$ | $\omega = 75°$ | $\omega = 90°$ |
|---|---|---|---|---|---|
| −0.99 | 0.0012 | 0.3211 | 0.4996 | 0.4365 | 0.3365 |
| −0.8 | 0.0058 | 0.2911 | 0.5168 | 0.4879 | 0.4334 |
| −0.5 | 0.0383 | 0.2789 | 0.5685 | 0.6010 | 0.6318 |
| −0.2 | 0.0579 | 0.2631 | 0.6550 | 0.7729 | 0.8954 |
| 0.0 | 0.0670 | 0.2500 | 0.7500 | 0.9330 | 1.0000 |
| 0.2 | 0.0735 | 0.2353 | 0.9190 | 1.0444 | 0.8954 |
| 0.5 | 0.0758 | 0.2130 | 1.5676 | 0.7269 | 0.6318 |
| 0.8 | 0.0689 | 0.1924 | 0.3074 | 0.3819 | 0.4334 |
| 0.99 | 0.0663 | 0.1884 | 0.1167 | 0.2312 | 0.3365 |

*Table 1(b).* Magnitude of $\mathbf{a}(\alpha, \beta = \alpha/4, \omega)$ for tension.

| $\alpha$ | $\omega = 15°$ | $\omega = 30°$ | $\omega = 45°$ | $\omega = 60°$ | $\omega = 75°$ | $\omega = 90°$ |
|---|---|---|---|---|---|---|
| −0.99 | 0.0010 | 0.0402 | 0.6410 | 0.6544 | 0.5647 | 0.4279 |
| −0.8 | 0.0152 | 0.2461 | 0.6422 | 0.6633 | 0.6189 | 0.5472 |
| −0.5 | 0.0374 | 0.2841 | 0.5948 | 0.6881 | 0.7400 | 0.7540 |
| −0.2 | 0.0572 | 0.2671 | 0.5372 | 0.7220 | 0.8469 | 0.9476 |
| 0.0 | 0.0670 | 0.2500 | 0.5000 | 0.7500 | 0.9330 | 1.0000 |
| 0.2 | 0.0736 | 0.2312 | 0.4648 | 0.7830 | 0.9828 | 0.9476 |
| 0.5 | 0.0752 | 0.2017 | 0.4143 | 0.8332 | 0.8313 | 0.7540 |
| 0.8 | 0.0664 | 0.1683 | 0.3452 | 0.4423 | 0.4654 | 0.5472 |
| 0.99 | 0.0622 | complex | 0.2572 | 0.0807 | 0.2652 | 0.4279 |

[1] Hibbitt, Karlsson and Sorenson, ABAQUS Users Manual, Version 5.2, HKS Inc. (1992).

*Table 1(c)*. Magnitude of $\mathbf{a}(\alpha, \beta = 0, \omega)$ for bending.

| $\alpha$ | $\omega = 15°$ | $\omega = 30°$ | $\omega = 60°$ | $\omega = 75°$ | $\omega = 90°$ |
|---|---|---|---|---|---|
| −0.99 | 0.00124 | 0.1121 | 0.1808 | 0.1945 | 0.2057 |
| −0.8 | 0.0160 | 0.2309 | 0.2576 | 0.2676 | 0.3003 |
| −0.5 | 0.0384 | 0.2600 | 0.3801 | 0.4188 | 0.5249 |
| −0.2 | 0.0573 | 0.2601 | 0.5556 | 0.6716 | 0.8685 |
| 0.0 | 0.0670 | 0.2500 | 0.7500 | 0.9330 | 1.0000 |
| 0.2 | 0.0730 | 0.2300 | 1.1118 | 1.1898 | 0.8685 |
| 0.5 | 0.0744 | 0.2147 | 2.6836 | 0.8299 | 0.5249 |
| 0.8 | 0.0641 | 0.1868 | 0.6210 | 0.3751 | 0.3003 |
| 0.99 | 0.0601 | 0.1550 | 0.2140 | 0.2081 | 0.2057 |

*Table 1(d)*. Magnitude of $\mathbf{a}(\alpha, \beta = \alpha/4, \omega)$ for bending.

| $\alpha$ | $\omega = 15°$ | $\omega = 30°$ | $\omega = 45°$ | $\omega = 60°$ | $\omega = 75°$ | $\omega = 90°$ |
|---|---|---|---|---|---|---|
| −0.99 | 0.0011 | 0.0302 | 0.2346 | 0.2719 | 0.2868 | 0.2933 |
| −0.8 | 0.0153 | 0.2311 | 0.3907 | 0.3694 | 0.3790 | 0.4209 |
| −0.5 | 0.0378 | 0.2731 | 0.4803 | 0.5012 | 0.5458 | 0.6734 |
| −0.2 | 0.0574 | 0.2665 | 0.5019 | 0.6409 | 0.7641 | 0.9348 |
| 0.0 | 0.0670 | 0.2500 | 0.5000 | 0.7500 | 0.9330 | 1.0000 |
| 0.2 | 0.0732 | 0.2314 | 0.4928 | 0.8843 | 1.0859 | 0.9348 |
| 0.5 | 0.0739 | 0.2015 | 0.4811 | 1.1891 | 0.9885 | 0.6734 |
| 0.8 | 0.0637 | 0.1703 | 0.5233 | 1.0651 | 0.5330 | 0.4209 |
| 0.99 | 0.0589 | complex | 0.3470 | 0.2952 | 0.2929 | 0.2933 |

### 2.2. Stress intensities and $T$-stresses for an interfacial edge crack

If the interface of the scarf joint is sufficiently brittle in comparison with the two substrates, it is expected that failure will occur by the initiation and growth of an interfacial crack from the interface-corner. Therefore, an analysis of the stress state for an interfacial edge crack in the bi-material strip is of practical relevance; the precise geometry is given in Figure 3(b). Again, we consider the two loading cases of remote uniform tension, of magnitude $\sigma_t$, and a bending stress field of magnitude $\sigma_b$.

In general, an interfacial crack between two dissimilar isotropic elastic solids suffers a singular stress field characterised by the complex interfacial stress intensity factor $K = K_1 + iK_2$, where $i = \sqrt{-1}$. In order to define $K$ we introduce the cylindrical polar co-ordinates $(r, \theta)$ placed at the crack tip as shown in Figure 3(b). Then, $K$ is defined such that at a distance $r$ directly ahead of the crack tip, the normal stress $\sigma_{\theta\theta}$ and shear stress $\sigma_{r\theta}$ are given by

$$\sigma_{\theta\theta} + i\sigma_{r\theta} = \frac{1}{\sqrt{2\pi}} K r^{-1/2+i\varepsilon} \tag{2.5}$$

where the oscillatory index $\varepsilon$ depends only upon the material elastic mismatch parameter $\beta$ via

$$\varepsilon = \frac{1}{2\pi} \ln\left(\frac{1-\beta}{1+\beta}\right). \tag{2.6}$$

In presenting the interfacial stress intensity solutions for both long and short edge cracks, it is convenient to define a phase angle $\psi$ in relation to the crack length $\ell$ by

$$\tan\psi = \frac{\mathrm{Im}(K\ell^{i\varepsilon})}{\mathrm{Re}(K\ell^{i\varepsilon})}. \tag{2.7}$$

In the series expansion for the interfacial crack tip stress field, the next highest order term to the $K$ field is given by in-plane direct stresses parallel to the crack plane. These stresses are of magnitude $T_1$ in material 1 and $T_2$ in material 2, and are referred to as the '$T$-stresses'. Since the strain component $\varepsilon_{rr}$ is the same on both sides of the interface, a direct relation exists between the $T$-stresses,

$$T_2 = \frac{1-\alpha}{1+\alpha}T_1. \tag{2.8}$$

When the cracked bi-material strip shown in Figure 3(b) is loaded by a stress $\sigma$ (equal to $\sigma_t$ for remote tension, and equal to $\sigma_b$ for remote bending) the stress field at the tip of the interfacial crack of length $\ell$ from the free-edge is governed by the complex stress intensity factor $K$ and the $T$-stresses, $T_1$ in material 1 and $T_2$ in material 2. Dimensional considerations require that the stress intensity factor and the $T$-stresses be related to the geometry and the applied stress $\sigma$ by

$$K\ell^{i\varepsilon} = \sigma\sqrt{\ell}\,\mathbf{b}(\ell/w, \alpha, \beta, \omega) \tag{2.9}$$

and

$$T_1 = \sigma\mathbf{c}(\ell/w, \alpha, \beta, \omega). \tag{2.10}$$

We emphasise that $\sigma \equiv \sigma_t$ for the case of remote tensile loading, $\sigma \equiv \sigma_b$ for the case of remote bending; $\mathbf{b}(= b_1 + ib_2)$ is a complex nondimensional function and $\mathbf{c}$ is a real nondimensional function of the elastic mismatch parameters $(\alpha, \beta)$, of the inclination $\omega$ and of the relative crack length $\ell/w$. We omit tables of values for **b** and **c** for the sake of brevity but will present relevant results in graphical form[2].

For the case where the interfacial edge crack of length $\ell$ lies within the zone of $H$-dominance the stress intensity factor and the $T$-stresses depend directly upon the magnitude of $H$. A boundary layer approach is used for evaluating the coupling between the crack tip parameters $(K, T)$ and the free-edge singularity parameter $H$. The boundary layer geometry shown in Figure 3(c) is loaded on the semi-circular boundary by the asymptotic free-edge displacement field characterised by $H$ (Equation (2.1)).

The interfacial stress intensity factor $K$ has the dimension (stress)(length)$^{1/2-i\varepsilon}$, while the free-edge intensity factor $H$ has the dimension (stress)(length)$^{1-\lambda}$. Dimensional considerations require that the interfacial stress intensity factor $K$ and the $T$-stress $T_1$ in material 1 be related to the geometry and the free-edge intensity factor $H$ by

$$K\ell^{i\varepsilon} = H\ell^{\lambda-1/2}\mathbf{d}(\alpha, \beta, \omega), \tag{2.11}$$

[2] We note in passing that the **b**-values for the butt joint presented in Table 2 of Akisanya and Fleck (1997) require correction: entries in the table should be multiplied by the factor $(10\ell/w)^{i\varepsilon-(1/2)}$.

and

$$T_1 = H\ell^{\lambda-1}\mathbf{e}(\alpha, \beta, \omega). \tag{2.12}$$

Here, $\mathbf{d}$ $(= d_1 + id_2)$ is a complex nondimensional function and $\mathbf{e}$ is a real nondimensional function of the elastic mismatch parameters $\alpha$ and $\beta$ and of the inclination $\omega$; the evaluation of these functions is described in the following sub-section.

By substituting the expression (2.3) for the free-edge intensity factor $H$ into (2.11) and (2.12), the interfacial stress intensity factor $K$ and the $T$-stress $T_1$ in material 1 for an interfacial crack embedded within the free-edge singularity zone are given in terms of the remote loading $\sigma$, crack length ratio $\ell/w$ and material parameters $(\alpha, \beta)$ by

$$K\ell^{i\varepsilon} = \sigma\sqrt{\ell}\left(\frac{\ell}{w}\right)^{\lambda-1} \mathbf{a}(\alpha, \beta, \omega) \,.\, \mathbf{d}(\alpha, \beta, \omega), \tag{2.13}$$

and

$$T_1 = \sigma\left(\frac{\ell}{w}\right)^{\lambda-1} \mathbf{a}(\alpha, \beta, \omega) \,.\, \mathbf{e}(\alpha, \beta, \omega). \tag{2.14}$$

### 2.3. Numerical evaluation of interfacial stress intensity factor and $T$-stress

The interfacial stress intensity factor $K$ and the $T$-stresses for a crack of length $\ell$ laying on the interface of the scarf joint are determined by:

(i) evaluating the path-independent $J$-integral for the cracked geometry, loaded by remote tension or bending, followed by
(ii) evaluating the $J$-integral for the same cracked geometry, with a suitably-chosen auxiliary elastic field superimposed on the remote loading.

Parks' (1974) virtual crack extension method is used to evaluate the path independent $J$-integral. For the evaluation of the components $K_1$ and $K_2$ of the interfacial stress intensity factor, the auxiliary field is taken to be the singular crack tip field for an interfacial crack, as described by Matos et al. (1989). In the evaluation of the $T$-stresses, the auxiliary elastic field consists of a point force placed at the tip of a semi-infinite interfacial crack and in a direction parallel to the crack faces. A detailed description of these methods is given by Akisanya and Fleck (1994).

Finite element analysis is again carried out using the finite element code ABAQUS. The finite element mesh contains 1354 to 1728 elements for the case of long crack (Figure 3(b)) and 638 elements for the boundary layer problem (Figure 3(c)). In all cases eight-noded plane strain isoparametric, quadrilateral elements are used. The numerically obtained values for the coefficients $\mathbf{d}$ and $\mathbf{e}$ are listed in Tables 2 and 3, for various values of material parameters $\alpha$ and $\beta(= 0$ and $\alpha/4)$, and interface orientation $\omega$. Numerical experimentation suggests that the results for $K$ and $T_1$ are accurate to within about 1 percent.

## 3. Results

### 3.1. MAGNITUDE OF THE FREE EDGE INTENSITY FACTOR $H$

The magnitude of the free-edge intensity factor $H$ is related to the applied stress $\sigma$ ($\sigma = \sigma_t$ for tension and $\sigma = \sigma_b$ for bending), the strip width $w$, the joint inclination $\omega$ and the material elastic properties $(\alpha, \beta)$ in the manner given by (2.3). The nondimensional real parameter $\mathbf{a}(\alpha, \beta, \omega)$ is determined by the contour integral method as described in Section 2.1 and detailed in Appendix B of Akisanya and Fleck (1997). Values for **a** are listed in Table 1 for the tensile and bending cases, over the full range of values for $\alpha$, and for the two values, $\beta = 0$ and $\alpha/4$. Contours of the nondimensional measure **a** of the intensity factor $H$ are plotted as a function of material parameter $\alpha$ and joint inclination $\omega$ in Figure 5(a) for the case $\beta = 0$ and in Figure 5(b) for the case $\beta = \alpha/4$.

For both remote tension and bending and for all $(\alpha, \beta)$ considered, **a** increases from zero as $\omega$ increases from zero, see Figures 5(a, b). For $\omega < 30°$, **a** remains small (less than 0.3) and is relatively insensitive to the value of $(\alpha, \beta)$ and to the nature of the remote loading; for larger values of $\omega$, the behaviour is more complicated and the precise value of **a** depends upon $(\alpha, \beta)$ and upon the type of remote loading. We note that **a** is complex (and has not been evaluated) within the same small domain of $(\alpha, \beta)$ values for which $\lambda$ is complex. For the limiting case of a butt joint under remote tension ($\omega = 90°$), Akisanya and Fleck (1997) showed that a unique correlation exists between **a** and $\lambda$ for all $(\alpha, \beta)$ considered. No such correlation is observed here for $\omega \neq 90°$. Numerical experimentation and a comparison of the results given by Qian and Akisanya (1997a) for the scarf joint under remote tension suggest that the values of **a** in the current study are accurate to within 1 percent.

### 3.2. ZONE OF DOMINANCE OF THE FREE-EDGE SINGULARITY

It is important to determine the physical size of the singularity zone in relation to the width $w$ of the bi-material strip in order to appraise whether the intensity of singularity $H$ can be used as a failure criterion. The intensity $H$ is expected to be a useful parameter provided the fracture process zone and any nonlinear plastic zone at the interface-corner are fully embedded within the elastic $H$-field. The finite element results are compared with the asymptotic solution in Figure 6 by plotting the stress components $(\sigma_{\theta\theta}, \sigma_{r\theta})$ along the interface against the radial distance $r$ from the left hand interface-corner A, for the selected cases $\omega = 30°$ and $\omega = 90°$, and $\alpha = 0.5, \beta = \alpha/4$. First, consider the butt joint, $\omega = 90°$; the asymptotic results are adequate (i.e. lay within 5 percent of the finite element values) for $r/w$ extending out to 0.01 in the bending case, and to 0.03 in the tensile case. Second, consider the inclination $\omega = 30°$; the asymptotic results are within 5 percent of the finite element results for $r/w$ extending to 0.07 in both the bending and tensile cases. Numerical experimentation reveals that these findings are relatively insensitive to the material mismatch parameters $(\alpha, \beta)$. We conclude that the singular zone extends across a significant fraction of the strip, particularly for the loading case of remote tension. Akisanya and Fleck (1997) and Qian and Akisanya (1997a) drew similar conclusions from their studies on the free-edge singularity at a butt joint ($\omega = 90°$) under remote tension, and at a scarf joint under remote tension, respectively.

*Table 2(a)*. Values of $\mathbf{d} = d_1 + id_2$ for $\beta = 0$.

| $\alpha$ | | $\omega = 15°$ | $\omega = 30°$ | $\omega = 60°$ | $\omega = 90°$ |
|---|---|---|---|---|---|
| −0.99 | $d_1$ | 6.2312 | 2.7684 | 2.6714 | 2.2803 |
| | $d_2$ | 4.6028 | 1.1213 | 0.9539 | 0.6468 |
| −0.8 | $d_1$ | 6.3063 | 3.0009 | 2.5898 | 2.2639 |
| | $d_2$ | 4.8353 | 1.4391 | 0.9203 | 0.5991 |
| −0.5 | $d_1$ | 6.2465 | 3.1601 | 2.4383 | 2.0957 |
| | $d_2$ | 5.1711 | 1.7968 | 0.8616 | 0.4109 |
| −0.2 | $d_1$ | 6.1762 | 3.2460 | 2.2730 | 1.9611 |
| | $d_2$ | 5.6044 | 2.1391 | 0.7852 | 0.1831 |
| 0.0 | $d_1$ | 6.1156 | 3.2600 | 2.1582 | 1.9241 |
| | $d_2$ | 5.9673 | 2.3702 | 0.7184 | 0.0000 |
| 0.2 | $d_1$ | 6.0540 | 3.2637 | 2.0231 | 1.9611 |
| | $d_2$ | 6.4357 | 2.6329 | 0.6192 | −0.1831 |
| 0.5 | $d_1$ | 5.8485 | 3.1658 | 1.8281 | 2.0957 |
| | $d_2$ | 7.3701 | 3.0659 | 0.2821 | −0.4109 |
| 0.8 | $d_1$ | 5.1891 | 2.8355 | 2.2328 | 2.2639 |
| | $d_2$ | 8.8061 | 3.6076 | −0.2660 | −0.5991 |
| 0.99 | $d_1$ | 3.0292 | 2.0912 | 2.3127 | 2.2803 |
| | $d_2$ | 10.3894 | 4.1046 | −0.4994 | −0.6468 |

*Table 2(b)*. Values of $\mathbf{d} = d_1 + id_2$ for $\beta = \alpha/4$.

| $\alpha$ | | $\omega = 15°$ | $\omega = 30°$ | $\omega = 60°$ | $\omega = 90°$ |
|---|---|---|---|---|---|
| −0.99 | $d_1$ | 4.6127 | 2.2327 | 2.2176 | 2.3259 |
| | $d_2$ | 6.3703 | 1.6322 | 1.0330 | 0.6793 |
| −0.8 | $d_1$ | 4.8996 | 2.4707 | 2.2125 | 2.1911 |
| | $d_2$ | 6.2074 | 1.8432 | 0.9619 | 0.5620 |
| −0.5 | $d_1$ | 5.6410 | 2.7821 | 2.2570 | 2.0343 |
| | $d_2$ | 5.9206 | 2.0482 | 0.8617 | 0.3558 |
| −0.2 | $d_1$ | 5.7932 | 3.0788 | 2.1779 | 1.9426 |
| | $d_2$ | 5.9303 | 2.2371 | 0.7702 | 0.1404 |
| 0.0 | $d_1$ | 6.1156 | 3.2600 | 2.1582 | 1.9241 |
| | $d_2$ | 5.9673 | 2.3702 | 0.7184 | 0.0000 |
| 0.2 | $d_1$ | 6.4834 | 3.4516 | 2.1257 | 1.9426 |
| | $d_2$ | 6.1322 | 2.5464 | 0.6668 | −0.1404 |
| 0.5 | $d_1$ | 7.0342 | 3.6711 | 2.0893 | 2.0343 |
| | $d_2$ | 6.6811 | 2.9049 | 0.5938 | −0.3558 |
| 0.8 | $d_1$ | 7.2927 | 3.6697 | 2.0563 | 2.1911 |
| | $d_2$ | 8.0678 | 3.6275 | 0.2955 | −0.5620 |
| 0.99 | $d_1$ | 5.5262 | complex | 2.2478 | 2.3259 |
| | $d_2$ | 14.9905 | complex | −0.1492 | −0.6793 |

*Table 3(a)*. Magnitude of $\mathbf{e}(\alpha, \beta = 0, \omega)$.

| $\alpha$ | $\omega = 15°$ | $\omega = 30°$ | $\omega = 60°$ | $\omega = 90°$ |
|---|---|---|---|---|
| −0.99 | 9.0868 | 0.0703 | −0.0005 | −0.0049 |
| −0.8 | 9.8381 | 0.7260 | −0.0106 | −0.0984 |
| −0.5 | 10.9850 | 1.4867 | −0.0363 | −0.2511 |
| −0.2 | 12.4639 | 2.2471 | −0.0814 | −0.4100 |
| 0.0 | 13.6886 | 2.7836 | −0.1311 | −0.5109 |
| 0.2 | 15.2441 | 3.3916 | −0.2117 | −0.6150 |
| 0.5 | 18.3923 | 4.4427 | −0.5514 | −0.7532 |
| 0.8 | 23.4519 | 5.8481 | −1.1748 | −0.8852 |
| 0.99 | 30.4237 | 7.2838 | −1.4929 | −0.9741 |

*Table 3(b)*. Magnitude of $\mathbf{e}(\alpha, \beta = \alpha/4, \omega)$.

| $\alpha$ | $\omega = 15°$ | $\omega = 30°$ | $\omega = 60°$ | $\omega = 90°$ |
|---|---|---|---|---|
| −0.99 | 9.7696 | 0.4853 | −0.0018 | −0.0050 |
| −0.8 | 10.4107 | 0.9419 | −0.0344 | −0.1000 |
| −0.5 | 11.4073 | 1.5610 | −0.0770 | −0.2521 |
| −0.2 | 12.5068 | 2.2519 | −0.1132 | −0.4075 |
| 0.0 | 13.6886 | 2.7836 | −0.1311 | −0.5109 |
| 0.2 | 15.2893 | 3.4308 | −0.1437 | −0.6112 |
| 0.5 | 18.7599 | 4.6783 | −0.1674 | −0.7564 |
| 0.8 | 25.0216 | 6.7842 | −0.4127 | −0.8998 |
| 0.99 | 49.0327 | complex | −0.8623 | −0.9947 |

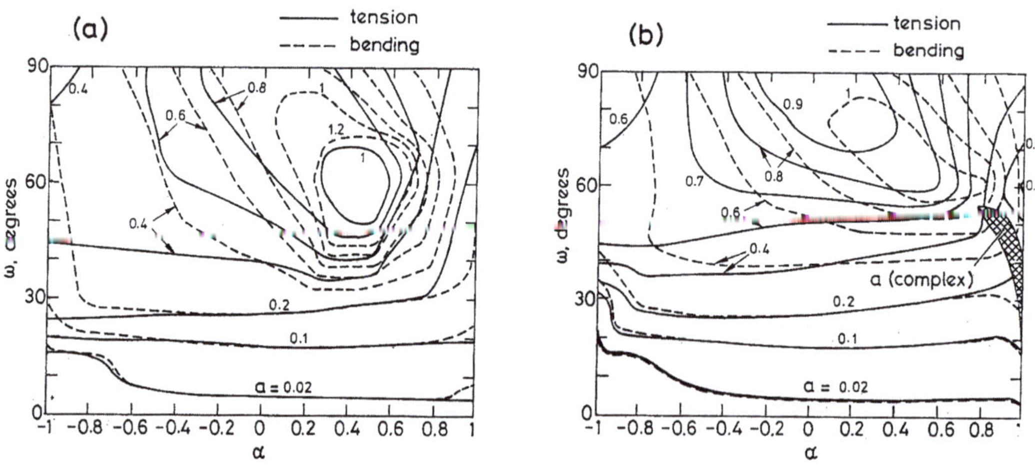


*Figure 5.* Contours of **a** as a function of $\alpha$ and joint inclination $\omega$, for both remote tension and remote bending. (a) $\beta = 0$ and (b) $\beta = \alpha/4$.

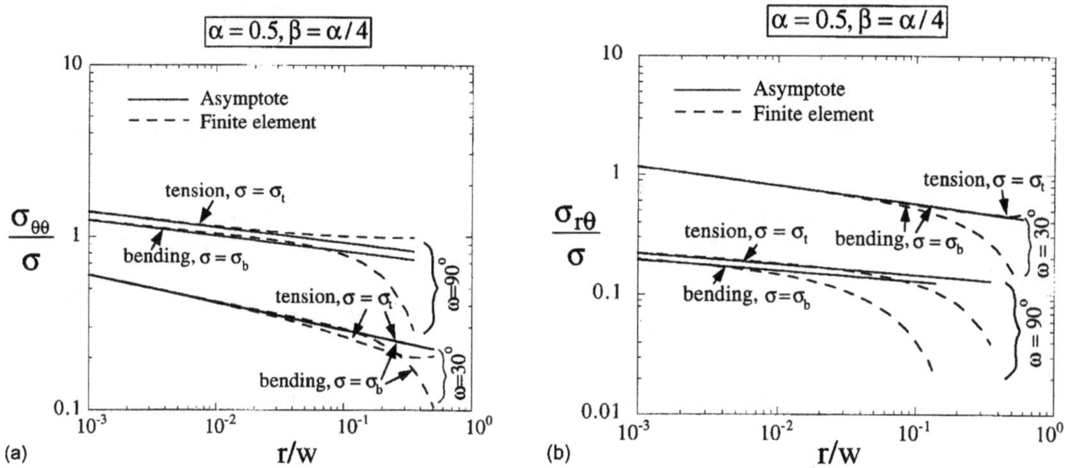


*Figure 6.* A comparison of the finite element and asymptotic solutions for the traction on the interface, as a function of distance $r$ from the interface-corner. $\alpha = 0.5$ and $\beta = \alpha/4$. (a) $\sigma_{\theta\theta}/\sigma$ versus $r/w$, and (b) $\sigma_{r\theta}/\sigma$ versus $r/w$, where $\sigma \equiv \sigma_t$ for remote tension and $\sigma \equiv \sigma_b$ for remote bending.

## 3.3. Interfacial stress intensity factor

The nondimensional real and imaginary parts of the complex interfacial stress intensity factor defined by (2.9) and (2.13) are plotted in Figures 7(a, b) for $\alpha = 0.5$, $\beta = \alpha/4$, and in Figures 7(c, d) for $\alpha = -0.5$, $\beta = \alpha/4$. Results are shown for selected values of joint inclination, $\omega = 15°, 30°, 60°$ and $90°$. The values for $K$ have been normalised by $\sigma\sqrt{\ell}$ where $\sigma = \sigma_t$ for the tensile case and $\sigma = \sigma_b$ for the bending case, in order to elucidate the significance of the corner singularity on $K$. The stress intensity factor values shown for $\ell/w \leqslant 0.01$ are the asymptotic results for a crack embedded within the free-edge singularity zone (see Figure 3(c)); results for $\ell/w \geqslant 0.06$ correspond to a long interfacial crack and are determined by finite element analysis of the cracked strip shown in Figure 3(b). Linear interpolation is used within the intermediate regime, $0.01 < \ell/w < 0.06$.

In order to provide a baseline reference for the magnitude of $K$ recall that for the homogeneous case ($\alpha = \beta = 0$) with $\omega = 90°$, we have $K_1 = \text{Re}(K\ell^{i\varepsilon}) = 1.12\sigma\sqrt{\pi\ell} = 1.985\sigma\sqrt{\ell}$ and $K_2 = \text{Im}(K\ell^{i\varepsilon}) = 0$. It is evident from Figures 7(a–d) that the interfacial stress intensity factor for short crack lengths ($\ell/w < 0.1$) is amplified when the stress field at the interface-corner has an unbounded singularity. The magnitude of the normalised stress intensity factor increases with decreasing crack length within the singularity region. This result can be deduced directly from (2.13): the magnitude of $K\ell^{i\varepsilon}/\sigma\sqrt{\ell}$ scales with $(\ell/w)^{\lambda-1}$. Now consider the opposite limit, $\ell/w \to 1$. In all cases there is a sharp up-turn in the magnitude of $K\ell^{i\varepsilon}/\sigma\sqrt{\ell}$ as $\ell/w \to 1$; it is clear from a separate asymptotic analysis that $(K\ell^{i\varepsilon}/\sigma\sqrt{\ell}) \to \infty$ as $\ell/w \to 1$.

There is no simple monotonic dependence of $K\ell^{i\varepsilon}$ upon the inclination $\omega$. It is clear, however, that $K\ell^{i\varepsilon}$ has the largest real values for the limiting case of a butt joint, $\omega = 90°$. If the inclination of the interface $\omega$ is less than about 30°, both the real and imaginary components of the interfacial stress intensity factor decrease with decreasing $\omega$ at a constant $\ell/w$: the inter-

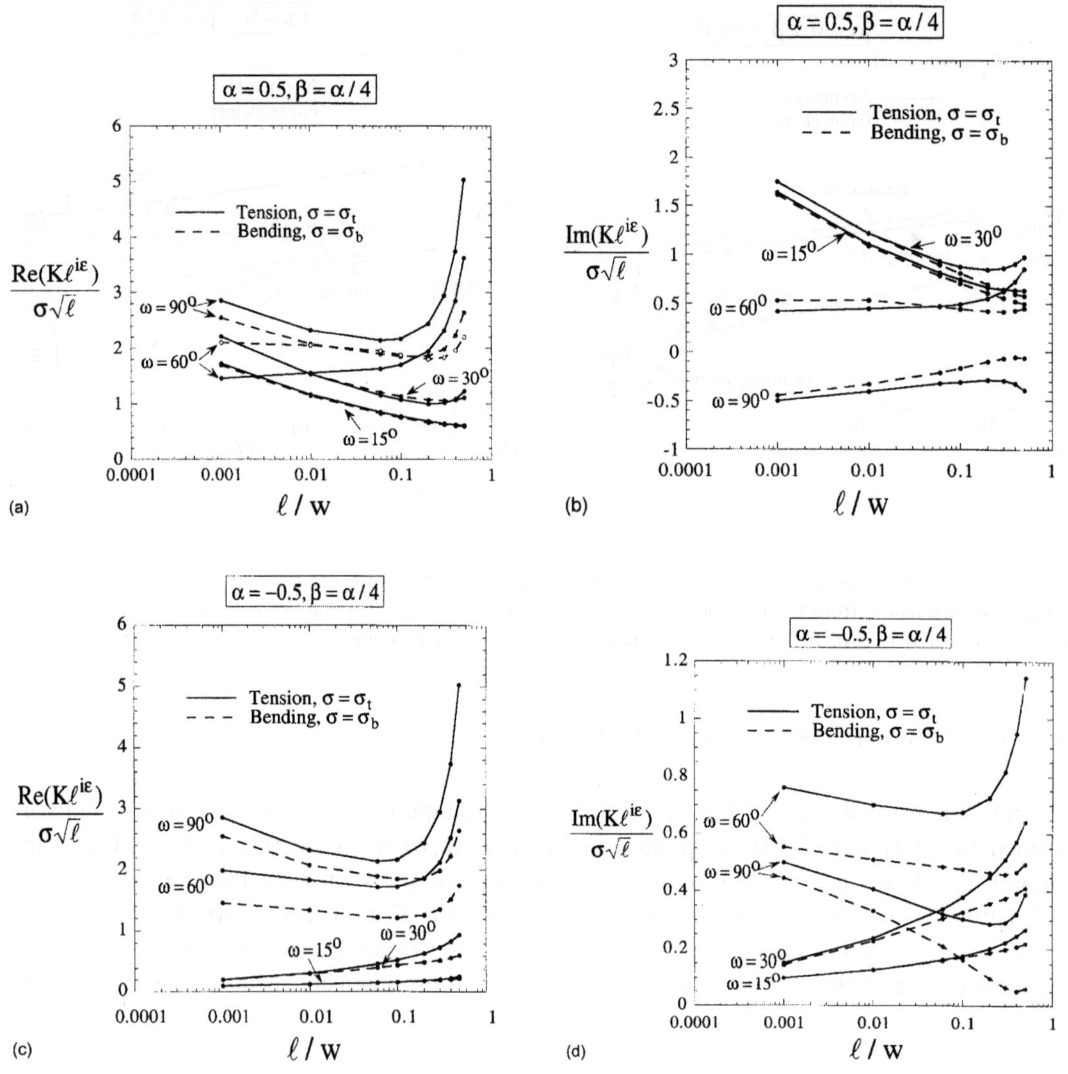


*Figure 7.* The effect of the interface orientation $\omega$ and relative crack length $\ell/w$ upon the interfacial stress intensity factor, for the case $\beta = \alpha/4$ (a) $[\mathrm{Re}(K\ell^{i\varepsilon})]/\sigma\sqrt{\ell}$ versus $\ell/w$ for $\alpha = 0.5$; (b) $[\mathrm{Im}(K\ell^{i\varepsilon})]/\sigma\sqrt{\ell}$ versus $\ell/w$ for $\alpha = 0.5$; (c) $[\mathrm{Re}(K\ell^{i\varepsilon})]/\sigma\sqrt{\ell}$ versus $\ell/w$ for $\alpha = -0.5$; and (d) $[\mathrm{Im}(Kl^{i\varepsilon})]/\sigma\sqrt{l}$ versus $\ell/w$ for $\alpha = -0.5$.

facial stress intensity factor tends to zero as $\omega$ tends to zero. Examination of the representative cases shown in Figure 7(a–d) shows that the nature of the remote loading (bending versus tension) has a greater effect on the interfacial stress intensity factor at larger values of $\omega$, for both short and long cracks.

The real and imaginary components of the stress intensity factor $K\ell^{i\varepsilon}$ are positive in all cases shown in Figure 7(a–d) with the following exception: the imaginary part of $K\ell^{i\varepsilon}$ is negative for $\omega = 90°$, $\alpha = 0.5$, $\beta = \alpha/4$ for both tensile loading and bending, as shown in Figure 7(b). A positive value of the real part of $K\ell^{i\varepsilon}$ implies that the crack is open (except over a very small segment behind the crack tip due to the oscillatory nature of the crack singularity),

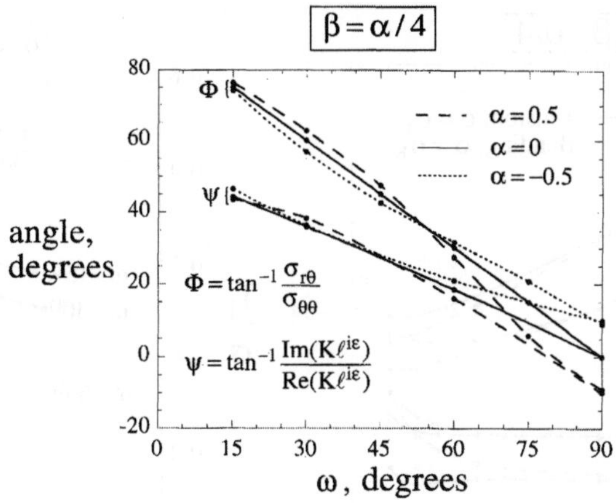


*Figure 8.* The phase angle $\psi \equiv \arctan[\mathrm{Im}(K\ell^{i\varepsilon})/\mathrm{Re}(K\ell^{i\varepsilon})]$ for an interfacial crack embedded within the interface-corner singular field, and the phase angle $\Phi \equiv \arctan[\sigma_{r\theta}/\sigma_{\theta\theta}]$ for the corner singularity in the absence of a crack, as a function of joint inclination $\omega$. $\alpha = -0.5$, 0 and 0.5, with $\beta = \alpha/4$.

and a positive value of the imaginary part of $K\ell^{i\varepsilon}$ implies that the interfacial crack has a tendency to kink out of the interface into the lower material 2, provided the toughness of material 2 is sufficiently low in comparison with the interfacial toughness. The phase angle $\psi \equiv \arctan[\mathrm{Im}(K\ell^{i\varepsilon})/\mathrm{Re}(K\ell^{i\varepsilon})]$ for an interfacial crack embedded within the interface-corner singular field is shown in Figure 8 as a function of inclination $\omega$, for $\alpha = -0.5$, 0 and 0.5, and $\beta = \alpha/4$. For the case $\alpha = \beta = 0$, the phase angle decreases with increasing $\omega$ in an almost linear manner, from a value $\psi = 44.3°$ at $\omega = 15°$ to a value $\psi = 0°$ at $\omega = 90°$. The effect of the value of the material mismatch parameters on the phase angle is minor provided $\omega$ is less than about 45°, but becomes significant for larger values of $\omega$.

It is of interest to compare the phase angle of the interfacial stress intensity factor with the phase angle of the traction on the interface, associated with the corner singularity (in the absence of a crack). For the corner-singularity, the traction on the interface comprises a shear traction $\sigma_{r\theta}$ and a normal traction $\sigma_{\theta\theta}$, and the ratio of shear traction to normal traction $\sigma_{r\theta}/\sigma_{\theta\theta}$ of the asymptotic solution can be characterised by the phase angle $\Phi \equiv \arctan(\sigma_{r\theta}/\sigma_{\theta\theta})$. The dependence of the angle $\Phi$ upon the inclination $\omega$ is included in Figure 8, for selected values of $\alpha$ with $\beta = \alpha/4$. For the case $\alpha = \beta = 0$, the singularity vanishes and $\Phi = 90° - \omega$. The value of $\Phi$ changes slightly for the general bi-material case with $\alpha \neq 0$. For $\omega = 90°$, geometric symmetry dictates that the phase angle $\psi$ of the stress intensity factor for a short interfacial crack equals the phase angle $\Phi$ of the traction on the interface; in the general case, $\omega \neq 90°$, $\Phi$ increases with decreasing $\omega$ at a faster rate than does $\psi$. Thus, for small values of $\omega$ the relative amount of shear to normal traction on a crack-free interface is significantly higher than the relative mode II to mode I stress intensity for a short interfacial crack.

### 3.4. THE INTERFACIAL $T$-STRESS

The magnitude and sign of the $T$-stress can affect the kinking behaviour of an interfacial crack out of the interface (He et al., 1991; Akisanya and Fleck, 1994). A positive value of $T$-stress both encourages kinking of an interfacial crack and causes the kinked crack to grow unstably under fixed remote load. In addition, the subsequent trajectory of the kink is expected

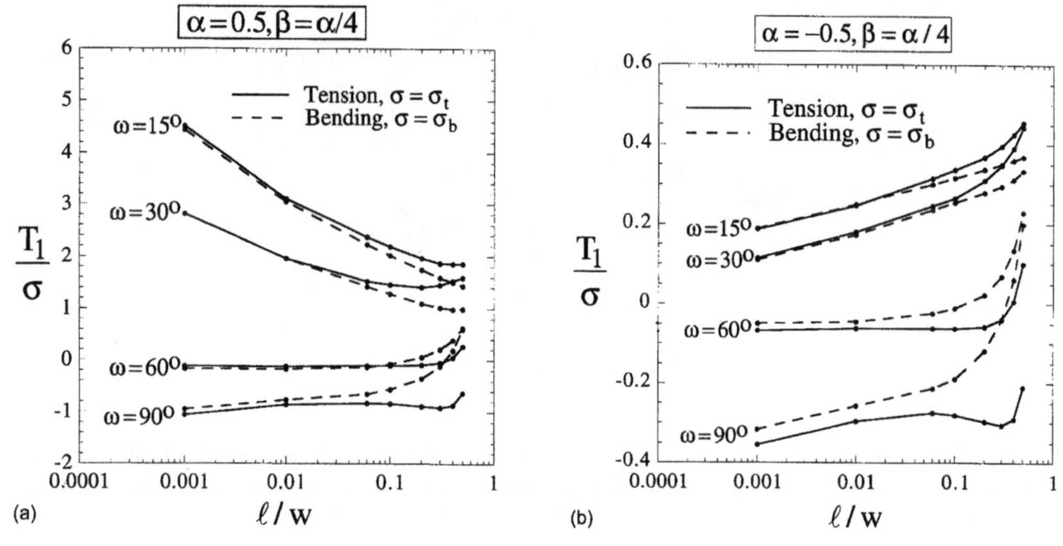


*Figure 9.* The effect of the interface orientation $\omega$ and relative crack length $\ell/w$ upon the $T$-stress in material 1, for $\beta = \alpha/4$. (a) $\alpha = 0.5$ and (b) $\alpha = -0.5$.

to diverge from the interface if the $T$-stress is positive (Cotterell and Rice, 1980). A negative value of the T-stress has the following effects:

(i) it stabilises an interfacial crack against kinking,
(ii) it may cause a kinked crack to arrest, and
(iii) crack growth from an initial kink is expected to converge towards the interface.

Figure 9 shows the normalised $T$-stress, $T_1/\sigma$, in material 1 due to remote tensile loading $\sigma = \sigma_t$, and due to remote bending of amplitude $\sigma = \sigma_b$, as a function of relative crack length $\ell/w$. Results are shown for $\alpha = \pm 0.5$ and $\beta = \alpha/4$. Values are calculated directly by the finite element method for long interfacial cracks of geometry shown in Figure 3(b), and as described in Section 2.3. For the case of short interfacial cracks ($\ell/w \leqslant 0.01$), embedded within the corner singularity, the magnitude of $T_1/\sigma$ is calculated in two steps as detailed in Section 2.2:

(i) calculation of the magnitude of $H$ for the geometry of Figure 3(a), and
(ii) calculation of the coupling between $H$ and $T_1$ for the boundary layer geometry of Figure 3(c).

Thus, results for $T_1$ in Figure 9 are taken from (2.10) for $\ell/w \geqslant 0.06$ and are taken from (2.14) for $\ell/w \leqslant 0.01$. Intermediate values of $\ell/w$, $0.01 < \ell/w < 0.06$, are derived by linear interpolation.

The level of $T$-stress is relatively insensitive to the nature of the remote loading for short interfacial cracks and $\omega \leqslant 30°$, as shown in Figure 9. This is obvious from an examination of (2.10): for short interfacial cracks the only dependence of $H$ upon type of loading is via the nondimensional parameter **a**. Provided $\omega \leqslant 30°$, **a** is not strongly influenced by the nature of the remote loading (see Figure 5(a, b)). When the interfacial crack is of comparable length to the specimen width, the $T$-stress increases with increasing $\ell/w$, and depends significantly upon the nature of the remote loading. For all lengths of interfacial crack considered, $T_1$ increases with diminishing $\omega$. For example, for a short interfacial crack $\ell/w = 10^{-3}$ and $\alpha = 0.5$, $T_1/\sigma$ increases from a value of about $-1$ at $\omega = 90°$ to $T_1/\sigma = 4.5$ at $\omega = 15°$, for both remote tension and remote bending. We further note from Figures 9(a, b) that, for short cracks, the sign of the $T$-stress switches from a negative value for joint inclinations $\omega$ close to the butt joint geometry ($\omega = 90°$) to a positive value at smaller values of $\omega$.

### 3.5. Strain energy release rate and phase angle

The normalised strain energy release rate $E^*G_I/\sigma^2 w$ is given by

$$\frac{E^*G_{\mathrm{I}}}{\sigma^2 w} = \left(\frac{\ell}{w}\right)^{2\lambda-1} \mathbf{a}^2(d_1^2 + d_2^2) \tag{3.1}$$

for an interfacial crack within the free-edge singularity zone, and by

$$\frac{E^*G_{\mathrm{I}}}{\sigma^2 w} = \left(\frac{\ell}{w}\right)(b_1^2 + b_2^2) \tag{3.2}$$

for an interfacial crack whose tip is outside the singularity region. Here, $E^* \equiv (1+\alpha)E_2/[(1-\beta^2)(1-\nu_2^2)]$, where $E_2$ and $\nu_2$ are the Young's modulus and the Poisson's ratio of material 2 below the interface. The corresponding phase angle of loading at the crack tip $\psi$ defined by (2.7) is obtained from (2.11) or (2.13). Once the calibration functions $\mathbf{a}(\alpha, \beta)$, $\mathbf{b}$ $(\ell/w, \alpha, \beta)$ and $\mathbf{d}$ $(\alpha, \beta)$ have been deduced from the finite element analysis, the quantities $G_{\mathrm{I}}$ and $\psi$ are known for any prescribed remote loading $\sigma$ and strip width $w$. Recall that the coefficient $\mathbf{a}$ is real while the coefficients $\mathbf{b}$ and $\mathbf{d}$ are complex functions: we restrict our attention to real values of the singularity level $\lambda$.

Results for $E^*G_{\mathrm{I}}/\sigma^2 w$ and $\psi$ are given in Figures 10(a) and (b) for the case $\alpha = 0.5$, and in Figures 10(c) and (d) for $\alpha = -0.5$. Each curve is for a prescribed value of inclination $\omega$ and results are shown for both remote tension and remote bending. The inclination $\omega$ has a major effect upon both the level of energy release rate $G_{\mathrm{I}}$ for a given crack length and upon the phase angle $\psi$. The details of the remote loading become increasingly important with crack extension, as commented upon previously.

The normalised strain energy release rate $E^*G_{\mathrm{I}}/\sigma^2 w$ increases monotonically from zero as the crack length is increased, with a relatively small attendant change in the phase angle $\psi$. This suggests that interfacial crack growth from the free edge is unstable under fixed remote loading. The effect of the material elastic mismatch parameter $\beta$ upon the values of $E^*G_{\mathrm{I}}/\sigma^2 w$ and $\psi$ is minor.

## 4. Concluding remarks

A singularity of the type $Hr^{\lambda-1}$ exists at the free-edge of a scarf joint between two dissimilar elastic solids under remote tension and remote bending. In this paper we have presented a finite element analysis for the evaluation of the intensity $H$ and of the order of the stress singularity $\lambda - 1$. For almost all values of material mismatch in the practical range, the eigenvalue $\lambda$ is real. The intensity $H$ is evaluated by a contour integral method. The singular zone extends to a distance of about $0.03w$ from the interface corner, where $w$ is the width of the bi-material

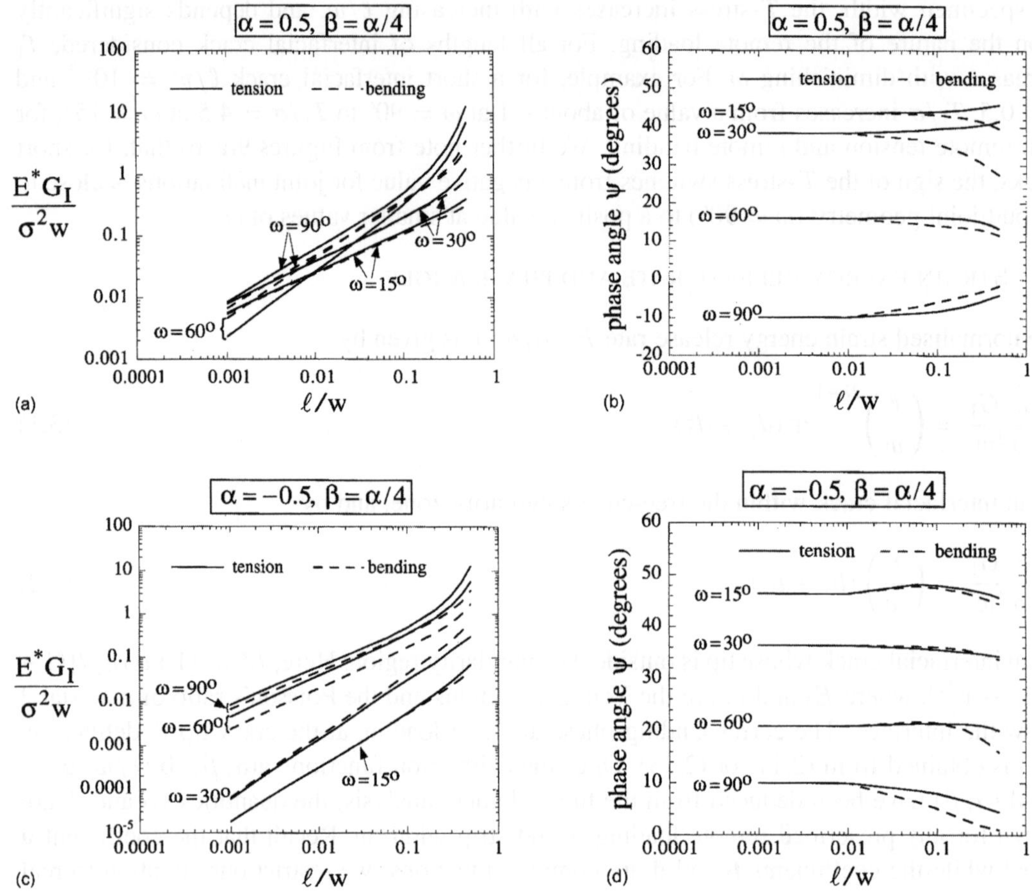


*Figure 10.* The effect of the interface orientation $\omega$ and relative crack length $\ell/w$ upon the nondimensional energy release rate $E^*G_\mathrm{I}/\sigma^2 w$ and the phase angle $\psi = \arctan[\mathrm{Im}(K\ell^{i\varepsilon})/\mathrm{Re}(K\ell^{i\varepsilon})]$ for $\beta = \alpha/4$. (a) $E^*G_\mathrm{I}/\sigma^2 w$ vs. $\ell/w$, and (b) $\psi$ vs. $\ell/w$ for $\alpha = 0.5$; (c) $E^*G_\mathrm{I}/\sigma^2 w$ vs. $\ell/w$, and (d) $\psi$ vs. $\ell/w$ for $\alpha = -0.5$.

strip. It is expected that $H$ provides a useful correlating parameter for fracture initiation at the free edge of the scarf joint.

Experimental data are not yet available to verify (or refute) the use of a critical value for $H$ as a fracture initiation criterion for scarf joints. This approach suggests that the strength of a scarf joint scales with the width $w$ of the panel to the power ($\lambda$-1) where $\lambda$ depends upon both the material mismatch and the inclination of the scarf joint. In general $\lambda$ is greater than 0.5 and so the expected size dependence is weaker than that for a cracked structure.

In the finite element evaluation of the free-edge intensity factor $H$ for the uncracked bi-material strip we have taken the length $L$ to equal $20w$, where $w$ is the width of the strip. Numerical experimentation shows that this ratio of $L/w$ is sufficiently large for the results to be independent of $L$: the numerical results pertain to the case of a long strip, with $w$ as the only length-scale in the problem.

A calibration of the interfacial stress intensity factor $K$ and the $T$-stress is given for a crack lying within the singularity region and also for a crack whose tip is outside the singularity

*International Journal of Fracture* **95:** 89–101, 1999.


# Size effect in fracture of unidirectional composite plates

G.J. DVORAK and A.P. SUVOROV
*Department of Mechanical Engineering, Aeronautical Engineering and Mechanics, 5003JEC, Rensselaer Polytechnic Institute, Troy, NY 12180-3590. e-mail: dvorak@rpi.edu*

Received 24 August 1998; accepted in revised form 18 December 1998

**Abstract.** Fracture of notched, unidirectionally reinforced composite plates with well-bonded ductile matrices is typically preceded by the formation of long, discrete plastic shear zones aligned in the fiber direction. Onset of fracture is associated with a critical tension stress in a certain small process zone ahead of the notch tip; the critical stress is often equal to the tensile strength of the unnotched composite plate. Length of the shear zones can be estimated by plastic limit-analysis, and the local tension stress ahead of the notch found by superposition of the stress caused by remotely applied loads with the stress induced by the shear tractions in the plastic zone, which is shown to be dominated by a logarithmic singularity. In as-fabricated boron-aluminum composites, this fracture mechanism was analyzed and confirmed by numerous experiments (G.J. Dvorak, J. Zarzour and Y. Benveniste, *Engineering Fracture Mechanics* **42**, 501–517, 1992). Since the notch tip field is not described by a stress intensity factor, experimental notched strength data cannot be interpreted in terms of a single material property, such as toughness. An alternative scaling procedure is outlined for prediction of notched strength of wide plates on the basis of data obtained from small size specimens.

## 1. Introduction

It is well known that fracture of notched fibrous composite laminates is usually preceded by small scale inelastic deformation and/or damage at the tip of the notch. Experimental observations suggest that fracture may start at loads that cause either a critical average normal stress within a certain process zone ahead of a notch, or a critical normal stress at a certain distance ahead of the notch (Whitney and Nuismer, 1974), while the underlying stress field is derived for a crack in an anisotropic elastic solid with effective properties of the composite laminate. As pointed out recently by Erdogan (1998), the average stress criterion was introduced by Weighardt (1907) in order to reconcile the existence of a crack tip stress singularity with finite strength or real materials. Both Whitney–Nuismer criteria provide expressions for modified stress intensity factors that reach a certain critical magnitude at the onset of fracture, and that can be applied in the usual fashion in structural design.

However, this approach may not be applicable to materials that exhibit large scale yielding or damage. The local stresses are no longer dominated by the elastic singularity, instead they depend on the actual geometry of the notch tip deformation. For example, such large notch tip deformations, in discrete shear zones aligned with the fiber direction, have been observed in unidirectional metal matrix composites with ductile matrices. In this case, the notch tip field is dominated by a logarithmic stress singularity, but the strength of the singular field depends on several material parameters that cannot be combined and replaced by a single intensity factor. Therefore, a procedure is needed for interpretation of experimental notch strength data and for their use in predicting notched strength of wide composite plates. This is addressed in the present paper. Section 2 reviews some of the experimental evidence and models of fracture in unidirectional B/Al plies with center notches. Section 3 outlines evaluation of the stress field at notches blunted by long discrete shear zones in notched plies of finite width; Section 4 gives the asymptotic estimate of notched strength of very wide plates with long notches, and Section 5 extends the analysis and interpretation of experiments to notched composite plates of any size. Our ealier work is relevant here; it will be referred to as (I): (Dvorak et al., 1989); (II): (Bahei-El-Din et al., 1989); (III): (Bahei-El-Din and Nigam, 1990); and (IV): (Dvorak et al., 1992).

## 2. Plastic zones in notched plates

Experimental evidence and associated modeling of the deformation mechanisms that precede the onset of rapid fracture from a crack or notch in a unidirectional composite plate with a ductile matrix suggest the existence of long, discrete plastic shear zones aligned with the fiber direction. Tirosh (1973) observed such zones in polymer matrix systems. In metal matrix B/6061-Al and FP/Al systems, zone length and plastic strain measurements were reported, together with references to related work, in (I) and (II) for plate specimens with sharp center notches, and in (III) for specimens with inclined edge notches, or circular and rectangular holes. Post et al. (1987), detected such zones in surface layers of laminated metal matrix composite plate specimens with center slots. However, no such zones or other evidence of extensive plastic deformation have been observed at fracture surfaces that cross the fibers (Dvorak et al., 1989).

In the as-fabricated B/6061-Al plates examined in (I), the width of the plastic zones was found to range from less that 0.5 mm at the notch tip, to at most 2.0 mm at the far ends; therefore the plastic zones can be regarded as discrete shear bands aligned with the fiber direction. The length of the zones was measured directly in (I) and also evaluated by a crack interaction scheme (Benveniste et al., 1989; Dvorak et al., 1992) and by finite element analysis (Bahei-El-Din et al., 1989). Figure 1 summarizes the results obtained for the two sets of center notch specimens tested by Dvorak et al. (1989), and Poe and Sova (1980); tables of the test appeared in (I) and (IV). A good estimate of the zone length $R$ in a center-notched plate under applied tension $p_0$ can be obtained from a simple limit analysis-based formula,

$$R \doteq cp_0/\tau^*. \tag{1}$$

where $2c$ is the total length of the notch, and $\tau^*$ is the longitudinal shear flow stress of the ply. The latter is best measured in a small rail-shear testing device, such as that described in (I).

A typical configuration of a center-notched tension test specimen with the plastic shear zones is shown schematically in Figure 2, together with the load superposition scheme used in evaluation of the local stress ahead of the notch tip. At the onset of fracture, the remotely applied tension $p_0$ defines the notched strength, $p_0 = \sigma_N$, and contributes to a uniform ligament stress $\sigma_{\text{lig}} = \sigma_N/(1-2c/W)$, where $2c$ is the notch length and $W$ is specimen width. The zone-induced stress $\Delta\sigma$ is obtained by averaging the normal stress $\sigma_{xx}(0, y)$, contributed by the shear stress $\sigma_{yx}$ in the plastic zone, over a certain small process zone $0 \leqslant y \leqslant \omega$; where $y$ is measured ahead of the notch tip. This zone-induced stress is evaluated from a separate analysis presented in Section 3 below.

Extensive experimental evidence supports the fracture initiation criterion,

$$\sigma_{\text{lig}} + \Delta\sigma = \sigma_N/(1 - 2c/W) + \frac{1}{\omega}\int_0^{\omega} \sigma_{xx}(0, y)\,\mathrm{d}y = \sigma^0, \tag{2}$$

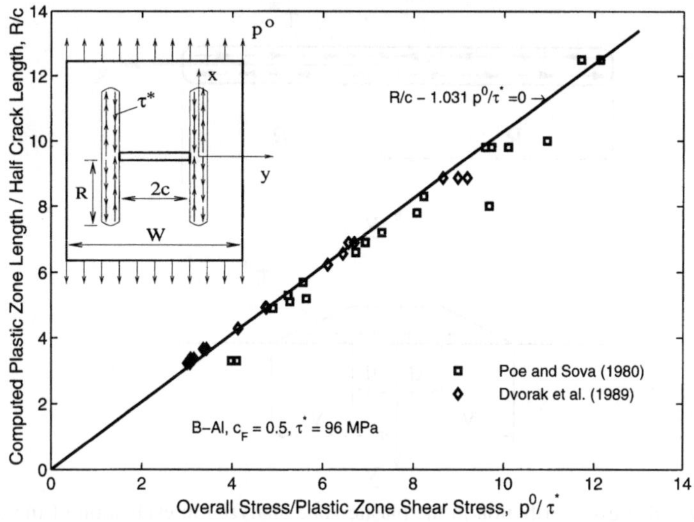


*Figure 1.* Normalized relationship between plastic zone length and applied overall stress computed at failure loads for center-notched B-Al specimens (Bahei-El-Din et al., 1989); (II).

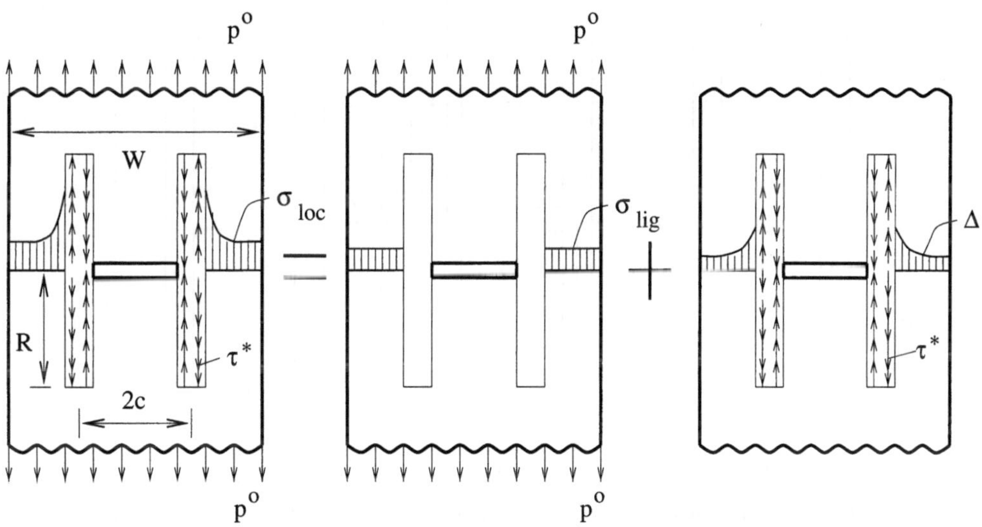


*Figure 2.* Schematic decomposition of stress distribution in the ligaments of center-notched specimens (Dvorak et al., 1989, 1992).

where $\omega$ is the length of the process zone ahead of the notch tip, and $\sigma^0$ is the ultimate tensile strength of the unnotched fibrous ply. The length of the process zone is the only quantity in the present analysis that needs to be evaluated empirically for each composite system, by comparing (1) with a set of measured magnitudes for specimens of different notch length. For example, in our earlier papers (I) and (III), we found a good fit with experimental data by selecting $\omega = 0.22$ mm in as-fabricated B/6061Al plates, and $\omega = 2.18$ mm in the FP/Al plates.

The main objective of fracture strength prediction is the evaluation of the plastic zone-induced stress in (2), in the small interval $0 \leqslant y \leqslant \omega$. A closed-form solution is sought for convenience in applications. In a wide plate, where the notch tip is far removed from free edges, the contribution from the zone adjoining the crack tip is found by finding the normal stress $\sigma_{xx}$ at $x = 0$ caused by a single zone of length $2R$, loaded according to Figure 3,

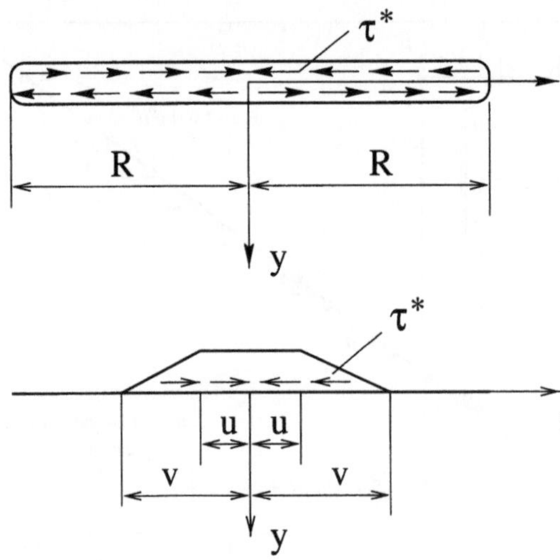


*Figure 3.* Geometry and stress distribution in the plastic zone analyzed in evaluation of the zone-induced stress ahead of the notch.

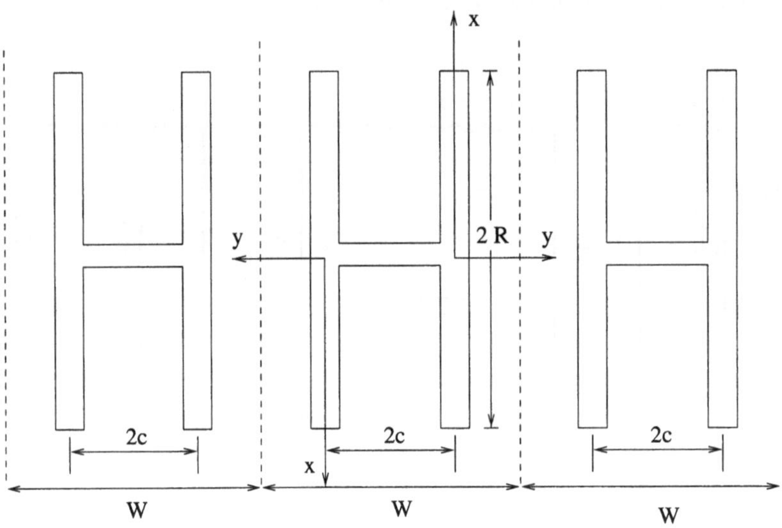


*Figure 4.* Model for estimating the effect of free edges on stresses ahead of a center notch.

where the uniform shear stress $\tau^*$ is applied everywhere, except in the vicinity of the notch tip. In finite width specimens, the interaction of the notch with the free edges is simulated by considering a row of H-cracks, as indicated in Figure 4. On the plane of symmetry $y = (W-2c)/2$, this cancels the shear stress and thus does not affect the zone length estimate (1); the transverse normal stress is still present, but does not influence (1). The arrangement provides for contributions to $\sigma_{xx}$ at $x = 0$ by both the plastic zone at the current notch tip, and by the image zone located at the distance $y = (W-2c)$. These contributions are approximated by replacing the integral on the right-hand side of (2) with the expression,

$$\Delta\sigma = \frac{1}{\omega}\left[\int_0^{\omega} \sigma_{xx}(0, y)\,\mathrm{d}y + \int_{y_1-\omega}^{y_1} \sigma_{xx}(0, y)\,\mathrm{d}y\right], \tag{3}$$

where $y_1 = (W-2c)$.

*Table 1.* Elastic constants and geometrical parameters

| Material | B/Al | FP/Al |
|---|---|---|
| $E_L$ (MPa) | $237.3 \times 10^3$ | $167.08 \times 10^3$ |
| $G_L$ (MPa) | $55 \times 10^3$ | $44.2 \times 10^3$ |
| $E_T$ (MPa) | $143.1 \times 10^3$ | $114.7 \times 10^3$ |
| $\nu_L$ | 0.205 | 0.279 |
| $\sigma_0$ (MPa) | 1650.0 | 450.0 |
| $\tau^*$ (MPa) | 96.0 | 130.0 |
| $\omega$ (mm) | 0.22 | 2.18 |
| $u$ (mm) | 0.1 | 0.1 |
| $v$ (mm) | 0.2 | 0.2 |
| $c_F$ | 0.5 | 0.35 |

## 3. Local stress in the process zone

We now proceed to evaluate the zone-induced contribution to the stress ahead of the notch. Considering the extensive stress shielding that the long plastic zones exercise over the integration path in (3), and the actual geometry of the blunted notch tip region, we evaluate $\Delta\sigma$ from the solution of the crack problem indicated in Figure 3, where a single crack of length $2R$ is loaded by a piecewise uniform shear stress $\tau^*$, equal to the measured flow shear stress of the matrix. The effect on the shear stress by the initial width of the machined slot and the opening displacement at notch tip is accounted for by local reduction of the shear stress in the interval $-v \leqslant x \leqslant v$; Table 1. The solution procedure outlined in (IV) is reviewed for completeness.

The plastic zone is assumed to be embedded in an orthotropic elastic solid, which has the effective elastic properties of the composite material. A fundamental solution is used, for concentrated shear forces $+Q$ and $-Q$ applied at $x = q$ to either edge of the otherwise stress-free crack $-R \leqslant x \leqslant +R$. This plane problem of orthotropic elasticity can be reduced to finding two complex stress functions $\Phi(z_1)$ and $\Psi(z_2)$ where the complex variables $z_j$ ($j = 1, 2$) are,

$$z_1 = x + s_1 y, \qquad z_2 = x + s_2 y \tag{4}$$

with the parameters $s_1$ and $s_2$ being the roots of the characteristics equation (Lekhnitskii, 1963),

$$a_{11}s^4 - 2a_{16}s^3 + (2a_{12} + a_{66})s^2 - 2a_{26}s + a_{22} = 0, \tag{5}$$

where $a_{ij}$ are the compliance coefficients in generalized plane stress defined by

$$\begin{Bmatrix} \varepsilon_{xx} \\ \varepsilon_{yy} \\ 2\varepsilon_{xy} \end{Bmatrix} = \begin{bmatrix} a_{11} & a_{12} & a_{16} \\ a_{12} & a_{22} & a_{26} \\ a_{16} & a_{26} & a_{66} \end{bmatrix} \begin{bmatrix} \sigma_{xx} \\ \sigma_{yy} \\ \sigma_{xy} \end{bmatrix}. \tag{6}$$

The general expressions for the stresses then are

$$\begin{aligned} \sigma_{xx} &= 2\,\mathrm{Re}[s_1^2\Phi(z_1) + s_2^2\Psi(z_2)], \\ \sigma_{yy} &= 2\,\mathrm{Re}[\Phi(z_1) + \Psi(z_2)], \\ \sigma_{xy} &= -2\,\mathrm{Re}[s_1\Phi(z_1) + s_2\Psi(z_2)]. \end{aligned} \tag{7}$$

For the concentrated force at $x = q$ on an isolated crack $-R \leqslant x \leqslant +R$, the potential functions $\Phi(z_1)$, $\Psi(z_2)$ for shear forces $Q$ were derived by Sih and Liebowitz (1968) as

$$(s_2 - s_1)\Phi(z_1) = Q(q)f(z_1), \qquad (s_1 - s_2)\Psi(z_2) = Q(q)f(z_2), \qquad Q > 0, \tag{8}$$

where

$$f(z) = [2\pi(z-q)]^{-1}[(R^2-q^2)/(z^2-R^2)]^{1/2}. \tag{9}$$

These forms can be used to generate the appropriate potentials for any loading distribution on the crack line.

In the present case, the local resolution of the shear stresses indicated in Figure 3 is represented by superimposing the following $Q(q)$ distributions,

$$\begin{aligned}
&Q'(q) = \tau^* \quad \text{for } 0 \leqslant q \leqslant R, && Q'(q) = -\tau^* \quad \text{for } -R \leqslant q \leqslant 0,\\
&Q''(q) = -\tau^* \quad \text{for } 0 \leqslant q \leqslant u, && Q''(q) = \tau^* \quad \text{for } -u \leqslant q \leqslant 0,\\
&Q'''(q) = \frac{-\tau^*(q-v)}{(u-v)} \quad \text{for } u \leqslant q \leqslant v, && Q'''(q) = \frac{\tau^*(q+v)}{(-u+v)} \quad \text{for } -v \leqslant q \leqslant -u.
\end{aligned} \tag{10}$$

When applied to $(7_1)$, (8), (9), these boundary conditions produce the respective contributions $\sigma'_{xx}(0, y) + \sigma''_{xx}(0, y) + \sigma'''_{xx}(0, y) = \sigma_{xx}(0, y)$, the stress in (3).

Consider first the uniform shear stress loading in $(10_1)$. The solution can be obtained by integrating the potential functions (8), (9) along the crack length,

$$\begin{aligned}
(s_2 - s_1)\Phi(z_1) &= \frac{-\tau^*}{2\pi(z_1^2 - R^2)^{1/2}}\left\{\int_{-R}^{0} \frac{(R^2-q^2)^{1/2}}{z_1 - q}\,dq - \int_{0}^{R} \frac{(R^2-q^2)^{1/2}}{z_1 - q}\,dq\right\},\\
(s_1 - s_2)\Psi(z_2) &= \frac{-\tau^*}{2\pi(z_2^2 - R^2)^{1/2}}\left\{\int_{-R}^{0} \frac{(R^2-q^2)^{1/2}}{z_2 - q}\,dq - \int_{0}^{R} \frac{(R^2-q^2)^{1/2}}{z_2 - q}\,dq\right\}.
\end{aligned} \tag{11}$$

Explicit integration can be performed using the formula

$$\begin{aligned}
&\int \frac{(R^2-q^2)^{1/2}}{q-z}\,dq\\
&\quad = (R^2-q^2)^{1/2} + iz\ln[-2q + 2i(R^2-q^2)^{1/2}]\\
&\qquad + i(z^2-R^2)^{1/2}\ln\left[\frac{2R^2 - 2zq - 2i(z^2-R^2)^{1/2}(R^2-q^2)^{1/2}}{(q-z)}\right].
\end{aligned} \tag{12}$$

This provides,

$$\begin{aligned}
&(s_2 - s_1)\Phi(z_1)\\
&\quad = \frac{\tau^*}{2\pi(z_1^2-R^2)^{1/2}} \times \left\{2R + i(z_1^2-R^2)^{1/2}\ln\left[\frac{z_1^2 - 2R^2 + 2iR(z_1^2-R^2)^{1/2}}{z_1^2}\right]\right\},\\
&(s_1 - s_2)\Psi(z_2)\\
&\quad = \frac{\tau^*}{2\pi(z_2^2-R^2)^{1/2}} \times \left\{2R + i(z_2^2-R^2)^{1/2}\ln\left[\frac{z_2^2 - 2R^2 + 2iR(z_2^2-R^2)^{1/2}}{z_2^2}\right]\right\}.
\end{aligned} \tag{13}$$

This result and $(7_1)$ yield

$$\sigma'_{xx}(x,y)/\tau^* = \mathrm{Re}\left\{\frac{s_1^2}{\pi(s_2-s_1)}\left[\frac{2}{((z_1/R)^2-1)^{1/2}}+i\ln(\alpha_1)\right]\right.$$
$$\left.+\frac{s_2^2}{\pi(s_1-s_2)}\left[\frac{2}{((z_2/R)^2-1)^{1/2}}+i\ln(\alpha_2)\right]\right\}, \tag{14}$$

where

$$\alpha_{1,2} = \frac{(z_{1,2}/R)^2+2i[(z_{1,2}/R)^2-1]^{1/2}-2}{(z_{1,2}/R)^2}.$$

In a unidirectional composite where the material principal axes coincide with the cartesian axes, the roots of (5) are purely imaginary, hence (4) becomes,

$$z_1 = ia_1 y, \qquad z_2 = ia_2 y$$

where $a_1$ and $a_2$ denote the imaginary parts of $s_1$ and $s_2$ respectively; for example, in B/Al and SiC/Al composites, with ply elastic moduli listed in Table 1, there is $a_1 = 0.6962$, $a_2 = 1.8491$.

Then, (14) can be reduced to

$$\sigma'_{xx}(0,y)/\tau^* = \frac{2}{\pi(a_2-a_1)}\left\{a_1^2\left[\frac{1}{[1+a_1^2(y/R)^2]^{1/2}}-\ln(\alpha_1)\right]\right.$$
$$\left.+a_2^2\left[\frac{-1}{[1+a_2^2(y/R)^2]^{1/2}}\quad \ln(\alpha_2)\right]\right\}, \tag{15}$$

where

$$\alpha_{1,2} = \frac{1+[1+a_{1,2}^2(y/R)^2]^{1/2}}{a_{1,2}(y/R)}.$$

In the final step, (15) is integrated as suggested by (3). Of particular interest is the first integral in the interval $0 \leqslant y \leqslant \omega$, which provides a dominant contribution to the first term of the superposition suggested by (10),

$$\frac{1}{\omega}\int_{\omega}^{0}\sigma'_{xx}(0,y)\,\mathrm{d}y/\tau^* = \frac{2}{\pi(a_2-a_1)}\left\{-a_1^2\ln\left[\frac{1+\sqrt{1+(a_1/t)^2}}{(a_1/t)}\right]\right.$$
$$\left.+a_2^2\ln\left[\frac{1+\sqrt{1+(a_2/t)^2}}{(a_2/t)}\right]\right\}, \tag{16}$$

where $t = R/\omega = (c/\omega)(\sigma_N/\tau^*)$, according to (1). Evaluation of the second term in (3) done in a similar manner.

Next, we outline the evaluation of the contributions to $\Delta\sigma$ by the second and third shear stress distributions in (10), which are localized at the notch tip and thus do not enter into the second integrand in (3). Evaluation of stress potentials in (7) is made now for an anisotropic elastic half-plane (Lekhnitskii, 1963). In particular, unloading within $-u \leqslant x \leqslant u$ contributes to the first term in (3),

$$\sigma''_{xx}(x,y)/\tau^* = \mathrm{Re}\left[\frac{s_1^2}{\pi i(s_2-s_1)}(-2\ln(-z_1)+\ln(u-z_1)+\ln(-u-z_1)) + \frac{s_2^2}{\pi i(s_1-s_2)}(-2\ln(-z_2)+\ln(u-z_2)+\ln(-u-z_2))\right] \tag{17}$$

and the linearly changing shear stress in ($10_3$) contributes

$$\begin{aligned}\Delta\sigma'''_{xx}(x,y)/\tau^* &= \mathrm{Re}\left[\frac{s_1^2}{\pi i(s_2-s_1)}\left\{\frac{1}{u-v}[v-u+(-v+z_1)\ln(v-z_1)-(-v+z_1)\ln(u-z_1)]\right.\right.\\ &\quad -\frac{1}{-u+v}[-u+v+(v+z_1)\ln(-u-z_1)\\ &\quad \left.-(v+z_1)\ln(-v-z_1)]\right\}\\ &\quad +\frac{s_2^2}{\pi i(s_1-s_2)}\left\{\frac{1}{u-v}[v-u+(-v+z_2)\ln(v-z_2)-(-v+z_2)\ln(u-z_2)]\right.\\ &\quad -\frac{1}{-u+v}[-u+v+(v+z_2)\ln(-u-z_2)\\ &\quad \left.\left.-(v+z_2)\ln(-v-z_2)]\right\}\right].\end{aligned} \tag{18}$$

Superposition of (14), (17) and (18) provides the total stress component $\sigma_{xx}(x,y)/\tau^*$ in the close vicinity of the notch tip. Of particular interest are the distributions at $x=0$, such as that in (15), which can be obtained from (17) and (18) as

$$\Delta\sigma''_{xx}(0,y)/\tau^* = \frac{a_1^2}{\pi(a_2-a_1)}\left(-2\ln(a_1y)+2\ln\sqrt{u^2+a_1^2y^2}\right) + \frac{a_2^2}{\pi(a_1-a_2)}\left(-2\ln(a_2y)+2\ln\left(\sqrt{u^2+a_2^2y^2}\right)\right). \tag{19}$$

$$\Delta\sigma'''_{xx}(0,y)/\tau^* = \frac{a_1^2}{\pi(a_2-a_1)}\left\{\frac{1}{u-v}\left(2(v-u)-2v\ln\sqrt{v^2+a_1^2y^2}+2v\ln\sqrt{u^2+a_1^2y^2}\right.\right.$$

$$+ 2a_1 y \arctan\left(\frac{a_1 y}{v}\right) - 2a_1 y \arctan\left(\frac{a_1 y}{u}\right)\Bigg)\Bigg\}$$

$$+ \frac{a_2^2}{\pi(a_1 - a_2)}\left\{\frac{1}{u - v}\left(2(v - u) - 2v \ln\sqrt{v^2 + a_2^2 y^2} + 2v \ln\sqrt{u^2 + a_2^2 y^2}\right.\right.$$

$$\left.\left. + 2a_2 y \arctan\left(\frac{a_2 y}{v}\right) - 2a_2 y \arctan\left(\frac{a_2 y}{u}\right)\right)\right\}. \tag{20}$$

The following integrals are required for finding the averages of these stresses over the small process zone length $\omega$,

$$\int_0^\omega \ln(ay)\,\mathrm{d}y = \omega \ln(a\omega) - \omega,$$

$$\int_0^\omega \ln\left(\sqrt{d^2 + a^2 y^2}\right)\mathrm{d}y = \omega \ln(\sqrt{d^2 + a^2\omega^2}) - \left(\omega - \frac{d}{a}\arctan\left(\frac{a\omega}{d}\right)\right),$$

$$\int_0^\omega ay \arctan\left(\frac{ay}{d}\right)\mathrm{d}y = \tfrac{1}{2}a\left(\omega^2 + \frac{d^2}{a^2}\right)\arctan\left(\frac{a\omega}{d}\right) - (\tfrac{1}{2}d)\omega, \tag{21}$$

where $a$ and $d$ are constants.

To summarize, the above contributions of the loads (10) to the normal stress ahead of the notch are superimposed as

$$\sigma_{xx}(0, y) = \sigma'_{xx}(0, y) + \sigma''_{xx}(0, y) + \sigma'''_{xx}(0, y) \tag{22}$$

and then integrated as required by (3). According to (16), the resulting zone-induced stress $\Delta\sigma$ is a function of the as yet unknown notched strength $\sigma_N$. Substituting $\Delta\sigma$ into (2) provides an implicit equation for evaluation of $\sigma_N$ in terms of the specimen or plate geometry and the material properties listed in Table 1, i.e., elastic moduli, unnotched strength $\sigma^0$, longitudinal shear flow stress $\tau^*$ and the empirical process zone length $\omega$. A more refined prediction of $\sigma_N$ can be found by comparing the maximum principal stress at the notch tip to the unnotched strength $\sigma^0$, as described in Section 5 of (II).

## 4. Notched strength of wide plates

The dominant contribution to the zone-induced normal stress ahead of the notch can be recast using the relation,

$$a^2 \ln[t/a + [1 + (t/a)^2]^{1/2}] = a^2\,\mathrm{Arsh}(t/a) \tag{23}$$

in (16). Substituting the result into (2) yields,

$$\sigma^0/\tau^* = \sigma_N/(1 - 2c/W)\tau^* + 2[\pi(a_2 - a_1)]^{-1}[-a_1^2\,\mathrm{Arsh}(t/a_1) + a_2^2\,\mathrm{Arsh}(t/a_2)]. \tag{24}$$

Note that for short cracks, $c/\omega \to 1$ and $t \to (\sigma_N/\tau^*) \gg 1 \Rightarrow t/a \gg 1$, while for long cracks, $(\sigma_N/\tau^*) \to 1$ and $t \to c/\omega \gg 1 \Rightarrow t/a \gg 1$. Therefore, in either case one can use the expansion

$$\mathrm{Arsh}(t/a) = \ln(2t/a) + (2t/a)^{-2} - \cdots \tag{25}$$

which gives (24) in the form

$$\sigma^0/\tau^* = \sigma_N/(1-2c/W)\tau^* + 2[(a_1+a_2)/\pi][\ln(\sigma_N/\tau^*) + \ln(c/\omega)] + C_1,$$

$$C_1 = 2[\pi(a_2-a_1)]^{-1}[-a_1^2\ln[2/a_1] + a_2^2\ln[2/a_2]]. \tag{26}$$

This is a useful expression for finding simple estimates of $\sigma_N$ in plies of any width but with small $2c/W$ ratios such that the second integral in (3) can be neglected.

Turning now to the case of wide plates where $2c/W \to 0$, and noting that $\sigma^0/\tau^*$ is a known constant, we rewrite (26),

$$\exp\{\ln(\sigma_N/\tau^*)\} + C_2\ln(\sigma_N/\tau^*) = -C_2\ln(c/\omega) + C_3. \tag{27}$$

For example, for B/Al and SiC/Al composite ply moduli specified ahead of (15), there is $C_1 = -0.1345$, $C_2 = 2[(a_1+a_2)/\pi] = 1.6204$ and $C_3 = 17.322$.

This leads to the important conclusion that for the case of very long cracks, where $(\sigma_N/\tau^*) \to 0$, the slope of the normalized notch strength vs. crack length line in logarithmic coordinates is

$$\frac{d[\ln(\sigma_N/\tau^*)]}{d[\ln(c/\omega)]} = -\frac{C_2}{(\sigma_N/\tau^*) + C_2} \to -1 \tag{28}$$

which is consistent with the logarithmic singularity in the normal stress (16) ahead of the notch. In contrast, the inverse square root singularity in linear fracture mechanics yields the asymptotic slope of $-1/2$.

## 5. Interpretation of experimental notched strength data

The above distinction between linear fracture mechanics and the mechanics of fracture in ductile fibrous plies is also reflected in the absence of a single parameter, such as a stress intensity factor or a $J$-integral, that would provide a basis for a simple fracture criterion. Therefore, it is not immediately obvious how the fracture strengths measured on relatively narrow specimens of different geometries can be correlated and used to predict strength of wide notched plates.

To resolve this problem, we first postulate that if in a given material system, there is agreement between predicted and measured fracture strengths of narrow specimens, then a similar agreement will hold for wide plates, where $W \gg c$.

If

$$(\sigma_{\text{lig}}^n)^{\text{pr}} + (\Delta\sigma^n)^{\text{pr}} = (\sigma_{\text{lig}}^n)^{\text{exp}} + (\Delta\sigma^n)^{\text{exp}} = \sigma^0,$$

then

$$(\sigma_{\text{lig}}^w)^{\text{pr}} + (\Delta\sigma^w)^{\text{pr}} = (\sigma_{\text{lig}}^w)^{\text{exp}} + (\Delta\sigma^w)^{\text{exp}} = \sigma^0 \tag{29}$$

where the superscripts $n$ and $w$ denote narrow and wide plates, and pr and exp indicate predicted and experimentally measured quantities.

Figure 5 illustrates the significance of these relations, by presenting the variation of the predicted normalized ligament stresses as functions of the normalized notch length for different ratios of $2c/W$, including $2c/W \to 0$, for both B/Al and FP/Al plies. According to (29),

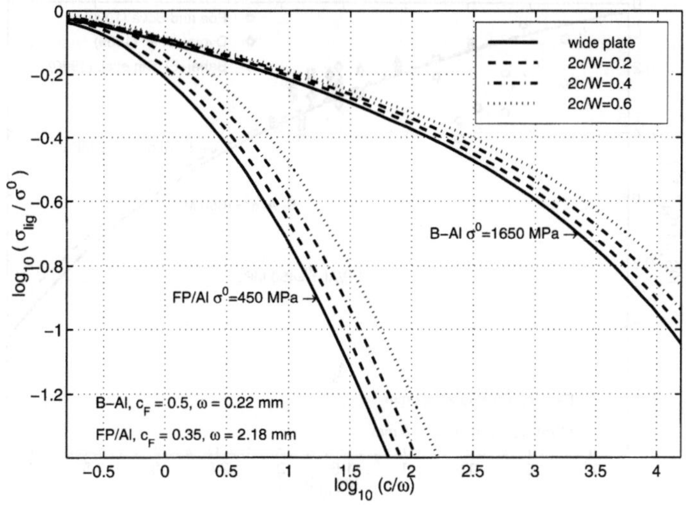


*Figure 5.* Estimates of the average ligament stress $\sigma_{\text{lig}} = \sigma_N/(1 - 2c/W)$ at fracture of center-notched B/Al and FP/Al plies.

experimental values of the ligament stress obtained on specimens with certain $2c/W > 0$ translate into different values of ligament stress on specimens with other $2c/W$ ratios, or on wide plates with $2c/W \to 0$.

Therefore, the difference in the experimental ligament stresses of narrow and wide plates can be evaluated as equal to the difference in the corresponding theoretical values,

$$(\sigma_{\text{lig}}^w)^{\text{exp}} - (\sigma_{\text{lig}}^n)^{\text{exp}} = (\sigma_{\text{lig}}^w)^{\text{pr}} - (\sigma_{\text{lig}}^n)^{\text{pr}}. \tag{30}$$

This gives the ligament stress $(\sigma_{\text{lig}}^w)^{\text{exp}}$ that would be found in wide plate experiments,

$$(\sigma_{\text{lig}}^w)^{\text{exp}} = (\sigma_{\text{lig}}^n)^{\text{exp}} - [(\sigma_{\text{lig}}^n)^{\text{pr}} - (\sigma_{\text{lig}}^w)^{\text{pr}}]. \tag{31}$$

The implication is that ligament stresses found from experiments on narrow specimens can be translated into wide plate data by subtracting from the former the indicated theoretical terms that can be found by appealing to (29) and using the connection

$$(\sigma_{\text{lig}}^n)^{\text{pr}} - (\sigma_{\text{lig}}^w)^{\text{pr}} = -((\Delta\sigma^n)^{\text{pr}} - (\Delta\sigma^w)^{\text{pr}}). \tag{32}$$

In wide plates, the ligament and fracture strengths coincide

$$(\sigma_{\text{lig}}^w)^{\text{pr}} \to \sigma_N^w \quad \text{for } 2c/W \to 0.$$

Figure 6 shows a comparison of the predicted ligament stresses or fracture strengths from Figure 5 with the experimental data obtained in (I) and (III), and by Poe and Sova (1980), on B/Al and FP/Al finite width specimens, where the data have been adjusted as required by (31). The agreement is satisfactory and thus provides a confirmation of the rule for translation of narrow plate to wide plate fracture strength data that follows from (31), (32)

$$(\sigma_{\text{lig}}^w)^{\text{exp}} = (\sigma_{\text{lig}}^n)^{\text{exp}} + [(\Delta\sigma^n)^{\text{pr}} - (\Delta\sigma^w)^{\text{pr}}] \tag{33}$$

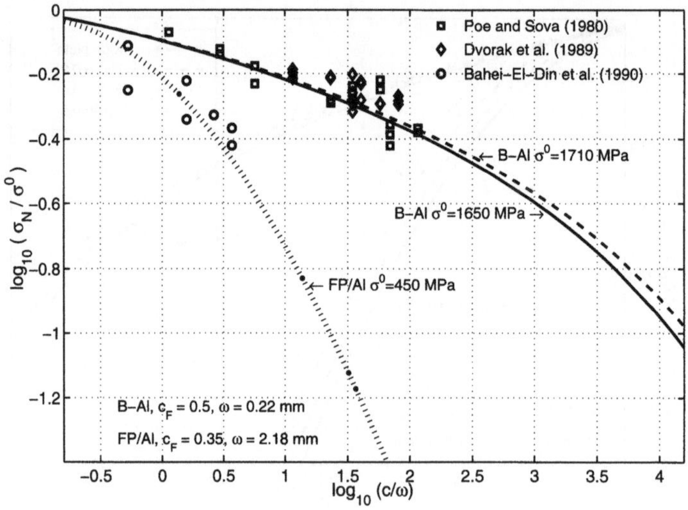


*Figure 6.* Estimates of the fracture strength $\sigma_N$ of wide, notched fibrous plies, compared with experimental data for 0.75–4 in. wide specimens, adjusted according to (33).

where the zone-induced stresses $\Delta\sigma$ are evaluated for the indicated geometries from (22), (3), (2). Obviously, the same rule would apply to any pair of specimen width.

## 6. Closure

The main conclusion obtained is that fracture strength of the B/Al and FP/Al plies considered can be estimated using the superposition (2) or (3), where the dominant part of the zone-induced stress follows from (16). A more direct estimate of the fracture strength can be found by solving (26). Further refinement of the results can be made by including the terms in (17) to (21), and using them in (22) and then in (3).

Since fracture of the composite plies can not be predicted in terms of a single parameter, such as a stress intensity factor, each notched geometry needs to be analyzed separately. However, with such analysis in place, experimental data obtained on a notched ply of a known geometry $(n)$ can be used to estimate the strength of a ply of a different geometry $(w)$ with the rule (33).

The results suggest a relatively simple method for estimating the fracture strength of both wide and narrow fibrous composite plies with ductile matrices, which show clear evidence of discrete plastic shear zone formation prior to fracture. In addition to the B/Al and FP/Al plies that provided experimental verification of the present models, such shear zones can be found in polymer matrix systems, and also in other metal systems, such as SiC/Al. Since the shear zones are sometimes observed in plies bonded to laminated plates, the present results may be useful in estimating the fracture strength of such structures. The present approach could also be extended to ceramic and other systems where matrix ductility is absent, but where discrete shear zones at notches may form due to fiber debonding and sliding. In addition to the center-notched plies analyzed here, the superposition scheme of Figure 2 and the estimates of zone-induced stresses can be applied in modified form to plies with other notch configurations, and to various shape holes, etc; this was illustrated in (III).

We note in passing that the analysis overestimates the fracture strength of B/Al composites with T6-tempered matrices which exhibit high yield stress and very limited ductility; behavior of such systems seems to conform to modified linear fracture mechanics predictions (Bahei-El-Din and Nigam, 1990).

## References

Bahei-El-Din, Y.A., Dvorak, G.J. and Wu, J.F. (1989). Fracture of fibrous metal matrix composites – II. Modeling and numerical analysis. *Engineering Fracture Mechanics* **34**, 105–123. (Paper II)

Bahei-El-Din, Y.A. and Nigam, H. (1990). Fracture of fibrous metal matrix composites – III. Effect of imperfection geometry and heat treatment. *Engineering Fracture Mechanics* **37**, 1207–1232. (Paper III)

Benveniste, Y., Dvorak, G.J., Zarzour, J. and Wung, E.C.J. (1989). On interacting cracks and complex crack configurations in linear elastic media. *International Journal of Solids and Structures* **25** (11) 1279–1293.

Dvorak, G.J., Bahei-El-Din, Y.A. and Bank, L.C. (1989). Fracture of fibrous metal matrix composites – I. Experimental results. *Engineering Fracture Mechanics* **34**, 87–104. (Paper I)

Dvorak, G.J., Zarzour, J. and Benveniste, Y. (1992). Fracture of fibrous metal matrix composites – IV. Plastic zones, local stresses and fracture strength. *Engineering Fracture Mechanics* **42**, 501–517. (Paper IV)

Erdogan, F. (1998). Private communication.

Lekhnitskii, S.G. (1963). *Theory of Elasticity of an Anisotropic Elastic Body* (Translated by P. Fern), Holden-Day, San Francisco.

Poe, C.C. and Sova, J.A. (1980). Fracture toughness of boron/aluminum laminates with various proportions of 0 and 45 plies. NASA Technical Paper 1707.

Post, D., Czarnek, R., Joh, D., Jo, J. and Guo, Y. (1987). Deformation of a metal matrix tensile coupon with a central slot: An experimental study. *Journal of Composite Technology and Research* **9**, 3–9.

Sih, G.C. and Liebowitz, H. (1968). Mathematical theories of brittle fracture. *Fracture* (Edited by H. Liebowitz), Vol. II, Chapter 2, Academic Press, New York, 67–190.

Tirosh, J. (1973). The effect of plasticity and crack blunting on stress distribution in orthotropic composite materials. *Journal of Applied Mechanics* **40**, 785–790.

Whitney, J. and Nuismer, R. (1974). Stress fracture criteria for laminated composites containing stress concentrations. *Journal of Composite Materials* **8**, 253–265.

Wieghardt, K. (1907). Über das Spalten und Zerreissen elastischer Körper. *Z. Mathematik und Physik* **55**, 60–103. (See also: Wieghardt, K. (1995). On splitting and cracking of elastic bodies. *Fatigue and Fracture of Engineering Materials and Structures* **18**, 1371–1405, Translated by H.P. Rossmanith.)

Zarzour, J. (1990). Ph.D. Thesis. Rensselaer Polytechnic Institute.

*International Journal of Fracture* **95:** 103–141, 1999.


# Size effect on compression strength of fiber composites failing by kink band propagation

ZDENĚK P. BAŽANT[1], JANG-JAY H. KIM[2], ISAAC M. DANIEL[3], EMILIE BECQ-GIRAUDON[4] and GOANGSEUP ZI[4]

[1] *Walter P. Murphy Professor of Civil Engineering and Materials Science, Northwestern University, Evanston, Illinois 60208; e-mail: z-bazant@nwu.edu.*

[2] *Graduate Research Assistant, Northwestern University; currently Research Engineer, Sandia National Laboratories, Albuquerque, New Mexico*

[3] *Walter P. Murphy Professor of Civil and Mechanical Engineering, Northwestern University.*

[4] *Graduate Research Assistant, Northwestern University.*

Received 16 July 1998; accepted in revised form 31 December 1998

**Abstract.** The effect of structure size on the nominal strength of unidirectional fiber-polymer composites, failing by propagation of a kink band with fiber microbuckling, is analyzed experimentally and theoretically. Tests of novel geometrically similar carbon–PEEK specimens, with notches slanted so as to lead to a pure kink band (not accompanied by shear or splitting cracks), are conducted. They confirm the possibility of stable growth of long kind bands before the peak load, and reveal the existence of a strong (deterministic, non-statistical) size effect. The bi-logarithmic plot of the nominal strength (load divided by size and thickness) versus the characteristic size agrees with the approximate size effect law proposed for quasibrittle failures in 1983 by Bažant. The plot exhibits a gradual transition from a horizontal asymptote, representing the case of no size effect (characteristic of plasticity or strength criteria), to an asymptote of slope $-\frac{1}{2}$ (characteristic of linear elastic fracture mechanics, LEFM). A new derivation of this law by approximate (asymptotically correct) $J$-integral analysis of the energy release, as well as by the recently proposed nonlocal fracture mechanics, is given. The size effect law is further generalized to notch-free specimens attaining the maximum load after a stable growth of a kink band transmitting a uniform residual stress, and the generalized law is verified by Soutis, Curtis and Fleck's recent compression tests of specimens with holes of different diameters. The nominal strength of specimens failing at the initiation of a kink band from a smooth surface is predicted to also exhibit a (deterministic) size effect if there is a nonzero stress gradient at the surface. A different size effect law is derived for this case by analyzing the stress redistribution. The size effect law for notched specimens permits the fracture energy of the kink band and the length of the fracture process zone at the front of the band to be identified solely from the measurements of maximum loads. The results indicate that the current design practice, which relies on the strength criteria or plasticity and thus inevitably misses the size effect, is acceptable only for small structural parts and, in the interest of safety, should be revised in the case of large structural parts.

**Key words:** Fracture, compression, fiber composites, testing, kink bands, microbuckling, size effect, scaling, asymptotic analysis, $J$-integral, equivalent LEFM, cohesive crack model.

## 1. Introduction

In the early 1980's it became clear that the size effect on the nominal strength of quasibrittle materials failing after large stable crack growth is caused principally by energy release (Bažant 1984) and cannot be explained by Weibull-type statistics of random microdefects. Ever since, the problem of size effect has received increasing attention (see the reviews in Bažant and Chen, 1997, and Bažant and Planas, 1998). Size effects caused by energy release into a finite-size FPZ (or damage localization zone) have been extensively demonstrated and mathematically described for concrete, mortar, rocks, ceramics and sea ice. Description of such a size effect requires an energy analysis of fracture mechanics type.

At present, all the textbooks and practical procedures for fiber composites characterize the failure in terms of strength or plasticity type criteria, which are inherently incapable of capturing the size effect. Recently, though, the existence of size effect has been demonstrated by tests of notched orthotropic and quasi-isotropic carbon-epoxy laminates under tensile (Mode I) loading (Bažant, Daniel and Li, 1996). A size effect of thickness in laminates was experimentally shown by Daniel and Hsiao (1996).

The present study will focus attention on the compression failure of unidirectionally reinforced fiber composites. This is a particularly complex type of material failure, which can involve two distinct mechanisms:

(1) delamination fracture, and
(2) a row of parallel axial shear cracks combined with microbuckling of fibers in a so-called kink band.

Only the latter mechanism of failure will be considered in this study.

Compression microbuckling in kink bands has been studied extensively for over thirty years and a large body of knowledge has been accumulated. Rosen (1965) presented a simple formula for compression strength based on the idea of buckling of parallel fibers embedded in an elastic matrix. Similar more refined formulae for elastic composites with wavy fibers were presented by Bažant (1968). Argon (1972) extended Rosen's formula by considering plastic yielding.

A further important refinement was introduced by Budiansky (1983) who took into account the initial misalignment of the axial fibers and showed its pronounced effect. Various subsequent refinements within the framework of elastoplastic analysis were made by Budiansky and Fleck (1993), Kyriakides et al. (1995), Christensen and DeTereza (1997), Soutis et al. (1991, 1993), Moran et al. (1995), Jelf and Fleck (1992), Fleck and Jelf (1995), Fleck and Shu (1995), Fleck et al. (1996), Kyriakides and Ruff (1997), and others; see the excellent reviews by Fleck (1997), Budiansky and Fleck (1994), Schultheisz and Waas (1996), Waas and Schultheisz (1996), and Sutcliffe and Fleck (1994). These studies included the analysis of post-buckling behavior of the fibers, inelastic behavior of matrix, and various imperfections. They provided understanding of the factors governing the inclination of the kink band with regard to the direction of the fibers, including the role of lateral expansion in the kink band and the shape of the assumed yield surface. Analysis of microbuckle initiation by small-scale yielding fracture mechanics was presented by Sutcliffe et al. (1996)

None of the existing formulae for the nominal strength in compression, however, predicts any size effect. Omission of the size effect in compression has seemed natural because small-scale laboratory tests indicated no size effect and because the maximum load has been thought to occur at the very beginning of microbuckling, before the size or the length of the kink band becomes macroscopically significant.

Recently, though, various fracture mechanics aspects of the kink band failure came to light. Based on strain measurements at the flank of the kink band, the experimental studies of Sutcliffe and Fleck (1994) and Fleck et al. (1996, 1997) demonstrated that the axial normal stress across the band decreases with the distance from the band front, and reaches a plateau equal to about 50 percent of the maximum stress. This means that the diagram of the axial normal stress versus the axial relative displacement across the band exhibits softening and then reaches a yield plateau at about 50 percent stress reduction. Such a crack-like behavior of the band (Fleck et al., 1996) was further confirmed by the study of Moran et al. (1995), who discovered the phenomenon of band broadening. In retrospect, the band broadening appears natural to expect because it is required to accommodate the increase of relative displacement across the band with the distance from the front, similar (but opposite in sign) to the increase of crack opening.

Fleck (1996) and Sutcliffe and Fleck (1996) conducted two-dimensional fracture analyses, adapting the cohesive crack model for compression, and estimated the fracture energy associated with the kink band propagation. Soutis et al. (1991) reported extensive numerical studies with a crack-bridging model and analyzed the effect of the size of a hole on the strength of a composite specimen. They calculated and experimentally verified how the apparent strength in the vicinity of a hole decreases with an increasing diameter of the hole. Although geometric similarity of the hole with the specimen dimensions was not maintained in these tests, the results nevertheless hint at the likelihood of size effect.

Recently, Budiansky, Fleck and Amazigo (1997) analyzed the propagation of a semi-infinite out-of-plane kink band, approximating the band with a crack whose face is allowed to overlap in compression. They assumed the FPZ to be lumped into a point, and analyzing energy balance derived the formula

$$\sigma_u = \sigma_b + \sqrt{G_b/L} \tag{1}$$

in which $\sigma_u$ = axial normal stress 'upstream' in the direction of propagation (far enough ahead of the front), $\sigma_b$ = axial normal stress transmitted across the band far enough behind the front, $L$ = specimen height, and $G_b$ = material constant playing the role of fracture energy and consisting of the work of sliding shear stresses in the band, the work of fiber debonding, and the work of the axial normal stress on the relative displacement across the band. Sutcliffe et al. (1996) examined the energy release rate required for microbuckle initiation by small-scale yielding fracture mechanics approach. required for microbuckle initiation

Although this formula does not give the effect of size $L$ on the nominal strength of geometrically similar specimens and does not take into account the effect of the size of the FPZ at the front of the kink band, it clearly suggests the existence of a size effect. More importantly, the type of analysis that has led to this formula gives an inspiration for taking a fracture mechanics approach.

The present study, involving specimens of geometrically similar shapes, reports experimental results that reveal the existence a size effect in kink band compression failure and permit an approximate calibration of size effect theory. A simplified analysis of the energy release, briefly outlined at a recent conference (Bažant 1998(a)), is used to obtain closed-form formulae for the size effect in failures that occur either after a large stable growth of kink band, or at the initiation of the kink band. These formulae represent a special application of the general asymptotic analysis of size effect proposed for compression fracture in Bažant (1997) and Bažant and Chen (1997).

## 2. Size effect tests of notched specimens

The kink band failure is often combined with axial splitting-shear cracks and delaminations. Such combined failures are difficult to analyze because the contributions of the microbuckling in the kink band and of the shear, splitting and delamination failures are hard to separate.

Therefore, the objective of experimental investigation aimed at verifying a theory should be to find the type of fiber composite and the shape of specimen that would lead to a pure kink band failure and would do so even for very large sizes.

After experimenting with various types of composites, the carbon fiber composite with a PEEK (poly-ether-ether-keton) thermoplastic polymer matrix was selected. The advantage of PEEK is that it is less brittle than carbon epoxy composites, which leads to a more stable type of failure. It follows that, if a size effect is revealed in this type of composite, it should exist and be in fact more pronounced in more brittle composites, such as the carbon fiber-epoxy composites.

To exclude the effect of random variation of material strength over the specimen volume (known to cause Weibull-type size effect), the specimens (shown in Figure 1(a,c) and 2) need to be provided with notches. The notches ensure the failure to begin in one desired place, preventing the failure from starting at diverse locations where the material is statistically weakest.

Why a one-sided notch, rather than two symmetric notches? The reason is that a bifurcation of the equilibrium path would have to be expected to occur, according to the analysis in Bažant and Cedolin (1991, Sec. 12.5). Symmetric growth of two interacting kink bands would surely represent an unstable path. The stable path is a one-sided growth of one kink band even if there are two notches.

Notches that are orthogonal to the surface have normally been used in fracture testing. At the beginning of this study, however, such notches were found to engender failures that begin by an axial splitting-shear crack (Figure 1(b)), which is only later followed by the development of an out-of-plane kink band. Such a band typically has a transverse inclination (such inclination was previously observed by Fleck et al. (1996), Sutcliffe and Fleck (1994), Kyriakides et al. (1995)). For this reason, the starter notch has been made inclined (Figure 1(c), 2) with the same angle as the out-of-plane kink band. This inclination was found by trial tests to be 25.4°. It made it possible to eliminate in most tests the axial splitting-shear cracks and thus obtain pure kink band failures.

The test specimens, shown in Figure 1(a,c) and 2, are scaled in two dimensions. the thickness $b$ in the third dimension being constant. The specimen dimensions and the notch depths of the specimens of different sizes are scaled in the ratio 1:2:4. The notch of length $a_0 = 0.3\,\mathrm{D}$, where D is the specimen width, is machined with a diamond bladed band saw up to 95 percent of its length. Then the notch is sharpened at the tip by a cut whose depth is 5 percent of the notch depth. The cut is machined with a 0.2 mm diameter diamond-studded wire, and thus the crack tip radius is 0.1 mm in all the specimens. The depth $a_0$ of the notch considered for scaling and in the analysis includes the depth of the wire saw cut.

The content of polymer resin in the specimen (supplied by Fiberite, Inc., Orange, California) was $32 \pm 3$ percent. The specimens have been molded at Northwestern University from 100 plies of sheets $304.8 \times 304.8$ mm, 0.05 mm thick. The molding was carried out under temperature 391°C (735°F) and pressure 0.69 MPa, using the standard time sequence of the curing process. After the specimens had been cut from the molded sheets, they were provided with massive end caps made of 1040 hot rod steel, to which they were glued by epoxy. To ensure proper alignment, the end caps were glued only after the specimen had been installed under the loading platens of the testing machine. The end plates were restrained to prevent any rotations. All the specimens have been tested under a controlled stroke rate of $1.27 \times 10^{-4}$ mm/s. After the kink band had initiated at the notch tip, it was seen to propagate stably on both the front and back sides of the specimen. On one side, the kink band was usually

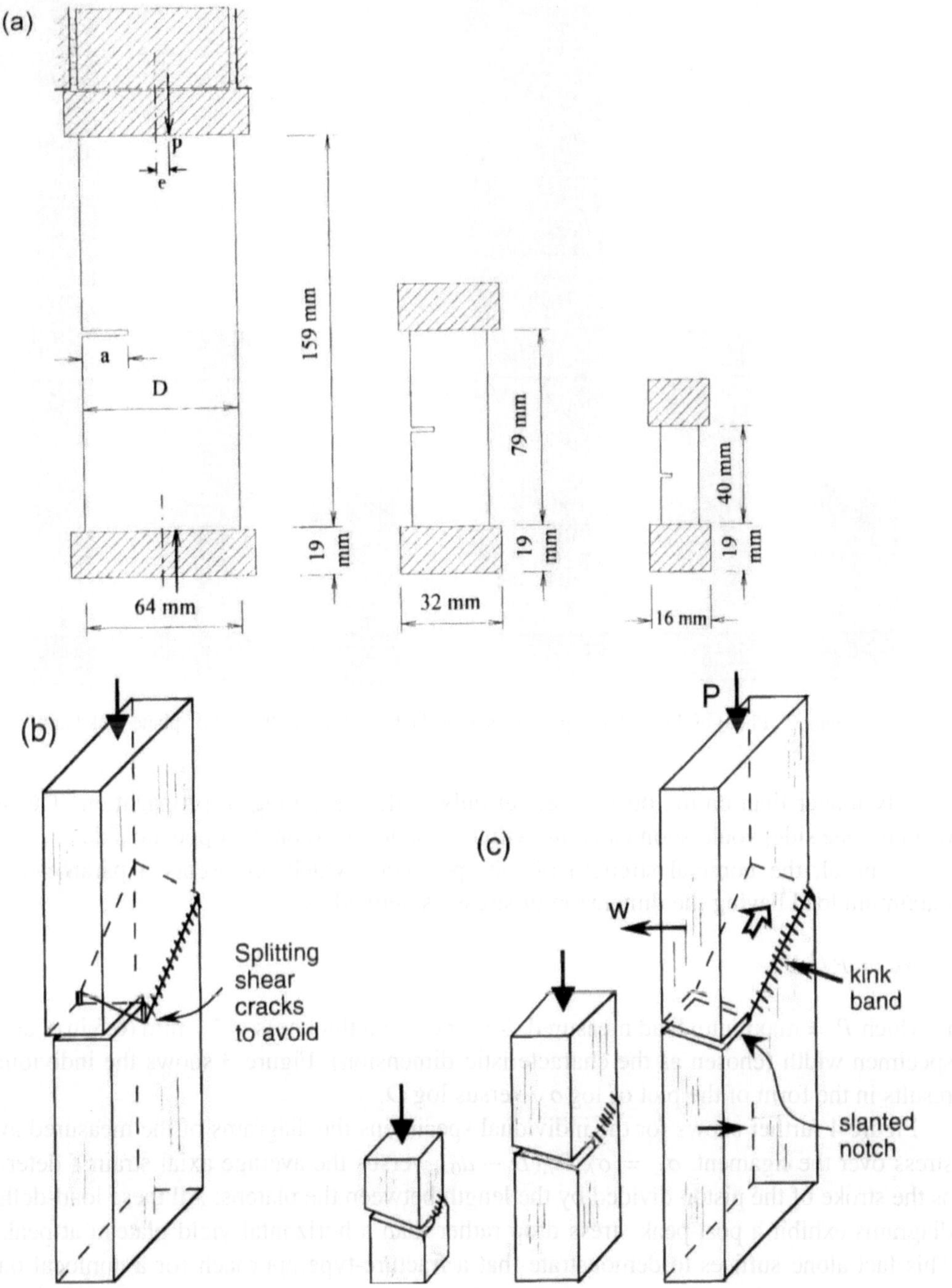


*Figure 1.* (a) Geometrically similar single-edge notched carbon-PEEK (poly-ether-ether-keton) specimens tested, and scheme of loading, (b) specimen with an orthogonal notch exhibiting undesirable failure (splitting-shear cracks), and (c) transversely slanted notches used in present tests (Figures 3 and 4), achieving pure kink band failure.

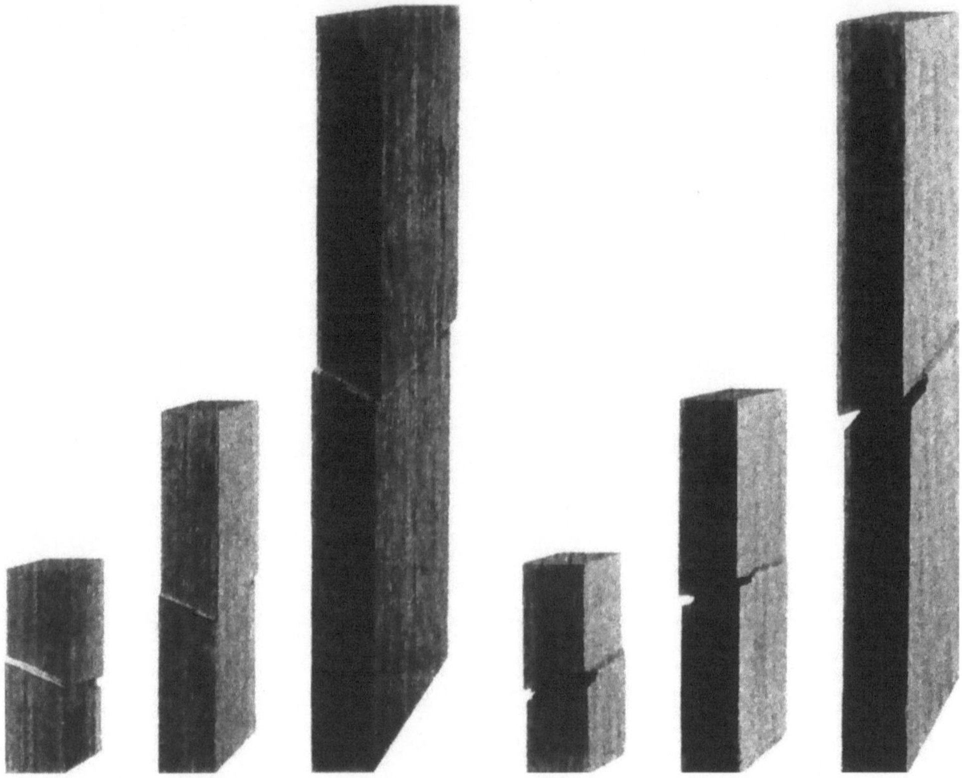

*Figure 2.* Photo of three of the specimens after the test, showing the out-of-plane kink band.

slightly longer than on the other side, but only at the beginning of propagation. The shorter band on one side would soon catch up with the longer band on the opposite side.

As usual, the nominal strength of the specimen, which represents a parameter of the maximum load having the dimension of stress, is defined as

$$\sigma_N = P/bD, \tag{2}$$

in which $P$ = maximum load measured, $b$ = specimen thickness (12.7 mm (0.5 in)), and $D$ = specimen width (chosen as the characteristic dimension). Figure 3 shows the individual test results in the form of the plot of $\log \sigma_N$ versus $\log D$.

Figure 4 further shows for the individual specimens the diagrams of the measured average stress over the ligament, $\sigma_L = \sigma_N D/(D - a_0)$, versus the average axial strain $\bar{\varepsilon}$ determined as the stroke of the piston divided by the length between the platens. All these load-deflection diagrams exhibit a post-peak stress drop rather than a horizontal yield plateau at peak load. This fact alone suffices to demonstrate that a fracture-type approach (or a nonlocal damage approach) is required. Furthermore, these diagrams reveal the existence of a terminal yield plateau, which confirms the existence of a finite residual stress across the kink band.

If there were no size effect, as currently assumed in design and exhibited by the existing formulae for the maximum load expressed in terms of stress or strain, the plot in Figure 3 would have to be horizontal. Every theory based on plasticity or on some critical values of stress or strain predicts a horizontal plot. However, the trend is clearly seen to be downward. The downward slope is quite steep and closer to the slope $-\frac{1}{2}$ corresponding to LEFM than

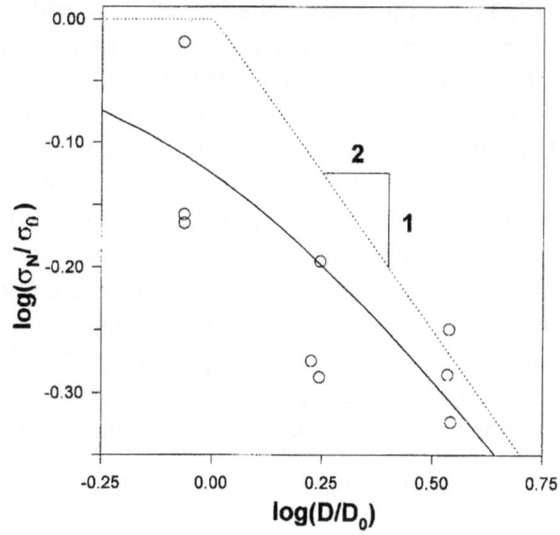


*Figure 3.* Results of the present tests of nominal strength of carbon-PEEK specimens (data points) and their optimum fit by equation (14).

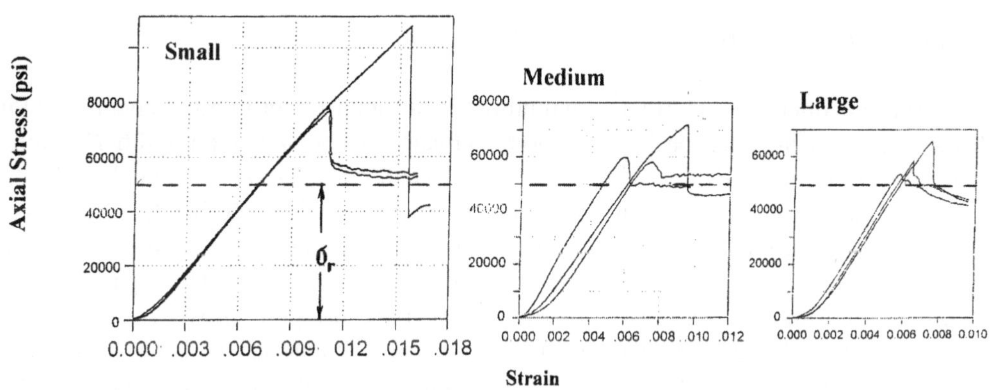


*Figure 4.* Load-deflection diagrams of carbon-PEEK specimens tested.

to the horizontal line for the strength theory. So, despite a relatively ductile PEEK matrix, the size effect for these large, albeit not extremely large, specimens is strong.

One of the small specimens, corresponding to the upper left data point in Figure 3, did not develop a kink band but failed by a vertical shear crack that started from the notch tip and produced axial splitting. Argued though it could be that this data point should be excluded, it has nevertheless been retained in the data set, because

(1) it is the highest point in the data set, and
(2) the load that would cause the kink band failure of this specimen must have been at least as high as this point.

In other words, the size effect could not be milder than it is when this point is included in the analysis.

The test results exhibit considerable random scatter. This is, however, typical of compression failure of fiber composites (because of their strong sensitivity to fiber misalignment;

Fleck 1997, Budiansky 1983). For this reason, the size effect would not have been revealed clearly if the size range were less than 1:4. The present size range of 1:4 appears just about the minimum for being able to clearly demonstrate the size effect. To reduce the ratio of the inevitable scatter band width to the range of sizes, the ratio of the sizes of the smallest and largest specimens should be at least 1:8 in future testing. For concrete, tests with size ranges up to 1:32 have been made (Bažant and Planas 1998). For sea ice, tests of a size range of 1:162 have been carried out by Dempsey's team in the Arctic Ocean (Dempsey et al., 1999), providing a clear picture of the size effect despite strong random scatter.

Compared to the loading through a pin, which would allow free rotations at the ends, the present boundary condition of restraint against rotation complicates a little bit the test evaluation because an unknown bending moment develops at the ends causing the compression resultant to shift laterally during kink band propagation. However, the end restraint has the advantage that the test becomes much more stable. This has made it possible to

(1) demonstrate that a stable propagation of a long kink band is possible (which has so far been doubted by most experts),
(2) observe the post-peak deformations well beyond the peak load, and
(3) approach the residual stress plateau and observe the residual stress (Figure 4).

## 3. Simple size effect analysis via $J$-integral

We analyze a specimen (Figure 5(a)) with unidirectional (axial) fiber reinforcement. The kink band has length $a$ which can be long or short compared to the specimen width $D$ taken as the characteristic dimension. The width of the kink band, considered to be small, is denoted as $w$, and its inclination as $\beta$ (Figure 5(a), 6). Although tractable, the bending stiffness of the fibers is neglected, for the sake of simplicity. The loading is assumed to produce cohesive shear cracks that are parallel to the fibers and have a certain characteristic spacing $s$. The axial normal stress transmitted across the kink band (band-bridging stress) is denoted as $\sigma$ (Figure 6). Although Figure 5(a) depicts an in-plane fiber inclination, the behavior is similar for the out-of-plane fiber inclination in the test specimen used because what matters for the analysis is the reduction of axial stress across the kink band, which is the same for both cases.

The diagram of the shear stress $\tau$ transmitted across the shear cracks versus the slip displacement $\eta_{fr}$ on these cracks must exhibit post-peak softening (Figure 7 top left). This is confirmed by two important recent experimental findings. First, Fleck and Shu (1995) placed strain gauges at the flanks of the kink band and, as the kink band grew, observed the strain in the gages to decrease, rather remain constant (see also Fleck, 1997). Second, Moran et al. (1995) recently discovered the phenomenon of band broadening (see also Sutcliffe et al., 1996), which implies that the relative displacement across the band increases as the band grows, and thus indicates that the kink band plays a role similar to a crack (whose opening width grows with the distance from the front and the transmitted stress decreases), rather than to a dislocation line (on which the relative displacement as well as the transmitted stress remains constant).

For the sake of simplicity, the stress-displacement diagram of the axial shear cracks is considered to be bilinear, as shown in Figure 7 (top right) where $\tau_p$ = peak stress or shear strength = shear stress parallel to fibers at which the cohesive crack initiates, and $\tau_r$ = the residual shear strength, representing the final yield plateau. According to the analysis of mode II slip bands by Palmer and Rice (1973), the area of the diagram above the yield plateau is

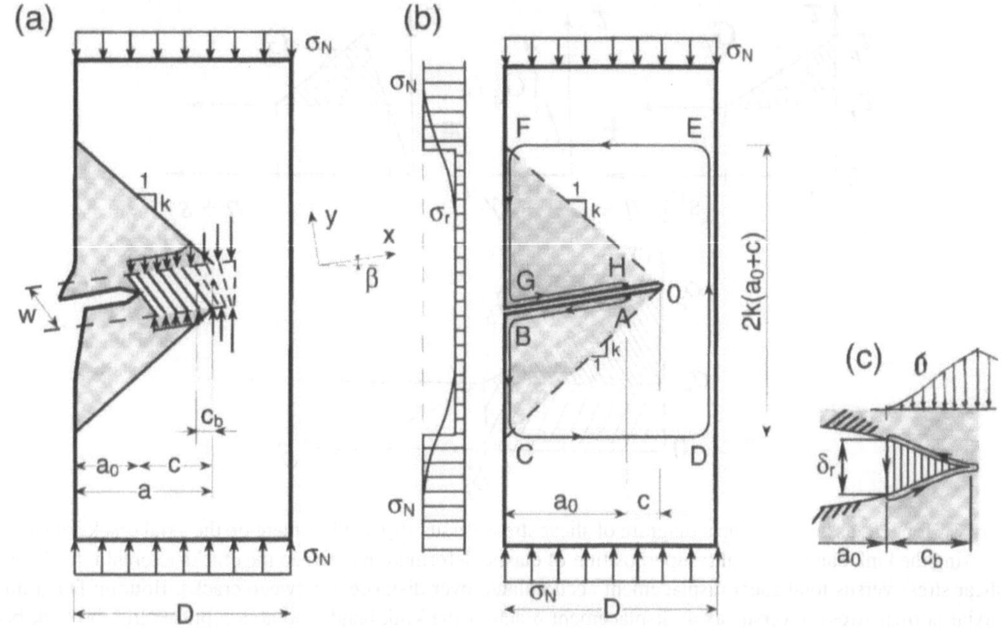


*Figure 5.* (a) Idealized kink band of width $w$ in a notched specimen, with a fracture process zone of effective length $c_b$; (b) path of $J$-integral, with energy release (stress relief) zones OFGO, OBCO; (c) fracture process zone of equivalent cohesive crack.

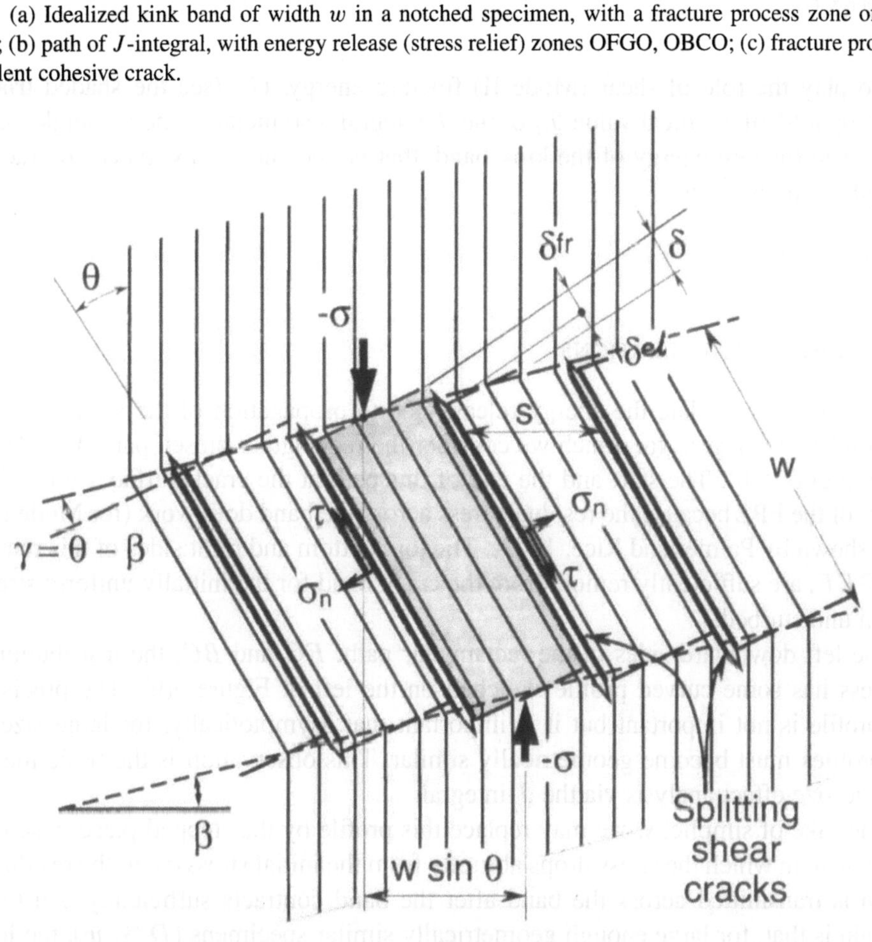


*Figure 6.* Idealized microbuckling of fibers in the kink band and axial shear cracks.

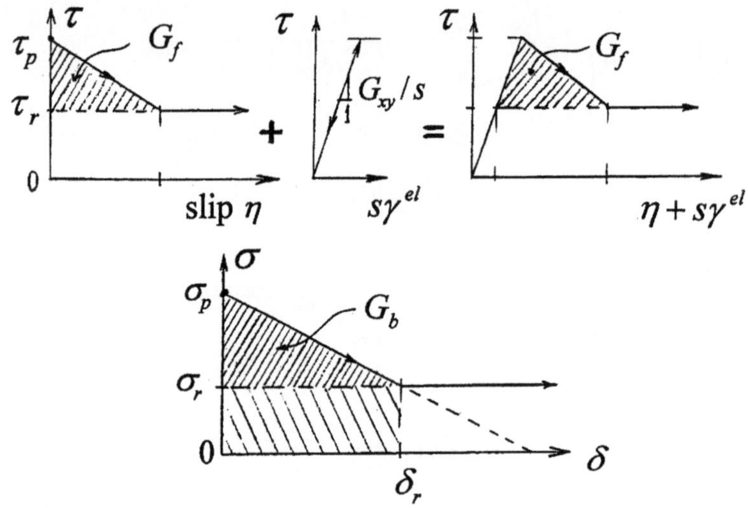


*Figure 7.* Top left: assumed bilinear diagram of shear stress versus slip displacement on the axial cracks of spacing $s$, crossing the kink band. Top right: superposition of elastic deformation between the cracks to obtain the diagram of shear stress versus total shear displacement accumulated over distance $s$ between cracks. Bottom: Diagram of the axial normal stress $\sigma$ versus axial displacement $\delta$ across the kink band, and area representing the kink band fracture energy $G_b$.

known to play the role of shear (Mode II) fracture energy, $G_f$ (see the shaded triangle in Figure 7 top left) (the critical value $J_{\rm cr}$ of the $J$-integral also includes the rectangle below the triangle). The fracture energy of the kink band, that is, the energy dissipated by fracture per unit length of the band, is

$$G_b = G_f\, w / s. \tag{3}$$

### 3.1. $J$-INTEGRAL CALCULATIONS

To approximately calculate the energy release due to propagation of the kink band, we use Rice's (1968a) $J$-integral, for which we consider the rectangular closed path $ABCDEFGH$ shown in Figure 5(b). The start and the end of this path at the crack surfaces must lie at the boundary of the FPZ because the residual stress across the band does work (for Mode II cracks this was shown by Palmer and Rice, 1973). The top, bottom and right sides of this rectangular path, $CDEF$, are sufficiently remote from the crack band for the initially uniform stress state to remain undisturbed.

On the left downward sides of the rectangular path, $FG$ and $BC$, the distribution of the axial stress has some curved profile sketched on the left of Figure 5(b). The precise shape of this profile is not important but it is important that asymptotically, for large sizes $D \gg w$, the profiles must become geometrically similar. This observation is the basic idea of the asymptotic size effect analysis via the $J$-integral.

For the sake of simplicity, we may replace this profile by the stepped piece-wise constant profile shown, in which the stress drops abruptly from the initial stress $\sigma_N$ to the residual stress $\sigma_r$ which is transmitted across the band after the band contracts sufficiently. An important point again is that, for large enough geometrically similar specimens ($D \gg w$), the locations of the stress steps in this replacement profile must also be similar, that is, points $F$ and $C$, must lie on inclined rays of a certain constant slope $k$ shown dashed in Figure 5(b). These rays may be imagined to emanate from the tip of the equivalent crack of length $a = a_0 + c_b$ (Figure 5(b)) where $c_b$ characterizes the length of the FPZ of the kink band and represents approximately the distance from the center of the FPZ of kink band to the point where the stress is reduced to its residual value $\sigma_r$ (Figure 7 top; the length of the FPZ is about $2c_b$). Slope $k$ depends on the structure geometry and on the orthotropic elastic constants.

The area between these rays and the kink band roughly represents the zone of stress relief caused by the drop of axial stress transmitted by the kink band. The strain energy contained within this area is released and is dissipated by the axial shear cracks forming at the front of the kink band. Noting that this area, and thus the energy release, increases in proportion to $D^2$, while the energy dissipated at the kink band front increases linearly with $D$, one immediately concludes that there must be size effect.

The zone at kink band front in which the axial shear cracks are forming represents the FPZ of the kink band. Its length $c_b$ may be regarded as a material property, almost independent of the specimen dimensions and geometry. It may be considered to be of the same order of magnitude as the width $w$. Throughout this zone, the fiber inclination increases from the initial misalignment angle $\overline{\varphi}$ up to the value $\overline{\varphi} + \varphi$ corresponding to the residual cracks. To make test evaluation simple, the specimens must be notched and the FPZ at maximum load must still be attached to the notch (i.e. $c = c_b$).

Referring to the sketch in Figure 5(b), the crack band of length $a_0 + c_f$ is approximately equivalent to a mode I crack whose faces are imagined to interpenetrate. The length of this crack is $a_0 + c$ where $c = c_f + (w/2k)$, which may again be assumed to be approximately a constant when the size is varied. Consequently, the height $FC$ of the rectangular path in Figure 5b is approximately $2k(a_0 + c)$, as labeled in the figure.

In view of these considerations, the first part of the $J$-Integral may be approximately expressed as follows

$$\oint \overline{W}\,\mathrm{d}y = 2k(a_0 + c)\left(\frac{\sigma_N^2}{2E_y} - \frac{\sigma_r^2}{2E_y}\right), \tag{4}$$

in which $\overline{W}$ = strain energy density, and $y$ = coordinate normal to the direction of propagation (Figure 5(b)), and $E_y$ = effective elastic modulus of the orthotropic fiber composite in the fiber direction $y$ (with different values for plane strain and plane stress). In (4) we have considered that the parts of the integral over the horizontal segments are 0, and that the stress on the vertical segment $DE$ may be assumed undisturbed by the kink band, i.e., equal to $\sigma_N$. The portions of the integral over the crack surface segments $GH$ and $AB$ are, likewise, 0.

The second part of the $J$-Integral may be calculated in a similar manner as that introduced by Palmer and Rice (1973) for the propagation of Mode II shear fracture with residual friction;

$$\oint \boldsymbol{\sigma}\cdot\frac{\partial \boldsymbol{u}}{\partial x}\,\mathrm{d}x = \int_{AB} \sigma_r \frac{\mathrm{d}}{\mathrm{d}x}\left[\frac{1}{2}\delta(x)\right]\mathrm{d}x - \int_{GH} \sigma_r \frac{\mathrm{d}}{\mathrm{d}x}\left[\frac{1}{2}\delta(x)\right]\mathrm{d}x$$

$$= \int_{x=0}^{a_0} \sigma_r\,\mathrm{d}\delta(x) = \sigma_r \int_{x=0}^{a_0} \mathrm{d}\delta(x) = \sigma_r \delta_{BG}, \tag{5}$$

in which $\boldsymbol{\sigma}$ = stress vector acting from the outside on the domain enclosed by the path, $\boldsymbol{u}$ = displacement vector, $\delta$ = relative displacement across the band, and $\delta_{BG}$ = relative displacement between points $B$ and $G$. That displacement can be estimated as the difference between the changes of length $\overline{ED}$ and length $\overline{FC}$;

$$\delta_{BG} = \Delta\overline{ED} - (\Delta\overline{FG} + \Delta\overline{BC}) \tag{6}$$

$$= 2k(a_0 + c)\frac{\sigma_N}{E_y} - 2k(a_0 + c)\frac{\sigma_r}{E_y} = 2k(a_0 + c)\frac{\sigma_N - \sigma_r}{E_y}. \tag{7}$$

Now the $J$-Integral may be readily evaluated as follows:

$$J = \oint\left(\overline{W}\,\mathrm{d}y - \boldsymbol{\sigma}\cdot\frac{\partial \boldsymbol{u}}{\partial x}\,\mathrm{d}s\right) = \frac{k}{E_y}(a_0 + c)\left[\sigma_N^2 - \sigma_r^2 - 2(\sigma_N - \sigma_r)\sigma_r\right]$$

$$= \frac{k}{E_y}(a_0 + c)(\sigma_N - \sigma_r)^2 \tag{8}$$

The energy consumed may be calculated again with the help of the $J$-Integral. Similar to Rice (1968b) and Palmer and Rice (1973), the integration path that runs along the equivalent crack surface and around the crack tip (Figure 5(c)) may be used;

$$J_{\mathrm{cr}} = \oint \boldsymbol{\sigma}\cdot\frac{\partial \boldsymbol{u}}{\partial x}\,\mathrm{d}x. \tag{9}$$

This represents the critical value $J_{cr}$, of the $J$-Integral required for propagation. This critical value may be subdivided into two terms

$$J_{\mathrm{cr}} = G_b + \sigma_r\delta_r \tag{10}$$

where $G_b$ is the fracture energy, i.e., the energy required to produce the axial shear cracks across the kink band, and $\sigma_r\delta_r$, represents the plastic work that is done by the residual stresses $\sigma_r$ within the FPZ of the kink band and is leaving the FPZ in its wake. This work corresponds in Figure 7 (top left) to the shaded rectangle lying under the shaded triangle. Following the way shown by Rice (1968b) and Palmer and Rice (1973) for shear bands, $J_{\mathrm{cr}}$ may be evaluated (Figure 5(c)) as follows

$$J_{\mathrm{cr}} = \oint \boldsymbol{\sigma}\cdot\frac{\partial \boldsymbol{u}}{\partial x}\,\mathrm{d}x$$

$$= -\int_{x=a_0}^{a_0+c} f[\delta(x)]\frac{\mathrm{d}}{\mathrm{d}x}[\tfrac{1}{2}\delta(x)]\,\mathrm{d}x + \int_{x=a_0+c}^{a_0} f[\delta(x)]\frac{\mathrm{d}}{\mathrm{d}x}[\tfrac{1}{2}\delta(x)]\,\mathrm{d}x$$

$$= -\int_{x=a_0}^{a_0+c} f[\delta(x)]\frac{\mathrm{d}\delta(x)}{\mathrm{d}x}\,\mathrm{d}x = \int_0^{\delta_r} f[\delta(x)]\,\mathrm{d}\delta(x). \tag{11}$$

This means that $J_{\mathrm{cr}}$ represents the sum of the shaded triangle and shaded rectangle in the stress-displacement diagram of Figure 7 (bottom). Therefore, according to (10), fracture energy $G_b$ is represented by the area under the descending stress-displacement curve and above the horizontal line for the residual stress.

### 3.2. Case of long kink band

Setting (8) equal to (10), and solving for the nominal strength $\sigma_N$ of the specimen, we obtain

$$\sigma_N = \sigma_R + \sqrt{\frac{E_y(G_b + \sigma_r \delta_r)/kc}{1 + D/D_0}} = \sigma_R + \frac{\sigma_0}{\sqrt{1 + D/D_0}}, \tag{12}$$

in which

$$D_0 = \frac{c}{\alpha_0}, \qquad \sigma_0 = \sqrt{\frac{E_y(G_b + \sigma_r \delta_r)}{kc}}, \qquad \alpha_0 = \frac{a_0}{D}, \qquad \sigma_R = \sigma_r \tag{13}$$

(for other geometries, $\sigma_R$ need not be equal to $\sigma_r$). The resulting formula (12) has the same form as that proposed by Bažant (1987) for the general case of quasibrittle failures with a residual plastic mechanism, and subsequently verified for several applications to concrete structure. This formula is valid when a long enough kink band transmitting constant residual stress $\sigma_r$ develops in a stable manner before the maximum load is reached. Because of $\sigma_r$, such stable propagation can happen even in specimens of positive geometry (i.e., for increasing $g(\alpha)$). Stable propagation is helped by rotational restraint of specimen ends.

Equation (12) with (13) bears some similarity to formula (1) derived by Budiansky, Fleck and Amazigo (1997). In contrast to that formula, however, (12) with (13) describes the size effect for geometrically similar specimens, takes into account the finiteness of the FPZ at the front of the kink band, and involves a stress quantity of a different meaning (namely the nominal strength, which is a parameter of the applied load) instead of the stress values in front of the kink band and on its flanks.

### 3.3. Failure at the start of kink band from a notch or stress-free crack

In the case of notched test specimens (of suitable geometry), the maximum load is achieved while the FPZ of the kink band is still attached to the notch. Except for the sign of the band-bridging stresses, the situation is analogous to tensile fracture of notched specimens. From experiments on concrete as well as analytical studies based on the cohesive crack model, it is known that only a short initial portion of the softening stress-displacement curve of the cohesive crack comes into play. It is only the initial downward slope of this curve which matters for the maximum load (the tail of the postpeak load-deflection diagram, of course, depends on the entire stress-displacement curve of the cohesive crack); see Bažant and Li (1995) or Bažant and Planas (1998).

A similar situation must be expected for kink bands in notched specimens. Since the shape of the softening stress displacement curve of the cohesive crack model is irrelevant for the maximum load, except for the initial downward slope of the curve, the maximum load must be the same as that for a linear stress-displacement diagram, shown by the descending dashed straight line shown in Figure 7 (bottom).

It follows that in this case the residual stress $\sigma_r$ should be disregarded and the fracture energy $G_B$ that mathematically governs the kink band growth at maximum load of a notched specimen corresponds to the entire area under the extended descending straight line in Figure 7. Obviously, $G_B > G_b$ if $\sigma_r > 0$. Consequently, setting $\delta_r = 0$ in (13) and replacing $G_b$ by $G_B$, we have the size effect law

$$\sigma_N = \frac{\sigma_0}{\sqrt{1 + D/D_0}}, \tag{14}$$

with

$$D_0 = \frac{c_0}{\alpha_0}, \qquad \sigma_0 = \sqrt{\frac{E_y G_B}{k c_0}}, \qquad c_0 = c_b + \frac{w}{2k}. \tag{15}$$

This coincides with the approximate size effect law proposed in Bažant (1983, 1984); Figure 8 (left, for $\sigma_R = 0$).

From experience with other materials, the length (at maximum load) of the crack band up to the beginning of the FPZ, $a_0$, may often be considered to be roughly proportional to the specimen size $D$, within a certain range of sizes. In other words, the ratio $D/a_0$ at maximum load of geometrically similar structures is often approximately constant. So is the value of $D_0$ in (14), provided that the specimens are geometrically similar.

### 3.4. Further considerations and generalization

The foregoing analysis is generally valid for any type of distributed or concentrated load applied on the top and bottom boundaries of the specimen, provided they are sufficiently remote. For the special case of a uniform load, the results can be more directly obtained by a simpler procedure that is based on the principle of superposition. As illustrated in Figure 9, the solution for a specimen with residual stress in the kink band consists of the solution of a specimen in which the distributed load at the top and bottom boundaries is equal to the residual stress (in which case the stress state is uniform, $\sigma = \sigma_r$) plus the solution of a specimen with a mode I stress-free crack loaded at top and bottom boundaries by $\sigma_N - \sigma_r$. In that case it suffices to take the $J$-integral only along the path $BCDEFG$, that is, omit the segments $AB$ and $HG$ along the crack surface. In such an approach, (5) vanishes. This leads directly to an expression of the type (14), but with $\sigma_N - \sigma_r$ on the left-hand side, which is evidently equivalent to (12).

In a complete analysis of the boundary value problem, the values of $\delta_c$ and of the kink band length are determined by the condition that the total stress intensity factor caused by the applied load and by the band-bridging stresses must vanish. As shown by Palmer and Rice (1973), by writing this conditions for the limiting case of an infinite body, an estimate of the length of the FPZ of the kink band can be obtained from the slope of the stress-displacement law of the cohesive shear cracks and their spacing $s$. From that length and the spacing, vice versa, one can obtain an estimate of the slope of the stress-displacement diagram.

Instead of expanding into Taylor series function $g(\alpha)$, it is equally justified to expand a power of $g(\alpha)$ as a function of a power of $\alpha$. In this way (same as in Bažant 1997 or Bažant and Planas 1998, Eq. 9.1.34), one can show that (12) may be replaced by the following more general formula (Figure 8(a,b)):

$$\sigma_N = \sigma_R + \sigma_0 \left[ 1 + \left( \frac{D}{D_0} \right)^r \right]^{-1/2r}. \tag{16}$$

Exponent $r$, a constant, controls the curvature of the size effect plot in Figure 8 (left). The optimal value of $r$ needs to be determined either experimentally or by some more refined theory (for concrete it is close to 1).

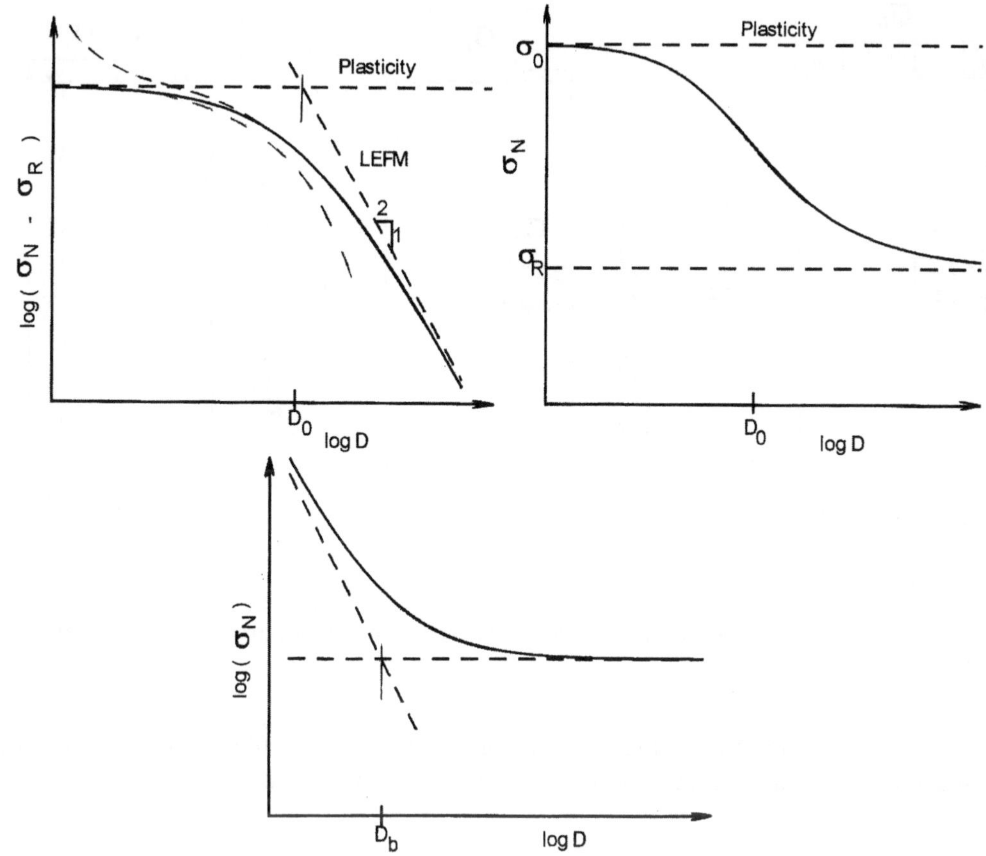


*Figure 8.* (a) Size effect law (solid curve) for specimens with a long kink band or notch (Eq.14) and asymptotic formulas (dashed curves). (b) Same but with $\sigma_N$ instead of $\log \sigma_N$ as the ordinate). (c) Size effect law when $P_{\max}$ occurs at kink band initiation.

### 3.5. Residual strength of specimens with notch or long kink band

The residual stress in the kink band may be estimated by a simple extension of the formula presented by Budiansky (1983). From the moment equilibrium condition of an element of the kink band between two adjacent shear cracks (shaded, in Figure 6), we have $(\sigma s)(w \sin \vartheta) = (\tau w)s$, which expresses the second-order nonlinear geometric effect of buckling. Assuming the fiber inclination $\theta$ at the onset of the slip plateau to be small, we thus have the approximations $\sigma = \tau/\vartheta$ and $\sigma_r = \tau_r/\vartheta$.

The total rotation of the fibers in the kink band may be expressed as

$$\vartheta = \overline{\varphi} + \varphi + \frac{\delta}{s} = \overline{\varphi} + \frac{\tau_r}{G_{xy}} + \frac{\delta_f}{s}, \tag{17}$$

in which $\overline{\phi}$ = initial inclination angle of the fibers (an imperfection), $G_{xy}$ = elastic shear modulus for planes parallel to the fibers, $\delta_f$ = slip displacement of the axial (shear) cracks

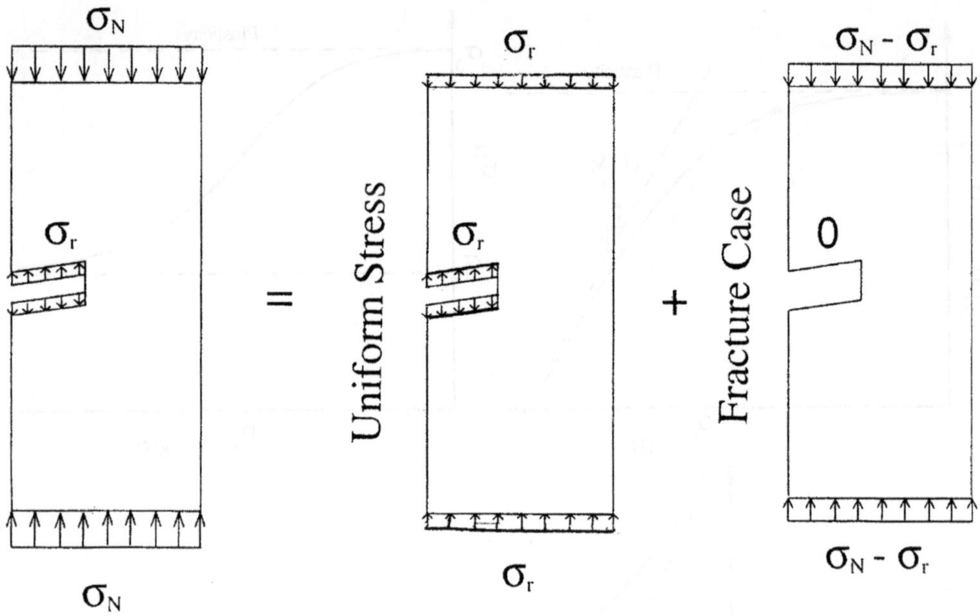


*Figure 9.* Application of principle of superposition to kink band with residual stress $\sigma_r$ (the stress field in the middle specimen is uniform, $K_I = 0$).

at the onset of the yield plateau (Figure 7), and $\phi$ = rotation of the fibers due to elastic shear deformation of the matrix in the kink band. From (17) it follows that

$$\sigma_r = \frac{\tau_r}{\vartheta} = \frac{\tau_r}{\overline{\varphi} + \frac{\tau_r}{G_{xy}} + \frac{\delta_f}{s}} = \frac{G_{xy}}{1 + \frac{\overline{\varphi} + (\delta_f/s)}{\varphi}}. \tag{18}$$

The last expression is the same as that obtained by Budiansky (1983) except for the additional term $\delta_f/s$, which expresses the increase of the shear angle due to crack slip. Budiansky's formula of course refers to the maximum load (of a plastic kink band without fractures), whereas the present formula applies to the residual stress in the kink band after fracturing.

### 3.6. Case of Kink Band Initiation from Smooth Surface

If there is no notch, the maximum load may often occur when the crack band initiates from a smooth surface. However, unless the size of the specimen is very large, a sizable FPZ, of a certain effective length $c_b$ (Figure 10), must form at the kink band front before the maximum load is reached. This must evidently cause significant stress redistribution, which may be seen as the source of the size effect.

This type of size effect, observed in the bending tests of the modulus of rupture, is well documented and well understood for concrete and rocks. It can also be explained on the basis of energy release. In this case, however, one must take into account the second-order derivatives of the energy release rate because the energy release rate for an initiating crack or kink band, still infinitely short, is 0. The present simple approach is not accurate enough for determining these higher-order derivatives. Therefore, the size effect can be more conveniently and perhaps more instructively explained and quantified by analyzing the stress redistribution (as already done for concrete in Bažant and Li, 1995).

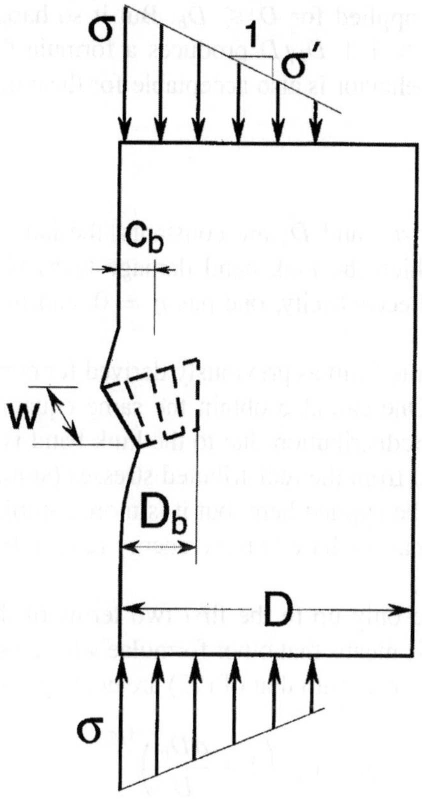


*Figure 10.* Kink band initiation from a smooth surface in a stress field with a stress gradient.

The size effect at crack initiation is known to occur only if the stress distribution has a significant stress gradient $\sigma_{,x}$ and reaches its maximum at the surface (Figure 10). The stress distribution near the surface is considered in the form

$$\sigma = \sigma_1 - \sigma_{,x} x = \sigma_N(1+\eta) - 2\eta\sigma_N \frac{x}{D}, \tag{19}$$

in which $x$ = distance from the surface; $\sigma_1, \sigma_{,x}$ = values of the stress and its gradient at the surface ($x = 0$) before the kink band has formed. If the stress distribution is calculated from the theory of bending and is caused by axial load $P = \sigma_N bD$ of eccentricity $e$, then $\eta = 6e/D$ (dimensionless stress gradient).

A simple solution may proceed as follows. Based on the experience with the analysis of size effect on the modulus of rupture in other quasibrittle materials (Bažant and Li 1995), the stress under maximum load at the front of the FPZ, lying at distance $x = c_b$ from the surface (Figure 10), may be assumed to be equal to the strength limit $\sigma_0$ at which the kink band begins to form, which corresponds to the maximum stress point on the diagram of band-bridging stress versus contraction of the band; $[\sigma]_{x=c_b} = \sigma_0$ or

$$\sigma_N = \sigma_\infty/(1 - D_b/D), \tag{20}$$

in which

$$D_b = 2c_b\eta/(1+\eta), \qquad \sigma_\infty = \sigma_0/(1+\eta). \tag{21}$$

Equation (20) cannot be applied for $D \leqslant D_b$. But it so happens that the large-size approximation $(1 - D_b/D)^{-1} \approx 1 + D_b/D$ produces a formula that has the same large-size asymptotic behavior yet its behavior is also acceptable for the entire range of $D$

$$\sigma_N = \sigma_\infty \left(1 + \frac{D_b}{D}\right), \tag{22}$$

(Figure 8(c)). The values of $\sigma_\infty$ and $D_b$ are constants, the latter representing the thickness of the boundary layer in which the kink band damage takes place. For a vanishing stress gradient or a vanishing load eccentricity, one has $\eta = 0$, and the (deterministic) size effect then disappears.

Equation (22) is of the same form as previously derived for concrete (Bažant and Li, 1995, Bažant and Planas, 1997). One can also obtain the same equation by a more sophisticated analysis in which the stress redistribution due to the kink band is actually calculated and the maximum load is determined from the redistributed stresses (similar to Bažant and Li, 1995). Such an analysis could also be applied here, but it is more complicated. A still more rational approach is an asymptotic analysis based on the energy release functions of LEFM, which is outlined next.

Equation (22) is accurate only up to the first two terms of the asymptotic power series expansion of $\sigma_N$ in $1/D$. This means that other formulae whose asymptotic expansion in $1/D$ coincides up to the first two terms with that of (22) are equally justified; for example

$$\sigma_N = \sigma_\infty \sqrt{1 + \frac{2D_b}{D}} \quad \text{or} \quad \sigma_N = \sigma_\infty \left(1 + \frac{qD_b}{D}\right)^{1/q}, \tag{23}$$

with any positive constant $q$.

## 4. LEFM of orthotropic materials

For an orthotropic material, the stress intensity factor of a sharp crack with a negligibly small FPZ may always be written in the form

$$K_\text{I} = \sigma_N \sqrt{\pi D \alpha}\, F(\alpha) \quad (\alpha = a/D), \tag{24}$$

where $\sigma_N$ = nominal stress, considered here at maximum load, $D$ = characteristic dimension, $a$ = crack length, $\alpha$ = relative crack length, and $F(\alpha)$ = function characterizing structure geometry and material orthotropy.

The energy release rate $\mathcal{G}$ may be related to $K_\text{I}$ using Bao et al.'s (1992) generalization of Irwin's (1958) relation for orthotropic materials

$$\mathcal{G} = \frac{K_\text{I}^2}{\bar{E}} = \frac{D}{\bar{E}} \sigma_N^2 g(\alpha), \quad g(\alpha) = \pi\alpha[F(\alpha)]^2, \tag{25}$$

where $g(\alpha)$ = dimensionless energy release function, characterizing the structure geometry and material orthotropy, and

$$\bar{E} = \frac{1}{Y(\rho)^2} \frac{(E_2/E_1)^{1/4}}{\sqrt{(1+\rho)/2E_1E_2}}, \quad \text{with } \rho = \frac{\sqrt{E_1E_2}}{2G_{12}} - \sqrt{\nu_{12}\nu_{21}}, \tag{26}$$

$$Y(\rho) = [1 + 0.1(\rho - 1) - 0.015(\rho - 1)^2 + 0.002(\rho - 1)^3][(\rho + 1)/2]^{-1/4}. \tag{27}$$

Subscripts 1 and 2 refer to Cartesian axes $x_1 \equiv x$ and $x_2 \equiv y$; $x_2$ coincides with the fiber direction; $E_1$, $E_2$, $G_{12}$, and $\nu_{12}$ are the orthotropic elastic constants; and parameters $E_2/E_1$ and $\rho$ characterize the degree of orthotropy. The formula is valid when the crack propagates in the direction $x_1$ orthogonal to the fibers, but it is used here as an approximation even for propagation directions forming a small angle with $x_1$.

For fracture specimens in the form of long notched strip or slender notched beams, function $g(\alpha)$ or $F(\alpha)$ may be taken approximately the same as for isotropic specimens.

## 5. Asymptotic size and shape effect laws via equivalent LEFM

### 5.1. Case of long kink band or long notch

Following a procedure analogous to Bažant (1997), we will now try to express the coefficients of the size effect law (12) in terms of the energy release functions of LEFM. This will further allow us to capture the effect of structure geometry (shape). Assuming that the FPZ at the kink band front is not so large as to spoil linearity, the stress intensity factor $K_{\rm I}$ at the tip of a Mode I crack equivalent to the kink band may be expressed according to the principle of superposition as follows

$$K_{\rm I} = K_{\rm I}^P - K_{\rm I}^r, \tag{28}$$

in which

$$K_{\rm I}^P = \sigma_N\sqrt{Dg(\alpha)}, \qquad K_{\rm I}^r = \sigma_r\sqrt{D\gamma(\alpha)} \qquad (\alpha = a/D). \tag{29}$$

Here $a$ is the length of the equivalent LEFM crack, whose tip is expected to lie roughly in the middle of the FPZ; $K_I^P$ or $K_I^r$ are the LEFM stress intensity factors caused by load $P$ acting alone (where $P$ can represent not only one applied load but also the parameter of a system of loads), or by a uniform normal traction $\sigma = \sigma_r$ applied on the crack faces (but not the notch faces; Figure 5(a)); $g(\alpha)$ and $\gamma(\alpha)$ are dimensionless LEFM energy release functions of the orthotropic specimen.

When the kink band is propagating, $\mathcal{G} = G_b$ = fracture energy of the kink band. Its meaning, when a residual stress is present, is

$$G_b = J_{\rm cr} - \sigma_r\delta_r = \int_\delta \sigma\, d\delta_r - \sigma_r\delta_r = \int_\delta (\sigma - \sigma_r)\, {\rm d}\delta_r \tag{30}$$

Thus, in view of (25) and (28) with (29), we have $K_{\rm I}^P - K_{\rm I}^r = \sigma_N\sqrt{Dg(\alpha)} - \sigma_r\sqrt{D\gamma(\alpha)} = \sqrt{\bar{E}G_b}$ ($= K_{{\rm I}b}$ = stress intensity factor of kink band). From this,

$$\sigma_N = \frac{\sqrt{\bar{E}G_b} + \sigma_r\sqrt{D\gamma(\alpha)}}{\sqrt{Dg(\alpha)}}. \tag{31}$$

This formula, however, is not acceptable for small sizes of notched specimens; it yields $\sigma_N \to \infty$ for $D \to 0$ because $\lim g(\alpha) = g(\alpha_0) > 0$.

If there is no notch, $a_0$ represents the length of the portion of the kink band length along which the normal stress $\sigma$ transmitted across the band has already been reduced to the residual value $\sigma_r$. This is the distance from the notch mouth or from the beginning of the kink band to the beginning of the FPZ (Figure 5(a)).

By cutting geometrically similar notches of depth $a_0$ in specimens of suitable geometry with a suitable type of loading, one can achieve that the FPZ at maximum load be still attached to the notch tip. In that case, $\alpha_0 = a_0/D$ = constant for all the specimen sizes $D$, which is convenient for evaluating test data.

A basic question is what is the value of $a - a_0$ at the maximum load, $P_{\max}$. In view of extensive experimental evidence and finite element results for other quasibrittle materials (Bažant and Planas 1998), it is reasonable to introduce the following simplifying hypothesis: *The effective length $c_b = a - a_0$ of the FPZ of the kink band at $P_{\max}$ (Figure 5) is approximately constant, governed essentially by the microstructure.*

5.1.1. *Large-size asymptotics*

To deduce the large-size asymptotic behavior, we write $g(\alpha) = g(\alpha_0 + \theta)$ and $\gamma(\alpha) = \gamma(\alpha_0 + \theta)$ where

$$\theta = c_b/D, \quad \alpha = \alpha_0 + \theta \tag{32}$$

and expand functions $g$ and $\gamma$ into a Taylor series as powers of $\theta$, centered at point $\alpha_0$ (or $D \to \infty$). Then we truncate each series after the second (linear) term. So, we may introduce into (31) the large-size approximations

$$g(\alpha) \approx g_0 + g'\theta, \quad \gamma(\alpha) \approx \gamma_0 + \gamma'\theta, \tag{33}$$

in which $g_0 = g(\alpha_0)$, $\gamma_0 = \gamma(\alpha_0)$, and the primes denote the derivatives with respect to $\alpha$; $g' = \mathrm{d}g(\alpha)/\mathrm{d}\alpha$, $\gamma' = \mathrm{d}\gamma(\alpha)/\mathrm{d}\alpha$ at $\alpha = \alpha_0$. With these approximations, one obtains the following size effect law for failures occurring only after a long stable kink band growth (Figure 8(a,b)))

$$\sigma_N = \frac{\sqrt{\bar{E}G_b} + \sigma_r\sqrt{\gamma' c_b + \gamma_0 D}}{\sqrt{g' c_b + g_0 D}}, \tag{34}$$

or, upon further algebraic rearrangement

$$\sigma_N = \frac{\sigma_0 + \sigma_Y\sqrt{1 + D/D_1}}{\sqrt{1 + D/D_0}}, \tag{35}$$

in which

$$D_0 = c_b\frac{g'}{g_0}, \qquad D_1 = c_b\frac{\gamma'}{\gamma_0}, \qquad \sigma_0 = \sqrt{\frac{\bar{E}G_b}{c_b g'}}, \qquad \sigma_Y = \sigma_r\sqrt{\frac{\gamma'}{g'}}. \tag{36}$$

Formula (35) has the following asymptotic values

$$\text{for large } D\text{:} \quad \sigma_N = \sigma_Y\sqrt{\frac{D_0}{D_1}} = \text{const.} = \sigma_r\frac{\gamma_0}{g_0}\sqrt{\frac{g'}{\gamma'}}, \tag{37}$$

$$\text{for small } D\text{:} \quad \sigma_N = \sigma_0 + \sigma_Y = \text{const.} \tag{38}$$

The size effect curve of $\log(\sigma_N - \sigma_Y)$ versus $\log D$ represents a smooth bridging between the size effect of plasticity (i.e. the case of no size effect, corresponding to a horizontal line) and the size effect of LEFM (i.e. the case of the strongest possible size effect, corresponding to a straight line of downward slope $-\frac{1}{2}$). The fact that the small-size asymptote is horizontal agrees with the fact that the strength theory or plasticity ought to be a good approximation for small sizes because the failure cannot localize.

Size $D_0$, called the transitional size, represents the intersection of the power scaling laws for plasticity and LEFM and characterizes the transition from plastic to brittle behavior. The ratio

$$\beta = D/D_0 \tag{39}$$

is a hypostatic characteristic called the brittleness number of a structure (Bažant and Planas 1998). For $\beta \to 0$, the structural behavior is perfectly ductile (plastic), and for $\beta \to \infty$ with $\sigma_R = 0$, the behavior is perfectly brittle, described by LEFM. Note that the shape effect on structural brittleness is included through function $g(\alpha)$ used to calculate $D_0$.

If the FPZ of the kink band is still attached to the tip at maximum load, $\gamma_0 = 0$ in (33), and so, for any structure geometry, the near-tip asymptotic behavior of function $\gamma(\alpha)$ for $\alpha \to \alpha_0$ is (see Appendix III)

$$\gamma(\alpha) = \gamma'\theta, \qquad \text{with } \gamma' = 8/\pi, \qquad \gamma' = \gamma'(\alpha - \alpha_0). \tag{40}$$

Thus (33) becomes a one-term linear approximation $\gamma(\alpha) \approx \gamma'\theta$ for $D \to 0$. This might not be sufficiently accurate for smaller $D$. For a broader range of sizes including small $D$, one must use at least a two-term approximation, $\gamma(\alpha) \approx \gamma'\theta + \frac{1}{2}\gamma''\theta^2$, or even better, not to replace $\gamma(\alpha)$ by an approximation. Thus, for notched specimens, the following formula should have a broader range of accuracy than (34)

$$\sigma_N = \frac{\sqrt{\bar{E}G_b} + \sigma_r\sqrt{D\gamma(\alpha)}}{\sqrt{g'c_b + g_0 D}} \quad \text{(FPZ at notch)}, \tag{41}$$

although it cannot be used if the structure is so small that $\alpha = \alpha_0 + c_b/D$ corresponds to a point that lies outside the structure or near its surface.

With the one-term linear approximation $\gamma(\alpha) = \gamma'\theta$, (31) and (34) reduce to the classical size effect law derived in 1984 for notched Mode-I (tensile) fracture specimens of positive geometry (i.e., those with $g' > 0$)

$$\sigma_N = \sqrt{\frac{\bar{E}G_B}{g'c_b + g_0 D}} = \frac{\sigma_0}{\sqrt{1 + D/D_0}} \quad \text{(FPZ at notch)}, \tag{42}$$

where

$$G_B = (K_b + \sigma_r\sqrt{\gamma' c_b})^2 / \bar{E} \quad \text{with } K_b = \sqrt{\bar{E}G_b}, \quad \gamma' = 8/\pi \tag{43}$$

Here $K_b$ is the fracture toughness of the kink band, and $G_B$ must be the same as in (15), representing the total fracture energy including the energy dissipation by $\sigma_r$. The value of $\gamma'$ is independent of the structure geometry. Equation (42) is the special case of (35) for $D_1 \to \infty$.

Since functions $g(\alpha)$ and $\gamma(\alpha)$ capture the effect of structure geometry, the foregoing formulae give not only the effect of size but also the effect of shape.

5.1.2. *Small-size asymptotics*

To deduce the small-size asymptotic behavior, we introduce new parameter $\xi = 1/\theta = D/c_b$ and new LEFM functions

$$p(\alpha_0, \xi) = \xi g(\alpha_0 + \xi^{-1}), \qquad \omega(\alpha_0, \xi) = \xi \gamma(\alpha_0 + \xi^{-1}). \tag{44}$$

Then we substitute $g(\alpha) = p(\alpha_0, \xi)/\xi$ and $\gamma(\alpha) = \omega(\alpha_0, \xi)/\xi$ into (31), expand these functions into a Taylor series in $\xi$ centered at point $\xi = 0$ (or $D \to 0$), and truncate the series after the second (linear) term, i.e.

$$p(\alpha_0, \xi) \approx p_0 + p'\xi, \qquad \omega(\alpha_0, \xi) \approx \omega_0 + \omega'\xi, \tag{45}$$

where $p_0, \omega_0, p', \omega' =$ constants. Expressing $g$ and $\gamma$ from (44) and substituting into (31), we obtain, after rearrangements,

$$\sigma_N = \frac{\sqrt{\bar{E}G_b} + \sigma_r\sqrt{\omega c_b + \omega' D}}{\sqrt{p c_b + p' D}}. \tag{46}$$

This expression can be brought to the form of the large-size asymptotic approximation (36) in which

$$D_0 = c_b \frac{p_0}{p'}, \qquad D_1 = c_b \frac{\omega_0}{\omega'}, \qquad \sigma_0 = \sqrt{\frac{\bar{E}G_b}{c_b p_0}}, \qquad \sigma_Y = \sigma_r \sqrt{\frac{\omega_0}{p_0}}. \tag{47}$$

5.1.3. *Asymptotic matching character*

For large enough sizes, the original formula (31) is more accurate than its subsequent approximations. But this formula is unacceptable for $D \to 0$, giving $\sigma_N \to \infty$. This conflicts with the small-size asymptotic form (46) which gives a finite $\sigma_N$. On the other hand, the approximations (35), (34), (41)and (42) of the large-size asymptotic formula happen to have also the correct small-size asymptotic behavior for $D \to 0$, approaching a horizontal asymptote. Moreover, they happen to have the same form as the small-size asymptotic approximation in (46). This means that these approximations serve as an interpolation between the opposite infinities (in the log $D$ scale), generally called asymptotic matching.

There is a difference, though. In asymptotic matching, solutions that are accurate for one and the other infinity serve as the starting point. Here, however, the small-size approximation cannot be accurate in the limit for $D \to 0$ because it is based on LEFM while the FPZ is in the limit larger than the structure (properly one would need to use for the small-size limit the nonlocal theory of damage localization, which however does not seem amenable to a simple analytical solution). So the numerical values of $D_0, D_1, \sigma_0, \sigma_1$ for the small-size approximation cannot be reliably predicted. Therefore, the asymptotic matching with the small-size asymptotic approximation (46) can only indicate the form of the matched asymptotic formula but not its coefficients (in particular, the horizontal asymptote $\sigma = \sigma_0$ could be higher or lower than indicated by (47)). For this reason, we simply assume $D_0, D_1, \sigma_0, \sigma_1$ to be the same as for the large-size asymptotic, and content ourselves with merely having the correct asymptotic form for $D \to 0$.

Since the justification of (35) rests in the matching of the asymptotic behaviors at opposite infinities, any smooth formula that satisfies the same small-size and large-size asymptotic properties should be equally valid; for instance,

$$\sigma_N = \frac{\sigma_0 + \sigma_Y[1 + (D/D_1)^r]^{1/2r}}{[1 + (D/D_0)^r]^{1/2r}}, \tag{48}$$

where $r$ is any positive constant. This formula may be derived similarly as (35) except that one needs to set $\zeta = \theta^r$, $\tilde{g}(\alpha) = [g(\zeta)]^{1/r}$, $\tilde{\gamma}(\alpha) = [\gamma(\zeta)]^{1/r}$, and use instead of (33) the approximations $\tilde{g}(\zeta) = \tilde{g}_0 + \tilde{g}'\zeta$ and $\tilde{\gamma}(\zeta) = \tilde{\gamma}_0 + \tilde{\gamma}'\zeta$.

5.1.4. *Length of kink band at maximum load*

In the classical LEFM corresponding to $\sigma_r = 0$, specimens of positive geometry, i.e. those with $g'(\alpha_0) > 0$, attain $P_{\max}$ while the FPZ is still attached to the notch. Then a sufficient characteristic of $P_{\max}$ is that $\mathcal{G} = G_b$, which implies that the FPZ length $2c$ has grown to its full value $2c_b$. If the geometry is negative, $g'(\alpha_0) < 0$, the kink band propagates stably, at increasing load $P$, and $P_{\max}$ occurs at kink band length $a$ such that $g'(\alpha) = 0$. When an R-curve, $R(c)$, is considered, the condition of $P_{\max}$ is that $\mathcal{G}'(a) = R'(c)$.

These conditions, however, do not apply if $\sigma_r > 0$. The total energy release rate due to $P$ and $\sigma_r$ may be written as $\mathcal{G} = K_I^2/\bar{E}$ where $K_I = [\sigma_N k(\alpha) - \sigma_r \kappa(\alpha)]\sqrt{D}$; $k(\alpha) = \sqrt{g(\alpha)}$ and $\kappa(\alpha) = \sqrt{\gamma(\alpha)}$ (dimensionless stress intensity factors of kink band). Under gravity load, the kink band is stable if $\mathcal{G}' > 0$ or $\mathrm{d}K_I/\mathrm{d}\alpha > 0$, and it is critical (i.e., the load $P = P_{\max}$) if $\mathrm{d}K_I/\mathrm{d}\alpha = 0$. This yields for gravity loading the conditions

$$k'(\alpha) < \kappa'(\alpha)\sigma_r/\sigma_N \text{........ stable, } P \text{ growing,} \tag{49}$$

$$= \text{........ critical, } P = \max. \tag{50}$$

The latter is of course the condition of maximum load for any type of loading (an equivalent equation without $\sigma_N$ and in terms of $g(\alpha)$ was derived from the condition $\mathrm{d}\sigma_N/\mathrm{d}\alpha = 0$ by Zi, 1999).

If condition (50) is violated already while FPZ is attached to the notch tip ($\alpha_0 = \alpha_n$), $P_{\max}$ occurs right at the beginning of kink band propagation, i.e., as soon as $\mathcal{G} = G_b$. Otherwise the value of $\alpha_0$ (location of the beginning of the FPZ at $P_{\max}$) along with $P_{\max}$ must be found by solving equations (42) and (50), in which $\alpha = \alpha_0 + c_b/D$ ($\approx$ coordinate of FPZ center).

Compared to propagation of cracks without residual stress, the range of stability of propagating kink bands is wider. According to (49), it includes not only negative but also positive values $g'(\alpha)$, up to a certain limit proportional to $\sigma_r$. So, a stable kink band growth can occur even for positive geometries (which is the case of the present tests). The previous difficulties in achieving stable pre-peak growth of a long kink band in experiments must have been due to unsuitable specimen and loading geometries (in post-peak, a stable growth further requires a sufficiently stiff loading).

5.1.5. *Other aspects*
Formula (12), which was derived for a long rectangular strip, is a special case of the general formula (35) for the case $D_1 = D_0$, i.e., for $\gamma(\alpha) = g(\alpha)$. The case of such an equivalence of the energy release functions means that the corresponding stress intensity factors are equal. This is immediately evident from the decomposition in Figure 10 based on the principle of superposition; indeed, the middle specimen is in a homogeneous stress state and thus has no stress singularity ($K_I = 0$).

Formula (34) is of the same form as the formula derived in general for compression failures under the assumption that the laterally propagating damage band causing compression failure applies a uniform pressure on its boundaries (Bažant and Chen 1997, Eq. 51).

In a similar manner as shown in Bažant (1997) for cracked specimens, the foregoing analysis could be refined by considering that the kink band ought to exhibit $R$-curve behavior. In other words, the critical energy release may be expected to follow an $R$-curve, such that $\mathcal{G}_b = R_b(c)$ where $R_b(c)$ is assumed to be a given function of the kink band extension $c$. The only change needed in the preceding formulation is to replace the value $\sqrt{\bar{E}G_b}$ in (31) by $\sqrt{\bar{E}R_b(c)}$. In addition to this, it becomes possible to implement in the calculations the condition of failure as a stability limit, which can be reduced to the condition of tangency of the curve of energy release rate at constant load to the $R$-curve (Bažant 1997) (this condition cannot be imposed when the critical energy release rate is constant, being equal to $G_b$, as assumed in the preceding).

Similar to (14), Equation (35) is accurate only up to the first two terms of the power series expansion in $(1/D)$. Therefore, it could be generalized in the same manner as (16).

## 5.2. Case of kink band initiation from smooth surface

For kink bands originating from a smooth surface, one may usually assume that $P_{\max}$ occurs at the beginning of propagation, while the FPZ is still attached to the surface. Accordingly, the initial crack length for the LEFM approximation $a_0 = 0$ or $\alpha_0 = 0$. But the energy release rate of a crack of zero length vanishes, $g(\alpha_0) = \gamma(\alpha_0) = 0$. Thus, if we truncated the Taylor series expansion of $g(\alpha_0 + \theta)$ and $\gamma(\alpha_0 + \theta)$ after the linear term, we could not capture the size effect. Therefore we need to truncate these series only after the quadratic term. In this manner, the large-size approximation of (31) takes the form

$$\sigma_N = \frac{\sqrt{\bar{E}G_b} + \sigma_r\sqrt{\gamma' c_b + \gamma''(c_b^2/2D)}}{\sqrt{g' c_b + g''(c_b^2/2D)}} \quad \text{(FPZ at surface).} \tag{51}$$

This may be rearranged as

$$\sigma_N = \frac{\sqrt{\bar{E}G_b/c_b} + \sigma_r\sqrt{\gamma'(1+2\chi_1)}}{\sqrt{g'(1-2\chi_0)}}, \quad \text{with } \chi_0 = -\frac{g'' c_b}{4g' D}, \quad \chi_1 = \frac{\gamma'' c_b}{4\gamma' D}, \tag{52}$$

with the notations $g' = \mathrm{d}g(\alpha)/\mathrm{d}\alpha$, $\gamma' = \mathrm{d}\gamma(\alpha)/\mathrm{d}\alpha$, $g'' = \mathrm{d}^2 g(\alpha)/\mathrm{d}\alpha^2$, $\gamma'' = \mathrm{d}^2\gamma(\alpha)/\mathrm{d}\alpha^2$, all evaluated at $\alpha = \alpha_0 = 0$; $\chi_0$ is defined with a minus sign because $g''$ is negative in bending specimens (Bažant and Li 1996).

Equation (52), however, is not acceptable for small $D \to 0$ because it gives $\sigma_N \to 0$. Therefore, it is expedient to transform it by asymptotic approximations that do not change the asymptotic behavior for $D \to \infty$ while making at the same time the formula acceptable for $D \to 0$. Such approximations are $\sqrt{1+2\chi_1} \approx (1+q\chi_1)^{1/q}$ and $1/\sqrt{1-2\chi_0} \approx (1+r\chi_0)^{1/r}$, with any positive constants $r$ and $q$. For the case $r = q = 1$, after noting that $(1+\chi_1)(1+\chi_0) \approx 1 + \chi_0 + \chi_1$ for large $D$, we obtain a size effect formula identical to (22) (Figure 8(c)), in which

$$\sigma_\infty = \sqrt{\frac{\bar{E}G_b}{g'c_b}} + \sigma_r\sqrt{\frac{\gamma'}{g'}}, \tag{53}$$

$$D_b = -\frac{c_b}{4\sigma_\infty}\left[\frac{g''}{g'}\sqrt{\frac{\bar{E}G_b}{g'c_b}} + \sigma_r\left(\frac{g''}{g'} - \frac{\gamma''}{\gamma'}\right)\sqrt{\frac{\gamma'}{g'}}\right]. \tag{54}$$

A generalization in the form of (23) is also possible.

With (53), we have thus obtained a more general confirmation of (22). Note that although the approximations that led from (52) to (53) are only first-order accurate in $1/D$, (51) is first-order accurate in $1/D$ as well. So, these approximations cause no loss in accuracy overall. But (53) can be applied more widely because it has a realistic behavior also for $D \to 0$ while (52) does not.

### 5.3. Transition between long and short kink bands and universal size effect law

For small $\alpha_0$, there must be a continuous transition between the size effect law for specimens with a notch or long kink band and that for specimens failing at kink band initiation from a smooth surface. A universal size effect law that describes this transition has already been developed for fracture of concrete ($\sigma_r = 0$; Bažant 1997). By a similar matching the asymptotic behaviors for long and short kink bands, a similar universal size effect law can be developed for kink bands.

## 6. Identification of fracture parameters from kink band tests

### 6.1. FPZ at notch tip

It is convenient to cut notches in the compression test specimens. The notch ensures the kink band to start in one place (which eliminates any possible statistical effect due to random spatial variation of local strength). Furthermore, if a suitable loading is used, it further ensures the FPZ at maximum load to be still attached to the notch tip. This brings about an important simplification—the value of $\alpha_0$ can be controlled and thus is known.

The size effect formula for notched specimens, (14) with (43), can be used for identifying the fracture energy of the kink band, $G_b$, and the FPZ length, $c_b$, from the measurements of size effect. The procedure may be based on determining $D_0$ and $\sigma_0$ by fitting the formula (42) to maximum load data for notched specimens of sufficiently different sizes; then

$$c_b = D_0\frac{g_0}{g'}, \qquad G_B = \frac{c_b}{\bar{E}}g'\sigma_0^2. \tag{55}$$

If geometrically similar specimens and notches are used, $\alpha_0$ = constant for all $D$. Geometrical similarity is not a requirement because the equivalent LEFM functions $g$ and $\gamma$ capture the shape effect as well. Geometrical similarity, however, simplifies the evaluation of test results and improves the accuracy because the shape effect is known only approximately and thus introduces some additional error.

The fitting of (41) or (34) to $\sigma_N$ data may be best accomplished the Levenberg–Marquardt nonlinear optimization algorithm. Alternatively, by algebraic rearrangement, equation (42) may be transformed to a linear regression plot $Y = AD + C$ for which

$$Y = \frac{1}{\sigma^2}, \qquad G_B = \frac{g_0}{A\bar{E}}, \qquad c_b = \frac{C}{A}\frac{g_0}{g'}. \tag{56}$$

The procedure is identical to that embodied in the RILEM Recommendation for concrete (whose statistical treatment and proper weighting is described in Bažant and Planas, 1998, Sec. 6.3).

According to (43), the kink band fracture energy may then be calculated as

$$G_b = (\sqrt{\bar{E}G_B} - \sigma_r\sqrt{\gamma' c_b}\,)\,/\,\bar{E}, \tag{57}$$

provided that $\sigma_r$ is determined separately.

An approximate way to estimate $\sigma_r$ may be to measure the residual load on very large specimens, provided that a post-peak plateau can be reached. Additional damage long after the peak load, however, may further reduce the stress in the kink band, making it hard to identify a plateau. An upper bound on $\sigma_r$ is provided by the compression strength of very small specimens with nearly uniform stress distribution at peak load.

Another way to determine $\sigma_r$ may be to test both notched and unnotched specimens, and use the Marquardt–Levenberg optimization algorithm to fit the maximum load data by (41) and (51), considering $G_b$, $c_b$ and $\sigma_r$ as three unknowns.

Still another way to determine $\sigma_r$ might be to measure under uniaxial compression the post-peak behavior of very small unnotched prismatic specimens – so small that the FPZ of the kink band would occupy the whole cross section, thus ensuring simultaneous (nonpropagating) failure.

It is instructive to relate $G_B$ to the $J$-integral. Consider the $J$-integral path starting at point $C$ in Figure 11 (top) of the imagined cohesive crack representing the kink band. Point $C$ lies immediately ahead of the tip of the notch, and so the $J$ integral along the path starting at $C$ represents the total energy flux into the cohesive crack equivalent to the kink band. When the extension of this crack (or kink band) is sufficient to reduce the stress at point $C$ to the residual stress $\sigma_r$, the full fracture energy $G_b$ of the kink band comes into play, and then, by a similar derivation as that which led to (10), the $J$-integral along the path starting at $C$, which must be equal to $G_B$, is $J_{cr} = G_B$, $G_B = G_b + \sigma_r\delta_r$ (Equation 10) where $\delta_r$ is the normal displacement (contraction) across the kink band at the notch tip (or the end of the FPZ), and $G_b$ is the fracture energy representing the cross-hatched triangular area in Figure 7 (bottom). Knowing $\sigma_r$, this relation can be used to estimate the value of $\delta_r$.

The value of $\delta_r$ ought to be equal to the opening displacement of the equivalent LEFM crack at the location of the notch tip (Figure 11). This displacement may be calculated from the parabolic asymptotic profile of the equivalent LEFM crack, whose tip lies at distance $c_B$ from the notch tip. According to the well-known formula (e.g. Eq. 5.5.11 in Bažant and Planas 1998), one may use the approximation

$$\delta_r = \sqrt{32 G_B c_b / \pi \bar{E}}. \tag{58}$$

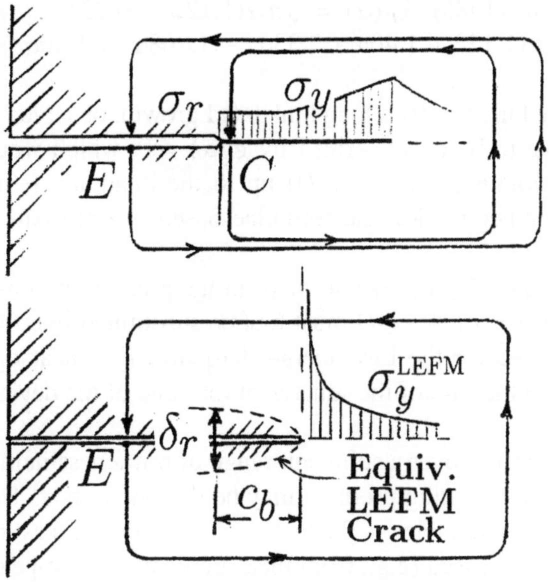


*Figure 11.* Equivalent cohesive crack emanating from the notch tip (top), equivalent LEFM crack (bottom), and $J$-integral paths.

### 6.2. FPZ REMOTE FROM NOTCH AND SPECIMENS WITH END RESTRAINT

While the rotational restraint at the ends of the present carbon-PEEK specimens has made it possible to demonstrate a long stable growth of kink band, it has slightly complicated the analysis of test results. The restraint causes the axial load resultant $P$ to gradually shift to the right in Figure 1(a) as the kink band propagates, i.e., the resultant eccentricity $e$ (Figure 1(a) and 13) increases as a function of $\alpha$. Function $e = e(\alpha)$ is determined by the condition that the relative rotation between the specimen ends $\phi = PC_{PM} + MC_{MM} + \sigma_r C_{Mr} = 0$ where $M = Pe$ = moment of the resultant. Consequently,

$$e(\alpha) = \frac{-[PC_{PM}(\alpha) + \sigma_r C_{Mr}(\alpha)]}{PC_{MM}(\alpha)} \quad (P = \sigma_N bD). \tag{59}$$

The compliances, according to well-known LEFM relations (e.g. Bažant and Planas 1998, Eq. 3.5.18), are

$$C_{PM}(\alpha) = \frac{2}{b\bar{E}} \int_0^\alpha k_P(\alpha')k_M(\alpha')\,\mathrm{d}\alpha', \qquad C_{MM}(\alpha) = \frac{12\,L}{bD^3\bar{E}} + \frac{2}{b\bar{E}} \int_0^\alpha [k_M(\alpha')]^2\,\mathrm{d}\alpha',$$

$$C_{Mr}(\alpha) = \frac{2}{b\bar{E}} \int_0^\alpha k_M(\alpha')k_r(\alpha')\,\mathrm{d}\alpha', \tag{60}$$

where $b$, $L$ = specimen thickness and length (Figure 1(a)), and the $k$'s are the dimensionless stress intensity factors for unit loading by $P$, $M$ or $\sigma_r$; $k_P = K_{\mathrm{I}}^P b\sqrt{D}/P$, $k_M = K_{\mathrm{I}}^M bD^{3/2}/M$ and $k_r = K_{\mathrm{I}}^r/\sigma_r\sqrt{D}$. Also,

$$g(\alpha) = \sigma_N^2 D\{k_P(\alpha) + k_M(\alpha)e(\alpha)/D\}^2. \tag{61}$$

According to Tada et al. (1985), $k_P(\alpha) = \sqrt{\pi\alpha}(1.122 - 0.231\alpha + 10.55\alpha^2 - 21.71\alpha^3 + 30.38\alpha^4)$, $k_M(\alpha) = \sqrt{\pi\alpha}(1.122 - 1.40\alpha + 7.33\alpha^2 - 13.08\alpha^3 + 14.0\alpha^4)$. For $k_r$ see Appendix IV.

If $\sigma_r$ were 0, then the failure after a long kink band growth, observed in the present carbon-PEEK tests, would have to be characterized by $g'(\alpha) = 0$, which would require the use of formulae based approximating $g(\alpha + c_b/D)$ up to the quadratic term in $c_b/D$. However, because $\sigma_r$ is not 0 (nor negligible), the formulae based on approximation up to the linear term are appropriate.

To identify the values of $G_b$, $c_b$, $\alpha_0$ and $\sigma_r$ from the present carbon-PEEK tests, equation (59) and the expression $\alpha = \alpha_0 + c_b/D$ must first be substituted into (42) and (50). Then the standard library subroutine for the Levenberg–Marquardt nonlinear optimization algorithm may be used to minimize the sum of the squares of (50) and of the deviations of (42) from the data points.

To succeed, the tests must cover a sufficient range of brittleness number $\beta = D/D_0$. If the range of $\beta$ is too limited, the fitting problem may be ill-conditioned. Then one needs to drop $\sigma_r$ from the set of unknown variables in the least-square fitting, and estimate $\sigma_r$ in advance by other means, as already discussed (e.g., from measurements of post-peak behavior), or use in data fitting various estimated fixed values of $\sigma_r$ and then compare the results.

In the present study, however, the values of $e$ and $\alpha_0$ at $P_{\max}$ have not been calculated by the aforementioned procedure. Rather, for the sake of simplicity, the values that were already available from a finite element simulation with the cohesive crack model carried out earlier by G. Zi have been used in fitting (42) to the data points in Figure 12 (top left). This furnished the following material parameters of the carbon-PEEK composite tested

$$G_b = 27.9\,\text{kN/m}, \qquad c_b = 0.70\,\text{mm}, \qquad \sigma_r/\sigma_0 = 0.436, \qquad \sigma_0 = 758\,\text{MPa}. \tag{62}$$

Figure 12 on top right shows also a comparison of the data with the size effect plot based on the material parameters obtained with the cohesive crack model. The plot is very close to that in Figure 12 on top left.

It is instructive to compare these values to those obtained by simple fitting of the data with the size effect law (14) shown in Figure 13. They are $\sigma_0 = 779$ MPa and $c_b = 18.5$ mm. Although $\sigma_0$ is about the same, $c_b$ is far too large, by an order of magnitude. This demonstrates that the differences in $e/D$ and $\alpha_0$ among specimens of different sizes cannot be neglected.

### 6.3. Test simulation by cohesive crack model

Simulations of the present and other tests with the cohesive crack model will be reported separately (Bažant and Zi, 1999). Two-dimensional elastic finite element analysis has been applied to obtain the compliance matrices for forces and displacements at the specimen ends and at the nodal points placed on the sides of the kink band. The condition that the kink band contraction indicated by the softening stress-displacement law must be compatible with the elastic deformation of the specimen leads to one integral equation, and the condition that the stress intensity factor at the kink band front must vanish leads to another. Based on the nodal point discretization, both integral equations are approximated by a system of algebraic equations, which is then solved by incremental loading.

A computer program for calculating the values of $P_{\max}$, $e(\alpha)$ and $\alpha_0$ from any given values of $G_b$, $c_b$ and $\sigma_r$ has been written. The values of $G_b$, $c_b$ and $\sigma_r$ have then been optimized so

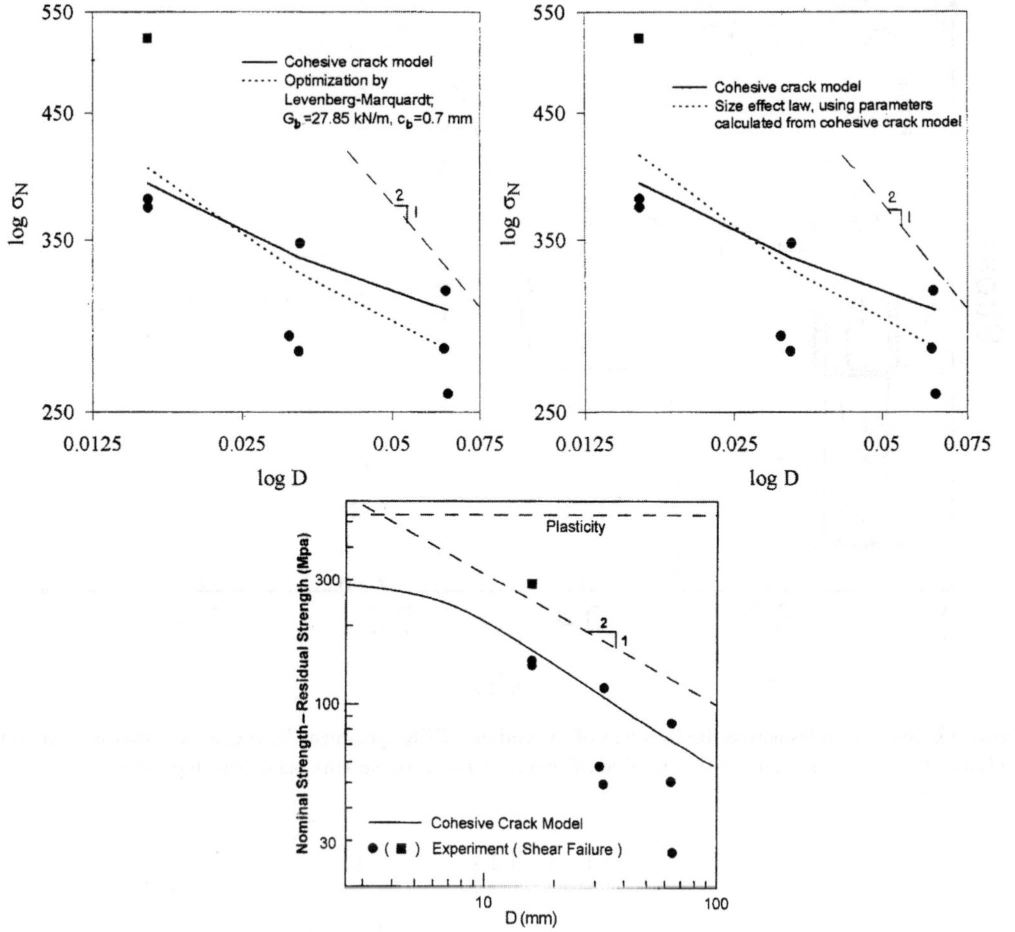


*Figure 12.* Optimum fits by size effect formulae and by numerical analysis with the cohesive crack model.

as to obtain the best possible fit of the test data. In the process, eccentricities $e$ have also been determined.

The calculated subsequent stress profiles throughout the ligament, along with the corresponding eccentricities $e$ of the load resultant, are shown in Figure 14. The optimum fit of the test data by the cohesive crack model, with $\log(\sigma_N - \sigma_r)$ as the ordinate, is shown by the curve in Figure 12 (bottom). The corresponding optimum value of fracture energy is $G_b = 28.2\,\text{kN/m}$, which is almost the same as in (62), but the length of the softening segment (Figure 14), which is about 2.7 mm for all $D$, is almost twice as large as $2c_b$. This suggests that the tip of the equivalent LEFM crack might not be at the center of the FPZ but farther behind its front. The optimum fit, however, is not very sensitive to $c_b$; this is documented by Figure 12 (top left and right), with $\log \sigma_N$ as the ordinate, which shows the optimum fit by the cohesive crack model (solid curve) and the results from (42) based on the optimum material parameters obtained with the cohesive crack model. Note that the dashed curves in Figure 12 on top left and right, corresponding to rather different $c_b$, are not very different. To eliminate this ambiguity, tests of a broader size range or fitting of post-peak response would be needed.

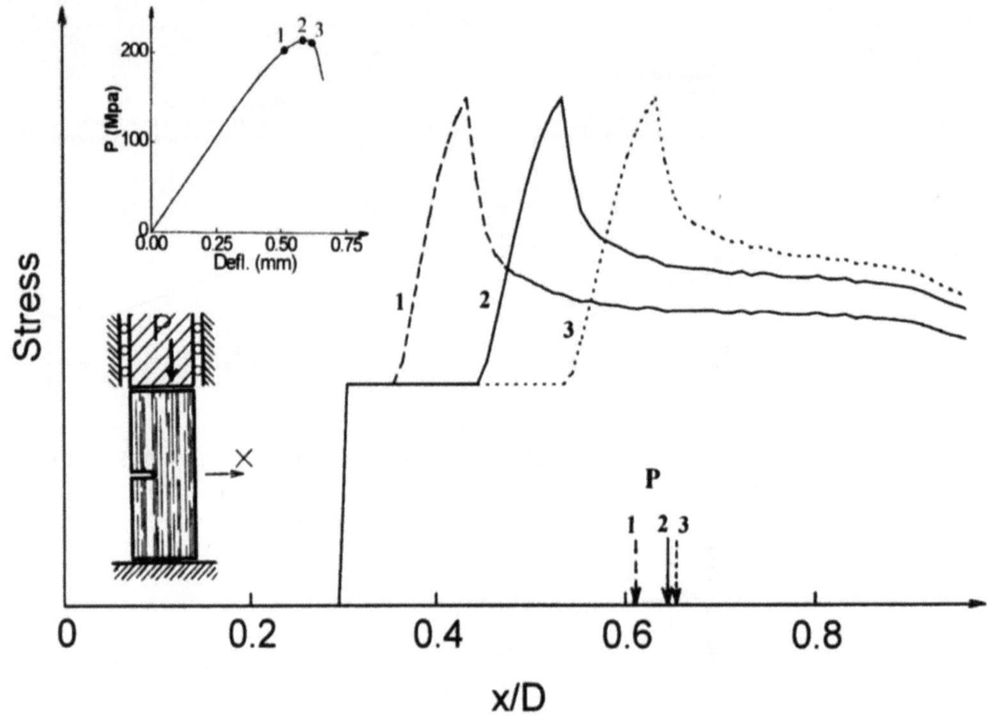


*Figure 13.* Stress profiles across the ligament of the carbon-PEEK specimens before, at and after the maximum load (note the shift of the compression resultant $P$ which makes a stable kink band growth possible).

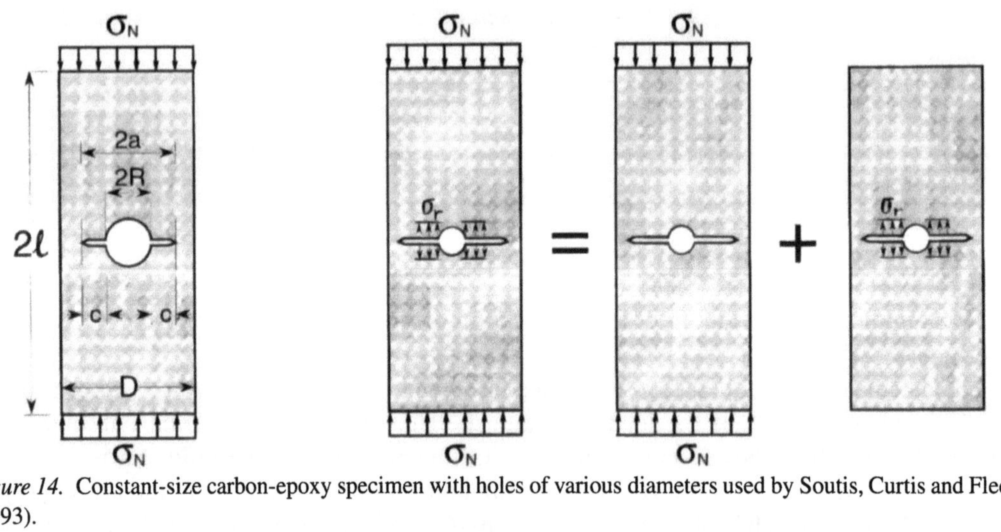


*Figure 14.* Constant-size carbon-epoxy specimen with holes of various diameters used by Soutis, Curtis and Fleck (1993).

## 7. Soutis et al.'s tests of carbon-epoxy laminates with holes of various sizes

The studies of kink band failures of composites have so far not been focused on the size effect. No data for geometrically similar specimens seem to be available in the literature. However, since functions $g(\alpha)$ and $\gamma(\alpha)$ characterize the shape effect, test data on nonsimilar specimens exhibiting the effect of shape can also be used to check and calibrate the theory provided that the range of the brittleness number $\beta = D/D_0$ in (12) is sufficient. The recent test data of

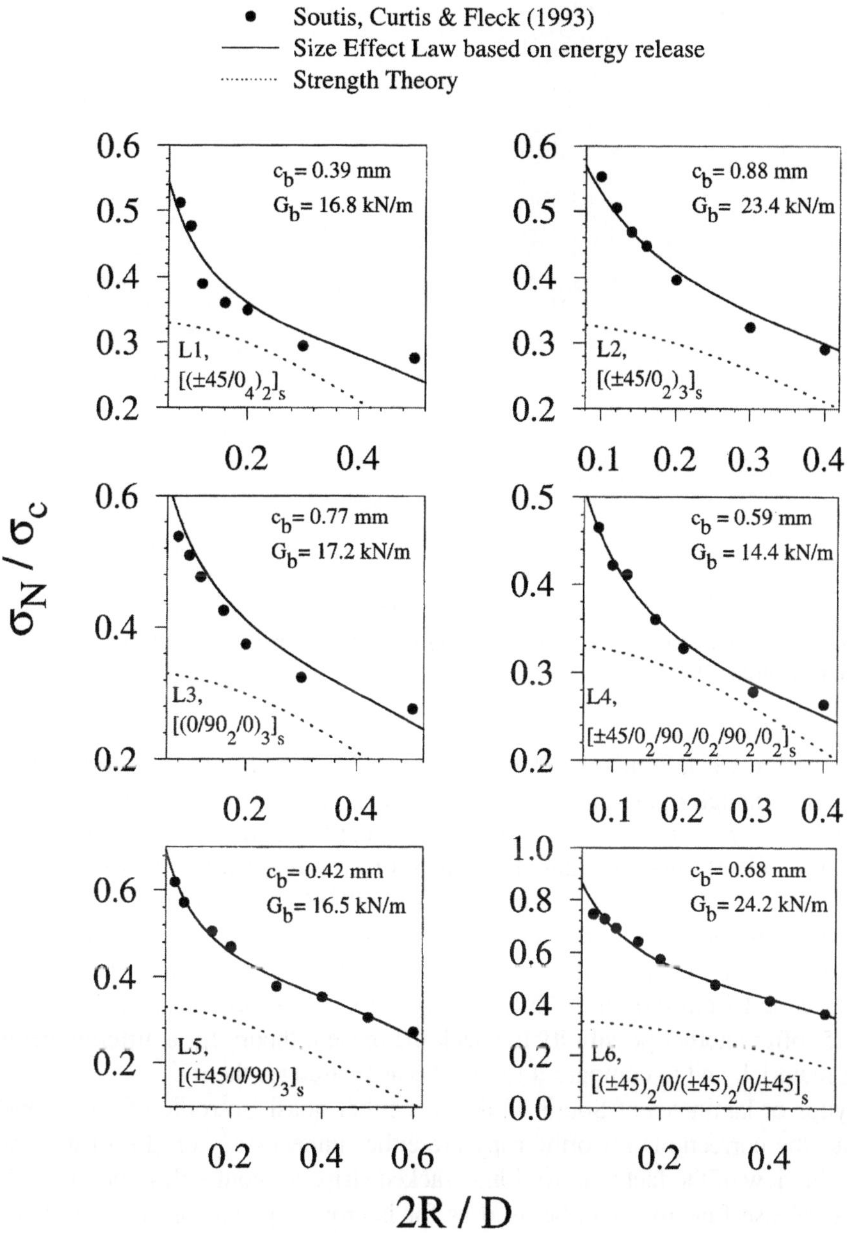


*Figure 15.* Soutis, Curtis and Fleck's (1993) test results for quasi-isotropic and orthotropic carbon–epoxy laminates of six different layups, with holes of various radii $R$ (data points) and constant width $D$. Solid curves: optimum fits by size-shape effect law in (31). Dashed curves: predictions of strength theory exhibiting no size effect.

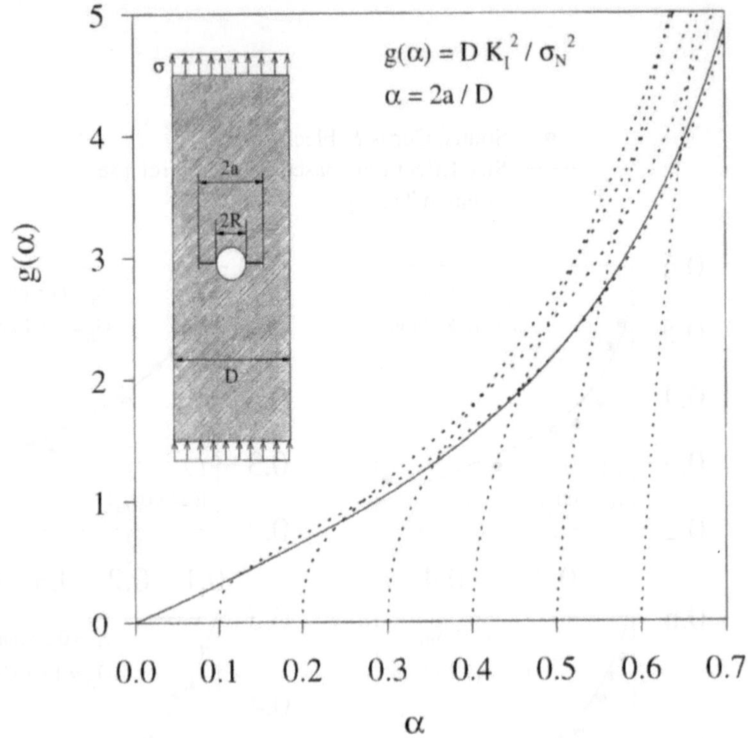


*Figure 16.* Dimensionless energy release function $g(\alpha)$ for isotropic specimens with centric holes of different radii (derived by Führing, 1973).

Soutis and Fleck (1991), Soutis, Fleck and Smith (1991), and Soutis, Curtis and Fleck (1993) exhibit a broad enough range of brittleness numbers, attained not through a variation of the overall specimen size $D$ (which was kept constant) but through a variation of $D_0$.

Soutis et al. (1991) used rectangular panels of constant width $D$ to study how centric holes of various radii $R$ affect the compressive strength of unidirectional and multidirectional carbon/epoxy (T800/924C) laminates of different layups, labeled as L1 to L6 (Figure 14 left). The specimens were 50 mm wide, 245 mm long and 3 mm thick, and the diameter–width ratio $2R/D$ varied from 0.08 to 0.60 for each layup. The results, shown by the data points in Figure 15, offer another possibility to check the present theory for a different geometry and a different material, and to compare it to the classical strength theory.

The layups of laminates of Soutis et al. were either quasi-isotropic of only weakly orthotropic. Thus the corrections for orthotropy are either unnecessary or almost unimportant for these tests. In view of the fact that, for long cracked strips without holes, the same dimensionless energy release functions can be used as for isotropic specimens (Bao et al. 1992), it is assumed that the same can be done for specimens with holes, at least as an approximation.

Therefore, the effect of the mild orthotropy of some of the Soutis et al.'s laminate layups is taken into account only through $\bar{E}$. With this assumption, the analysis of these data is made easy by the availability of the solution of the stress intensity factor $K_I$ (Führing 1973, and Murakami's 1987 handbook, p. 291) for an in finite strip with a hole, which may be used as an approximation for a rectangular specimen with a hole (Figure 14). This solution, used here for compressive instead of tensile loading, is shown in Figure 16 where $\alpha = (c + R)/D$. As seen, the curve of $K_I$ as a function of $\alpha$ first quickly rises as the horizontal cracks grow from the sides of the hole (Figure 14 left and Figure 16 insert). But soon the slope diminishes as this curve approximately joins the solid curve in Figure 16, which corresponds to $K_I(\alpha)$ for a specimen containing a centric horizontal crack (of relative depth $\alpha$) instead of a hole.

By virtue of similarity transformations, the slope of the initial rise of the curve must be the same as for a crack starting from the surface of an elastic halfspace. For such a crack, $K_I = 1.12\sigma\sqrt{\pi c}$, which implies that $g(\alpha) = K_I^2/\sigma^2 D = 1.12^2\pi c/D$ and $g'(\alpha) = 1.12^2\pi < \infty$. The formula plotted in Figure 16, however, gives an infinite (vertical) initial slope, which means that its initial portion cannot be accurate. Because of this asymptotic inaccuracy, $g'$ (at $c = 0$) in (51) for cracks initiating from a smooth surface is obtained, incorrectly, as infinite, which precludes using this equation and forces us to use the original equation (31) from which (51) was derived. Nevertheless we assume that for the present finite values of $c$ the formula plotted in Figure 16 is accurate enough.

The maximum load may be expected to be reached only after the slope of the curve in Figure 16 diminishes and the solid curve is approached, which means that the specimen behaves roughly as a panel with a large centric crack of length $2(R + c)$. This is another reason why the general original formula (31), rather than formula (51) derived from it for a kink band initiating from a smooth surface, appears to be appropriate.

Let us now assume the FPZ of the kink band at maximum load is still attached to the hole. In that case, the equivalent LEFM crack should have its tip roughly at the distance $c = c_b$ from the hole. This value of $c$ is large enough for approaching the solid curve in Figure 16, i.e., $a$ lies in Figure 16 beyond the sharp decrease of slope of the curves. But then the truncation of the series expansion after the second term that was made in deriving (12) would not be accurate enough. Therefore, the original formula (31) is used.

In (31), we now have $\alpha = 2(R + c)/D \approx 2(R + c_b)/D$. Functions $g(\alpha)$ and $\gamma(\alpha)$ are the dimensionless energy release functions corresponding to the case of a tensioned cracked plate with a centric hole and a pair of cracks, loaded either at the specimen ends or along a portion of each crack (Figure 14 right); see the Appendix. The value of the residual stress $\sigma_r$ needs to be considered as 0 because the FPZ of the kink band is still attached to the hole when the maximum load is reached. Considering $G_b$ and $c_b$ as unknowns, Equation (31) has been fit to the data of Soutis et al. (1993) using the Levenberg-Marquardt nonlinear optimization algorithm. The results, shown by the solid curves in Figure 15, are quite satisfactory. The optimum values of $G_b$ and $c_b$ corresponding to these fits are indicated in each plot in Figure 15 (and $\sigma_r = 0$).

To check whether a theory capable of representing the size effect is necessary, the test results are further compared to the simplest version of a theory that exhibits no size effect – the strength theory in which it is assumed that the specimen fails as soon as the vertical normal stress at the sides of the hole reaches the strength of the material. The dashed curves in Figure 15 show the predictions of the strength theory (obtained under the assumption that the material strength is the same as used in the kink band analysis). Obviously the trend of those curves does not agree with the trend of the data. These comparisons provide additional support for the present theory. They reveal that a theory exhibiting the size effect is necessary to describe Soutis et al.'s test results well.

The reason that the last data points on the right of the plots for specimens L1, L3 and L4 in Figure 15 systematically deviate slightly upwards from the solid curves for the present theory could be that (31) is not acceptable for large holes (because the cracks at maximum load are not long enough for approaching the solid curve in Figure 16).

## 8. Conclusions

(1) Transverse slanting of the notch in test specimens of carbon-PEEK laminate can achieve pure out-of-plane kink band failure, not contaminated by shear splitting cracks.

(2) Restraining specimen ends against rotation helps to stabilize kink band growth and makes it possible to demonstrate the possibility of a stable growth of long kink bands before the peak load.

(3) Compression tests of notched carbon-PEEK specimens show that the nominal strength of geometrically similar notched specimens failing purely by kink band propagation exhibits a strong (non-statistical) size effect.

(4) The size effect observed is transitional between the asymptotic case of no size effect, which is characteristic of plasticity or any theory whose failure criterion is expressed solely in terms of stress or strain, and the asymptotic case of size effect of linear elastic fracture mechanics, which is governed by energy release.

(5) The results of the present carbon-PEEK tests roughly agree with the approximate general size effect law proposed by Bažant (1983, 1984) and derived recently by asymptotic analysis of energy release (Bažant 1997).

(6) The present theory gives not only the effect of size but also the effect of shape. The theory is found to agree with the recent shape effect tests of Soutis et al. (1993), which utilized constant-size carbon epoxy specimens with centric holes of different diameters.

(7) The nominal strength of specimens failing at the initiation of a kink band from a smooth surface is also predicted to exhibit a size effect. A simple size effect formula is derived for this case, too. This formula is the same as that previously derived for the size effect on the modulus of rupture measured in bending tests of notch-free specimens.

(8) In addition to previous derivations, the size effect law for specimens with notch of long kink band can also be derived by J-integral analysis of energy release, as well as from the recently proposed nonlocal LEFM.

(9) The size effect law for notched specimens permits the fracture energy of the kink band and the length of the fracture process zone at the front of the band to be easily identified solely from the measurements of maximum loads.

(10) The results suggest that the current design practice, in which the compression failure is predicted on the basis of strength criteria (or plasticity), thus inevitably missing the size effect, is acceptable only for small specimens or structural parts. In the interest of safety, it should be revised for large structural parts.

*Remark on statistical size effect.* According to the arguments in Bažant and Planas (1991, ch. 12), this size effect, not addressed here, is overpowered by the present energetic size effect. It is significant only for

(1) very small notched specimens for which the FPZ involves most of the cross section;
(2) unnotched specimens so large that the energetic size effect approaches its horizontal asymptote (Figure 8(c)), and
(3) uniformly stressed specimens with negligible FPZ reaching peak load at fracture inititiation (probably relevant to Kyriakides and Ruff, 1997).

## Appendices

### I. $K_{\mathrm{I}}$ FOR ISOTROPIC PANEL WITH HOLE AND CRACKS

The stress intensity factor for an isotropic panel with a circular hole and two cracks (Figure 14 and 16) may be written as $K_{\mathrm{I}} = \sigma\sqrt{D}k(\alpha)$ (e.g. Bažant and Planas, 1998, Sec. 5.1) where $k(\alpha)$ = function of $\alpha$. Then $g(\alpha) = k^2(\alpha)$. The solution of $K_{\mathrm{I}}$ for a cracked rectangular panel with a centric hole (Figure 14 and 16 insert) was obtained by Führing (1973) (see also Murakami, 1987, p. 291). With the notations $\alpha = a/D$, $\bar{\alpha} = \pi\alpha$, $\delta = b/R$, $\gamma = 2R/D$, $\beta = (2\alpha - \gamma)(1 - \gamma)$, one gets $g(\alpha) = 2\pi\alpha F^2(\alpha)$ where $F(\alpha) = \phi\psi$ and

$$\phi = \frac{1}{\pi - 1}\left\{\pi\sqrt{\frac{1}{\bar{\alpha}}\tan\bar{\alpha} + g\sin(2\bar{\alpha})}\left(1 + \frac{\epsilon^2(2 - \epsilon^2)}{1 - \varepsilon}\right) + \sqrt{1 + 2g}\right\}, \tag{63}$$

$$\psi = \xi\left(3\beta^{2P/3} - 2\beta^P\sqrt{\xi}\right). \tag{64}$$

Here $g = 0.13\left[(2/\pi)\arctan\delta\right]^2$, $\epsilon = (\alpha/\pi)\arctan(0.6\sqrt[3]{\delta})$, $P = \log(\xi^{-3/2})/\log(\beta^*)$, $\beta^* = \gamma\delta/[\gamma(2\delta - 1) + 1]$ and $\xi = 1 + (2/\pi)\arctan(1.5\sqrt{\delta})$.

### II. $K_{\mathrm{I}}$ FOR ISOTROPIC PANEL WITH A HOLE LOADED ON A PORTION OF CRACK LENGTH

The stress intensity factor for an isotropic panel in which the crack lips are subjected to a distributed load of resultant $P$ in each crack was calculated by Newman (1982). With the notations $R$ = radius of the hole, $a$ = crack length, $\alpha = a/D$, $\lambda = R/a$, $\beta_1 = b_1/a$, $\beta_2 = b_2/a$, $b_1$ or $b_2$= distance between the initial or terminal points of the loaded crack segment (Figure 14 right), respectively,

$$\gamma(\alpha) = (8/\pi)\alpha(\arcsin\beta_2 - \arcsin\beta_1)F_3F_4, \tag{65}$$

where

$$F_3 = \frac{G(\beta,\lambda)}{\arcsin\beta_2 - \arcsin\beta_1}, \qquad F_4 = \frac{\arcsin B_2 - \arcsin B_1}{\arcsin\beta_2 - \arcsin\beta_1}\sqrt{\sec\pi\alpha}, \tag{66}$$

$$\begin{aligned} G(\beta,\lambda) = {} & \left[\left(1 + \frac{A_1}{1-\lambda} + \frac{3A_2}{2(1-\lambda)^2}\right)\arcsin\beta \right. \\ & \left. + \left(\frac{A_1}{1-\lambda} + \frac{(4-\beta)A_2}{2(1-\lambda)^2}\right)\sqrt{1-\beta^2}\right]_{\beta_1}^{\beta_2}, \end{aligned} \tag{67}$$

with $A_1 = -0.02\lambda^2 + 0.558\lambda^4$ and $A_2 = 0.221\lambda^2 + 0.046\lambda^4$, and $B_1 = \sin(\pi b_1/D)/\sin\pi\alpha$, $B_2 = \sin(\pi b_2/D)/\sin\pi\alpha$.

### III. PROOF OF EQUATION (40)

Equation (40) is proven by noting that the stress intensity factor $K_{\mathrm{I}}$ caused by $\sigma_r$ when $c = c_b \ll a_0$ is the same as for a crack of length $2a$ in an infinite body, loaded on the crack faces by $\sigma_r$ over distance $c_b \ll a$ from one tip. By superposition of a well known formula (e.g. Tada et al., 1985) used to derive Dugdale model

$$K_{\mathrm{I}} = \int_{a-c}^{a} k_P(x)(\sigma_r \mathrm{d}x) \approx \sigma_r \sqrt{\frac{8c}{\pi}}, \qquad k_P(x) = \frac{1}{\sqrt{\pi a}} \sqrt{\frac{a+x}{a-x}} \approx \sqrt{\frac{2}{\pi(a-x)}}, \tag{68}$$

where $k_P(x)$ = Green's function for the effect on $K_{\mathrm{I}}$ of a concentrated crack-face load pair at distance $x$ from the crack center. Then $\gamma = K_{\mathrm{I}}^2/\sigma_r^2 D \approx (8/\pi)c/D$.

## IV. Stress intensity factor due to $\sigma_r$

By superposition of the solution for crack-face concentrated load pair in infinite edge-cracked strip (Tada et al., 1985)

$$k_r(\alpha) = \int_{\alpha_{\text{notch}}}^{\alpha} (1-\alpha)^{-3/2}(1-t^2/\alpha^2)^{-1/2} \sum_{n=1}^{4} g_n(\alpha)(t/\alpha)^{n-1}\,\mathrm{d}t, \tag{69}$$

$$\begin{aligned}
g_1(\alpha) &= 0.46 + 3.06\alpha + 0.84(1-\alpha)^5 + 0.66\alpha^2(1-\alpha)^2,\\
g_2(\alpha) &= -3.52\alpha^2,\\
g_3(\alpha) &= 6.17 - 28.22\alpha + 34.54\alpha^2 - 14.39\alpha^3 - (1-\alpha)^{3/2}\\
&\quad -5.88(1-\alpha)^5 - 2.64\alpha^2(1-\alpha)^2,\\
g_4(\alpha) &= -6.63 + 25.16\alpha - 31.04\alpha^2 + 14.41\alpha^3 + 2(1-\alpha)^{3/2}\\
&\quad +5.04(1-\alpha)^5 + 1.98\alpha^2(1-\alpha)^2.
\end{aligned}$$

## V. Comment on application of nonlocal LEFM

All the size effect equations presented here, along with their large-size and small-size asymptotic forms, can alternatively be derived from nonlocal fracture mechanics – a new general model (Bažant 1998b) proposed as a simpler alternative to the cohesive crack model and equivalent to it asymptotically for large sizes. This model is based on the smeared-tip superposition method (Bažant and Planas 1998, Sec. 7.5.6) in which the stress tensor $\boldsymbol{\sigma}(\boldsymbol{x}, a)$ and displacement vector $\boldsymbol{v}(\boldsymbol{x}, a)$ at any point of coordinate vector $\boldsymbol{x}$ in a body with the FPZ centered at crack length $a$ (kink band length) is represented as a superposition of LEFM solutions for crack (kink band) tips within the FPZ corresponding to loads $Pq(\zeta)\mathrm{d}\zeta$, i.e.

$$\begin{aligned}
\boldsymbol{\sigma}(\boldsymbol{x}, a) &= \sigma_N \int_{-1}^{1} \boldsymbol{S}(\boldsymbol{\xi}, \alpha + \zeta c/D)\, q(\zeta)\,\mathrm{d}\zeta,\\
\boldsymbol{v}(\boldsymbol{x}, a) &= \sigma_N \frac{D}{\bar{E}} \int_{-1}^{1} \boldsymbol{V}(\boldsymbol{\xi}, \alpha + \zeta c/D)\, q(\zeta)\,\mathrm{d}\zeta.
\end{aligned} \tag{70}$$

Here $\alpha D = a$ = length of crack equivalent to kink band at FPZ center, $\alpha D + \zeta c$ = crack (kink band) lengths at various points along the FPZ; the crack (kink band) lengths in FPZ occupy the interval $(a - c, a + c)$ corresponding to $\zeta \in (-1, 1)$; $q(\zeta)$ = dimensionless load distribution function such that $\int_{-1}^{1} q(\zeta)\,\mathrm{d}\zeta = 1$. The dimensionless functions $\boldsymbol{S}(\boldsymbol{\xi}, \alpha')$ and $\boldsymbol{V}(\boldsymbol{x}, \alpha')$ represent the stress tensor and displacement vector at relative coordinate vector $\boldsymbol{\xi} = \boldsymbol{x}/D$ in a body with crack (kink band) length $a' = \alpha' D$, calculated by LEFM for $\sigma_N = 1$ (or for load $P = bD$). The half-length, $c$, of FPZ is taken as the smaller of $c_b$ and the distance from FPZ center to the notch tip or the surface of the body. In the smeared-tip method, the stresses and displacements on the crack line expressed from (70) are substituted into the stress-separation function of the cohesive crack model, which yields a singular integral equation for $q(\zeta)$. To solve the size effect from that equation is not easy. Therefore it was proposed to avoid prescribing a functional relation between the crack-bridging stress and the crack opening (separation), and introduce instead the *hypothesis* that the FPZ propagation is decided by the average of $g(\alpha)$ over the FPZ, i.e., by the condition

$$\frac{\sigma_N^2}{E} D\,\overline{g}(\alpha) = G_b, \qquad \text{with } \overline{g}(\alpha) = \int_{-1}^{1} g(\alpha + \zeta c/D)\, q(\zeta)\,\mathrm{d}\zeta. \tag{71}$$

## Acknowledgment

Grateful acknowledgment is due to the Office of Naval Research (monitor Yapa D.S. Rajapakse) for partial financial support under grants N00014-91-J-1109 (directed by Bažant) N00014-96-1-1219 (directed by Daniel) to Northwestern University. J.-J.H. Kim, E. Becq-Giraudon and M. Brocca wish to thank ONR for financial support under the former grant.

## References

Argon, A.S. (1972). Fracture of composites. *Treatise of Materials Science and Technology*, vol. 1, p. 79, Academic Press, New York.

Bao, G., Ho, S., Suo, Z. and Fan, B. (1992). The role of material orthotropy in fracture specimens for composites. *International Journal of Solid Structures*, **29**(9), 1105–1116.

Bažant, Z.P. (1976). Instability, ductility, and size effect in strain-softening concrete. *Journal of Engineering Mechanics Division*, Am. Soc. Civil Engrs., **102**, EM2, 331–344; disc. **103**, 357–358, 775–777, **104**, 501–502.

Bažant, Z.P. (1983). Fracture in concrete and reinforced concrete, Preprints, IUTAM Prager Symposium on *Mechanics of Geomaterials: Rocks, Concretes, Soils*, (Edited by Z.P. Bažant), Northwestern University, Evanston, Illinois, 281–316.

Bažant, Z.P. (1984). Size effect in blunt fracture: Concrete, rock, metal. *Journal of Engineering Mechanics*, ASCE, **110**, 518–535.

Bažant, Z.P. (1987). Fracture energy of heterogeneous material and similitude. Preprints, *SEM-RILEM International Conference on Fracture of Concrete and Rock* (held in Houston, Texas, June 1987), (Edited by S.P. Shah and S.E. Swartz), publ. by SEM (Soc. for Exper. Mech.) 390–402.

Bažant, Z.P. (1997). Scaling of quasibrittle fracture: Asymptotic analysis. *International Journal of Fracture* **83**(1), 19–40.

Bažant, Z.P. (1998a). Modeling of compressive strain-softening, fracture and size effect in concrete. vol. 1 (*Proceedings of the EURO-C 1998 Conference on Composite Modelling of Concrete Structures* held at Badgastein, Austria, March 31–April 3, 1998), (Edited by R. de Borst, N. Bićanić, H. Mang, G. Meschke) publ. by A.A. Balkema Publishers, Rotterdam, Netherlands, 249–264.

Bažant, Z.P. (1998b). *Nonlocal Fracture Mechanics – Proposal of a new Approach to Nonlinear or Quasibrittle Fracture*. Preliminary Internal Report, Northwestern University, Dec.

Bažant, Z.P., and Cedolin, L. (1991). *Stability of Structures: Elastic, Inelastic, Fracture and Damage Theories*, Oxford University Press, New York.

Bažant, Z.P., and Chen, E.-P. (1997). Scaling of structural failure. *Applied Mechanics Reviews ASME* **50**(10), 593–627.

Bažant, Z.P., and Kazemi, M. T. (1990). Determination of fracture energy, process zone length and brittleness number from size effect, with application to rock and concrete. *International Journal of Fracture*, **44**, 111–131.

Bažant, Z.P., and Li, Z. (1995). Modulus of rupture: size effect due to fracture initiation in boundary layer. *Journal of Structural Engineering ASCE* **121**(4), 739–746.

Bažant, Z.P., and Li, Yuan-Neng (1995). Stability of cohesive crack model: Part II – Eigenvalue analysis of size effect on strength and ductility of structures. Trans. ASME, *Journal of Applied Mechanics* **62** (Dec.), 965–969.

Bažant, Z.P., Daniel, I.M., and Li, Zhengzhi (1996). Size effect and fracture characteristics of composite laminates. ASME Journal of Engineering. *Materials and Technology* **118**(3), 317–324.

Bažant, Z.P., and Zi, G. (1999). *Cohesive Crack Analysis of Kink Band Propagation and Size Effect in Fiber Composites*. Report, Dept. of Civil Engrg., Northwestern University, Evanston, Illinois – in preparation.

Budiansky, B. (1983). Micromechanics. *Computers and Structures* **16**(1–4), 3–12.

Budiansky, B. and Fleck, N.A. (1993). Compressive failure of fiber composites. *Journal of Mechanics and Physics of Solids* **41**(1), 183–211.

Budiansky, B., Fleck, N.A., and Amazigo, J.C. (1997). On kink-band propagation in fiber composites. *Journal of Mechanics and Physics of Solids* **46**(9), 1637–1635.

Christensen, R.M. and DeTeresa, S.J. (1997). The kink band mechanism for compressive failure of fiber composite materials. *Journal of Applied Mechanics ASME* **64** (March), 1–6.

Daniel, I.M., and Hsiao, H.M. (1996). Failure mechanisms in thick composites under compressive loading. *Composites, Part B*, **27B** 543–552.

Dempsey, J.P., Adamson, R.M., and Mulmule, S.V. (1995), Large-scale in-situ fracture of ice. vol. 1 (Procedings of 2nd International Conference on Fracture Mechanics of Concrete Structures (FraMCoS-2), held at ETH, Zürich), edited by F.H. Wittmann, Aedificatio Publishers, Freiburg, Germany 575–684.

Dempsey, J.P. et al. (1999). *International Journal of Fracture* (present Special Issue on Fracture Scaling, edited by Z.P. Bažant and Y.D.S. Rajapakse), in press.

Fleck, N.A. (1997). Compressive failure of fiber composites. *Advances in Applied Mechanics* **33**, 43–117.

Fleck, N.A., and Jelf, P.M. (1995). Deformation and failure of a carbon fibre composite under combined shear and transverse loading. *Acta Metallurgica Materialia* **43**(8), 3001–3007.

Fleck, N.A., and Shu, J.Y. (1995). Microbuckle initiation in fibre composites: a finite element study. *Journal of the Mechanics and Physics of Solids* **43**, 1887–1918.

Fleck, N.A., Sivashanker, S., and Sutcliffe, M.P.F. (1996). The propagation of a microbuckle in undirectional and multidirectional composites. *Acta Metallurgica Materialia*, submitted.

Fleck, N.A., Sivashanker, S., and Sutcliffe., M.P.F. (1997). Compressive failure of composites due to microbuckle growth. *European Journal of Mechanics A / Solids* 16, 65–82.

Fleck, N.A., Sutcliffe, M.P.F., Sivashanker, S., and Xin, X.J. (1996). Compressive R-curve of a carbon fibre-epoxy matrix composite. *Composites, Part B: Engineering* **27B**(6), 531–541.

Führing, H. (1973). Approximation functions for K-factors of cracks in notches. *International Journal of Fracture* **9**, 328–331.

G.R. Irwin, Fracture, in *Handbuch der Physik*, vol. VI, (Edited by W. Flügge), Springer Verlag, Berlin (1958) 551–590.

Jelf, P.M., and Fleck, N.A. (1992). Compression failure mechanisms in unidirectional composites. *Journal of Composite Materials* **26**(18), 2706–2726.

Kyriakides, S., Ascerulatne, R., Perry, E.J., and Liechti, K.M. (1995). On the compressive failure of fiber reinforced composites. *International Journal of Solids and Structures* **32**(6/7), 689–738.

Kyriakides, S., and Ruff, A.E. (1997). Aspects of failure and postfailure of fiber composites in compression. *Journal of Composite Materials* **31**(20), 2000–2037.

Moran, P.M., Liu, X.H. and Shih, C.F. (1995). Kink band formation and band broadening in fiber composites under compressive loading. *Acta Metallurgical Materialia* **43**(8), 2943–2958.

Newman, J.C. (1982). A nonlinear fracture mechanics approach to the growth of small cracks. *Proceedings, AGARD Conference on Behavior of Short Cracks on Airframe Components*, France.

Palmer, A., and Rice, J.R. (1973). The growth of slip surfaces in the progressive failure of over-consolidated clay. *Proceedings of the Royal Society of London* **A332**, 527–548.

Rice, J.R. (1968a). A path independent integral and the approximate analysis of strain concentrations by notches and cracks. *Journal of Applied Mechanics ASME* **35**, 379–386.

Rice, J.R. (1968b). Mathematical analysis in the mechanics of fracture. In *Fracture – an Advanced Treatise*, vol. 2, (Edited by H. Liebowitz), Academic Press, New York, 191–308.

Rosen, B.W. (1965). Mechanics of composite strengthening. *Fiber Composite Materials*, Am. Soc. for Metals Seminar, Chapter 3, American Society for Metals, Metals Park, Ohio, 37–75.

Schultheisz, C.R. , and Waas, A.M. (1996). Compressive failure of composites, part I: testing and micromechanical theories. *Progresive Aerospace Science* **32**, 1–42.

Soutis, C. and Fleck, N.A. (1990). Static compression failure of carbon fibre T800/924C composite plate with a single hole. *Journal of Composite Materials* **24**, 536–558.

Soutis, C., Fleck, N.A., and Smith, P.A. (1991). Failure prediction technique for compression loaded carbon fibre-epoxy laminate with open holes. *Journal of Composite Materials* **25** (Nov.), 1476–1497.

Soutis, C., Curtis, P.T., and Fleck, N.A. (1993). Compressive failure of notched fibre composites. *Proceedings of the Royal Society of London A* **440**, 241–256.

Sutcliffe, M.P.F., and Fleck, N.A. (1994). Microbuckle propagation in carbon fibre-epoxy composites. *Acta Metallurgica et Materialia* **42**(7), 2219–2231.

Sutcliffe. M.P.F., Fleck, N.A., and Xin, X.J. (1996). Prediction of compressive toughness for fibre composites. *Proceedings of the Royal Society of London*, Series **A452**, 2443–2465.

Tada, H., Paris, P.C. and Irwin, G.R. (1985), *The Stress Analysis of Cracks Handbook*, Del Research Corp., Hellertown, PA.

Waas, A.M., and Schultheisz, C.R., (1996). Compressive failure of composites, part II: experimental studies. *Progressive of Aerospace Science* **32**, 43–78.

Zi, Goangseup (1998). Note privately communicated to Z.P. Bažant on Nov. 13.

*International Journal of Fracture* **95**: 143–158, 1999.


# Is there a thickness effect on compressive strength of unnotched composite laminates?

I.M. DANIEL[1] and H.M. HSIAO[2]
[1]*Robert R. McCormick School of Engineering and Applied Science, 2137 N. Campus Drive, Northwestern University, Evanston, IL 60208. e-mail: imdaniel@nwu.edu*
[2]*Hexcell Corporation, Dublin, CA 94568*

Received 17 July 1998; accepted in revised form 31 December 1998

**Abstract.** The effect of laminate thickness was investigated on the compressive behavior of unidirectional and crossply composites. A recently developed compression test method for thick composites was used to test specimens from 16 to 200-plies thick. In all cases the stress-strain behavior to failure is nonlinear and failure strength is matrix dominated. Longitudinal compressive failure is triggered by matrix failure accompanied by fiber microbuckling and the compressive strength is greatly degraded by initial fiber misalignment. The longitudinal compressive strength shows a mild trend of decreasing values with increasing thickness. It can be explained that, even if such a trend is significant, increasing size would have a diminishing effect on compressive strength for initial fiber misalignments greater than 1.5 to 2°.

## 1. Introduction

The design of some new composite structures requires advanced composite materials in thick sections subjected to compressive loading. Test results of thick composite cylinders loaded under hydrostatic pressure have shown premature failures at stresses below the compressive strengths measured in thin composite laminates (Garala, 1987). These results show that compressive behavior of thick composites is not well understood and is possibly influenced by numerous factors. Early compressive failures of thick composites have been attributed primarily to fiber misalignment/waviness and nonlinear behavior of the matrix.

Compressive behavior of composites has been studied extensively on the micromechanical scale. Failure modes proposed for longitudinal compression include global Euler buckling, fiber microbuckling (Rosen, 1965; Greszczuk, 1975; Wang, 1978; Hahn and Williams, 1986), kink banding (Argon, 1972; Berg and Selama, 1973; Weaver and Williams, 1973; Parry and Wronski, 1982), plastic kinking (Budiansky, 1983; Budiansky and Fleck, 1993), matrix yielding (Piggott, 1981) and buckling/kinking (Steif, 1990; Chung and Weitsman, 1994; Schapery, 1995; Häberle and Matthews, 1994; Daniel et al., 1996). In most cases models attempting to predict longitudinal compressive strength of composites assume the existence of some initial fiber misalignment. They differ in the way they treat the effects of this misalignment and in the assumed response of the material.

Kink banding, representing localization of failure, is more easily observed under some conditions, e.g., when initial notches or other large defects are present, under constraints near tabs and especially in materials with brittle fibers and quasi ductile matrices. In general, compressive failure in composites with semi-brittle matrices is a near instantaneous and catastrophic phenomenon and for this reason it is not easy to monitor experimentally the nature and sequence of failure mechanisms in determining the compressive strength.

In trying to explain early failures of thick composite structures, questions were raised on the thickness effect on compressive strength. It was argued that the compressive properties measured from thin laminates may not be applicable to thick composites. Strength reductions of up to 35 percent for thicknesses of 25 mm (1 in.) have been reported. A size effect would be expected, since the compressive strength is governed and is sensitive to initial defects such as fiber misalignment. However, the extent of any such size effect must be investigated and understood.

In this study, the effect of laminate thickness on the compressive behavior of composite materials was investigated. Composite laminates of various thicknesses were fabricated under uniform conditions without any major fabrication defects. A recently developed compression test method for thin and thick composites was used. The stiffness, strength, stress-strain behavior, and failure mechanisms of carbon/epoxy composite laminates were studied as a function of specimen thickness.

## 2. Prediction of compressive failure

Most theoretical models analyze compressive failure by assuming some initial fiber misalignment as discussed above. They differ mostly in the manner they treat the effects of this misalignment and in the assumed response of the material. For example, they may include the effects of matrix nonlinearity or plasticity and local transverse failure of the composite. Many of these studies emphasize the study of failure mechanisms, such as kink banding, over that of the ultimate strength property. Failure mechanisms have been investigated by many of these researchers including the authors (Daniel et al., 1996). In the present study a relatively simple model for compressive strength prediction based on an assumed initial fiber misalignment and the in-plane shear characteristics of the composite was adopted (Häberle and Matthews, 1994; Daniel et al., 1996). The model predictions are in good agreement with experimental results (Daniel et al., 1996).

The initiation and development of failure mechanisms cannot be observed easily, especially when the matrix is relatively brittle, and fracture is near instantaneous and catastrophic. Failure mechanisms were recorded only after the test (post mortem) by means of optical as well as scanning electron photomicrography. Failure mechanisms in the gage sections are very difficult to observe because of the catastrophic nature of failure. The easiest ones to observe are those occurring in constrained regions, such as the tabbed region of the specimen following premature failure initiation by end crushing. Early failures and discernible failure patterns occur in regions where an additional shear stress component is added (through the tabs) to the direct compressive normal stress component. Adding shear is tantamount to increasing the fiber misalignment and causes earlier failure. The possibility of end crushing was minimized in the present study by following previously established careful test procedures (Hsiao et al., 1995).

It was observed that the first failure mechanism appears to be shear yielding or shear failure of the matrix precipitated by pre-existing fiber misalignment. Matrix shear yielding is followed by fiber microbuckling and fracture. The resulting failure pattern takes the form of a kink band as shown in Figure 1. As the specimen thickness increases, the kink band

*Figure 1.* Kink band in unidirectional IM6/3501-6 carbon/epoxy under longitudinal compressive loading.

geometry becomes more complex, with multiple kink bands forming on one or more planes which, in some cases, rotate along the path of the kink band (Figures 2 and 3).

The longitudinal compressive strength can be defined as the stress at initiation of matrix shear failure and the accompanying fiber microbuckling and kink formation. A simple prediction of this compressive strength can be obtained based on maximizing the axial stress in an off-axis (misaligned) lamina (Häberle and Matthews, 1994; Daniel et al., 1996). The predicted compressive strength is matrix dominated and intimately related to the in-plane shear stress-strain behavior of the lamina and the initial misalignment. Referring to Figure 4, the longitudinal compressive strength is predicted as

$$F_{1c} = \frac{\tau^*}{\varphi + \gamma^*},$$

where $\varphi$ is the initial fiber misalignment and $\tau^*$ and $\gamma^*$ are the coordinates of the shear stress-strain curve where its tangent intersects the shear strain ($\gamma_{12}$) axis at $\gamma = -\varphi$. Figure 4 shows an actual experimental $\tau_{12}$ vs. $\gamma_{12}$ curve for the material investigated. Using this relationship, one can predict the variation of compressive strength as a function of fiber misalignment for the carbon/epoxy material investigated (Figure 5). It is seen that a measured compressive strength of 1,725 MPa (250 ksi), which is approximately one-fourth of the initial shear modulus $G_{12} = 7.0$ GPa (1.0 Msi), corresponds to an initial fiber misalignment of $\varphi \cong 1.5°$. Furthermore, the rate of compressive strength reduction decreases as the misalignment increases. This implies that, as the specimen thickness (or volume) increases and the probability of encountering a higher degree of misalignment increases (within an expected narrow range), the expected reduction in compressive strength with size becomes less pronounced.

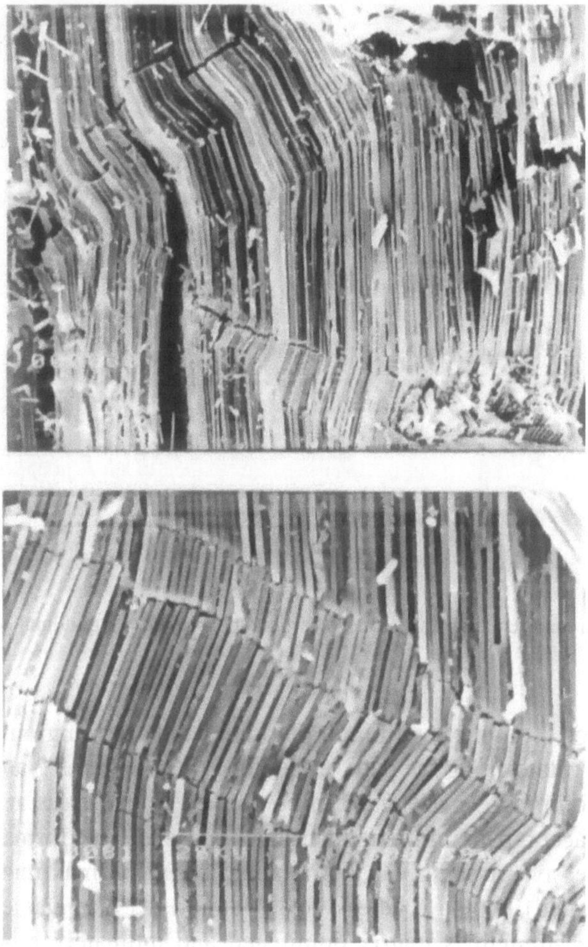

*Figure 2.* Multiple kink bands of various orientations and widths in carbon/epoxy under longitudinal compression.

## 3. Experimental procedures

### 3.1. Compression test method

Compression testing of thick composite materials is one of the most difficult types of testing, because of the tendency for premature failure due to global buckling or end crushing. The test itself is sensitive to many experimental parameters, such as alignment, specimen geometry, load introduction scheme, stress concentration, buckling stability, and specimen end-crushing. Over the years many test methods have been developed and applied to relatively thin laminates (Whitney et al., 1985; Berg and Adams, 1989; Camponeschi, 1991). The IITRI, Celanese, and sandwich test fixtures have been standardized (ASTM specification D3410-87). In contrast, test methods for thick composites have not been successfully developed and widely accepted.

The most commonly used test fixture today is the IITRI fixture, developed for relatively thin laminates (Hofer and Rao, 1977). Compressive loading is introduced into the composite laminate through shear by wedge grips. The limitation of the IITRI fixture is the maximum load that can be transmitted through shear by the tabs, which is controlled by the adhesive shear strength, the interlaminar shear strength of the tab and specimen materials, and the total

*Figure 3.* Illustration of kink band with changing plane.

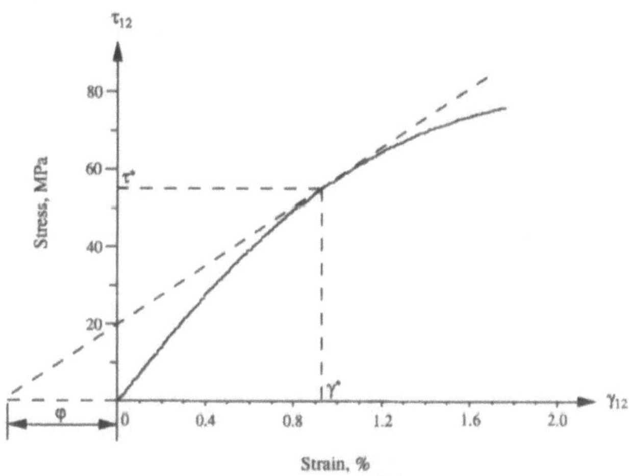


Compressive Strength:

$$F_{1c} = (\sigma_x)_{\max} = \frac{\tau^*}{\varphi + \gamma^*}$$

*Figure 4.* Measured shear stress-strain curve for carbon/epoxy composite (IM6G/3501-6) and graphical determination of longitudinal compressive strength. ($\varphi$ = initial misalignment.)

tabbed area. The capacity for testing thicker specimens by the IITRI fixture can be increased only by increasing the tabbed area of the specimen or the adhesive shear strength, factors that have practical limitations.

Recently a new compression test method for thick composites, incorporating both shear loading and end loading concepts, was successfully developed by the authors (Daniel et al., 1993; Hsiao et al., 1995) (Figure 6). This fixture is basically a modified IITRI fixture. The concept of this fixture is to transmit the initial load (up to 25 percent of the ultimate compress-

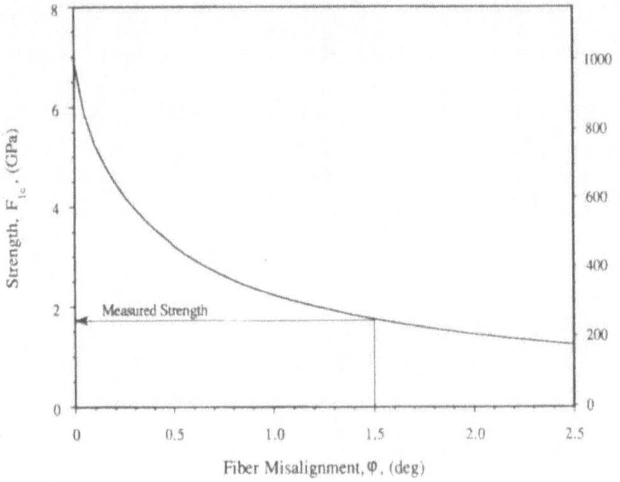


*Figure 5.* Variation of predicted longitudinal compressive strength with initial fiber misalignment for IM6G/3501-6 carbon/epoxy.

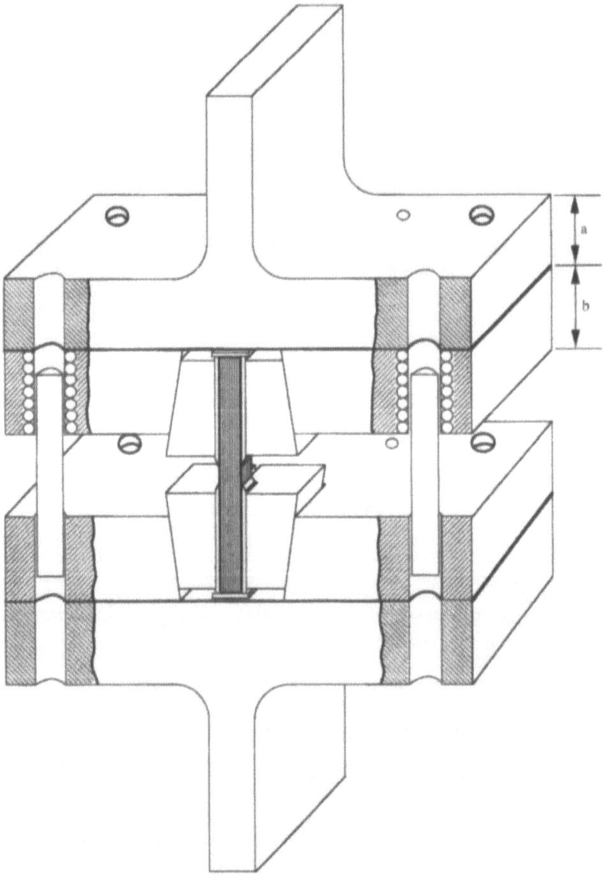


*Figure 6.* Schematic of NU compression test fixture. (Dimensions: adaptor $a = 25$ mm, base $b = 38$ mm.)

*Table 1.* Summary of specimen dimensions

| Layup | Number of plies | Specimen length cm | (in.) | Gage length mm | (in.) |
|---|---|---|---|---|---|
| $[0]_n$ | 16 | 11.4 | (4.50) | 12.7 | (0.50) |
| | 48 | 12.7 | (5.00) | 19.1 | (0.75) |
| | 72 | 13.3 | (5.25) | 25.4 | (1.00) |
| $[(0_n/90_n)_2/\bar{0}_n]_s$ | 27 | 11.4 | (4.50) | 12.7 | (0.50) |
| | 45 | 12.7 | (5.00) | 19.1 | (0.75) |
| | 72 | 13.3 | (5.25) | 25.4 | (1.00) |
| $[90]_n$ | 60 | 12.7 | (5.00) | 19.1 | (0.75) |
| | 200 | 14.0 | (5.50) | 25.4 | (1.00) |

ive strength) through the tabs by shear loading and thereafter engage the ends to apply the additional load to failure by end loading. The end loading is accomplished by inserting steel shims in the gap between the end of the specimen and the base of the fixture. Experimental results show that longitudinal compressive strengths of approximately 1,725 MPa (250 ksi) are obtainable for thick composites. In comparison with other compression test methods for thick composites reported to date (Shuart, 1989), the NU fixture produces the highest compressive strengths with acceptable scatter.

There are several important factors that influence the results of a compression test. Major factors include alignment, specimen geometry, stress concentration, specimen end-crushing, and timing of end-loading engagement (Hsiao et al., 1995). Testing becomes more difficult for thicker composites since all factors above become more serious and aggravated. Premature failure may be caused by either manufacturing defects or inappropriate testing methodology, and its probability increases with specimen thickness. Thus, in order to accurately assess the thickness effect, it is important to insure the validity of the test method.

### 3.2. Material selection and specimen configuration

The material used in this study was IM6G/3501-6 carbon/epoxy composite. The prepreg tape was obtained from Hercules Inc. (now Hexcell) as IM6G carbon fiber with 3501-6 177°C (350°F) epoxy resin. The prepreg layup was cured in a press/autoclave by a three-step curing cycle developed for thick composites (Daniel et al., 1993). Following fabrication, samples were cut from the plate, polished and optically observed under a microscope to examine the fabrication quality. Less than one percent void content was observed in all panels tested in this study.

Unidirectional and crossply laminates of $[0_n]$, $[(0_n/90_n)_2/\bar{0}_n]_s$ and $[90_n]$ layups were used to investigate the thickness effect. The respective specimen dimensions are summarized in Table 1. The specimen gage length was selected to prevent buckling failure. Its suitability was verified based on back-to-back strain gage readings from preliminary test results showing no evidence of bending or buckling. Each specimen was machined to 12.7 mm (0.5 in.) width by a water cooled diamond saw. Woven glass/epoxy tabs were adhesively bonded to the specimen with Hysol 9430 two-part adhesive. After tabbing, the specimens were machined to final

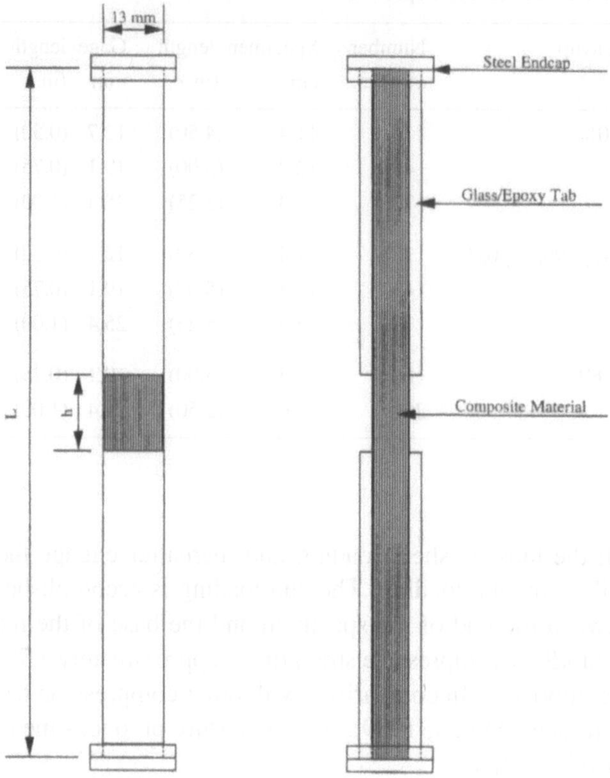


| | L, mm (in.) | h, mm (in.) |
|---|---|---|
| 16-ply | 114 (4.50) | 13 (0.50) |
| 48-ply | 127 (5.00) | 19 (0.75) |
| 72-ply | 133 (5.25) | 25 (1.00) |

*Figure 7.* Schematic of specimen configuration.

tolerances by grinding all sides parallel and perpendicular to within 0.025 mm (0.001 in.). Specimen ends must be flat and perpendicular to the loading axis to ensure uniform load distribution in the specimen. Steel endcaps were glued at both ends of the specimen to prevent premature crushing or longitudinal splitting near the specimen ends. The NU compression test specimen configuration is shown in Figure 7. Back-to-back strain gages were mounted at the center of the gage section to monitor the stress-strain behavior to failure.

### 3.3. Testing

All tests were carried out in an Instron universal servohydraulic testing machine with a 250 kN (56.2 kips) tension-compression load cell and a pair of hydraulic grips. The strain gages on the specimen were connected to a strain bridge conditioner (BC-888G) and amplified by the data acquisition interface (EXP-16, MetraByte Corporation). The measured load and strain were recorded by a personal computer.

Thin specimens were tested up to failure by shear loading only at a crosshead speed of 0.508 mm/min (0.02 in./min). For the thick specimens tested with combined shear and end loading, the initial crosshead speed of 0.508 mm/min (0.02 in./min) used for the shear-loading stage was subsequently reduced to 0.127 mm/min (0.005 in./min) for the end-loading stage of the test.

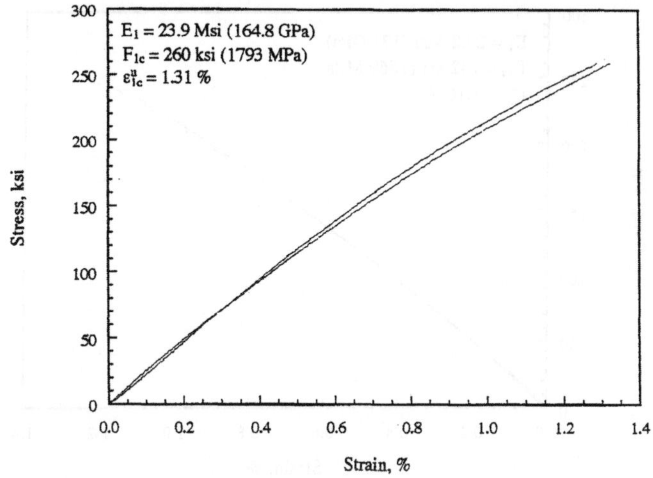


*Figure 8.* Stress-strain curve of $[0]_{16}$ carbon/epoxy under longitudinal compressive loading (IM6G/3501-6).

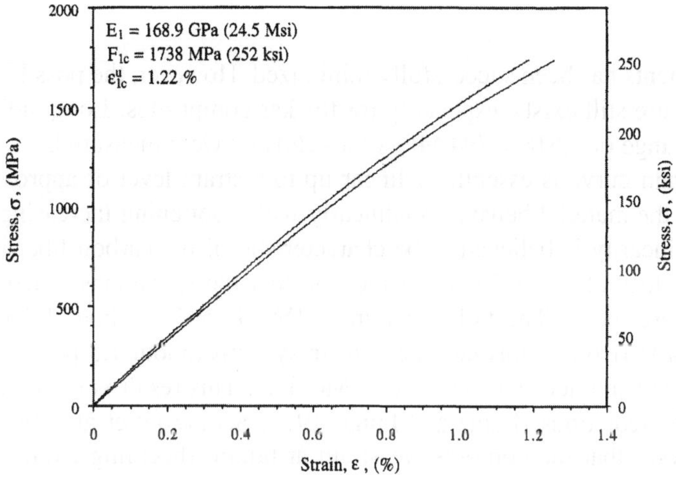


*Figure 9.* Stress-strain curve of unidirectional 48-ply carbon/epoxy under longitudinal compressive loading (IM6G/3501-6).

## 4. Results and discussion

The experimental results obtained include stress-strain curves to failure and post-failure examination of fracture patterns. The stiffness, strength, overall stress strain behavior and failure patterns were studied as a function of specimen thickness. The compressive elastic modulus was determined by averaging the initial slopes of the stress-strain curves from readings of back-to-back strain gages. Failure strains represent the average of the two back-to-back strain gages at the point of failure. In general, test results for all thicknesses were good and reproducible. The scatter in modulus and strength for all layup configurations was satisfactorily low. However, it was observed that the strain readings were more scattered. This suggests that the degree of nonlinearity increases rapidly as the loading approaches the ultimate value.

Typical longitudinal compressive stress-strain curves for 16-, 48-, and 72-ply unidirectional specimens are shown in Figures 8–10. Results from all tests are summarized in Table 2. The consistency of the two strain gage readings up to failure in each figure indicates that bending

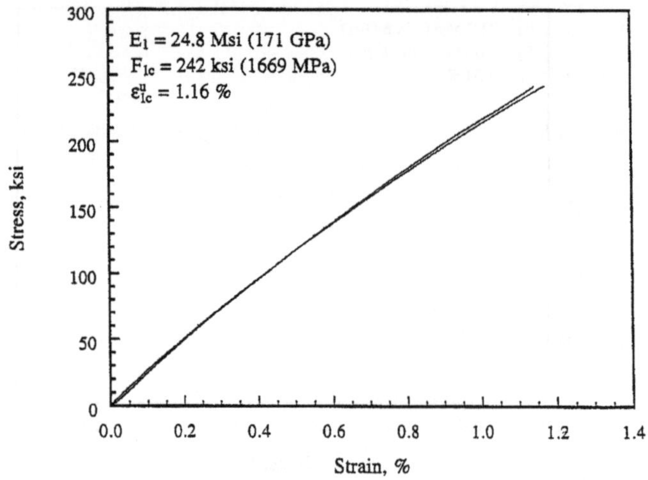


*Figure 10.* Stress-strain curve of unidirectional 72-ply carbon/epoxy under longitudinal compressive loading (IM6G/3501-6).

due to misalignments has been successfully minimized. However, the possibility of premature end-crushing failure still exists, especially for thicker composites. Longitudinal compressive strengths in the range of 1,518-1,794 MPa (220–260 ksi) were measured.

The stress-strain curve is essentially linear up to a strain level of approximately 0.5 percent. Thereafter, the material behaves nonlinearly with a softening increasing with increasing strain. This nonlinearity is believed to be characteristic of the carbon fiber itself. Consistent ultimate strain values of 1.1–1.3 percent indicate that the specimens failed primarily in the compressive failure mode. The failure strain of IM6G/3501-6 (about 1.3 percent) is lower than that of AS4/3501-6 or equivalent composite systems (about 1.5 percent) because of the higher modulus and smaller diameter of the IM6 fiber. This result is in good agreement with some theoretical predictions (Lanir and Fung, 1972; Sadowsky et al., 1967). It was shown from these analyses that the compressive strain at failure (buckling strain) is related to the moduli of fiber and matrix by the relationship

$$\varepsilon \propto (E_m/E_f)^{1/2},$$

where $\varepsilon$ is the failure strain, $E_f$ is the fiber modulus, and $E_m$ is the matrix modulus.

The measured compressive strength is plotted versus specimen thickness in Figure 11. A slight trend of decreasing strength with increasing thickness appears. An apparent slight increase in the rate of strength reduction with number of plies might be attributed to the fact that as the specimen thickness increases it becomes more difficult to minimize premature failures due to end crushing. Furthermore, the relatively small number of tests does not warrant any fast conclusions on the rate of strength reduction. Assuming a random distribution of flaws (fiber misalignment), one could apply Weibull distribution statistics to quantify the size (or thickness) dependence of the compressive strength. The probability of fracture at stress $\sigma$ of a uniformly loaded specimen is

$$P(\sigma) = 1 - \exp\left[-V\left(\frac{\sigma - \sigma_u}{\sigma_0}\right)^m\right],$$

*Table 2.* Longitudinal compressive properties of unidirectional carbon/epoxy (IM6G/3501-6)

| Number of plies | Specimen number | Modulus, $E_1$ GPa | (Msi) | Failure stress, $F_{1c}$ MPa | (ksi) | Ultimate strain, $\varepsilon_{1c}^u$ percent |
|---|---|---|---|---|---|---|
| 16 | 1 | 164.8 | (23.9) | 1538 | (223) | 1.06 |
| | 2 | 164.8 | (23.9) | 1793 | (260) | 1.31 |
| | 3 | 165.5 | (24.0) | 1662 | (241) | 1.17 |
| | 4* | 166.9 | (24.2) | 1820 | (264) | 1.58 |
| | 5 | 164.1 | (23.8) | 1607 | (233) | 1.17 |
| | Mean | 165.2 | (24.0) | 1684 | (244) | 1.26 |
| | Standard deviation | 1.1 | (0.2) | 121 | (18) | 0.20 |
| | $C_v$ | 0.7 | | 7.2 | | 15.9 |
| 48 | 1 | 168.9 | (24.5) | 1738 | (252) | 1.22 |
| | 2 | 165.5 | (24.0) | 1545 | (224) | 1.04 |
| | 3 | 171.0 | (24.8) | 1634 | (237) | 1.13 |
| | 4 | 172.4 | (25.0) | 1565 | (227) | 1.10 |
| | 5 | 164.8 | (23.4) | 1641 | (238) | 1.15 |
| | 6 | 175.8 | (25.5) | 1731 | (251) | 1.19 |
| | Mean | 169.7 | (24.5) | 1642 | (238) | 1.14 |
| | Standard deviation | 4.2 | (0.7) | 81 | (12) | 0.06 |
| | $C_v$ | 2.5 | | 4.9 | | 5.6 |
| 72 | 1 | 175.8 | (25.5) | 1586 | (230) | 1.09 |
| | 2* | 175.8 | (25.5) | 1524 | (221) | 1.04 |
| | 3 | 171.7 | (24.9) | 1593 | (231) | 1.11 |
| | 4 | 171.0 | (24.8) | 1669 | (242) | 1.16 |
| | Mean | 173.6 | (25.2) | 1593 | (231) | 1.10 |
| | Standard deviation | 2.6 | (0.4) | 59 | (9) | 0.05 |
| | $C_v$ | 1.5 | | 3.7 | | 4.5 |

*Results from one gage.

where $m$ is the flaw density exponent or shape parameter; $\sigma_0$ is the scale parameter; $\sigma_u$ is the zero probability stress or stress threshold below which the material never fails; and $V$ is the specimen volume.

The ratio of mean strengths of two specimens of volumes $V_1$ and $V_2$ is

$$F_{v_1}/F_{v_2} = \left(\frac{V_2}{V_1}\right)^{1/m}.$$

From the experimental data obtained and the range of sizes (thicknesses) considered an average shape parameter of $m = 40$ was computed. This is much higher than typical shape parameter values of $m = 15$ and $m \cong 25$ for tensile strengths of fibers and unidirectional specimens (Fukuda and Miyazawa, 1994; Bullock, 1974). This shows that if there is indeed

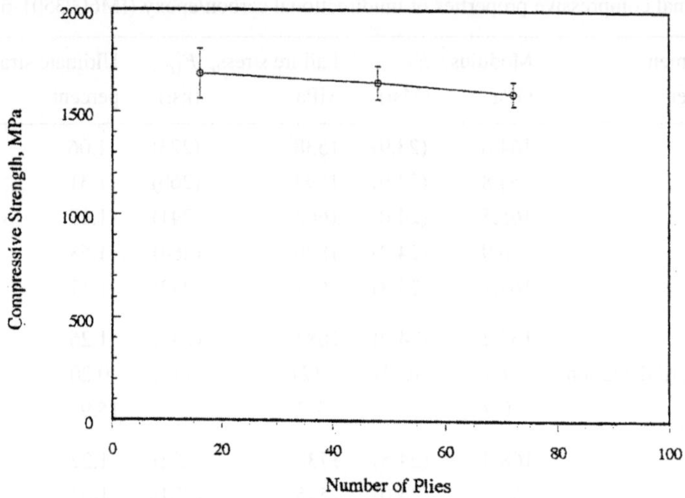


*Figure 11.* Thickness effect on longitudinal compressive strength.

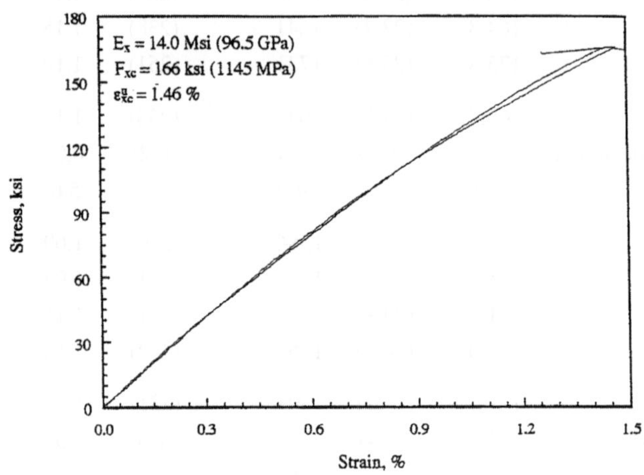


*Figure 12.* Stress-strain curve of $[(0_3/90_3)_2/\bar{0}_3]_s$ carbon/epoxy under compressive loading (IM6G/3501-6).

a size (or thickness) effect on compressive strength it is very mild. This is corroborated by the fact that the principal type of flaw that triggers longitudinal compressive failure, fiber misalignment, has a decreasing influence on strength as the fiber misalignment increases (see Figure 5). Another factor that mitigates against the apparent trend of decreasing strengths with increasing thickness is the fact that despite all care, compression testing becomes more difficult with increasing thickness and the probability of premature failure increases.

Typical stress-strain curves of 27-, 45- and 72-ply crossply specimens are shown in Figures 12–14. Results are summarized in Table 3. The stress-strain curves are linear up to a strain of approximately 0.3 percent with softening behavior to failure after that. The non-linearity of the crossply specimens is higher than that of the 0° longitudinal ones due to matrix nonlinearity in the 90° layers. The measured average compressive strength of 1,035 MPa (150 ksi) is in good agreement with a rule of mixtures prediction of 1,053 MPa (153 ksi), based on a longitudinal strength of 1,725 (250 ksi) and a transverse strength of 214 MPa (31 ksi). The average failure strain of the crossply specimens is only slightly higher than that of the

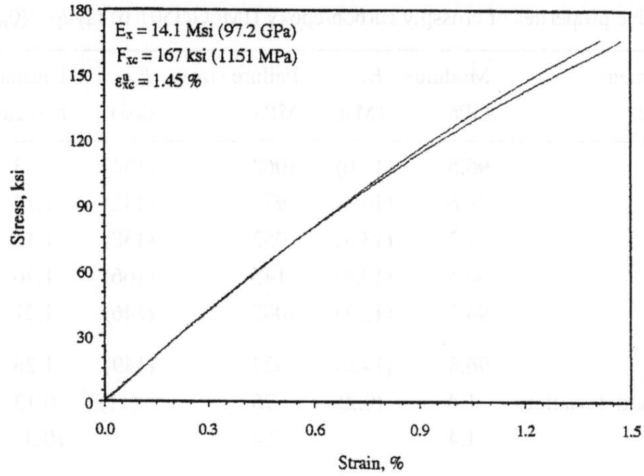


*Figure 13.* Stress-strain curve of $[(0_5/90_5)_2/\bar{0}_5]_s$ carbon/epoxy under compressive loading (IM6G/3501-6).

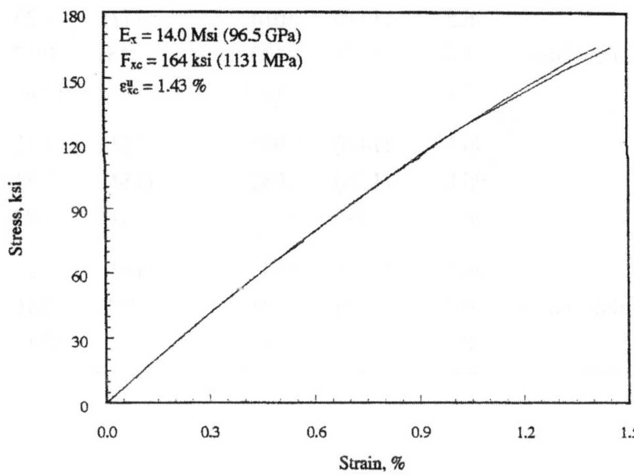


*Figure 14.* Stress-strain curve of $[(0_8/90_8)_2/\bar{0}_8]_s$ carbon/epoxy under compressive loading (IM6G/3501-6).

0° unidirectional ones, suggesting that failure is governed primarily by the 0° layers. The slightly higher ultimate strain may be attributed to the lateral support provided to the 0° layers by the adjacent 90° layers. Similar behavior has been observed in quasi-isotropic laminates (Sohi et al., 1987). Considering the scatter in the results, no significant thickness effect was observed.

Typical transverse compressive stress-strain curves for 60- and 200-ply $[90_n]$ unidirectional specimens are shown in Figures 15 and 16. Transverse compressive testing is relatively easier because of the lower strength of the material in the transverse direction. All tests exhibited nonlinear softening behavior which may be attributed to matrix nonlinearity. The measured transverse compressive strength was in the range of 200–221 MPa (29–32 ksi) with no apparent thickness effect.

*Table 3.* Compressive properties of crossply carbon/epoxy (IM6G/3501-6) layup: $[(0_n/90_n)_2/\bar{0}_n]_s$

| Number of plies | Specimen number | Modulus, $E_x$ GPa | (Msi) | Failure stress, $F_{xc}$ MPa | (ksi) | Ultimate strain, $\varepsilon^u_{xc}$ percent |
|---|---|---|---|---|---|---|
| 27 | 1 | 96.5 | (14.0) | 1062 | (154) | 1.33 |
| | 2 | 98.6 | (14.3) | 979 | (142) | 1.14 |
| | 3 | 95.2 | (13.8) | 952 | (138) | 1.16 |
| | 4 | 96.5 | (14.0) | 1145 | (166) | 1.46 |
| | 5 | 94.5 | (13.7) | 1007 | (146) | 1.21 |
| | Mean | 96.5 | (14.0) | 1027 | (149) | 1.26 |
| | Standard deviation | 1.4 | (0.2) | 76 | (11) | 0.13 |
| | $C_v$ | 1.4 | | 7.4 | | 10.3 |
| 45 | 1 | 95.8 | (13.9) | 1027 | (149) | 1.27 |
| | 2 | 94.5 | (13.7) | 924 | (134) | 1.11 |
| | 3 | 98.6 | (14.3) | 958 | (139) | 1.09 |
| | 4 | 97.2 | (14.1) | 1151 | (167) | 1.45 |
| | Mean | 96.5 | (14.0) | 1014 | (147) | 1.23 |
| | Standard deviation | 1.8 | (0.3) | 100 | (15) | 0.17 |
| | $C_v$ | 2.1 | | 10.2 | | 13.8 |
| 72 | 1 | 96.5 | (14.0) | 945 | (137) | 1.12 |
| | 2 | 95.8 | (13.9) | 1062 | (154) | 1.31 |
| | 3 | 96.5 | (14.0) | 1131 | (164) | 1.43 |
| | Mean | 96.5 | (14.0) | 1046 | (152) | 1.29 |
| | Standard deviation | 0.4 | (0.1) | 94 | (14) | 0.16 |
| | $C_v$ | 0.4 | | 9.0 | | 12.1 |

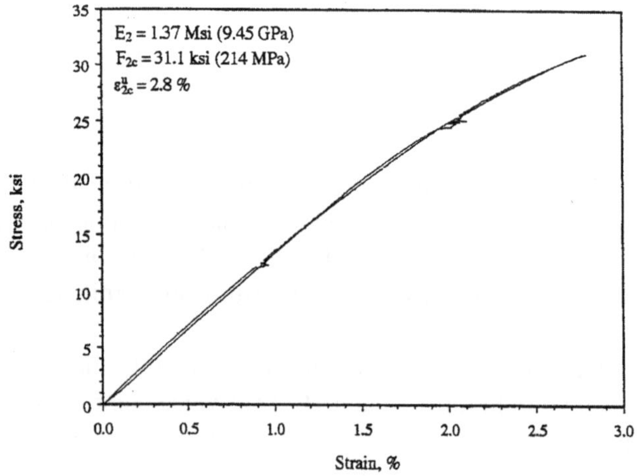


*Figure 15.* Stress-strain curve of $[90]_{60}$ carbon/epoxy under transverse compressive loading in the 2-direction (IM6G/3501-6).

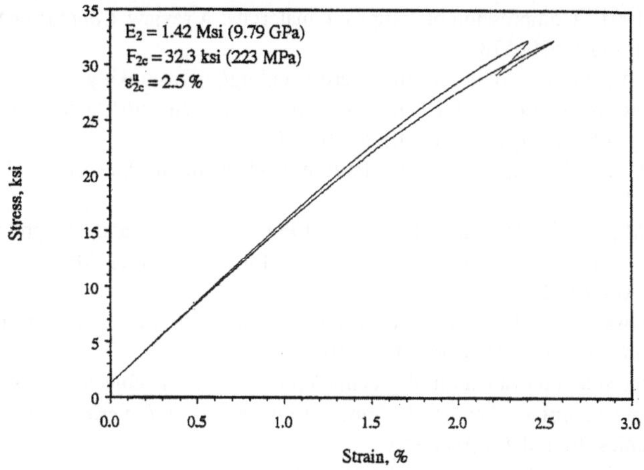


*Figure 16.* Stress-strain curve of $[90]_{200}$ carbon/epoxy under transverse compressive loading in the 2-direction (IM6G/3501-6).

## 5. Summary and conclusions

The effect of laminate thickness was investigated on the longitudinal and transverse compressive strength of unidirectional composites and on the axial compressive strength of crossply laminates. In all cases failure is matrix-dominated. In the case of longitudinal loading, composite failure is triggered by matrix failure accompanied by fiber microbuckling and is greatly dependent on initial fiber misalignment. The longitudinal compressive strength shows a mild trend toward decreasing values with increasing thickness. Even if such a trend is significant, it was explained that for initial fiber misalignments of the order of 1.5–2°, increasing size would have a diminishing effect on strength. No significant size effects were discerned in crossply and in transversely loaded specimens.

## Acknowledgements

The work described in this paper was sponsored by the Office of Naval Research. We are grateful to Dr. Y.D.S. Rajapakse for his encouragement and cooperation and to Mrs. Yolande Mallian for typing the manuscript.

## References

Argon, A.S. (1972). Fracture of composites. *Treatise on Materials Science and Technology* (Edited by H. Herman), Academic Press, New York, 79–114.

Berg, C.A. and Salama, M. (1973). Fatigue of graphite fiber-reinforced epoxy in compression. *Fibre Science and Technology* **6**, 79.

Berg, J.S. and Adams, D.F. (1989). An evaluation of composite material compression test methods. *Journal of Composites Technology and Research* **11**(2), 41–46.

Budiansky, B. (1983). Micromechanics. *Computers and Structures* **16**, 3.

Budiansky, B. and Fleck, N.A. (1993). Compressive failure of fibre composites. *Journal of Mechanics and Physics of Solids* **41**(1), 183.

Bullock, R.E. (1974). Strength ratios of composite materials in flexure and in tension. *Journal of Composite Materials* **8**, 200–206.

Camponeschi, E.T., Jr. (1991). Compression of composite materials: A review. *Composite Materials: Fatigue and Fracture*, ASTM STP 1110, 550–578.

Chung, I. and Weitsman, Y.J. (1994). Model for the micro-buckling/micro-kinking compressive response of fiber-reinforced composites. *Mechanics USA 1994, Proceedings of the Twelfth US National Congress of Applied Mechanics* (Edited by A.S. Kobayashi), ASME, S256–S261.

Daniel, I.M., Hsiao, H.M. and Wooh, S.C. (1996). Failure mechanisms in thick composites under compressive loading. *Composites Part B* **27B**, 543–552.

Daniel, I.M., Hsiao, H.M., Wooh, S.C. and Vittoser, J. (1993). Processing and compressive behavior of thick composites. *Mechanics of Thick Composites*, The First Joint Mechanics Meeting of ASME/ASCE/SES, Charlottesville, VA, AMD **162**, 107–126.

Fukuda, H. and Miyazawa, T. (1994). Micromechanical approach to the tensile strength of unidirectional composites. *Advanced Composite Materials* **4**(2), 101–110.

Garala, H.G. (1987). Experimental evaluation of graphite/epoxy cylinders subjected to external hydrostatic compressive loading. *Proceedings of 1987 SEM Spring Conference on Experimental Mechanics*, Society for Experimental Mechanics, Bethel, CT, 948–951.

Greszczuk, L. B. (1975). Microbuckling failure of circular fiber-reinforced composites. *AIAA Journal* **13**, 1311–1318.

Häberle, J.G. and Matthews, F.L. (1994). A micromechanics model for compressive failure of unidirectional fibre-reinforced plastics. *Journal of Composite Materials* **28**(17), 1618.

Hahn, H.T. and Williams, J.G. (1986). Compression failure mechanisms in unidirectional composites. *Composite Materials: Testing and Design (Seventh Conference)*, ASTM STP 893, American Society for Testing and Materials, Philadelphia, PA, 115–139.

Hofer, K.E., Jr. and Rao, P.N. (1977). A new static compression fixture for advanced composite materials. *Journal of Testing and Evaluation* **5**(4), 278–283.

Hsiao, H.M., Daniel, I.M. and Wooh, S.C. (1995). A new compression test method for thick composites. *Journal of Composite Materials* **29**(13), 1789–1806.

Lanir, Y. and Fung, Y.C.B. (1972). Fiber composite columns under compression. *Journal of Composite Materials* **6**, 387–401.

Parry, T.V. and Wronski, A.S. (1982). Kinking and compressive failure in uniaxially aligned carbon fibre composites tested under superposed hydrostatic pressure. *Journal of Materials Science* **17**, 893.

Pigott, M.R. (1981). A theoretical framework for the compressive properties of aligned fiber composites. *Journal of Materials Science* **16**, 2837.

Rosen, B.W. (1965). Mechanics of composite strengthening. *Fiber Composite Materials*, American Society of Metals, Metals Park, OH, 35–75.

Sadowsky, M.A., Pu, S.L. and Hussain, M.A. (1967). Buckling of microfibers. *Journal of Applied Mechanics* **34**, 1011–1016.

Schapery, R.A. (1995). Prediction of compressive strength and kink bands in composites using a work potential. *International Journal of Solids and Structures* **32**(6/7), 739.

Shuart, M.J. (1989). Failure of compression-loaded multidirectional composite laminates. *AIAA Journal* **27**(9), 1274–1279.

Sohi, M.M., Hahn, H.T. and Williams, J.G. (1987). The effect of resin toughness and modulus on compressive failure modes of quasi-isotropic graphite/epoxy laminates. *Toughened Composites*, ASTM STP 937, 37–60.

Steif, P.S. (1990). A model for kinking in fiber composites; I. Fiber breakage via microbuckling; II. Kink band formation. *International Journal of Solids and Structures* **26**, 549.

Wang, A.S.D. (1978). A non-linear microbuckling model predicting the compressive strength of unidirectional composites. *ASME Paper 78-WA/Aero 1*.

Weaver, C.W. and Williams, J.G. (1975). Deformation of a carbon/epoxy composite under hydrostatic pressure. *Journal of Materials Science* **10**, 1323.

Whitney, J.M., Daniel, I.M. and Pipes, R.B. (1985). *Experimental Mechanics of Fiber Reinforced Composite Materials*. Society for Experimental Mechanics, Bethel, CT, and Prentice-Hall, Englewood Cliffs, NJ.

*International Journal of Fracture* **95:** 159–174, 1999.


# Failure and scaling properties of a softening interface connected to an elastic block

ARNAUD DELAPLACE[1], STÉPHANE ROUX[2] and GILLES PIJAUDIER-CABOT[3,*]
[1]*LMT, ENS Cachan / CNRS / Université P. et M. Curie, 61, Avenue du Président Wilson, F-94235 Cachan, France.*
[2]*Laboratoire de Physique et Mécanique des Milieux Hétérogènes, Ecole Supérieure de Physique et de Chimie Industrielles, 10 rue Vauquelin, F-75231 Paris Cedex 05, France.*
[3]*LMT, ENS Cachan / CNRS / Université P. et M. Curie, 61, Avenue du Président Wilson, F-94235 Cachan, France, and Institut Universitaire de France.*

Received 13 July 1998; accepted in revised form 18 December 1998

**Abstract.** The damage growth in a softening interface connected to an elastic block is analysed. The elastic block, assumed to be infinite, is modelled as a two-dimensional continuum and the interface is one-dimensional with a constitutive response which follows a scalar damage model. The solution technique is based on the equilibrium of the interfacial forces resulting from the deformation of the elastic block and from the interface constitutive response. The interface failure process is compared to that of a hierarchical model which was obtained analytically (Delaplace et al., 1998). The two are found to be similar, without an internal length scaling the distribution of damage at the inception of macro-cracking. Finally, scale effects on the occurrence of bifurcation and instability are considered. It is shown that bifurcation may occur prior to or after the limit point under displacement control, depending on the elastic block height or stiffness.

## 1. Introduction

One of the important characteristics of quasi-brittle heterogeneous materials in tension is the existence of a strain softening regime. This peculiarity of the stress-strain response induces strain localisation which can be understood as a sudden increase of the strains in a very narrow region. In tension, the inception of strain localisation is often seen as a bifurcation from an homogeneous deformation regime to a non-homogeneous, continuous or discontinuous one, occurring in a band of material oriented perpendicularly to the applied tensile load. If the band is small with respect to the specimen size, it can be considered as a softening interface which is a precursor of the macro-crack observed at complete failure.

Among the popular techniques for describing this type of material and structural response is the approach where the material response in the softening band, collapsed onto a line, is expressed as a function of the stress versus the relative displacement across the band (the cohesive crack model). This function is, of course central in the constitutive model and there have been many proposals, assuming that the stress versus relative displacement relation is linear, bi-linear, or exponential (see for a recent review (Bažant and Planas, 1998)). Once a crack is initiated, that is when the material stress has been exhausted in the course of loading in a region of the band, another issue is the fracture process zone which occurs ahead of the crack due to progressive failure. The size of this process zone, here the length of the interface

* Correspondence Address: LMT ENS Cachan/CNRS/Université P. et M. Curie, 61 Avenue du President Wilson, F-94235 Cachan, France. e-mail: pijaudie@lmt.ens-cachan.fr

ahead of the crack where softening occurs, induces size effect which is typical of quasi-brittle structures. Hence it is critical to understand how the fracture process zone forms in the course of loading.

In the present contribution, we consider the response of a strain softening solid in tension from the viewpoint of an interface problem. Hence the width of the softening zone is assumed to be small compared to the specimen length. We will use a one-dimensional continuous damage model for describing the interface behaviour. Our intent is to understand how failure is initiated, what is the process of damage localisation in the interface, and what scaling parameters can be expected as outcomes of the redistribution of forces in the interface and of the interaction between the interface and the rest of the specimen.

In order to better understand how damage develops in the interface, a first study was made, based on a hierarchical model which allows to extract the interface behaviour from the global system response (Delaplace et al., 1998). The interface was composed of a set of elastic brittle fibres, and heterogeneity was included through a random variable, the threshold where the fibres break. With this model, we have shown that at the beginning of the loading, damage develops homogeneously *all over* the interface. Then damage becomes localised in the interface, first at a *macro scale*, that is the half system size, and further in a narrower and narrower region. In the limit of an infinite number of fibres corresponding to a continuous description of the interface, this process was understood as a cascade of bifurcations. At the initiation of the macro-crack, the damage profile in the interface was computed analytically, and damage was shown to spread all over the interface. In other words, the fracture process zone resulting from the progressive failure process considered did not exhibit a scale other than the entire length of the interface.

Although the hierarchical description is equivalent to the classical elastic one and provides a convenient setting for deriving analytical results and is more efficient for computations, it has two drawbacks: (i) because the interaction between two points on the interface occurs through hierarchical levels, correlation lengths are not straightforward to interpret physically; (ii) during the failure process, the discrete nature of the hierarchical description becomes more and more prominent and can result into unrealistic responses such as strength recoveries. In order to confirm the above results and to arrive at an easier interpretation, we will use here a more classical representation where two-dimensional elasticity is used for the description of the elastic block instead of a hierarchical model. The tensile specimen will be modelled by an elastic block connected to an interface with a continuous damage law for its behaviour, without any imperfection at the interface. In order to detect possible bifurcation points along the specimen response, we will implement classical criteria (see for instance (Pijaudier-Cabot and Huerta 1991; De Borst, 1987) in order to treat loss of uniqueness or loss of stability.

In the first part of the paper, we present the model geometry and the discretisation technique which yields to the basic equilibrium equations, paying specific attention to the boundary conditions. The bases of the stability and uniqueness criteria are recalled in section three. Then, the interface failure process is compared to the hierarchical model. The two are found to be similar. Finally scale effects on the occurrence of bifurcation and instability are considered.

## 2. Discretisation of the problem

We consider in this contribution the problem of an elastic block connected to a rigid substrate through an interface. The elastic block is a two-dimensional body, infinite in direction $x$ of

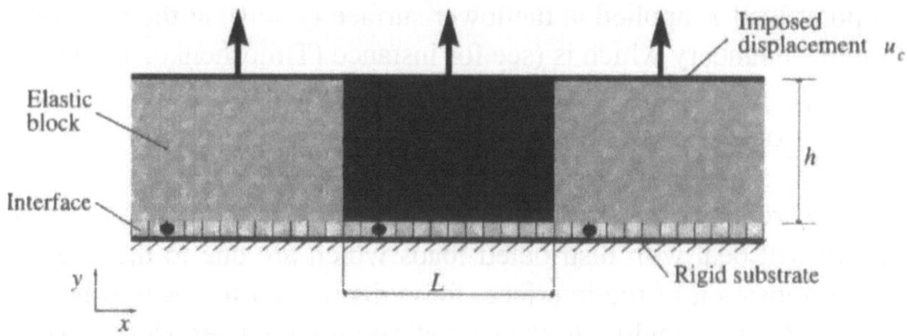


*Figure 1.* The interface, composed of $n = 12$ elements, coupled in series with the elastic block. Three cells (blocks) are represented. The black circles in the interface are just plotted in order to better understand the periodic boundary conditions.

height $h$ (Figure 1). At $y = h$, the boundary of the elastic block is constrained to follow a fixed displacement $u_c$. The material properties of the block are its Poisson ratio $\nu$, and its Young modulus $E$. Periodic loading conditions are used on the boundary $y = 0$ in the $x$ direction. Hence, the periodic cell is an elastic body connected to an interface at $x = 0$. $L$ and $h$ are the width and the height of the cell respectively. The interface is one-dimensional which means that the shear stiffness and tension-shear coupling terms are neglected. The interface is a row of $n$ identical uniaxial elements of width $l = L/n$ (its behaviour will be detailed in Section 2.4). As we will see later $L/l$ will be the discretisation parameter in the solution technique. Note that because the interface is one-dimensional, the rigid substrate on which the interface is connected can be seen as an axis of symmetry. Therefore, the problem considered is also equivalent to that of an infinite elastic band of width $2h$, with a softening interface in the middle, which is subjected to fixed displacements on its upper and lower boundaries.

In order to solve the equilibrium equations in the block and at the interface, one can use the finite element method. Another technique, which will be implemented here uses, to some extent, the boundary element method for computing the displacement and stresses in the elastic block and to write the equilibrium of each uniaxial element of the interface. In fact, one needs only to know the relationship between the boundary tractions and the boundary displacements in the upper ($y = h$) and lower ($y = 0$) contours of the periodic cell. In the following, we are going to derive from known Green functions these relationships, starting from a single elastic block and then implementing the periodicity conditions. Then, the equilibrium at the interface level will be introduced along with the nonlinear constitutive response of the interface.

### 2.1. The elastic block

We first consider the elastic block without periodic conditions. The vertical displacement of each point of the block is called $u(x, y)$. $x$ varies between $-\infty$ and $+\infty$, as $y$ varies between 0 and $h$. Note that for $y = h$, we have

$$u(x, h) = u_c \quad \forall x. \tag{1}$$

Because we are just focusing on the displacement along the interface, we will call in the following $u(x)$ the displacement $u(x, 0)$ in order to simplify the notation. For a perfectly elastic block, a point load $P$ applied at the lower surface ($y = 0$) at the position $x_0$ induces a displacement on this boundary which is (see for instance (Timoshenko, 1948))

$$u(x) - u_c = \frac{2P}{\pi E} \log \frac{h}{|x - x_0|} - \frac{(1+\nu)}{\pi E} P, \tag{2}$$

where the symbol $|x|$ refers to the absolute value of $x$.

In our case, we will deal with distributed loads which are due to the interface response. Because of the discretisation of the interface, these distributed forces will be piecewise constant over the length $l$ corresponding to the discretisation parameter. Hence, we consider now a constant pressure distribution $q$ over a distance $l$, which starts at position $x_0$. At points (with the coordinate $y = 0$) located outside the pressure distribution (Figure 2a), and for $x < x_0$, the displacement $u(x)$ is

$$\begin{aligned} u(x) - u_c \; = \; & \frac{2q}{\pi E}\left((l + |x - x_0|) \log \frac{h}{l + |x - x_0|} - |x - x_0| \log \frac{h}{|x - x_0|}\right) \\ & + \frac{(1-\nu)}{\pi E} q l. \end{aligned} \tag{3}$$

At points (with the coordinate $y = 0$) located outside the pressure distribution, and for $x > x_0 + l$, the displacement $u(x)$ is

$$\begin{aligned} u(x) - u_c \; = \; & \frac{2q}{\pi E}\left((l + |x - x_0|) \log \frac{h}{l + |x - x_0|} - |x - x_0 - l| \log \frac{h}{|x - x_0 - l|}\right) \\ & + \frac{(1-\nu)}{\pi E} q l. \end{aligned} \tag{4}$$

For the points located inside the pressure distribution, the displacement becomes (Figure 2b)

$$\begin{aligned} u(x) - u_c \; = \; & \frac{2q}{\pi E}\left((l - |x - x_0|) \log \frac{h}{l - |x - x_0|} + |x - x_0| \log \frac{h}{|x - x_0|}\right) \\ & + \frac{(1-\nu)}{\pi E} q l. \end{aligned} \tag{5}$$

Note that the two expressions in (3, 5) are obtained under the condition that the displacement on the upper boundary of the block at $y = h$ satisfies the prescribed fixed displacement $u_c$.

### 2.2. Periodicity of the Pressure Distribution

Let us come back to our problem with periodic loads. The periodicity condition for the displacements is $u(x) = u(x + L)$. The periodicity of the pressure distribution entering in (3, 5) needs also to be taken into account. This can be performed with a superposition technique where the influences of pressure distributions spaced out by $L$ are added (see appendix for the complete derivation). An other way to find this result is to consider a Fourier transform version of the Green function. Then, the mode $k = 0$ can be treated in real space as the mere uniaxial compression, like that introduced in the appendix. From (32, 36), we observe that for a large number $m$ of left and right elastic blocks (cells) in the superposition scheme, the displacement behaves as

$$u(m) = u_\infty + \frac{A}{m}, \tag{6}$$

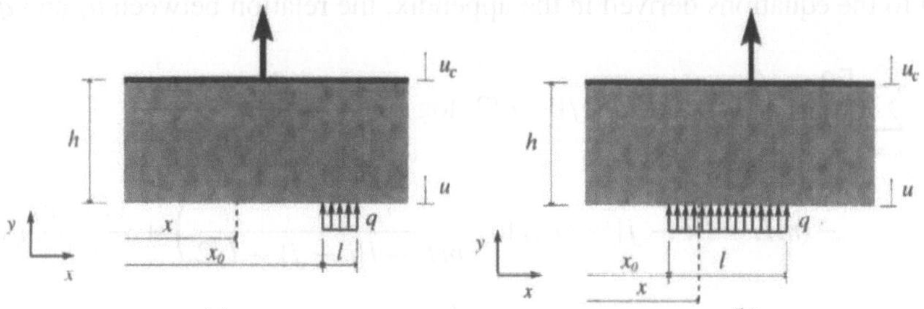


*Figure 2.* In the two schemes, just the elastic block is represented. On the left (a), displacement is measured *outside* the location of the distributed load. On the right (b), displacement is measured *inside* the location of the distributed load.

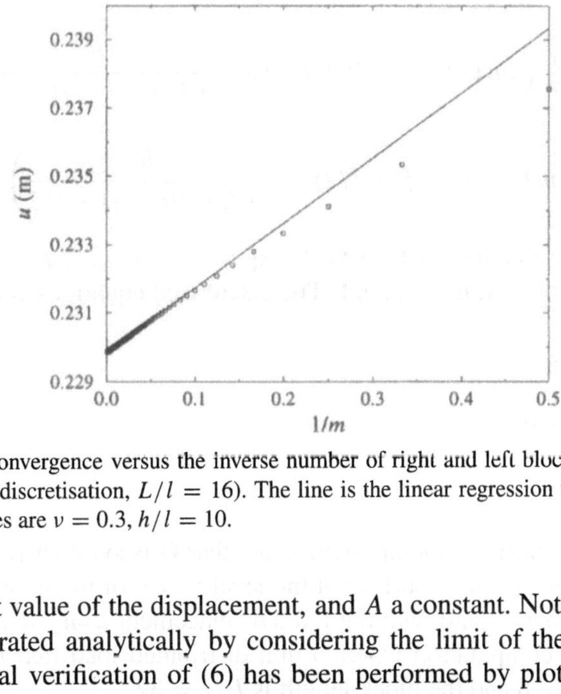


*Figure 3.* Displacement convergence versus the inverse number of right and left blocks for the element under a load $P = E$ (16-element discretisation, $L/l = 16$). The line is the linear regression with respect to equation 6. The elastic block properties are $\nu = 0.3$, $h/l = 10$.

where $u_\infty$ is the exact value of the displacement, and $A$ a constant. Note that this result could be probably demonstrated analytically by considering the limit of the series defined in the appendix. A numerical verification of (6) has been performed by plotting the displacement versus the inverse of number $m$ of blocks (Figure 3). The intersection with the vertical axis provides the exact displacement for an infinite periodic block resulting from the superposition of an infinite number of pressure distributions.

### 2.3. Discretisation of the elastic block response

The last step necessary to compute the displacement is to write the previous (32, 36) in a discrete form. We choose to compute the lower surface displacement in the middle of each part of width $l$. For a distributed pressure $q_j$ on the element $j$ and a measured displacement $u_i$ on element $i$, the continuous variable $x$ (the distance between the applied load and the measured displacement) is changed into

$$\begin{cases} x = l|i-j| - l/2 & \text{outside the distributed load (32)} \\ x = l/2 & \text{inside the distributed load (36).} \end{cases} \tag{7}$$

According to the equations derived in the appendix, the relation between $u_i$ and $q_j$ becomes

$$\begin{aligned}
u_i - u_c \;=\; & \sum_{m=1}^{+\infty}\left[\frac{2q_j}{\pi E}\left((mL - l|i-j| + l/2)\log\frac{h}{mL - l|i-j| + l/2}\right.\right.\\
& \left.\left. -(mL - l|i-j| - l/2)\log\frac{h}{mL - l|i-j| - l/2}\right) + \frac{(1-\nu)}{\pi E}q_j l\right]\\
& +\frac{2q_j}{\pi E}\left((l|i-j| + l/2)\log\frac{h}{l|i-j| + l/2}\right.\\
& \left. -\,(l|i-j| - l/2)\log\frac{h}{l|i-j| - l/2}\right) + \frac{(1-\nu)}{\pi E}q_j l\\
& +\sum_{m=1}^{+\infty}\left[\frac{2q_j}{\pi E}\left((mL + l|i-j| + l/2)\log\frac{h}{mL + l|i-j| + l/2}\right.\right.\\
& \left.\left. -(mL + l|i-j| - l/2)\log\frac{h}{mL + l|i-j| - l/2}\right) + \frac{(1-\nu)}{\pi E}q_j l\right] \qquad (8)
\end{aligned}$$

We introduce now the distributed load vector $\mathbf{q} = \{q_1, q_2, \ldots, q_n\}$ and the corresponding displacement vector $\mathbf{u} = \{u_1, u_2, \ldots, u_n\}$. The discretised equations can be cast in the form

$$\mathbf{q} = \mathbf{G}.(u_c \begin{pmatrix} 1 \\ 1 \\ \vdots \\ 1 \end{pmatrix} - \mathbf{u}), \qquad (9)$$

where $\mathbf{G}$ is the Green matrix of the problem. Note that $\mathbf{G}$ is symmetric and is independent of $\mathbf{u}$. The displacement of the block surface at the level $y = 0$ of the interface can be computed for any pressure distribution and with the fixed displacement $u{=}u_c$ at $y = h$. Figure 4 shows an example of such a computation where a unit distributed load (tension) is locally applied between $x = 7$ and $x = 8$ and the block length is $L/l = 32$.

### 2.4. The Interface Behaviour

The constitutive relation of the interface is assumed to derive from a damage model. Because the interface is one-dimensional, damage is a scalar variable, called $D$, that evolves between 0 (undamaged material) and 1 (broken material). For any interface element, the force-displacement relation is

$$f = C(1 - D)u, \qquad (10)$$

where $C$ is the stiffness of the undamaged material. We introduce the simplest form for the damage assuming it is simply proportional to the displacement

$$D = u/u_0. \qquad (11)$$

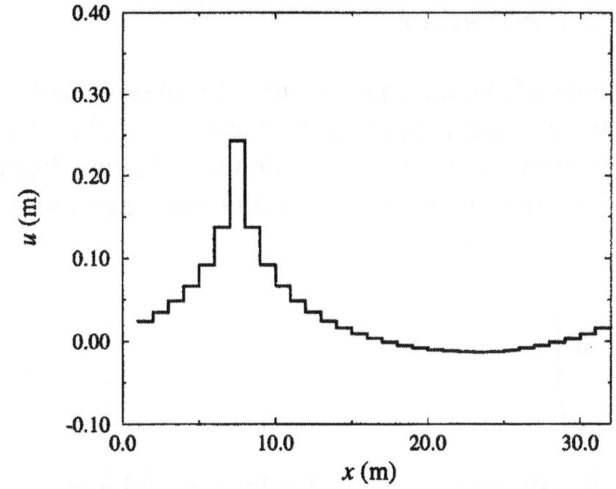


*Figure 4.* The deformation $u$ of one block of width $L/l = 32$ (32 elements) with periodic conditions, under a pressure applied on an unit length segment $P = E$. The elastic block properties are $\nu = 0.3$, $h/l = 10$.

This relation is in fact, the asymptotic behaviour of a fibre bundle model with an infinite number of fibres of unit stiffness, and with a uniform distribution between 0 and 1 for the thresholds where the fibres break irreversibly (Daniels 1945; Hemmer and Hansen, 1992). For simplicity, we introduce a dimensionless displacement $\tilde{u} = u/u_0$. With these notations, we have simply $D = \tilde{u}$.

In the following, geometrical distances will be made dimensionless by scaling them by the size of one element $l$. Forces will be scaled by $\mathrm{C}u_0/l^2$. Thus we introduce the reduced variable $\tilde{E} = El^2/\mathrm{C}u_0$, and $\tilde{f} = fl/(\mathrm{C}u_0)$. Note that $\tilde{E}$ is the only the pertinent parameter in our problem. Equation (10) becomes

$$\tilde{f} = l(1 - \tilde{u})\tilde{u}. \tag{12}$$

For an element $i$, the equivalent distributed load $q_i$ over the element width $l$ is

$$\tilde{q}_i = \frac{\tilde{f}_i}{l}. \tag{13}$$

Finally, if the element is undergoing loading, the relation between $\tilde{q}_i$ and $\tilde{u}_i$ is

$$\tilde{q}_i = \tilde{u}_i(1 - \tilde{u}_i). \tag{14}$$

In the situation of unloading, we have

$$\tilde{q}_i = K_i^{sc}\tilde{u}_i, \tag{15}$$

where the secant modulus $K_i^{sc} = 1 - \max(\tilde{u}_i)$ over the loading history. The corresponding rate relations are

$$\begin{cases} \dot{\tilde{q}}_i = (1 - 2\tilde{u}_i)\dot{\tilde{u}}_i & \text{loading } \dot{\tilde{u}}_i > 0 \\ \dot{\tilde{q}}_i = K_i^{sc}\dot{\tilde{u}}_i & \text{unloading } \dot{\tilde{u}}_i \leqslant 0. \end{cases} \tag{16}$$

From now on, we will omit the $\sim$ superscript for reduced variables.

### 2.5. Equilibrium of the interface

The set of (9) has been derived from the equilibrium of the elastic block. We have now to write the equilibrium of the interface, that is to write that the distributed forces applied on the lower surface of the block result from the interface constitutive relations. Because the interface has a nonlinear constitutive response, we write this condition in a rate form. Equation (9) is

$$\dot{\mathbf{q}} = \mathbf{G}.\left(\dot{u}_c \begin{pmatrix} 1 \\ 1 \\ \vdots \\ 1 \end{pmatrix} - \dot{\mathbf{u}}\right), \tag{17}$$

where $\dot{\mathbf{u}} = \{\dot{u}_1, \dot{u}_2, \ldots, \dot{u}_n\}$ the vector of nodal velocities and $\dot{\mathbf{q}} = \{\dot{q}_1, \dot{q}_2, \ldots, \dot{q}_n\}$ the vector of loading rate are introduced. In the same way, the interface response is written from equation (16)

$$\dot{\mathbf{q}} = \mathbf{K}(\mathbf{u})\dot{\mathbf{u}}, \tag{18}$$

where $\mathbf{K}(\mathbf{u})$ is just a diagonal matrix.

In these two last equations, the vectors of the rate of distributed forces should be equal in order to ensure equilibrium of the interface. We will define now

$$g_0 = \sum_{j=1}^{j=n} G_{ij} \ \forall i, \quad \text{and } \dot{\mathbf{u}}_g = \{g_0\dot{u}_c, g_0\dot{u}_c, \ldots, g_0\dot{u}_c\}. \tag{19}$$

Hence, the global rate equation that governs the system is

$$(\mathbf{K} + \mathbf{G})\dot{\mathbf{u}} = \dot{\mathbf{u}}_g. \tag{20}$$

Or

$$\mathbf{M}\dot{\mathbf{u}} = \dot{\mathbf{u}}_g, \tag{21}$$

where $\mathbf{M} = \mathbf{K} + \mathbf{G}$ is a symmetric matrix. Note that this equation has been derived assuming that the displacement at the upper surface of the elastic block is controlled with a rate $\dot{u}_c$. Load control conditions would provide a slightly different system of equations.

## 3. Bifurcation and stability analysis

### 3.1. Loss of uniqueness

The system has an unique solution unless $\mathbf{M}$ is singular, i.e.

$$\det \mathbf{M} = 0. \tag{22}$$

Because $\mathbf{M}$ is real and symmetric, therefore it can be diagonalised, this condition corresponds to the existence of at least one vanishing eigenvalue.

### 3.2. Loss of stability

For our problem, the condition that guarantees the stability of the system is (Hill, 1959)

$$\dot{\mathbf{u}}^t.\mathbf{M}.\dot{\mathbf{u}} > 0 \quad \forall \dot{\mathbf{u}} \tag{23}$$

The condition of critical equilibrium ($\dot{\mathbf{u}}^t.\mathbf{M}.\dot{\mathbf{u}} = 0$) is

$$\det \mathbf{M} = 0 \text{ or } \dot{\mathbf{u}} \text{ is orthogonal to } \mathbf{M}\dot{\mathbf{u}}. \tag{24}$$

### 3.3. Post-bifurcation response

When the condition $\det \mathbf{M} = 0$ is reached, i.e. when at least one eigenvalue, called here $\lambda_1$, vanishes, the solution of the rate equilibrium equations is no longer unique and a classical perturbation technique can be implemented in order to search for another solution in addition to the fundamental one $\dot{\mathbf{u}}$ (De Borst, 1987; 1988). We call $\mathbf{v}_1$ the eigenvector associated to $\lambda_1$. One can easily show that the following displacement rate vector $\dot{\mathbf{u}}^*$ also satisfies the equilibrium equation for any arbitrary scalar $\beta$

$$\dot{\mathbf{u}}^* = \dot{\mathbf{u}} + \beta.\mathbf{v}_1. \tag{25}$$

We have also to determine if the singularity of $\mathbf{M}$ corresponds to a limit point (under displacement controlled conditions) or to a bifurcation point. This can be performed following the steps in (De Borst, 1987; 1988). The solution verifies

$$\mathbf{M}\dot{\mathbf{u}} = \dot{\mathbf{u}}_g. \tag{26}$$

$\dot{\mathbf{u}}$ could be written in the $\mathbf{M}$-eigenvectors $\mathbf{v}_i$ base

$$\dot{\mathbf{u}} = \sum_{i=1}^{n} (\mathbf{v}_i^t.\dot{\mathbf{u}}).\mathbf{v}_i. \tag{27}$$

The equilibrium equation (21) becomes

$$\mathbf{M}\sum_{i=1}^{n} (\mathbf{v}_i^t.\dot{\mathbf{u}}).\mathbf{v}_i - \sum_{i=1}^{n} (\mathbf{v}_i^t.\dot{\mathbf{u}}_g).\mathbf{v}_i = 0 \tag{28}$$

or

$$\sum_{i=1}^{n} (\lambda_i(\mathbf{v}_i^t.\dot{\mathbf{u}}) - \mathbf{v}_i^t.\dot{\mathbf{u}}_g).\mathbf{v}_i = 0. \tag{29}$$

All the eigenvectors are independent, and then for all $i$ we have

$$\lambda_i(\mathbf{v}_i^t.\dot{\mathbf{u}}) - \mathbf{v}_i^t.\dot{\mathbf{u}}_g = 0. \tag{30}$$

For $i = 1$, it leads to $\mathbf{v}_i^t.\dot{\mathbf{u}}_g = 0$. Then, a limit point corresponds to $\dot{\mathbf{u}}_g = 0$ (loss of stability). A bifurcation point corresponds to $\mathbf{v}_i^t \perp \dot{\mathbf{u}}_g$ (loss of uniqueness).

For our system, the first eigenvalue that vanishes is unique for a limit point, and is at least double for a bifurcation point because of the periodic boundary conditions. In these equations, however, the eigenvalues depend on **M**, and then on **K**. But from (16), **K** is not a single valued matrix. It depends on the strain rate field in the interface, i.e. if damage grows or not in each element. It follows that a complete bifurcation and stability analysis would require the investigation of all the possible loading/unloading combinations in the rate constitutive relations of the interface. This can be hardly performed, except for a very coarse discretisation of the interface. Computations for coarse discretisations show that it may be sufficient to consider the tangent stiffness matrix computed under the assumption that all the interface elements which were loading (growth of damage) during the previous displacement step are still undergoing loading.

## 4. Results

### 4.1. Bifurcation

The post bifurcation analysis is applied to the system with a number $L$ of elements in the interface that varies between 4 and 32. The first singularity encountered is as just mentioned either a limit point (i.e. the maximum global displacement) or a bifurcation point. At the bifurcation point and according to the considerations in the previous section, we look at the case where the vanishing eigenvalue is at least double. Then, two independent eigenvectors are found. They have a $L$-periodic 'sine' form, and are shifted forward with a phase of $L/4$. Note that because of the periodic boundary conditions, all eigenvectors of this form could be associated to the vanishing eigenvalues. One of these eigenvectors is then added to the perturbation $\dot{\mathbf{u}}$ in accord with (25), and the loading could be continued according to the procedure defined in (De Borst, 1987; Pijaudier-Cabot and Huerta, 1991). In the all computations only one bifurcation point was found during the course of loading. After bifurcation occurred, the equilibrium equations had a single solution in the post-bifurcation regime.

The response of a 32-element system is shown in Figure 5. It is similar to the hierarchical model response (Delaplace et al., 1998). This result is also obtained if we introduce artificially an initial defect (initial nonzero damage for instance) in the interface. In this case, the damage profile in the interface is determined by the defect location and bifurcation does not occur, same as in the hierarchical model with initial disorder in the interface. In the presence of an initial defect, the global behaviour is only slightly affected.

### 4.2. Damage evolution

One of the main motivations of this work is to study the evolution of damage in the interface. This evolution is plotted in Figure 6. From the very beginning of the loading to the first bifurcation point, damage $D$ grows homogeneously in the interface. It condenses progressively into a narrower and narrower region down to just one element, which breaks (i.e. $D = 1$ in this element). Then, a macro-crack is initiated in this element and propagates in the interface. This last stage is not analysed in this paper. This damage evolution is very close to what was encountered with the hierarchical model. An interesting analytical result was obtained with

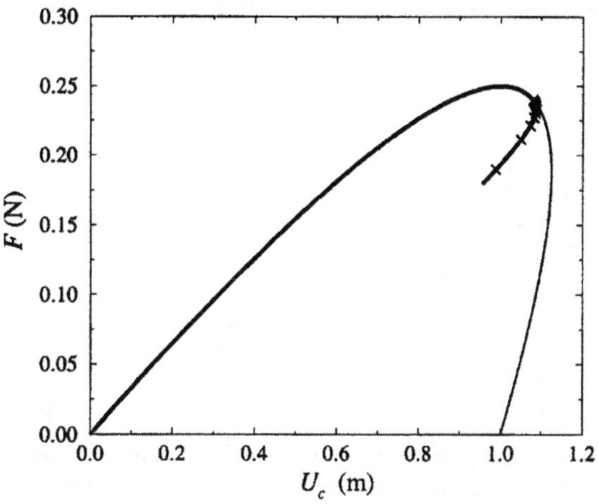


*Figure 5.* Global-Force displacement of a 32-element size, $L/l = 32$. Crosses correspond to the different damages profiles in Figure 6. The thin curve is the homogeneous behaviour. The elastic block properties are $\nu = 0.3$, $\tilde{E} = 2$, $h/l = 4$ m.

the hierarchical model about the damage profile at the initiation of the first macro-crack. This unusual profile was

$$D(x) = \frac{1}{2} + \frac{E}{2x}, \tag{31}$$

where $x$ is the distance, in the hierarchical sense, between the macro-crack location and the location at which damage is computed. This analytical result could not be found with the present approach, but, as shown in Figure 6, the damage profile at the initiation of the first crack is well fitted with such an inverse law with the distance. Note that the damage spreads all over the interface, and that there is no characteristic length which scales the damage distribution at the inception of macro-cracking. This is contrary to what is expected in the Hillerborg approach (Hillerborg et al., 1976), at least in the regime of inception of cracking. It is important to note the present conclusion does not pertain to the macro-crack propagation regime which is not analysed here.

### 4.3. SCALE EFFECT

Two scaling parameters are relevant in this problem. The first one is the block height, called $h$, and the second is the ratio between the block stiffness $E$ and the interface stiffness $C$. If we change these two parameters independently, we observe the same effect: the initial slope in the force displacement response changes (Figure 7). It can be easily understood by considering an equivalent system: a spring of stiffness $E$ and of length $h$ which is coupled in series with one damageable element influences the initial slope of the response, same as in the one-dimensional analysis devised in (Bažant and Cedolin, 1991).

A more relevant variable is the so called variable $g_0$ (Section 2.5). This variable depends of course on $E$ and $h$, but its influence is directly related to the global behaviour of the system through (21). Hence, a constant $g_0$ gives the same initial stiffness and the same characteristic if the interface displacement is homogeneous. Under this condition, the influence of $h$ and $E$ can be conveniently analysed: The main effect of $h$ and $E$ is to shift the position of the first

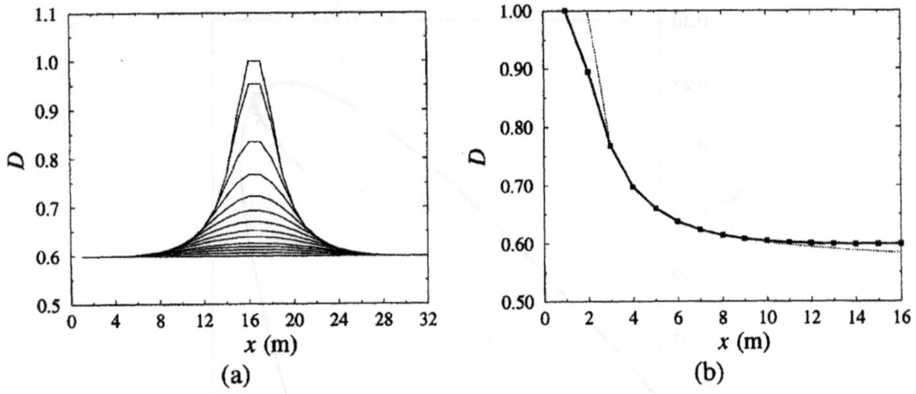


*Figure 6.* On the left (a) is plotted the post-bifurcation evolution of damage in the interface with 32 elements. The right side (b) show the half damage profile just at the first macro-crack tip, with a fitted curve that follows Equation 31.

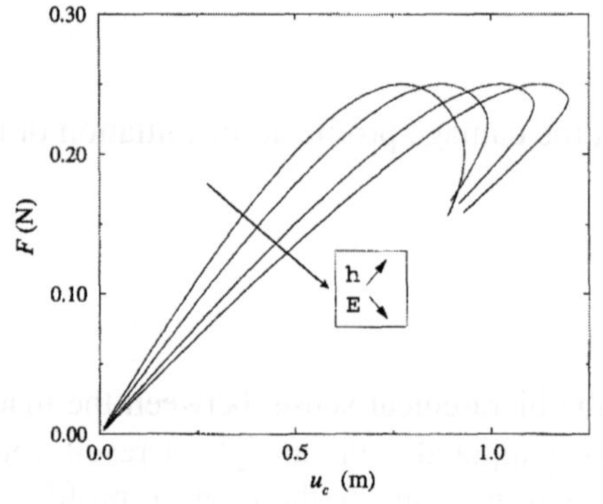


*Figure 7.* The force-displacement response of a system with 16 elements in the interface, $L/l = 16$. Initial stiffness decreases if $h$ increases (for a fixed $E$). Initial stiffness decreases if $E$ decreases (for a fixed $h$).

bifurcation point. More specifically, bifurcation can be encountered before or after the limit point. For a fixed $g_0$, if $h$ and $E$ increase, the first bifurcation is delayed during the loading. On the lower left zoom of the Figure 8, the first post-bifurcation branch (dashed-bold) is in the stable regime. The solid becomes unstable after the maximum displacement is reached (fine-dashed). Upon an increase of $h$ and $E$ (lower right zoom) the response of the system becomes unstable before bifurcation occurs (fine continuous curve).

## 5. Conclusion

We have analysed the growth of damage in a system that is composed of a softening interface coupled in series to an elastic block. This study follows a first one where a hierarchical model was used in order to obtain main results analytically. We have used here a more classical representation (two-dimensional continuum), instead of a hierarchical decomposition. We have focused attention on the post bifurcation response, where damage localises progressively. The following conclusions can be drawn

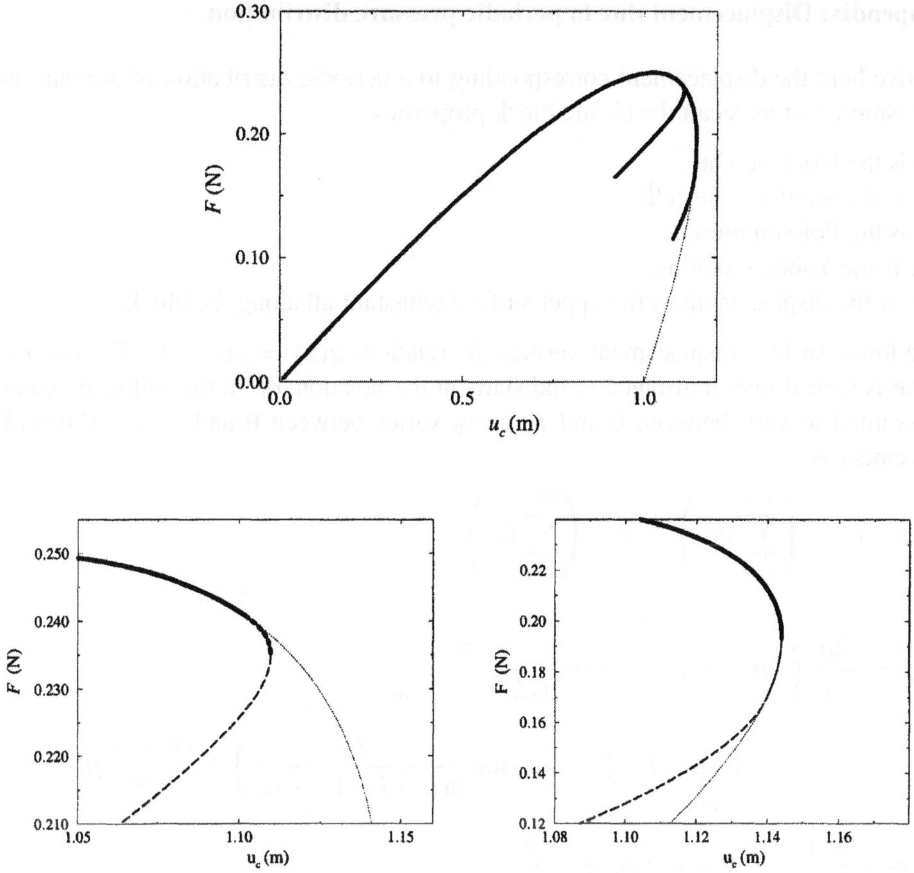


*Figure 8.* The force-displacement response of a system with 32 elements in the interface, $L/l = 32$. The elastic block Poisson ratio is 0.3. The two lower graphs are zooms on the bifurcation point. The left one corresponds to $\tilde{E} = 4$ and $h/l = 2$. The right one corresponds to $\tilde{E} = 14$ and $h/l = 7$. The stable states are represented by bold (continuous or dashed) lines, whereas the unstable states are represented by thin (continuous or dashed) lines.

- Damage grows first homogeneously in the interface. At the first bifurcation point, damage localises continuously from a macroscopic scale to a narrower and narrower region. Then, just one element fails and a macro-crack is initiated.
- The damage profile is computed at the initiation of the macro-crack. For the chosen interface behaviour, an inverse power law with the distance was obtained for the hierarchical model and the numerical result is very close the analytical one.
- The scaling parameters in the problem are the block height and the relative stiffnesses of the elastic block and of the interface. These two parameters influence the relative location of the bifurcation point and of the limit point on the response of the solid. There is no length scale in the damage profile at the inception of macro-cracking.

Finally, it should be noted that obtaining experimentally the response of such a softening interface should be quite difficult, due to the loss of stability and most of all due to the bifurcation of equilibrium states. The same problem is encountered when the constitutive response of a material is sought from a tensile test.

## 6. Appendix: Displacement due to periodic pressure distribution

We derive here the displacement corresponding to a periodic distribution of pressure using a superposition. Let us recall the elastic block properties

- $h$ is the block height;
- $L$ is the width of one cell;
- $\nu$ is the Poisson ratio;
- $E$ is the Young modulus;
- $u_c$ is the displacement of the upper surface (constant all along the block).

The lower surface displacement verifies the relation: $u(x) = u(x+L)$. For one cell, the pressure is spread over a distance $l$, and starts at the position $x_0$. In the following equations, $x$ is assumed to vary between 0 and $L$, as $x_0$ varies between 0 and $L-l$. From (3), the displacement is

$$u(x) - u_c = \left(\sum_{m=1}^{+\infty} \mathcal{A}_m\right) + \mathcal{B} + \left(\sum_{m=1}^{+\infty} \mathcal{C}_m\right) \tag{32}$$

with

$$\begin{aligned}\mathcal{A}_m &= \frac{2q}{\pi E}\left((mL - |x-x_0|)\log\frac{h}{mL-|x-x_0|}\right.\\ &\quad \left. - (mL - l - |x-x_0|)\log\frac{h}{mL-l-|x-x_0|}\right) + \frac{(1-\nu)}{\pi E}ql,\end{aligned} \tag{33}$$

$$\begin{aligned}\mathcal{B} &= \frac{2q}{\pi E}\left((l + |x-x_0|)\log\frac{h}{l+|x-x_0|}\right.\\ &\quad \left. - |x-x_0|\log\frac{h}{|x-x_0|}\right) + \frac{(1-\nu)}{\pi E}ql,\end{aligned} \tag{34}$$

$$\begin{aligned}\mathcal{C}_m &= \frac{2q}{\pi E}\left((mL + l + |x-x_0|)\log\frac{h}{mL+l+|x-x_0|}\right.\\ &\quad \left. - (mL + |x-x_0|)\log\frac{h}{mL+|x-x_0|}\right) + \frac{(1-\nu)}{\pi E}ql.\end{aligned} \tag{35}$$

$\mathcal{A}_m$ is the left load contribution, $\mathcal{B}$ refers to the pressure distribution applied on the block where the displacement is measured, and $\mathcal{C}_m$ is the right load contribution. In the same way, if the measured displacement is inside the pressure distribution, the displacement is

$$u(x) - u_c = \left(\sum_{m=1}^{+\infty} \mathcal{D}_m\right) + \mathcal{E} + \left(\sum_{m=1}^{+\infty} \mathcal{F}_m\right) \tag{36}$$

with

$$\begin{aligned}\mathcal{D}_m &= \frac{2q}{\pi E}\left((mL + |x-x_0|)\log\frac{h}{mL+|x-x_0|}\right.\\ &\quad \left. - (mL - l + |x-x_0|)\log\frac{h}{mL-l+|x-x_0|}\right) + \frac{(1-\nu)}{\pi E}ql,\end{aligned} \tag{37}$$

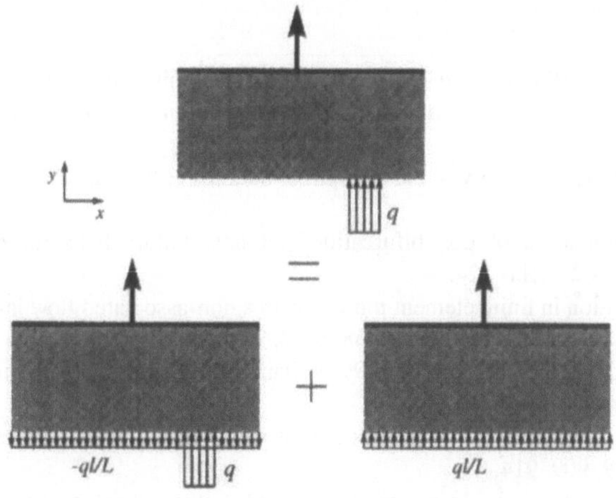


*Figure 9.* Superposition of the two problems.

$$\mathcal{E} = \frac{2q}{\pi E}\left((l - |x - x_0|)\log\frac{h}{l - |x - x_0|}\right.$$
$$\left. - |x - x_0|\log\frac{h}{|x - x_0|}\right) + \frac{(1-\nu)}{\pi E}ql, \quad (38)$$

$$\mathcal{F}_m = \frac{2q}{\pi E}\left((mL + l - |x - x_0|)\log\frac{h}{mL + l - |x - x_0|}\right.$$
$$\left. - (mL + |x - x_0|)\log\frac{h}{mL + |x - x_0|}\right) + \frac{(1-\nu)}{\pi E}ql, \quad (39)$$

$\mathcal{D}_m$ is the left pressure contribution, $\mathcal{E}$ refers to the pressure distribution applied on the block where the displacement is measured, and $\mathcal{F}_m$ is the right pressure contribution.

This superposition scheme leads to an infinite displacement, due to the constant expression $(1-\nu)/(\pi E)ql$. Of course, this unrealistic physical feature is due to the periodic conditions. In order to avoid this divergence, one solution is to superpose two problems (Figure 9)

- the first problem is the infinite elastic block, loaded by a pressure distribution $q$ over a length $l$, that starts at position $x_0$, and an uniform pressure distribution $-ql/L$ all over the surface. Hence, the load resultant along the surface is zero, and the divergent term vanishes in (8).
- The second problem is the same elastic block, loaded by just an uniform pressure distribution $ql/L$. In this case, the induced surface displacement is finite and well known for all $x$

$$u(x) - u_c = h\left(\frac{1-\nu^2}{E}\right)\frac{ql}{L} \quad (40)$$

Finally, the addition of these two displacements provides a finite displacement as expected on physical grounds.

## References

Bažant, Z.P. and Cedolin, L. (1991). *Stability of Structures*, Oxford University press.

Bažant, Z.P. and Planas, J. (1998). *Fracture and Size Effect in Concrete and Other Quasi-Brittle Materials*, CRC Press.

Daniels, H.E. (1945). The statistical theory of the strength of bundles of threads, *Proceedings of the Royal Society*, London, **A183**, 405–435.

De Borst, R. (1987). Computation of post-bifurcation and post failure behavior of strain softening solids. *Computers and Structures* **25**, 211–224.

De Borst, R. (1988). Bifurcation in finite element models with a non-associated flow law. *International Journal of Numerical and Analytical Methods in Geomechanics* **12**, 99–116.

Delaplace, A., Roux, S. and Pijaudier-Cabot, G. (1998). Damage cascade in a softening interface. *International Journal of Solids and Structures*, In press.

Hemmer, Per C. and Hansen, A. (1992). The distribution of simultaneous fiber failures in fiber bundles. *Journal of Applied Mechanics* **59**, 909–914.

Hill, R. (1959). Some basic principles in the mechanics of solids without a natural time. *Journal of the Mechanics and Physics of Solids* **7**, 209–225.

Hillerborg, A., Modeer, M. and Petersson, P.E. (1976). Analysis of crack formation and crack growth in concrete by means of fracture mechanics and finite elements. *Cement and Concrete Research* **6**, 773–782.

Pijaudier-Cabot, G. and Huerta, A. (1991). Finite element analysis of bifurcation in nonlocal strain softening solids. *Computer Methods in Applied Mechanics and Engineering* **90**, 905–919.

Timoshenko, S. (1948). *Théorie de l'Élasticité*, Librairie Polytechnique, Paris and Liege.

*International Journal of Fracture* **95:** 175–194, 1999.


# A model for predicting grain boundary cracking in polycrystalline viscoplastic materials including scale effects

K.L.E. HELMS[1], D.H. ALLEN[1] and L.D. HURTADO[2]
[1]*Aerospace Engineering Department, Texas A&M University, College Station, TX 77843*
*e-mail: dhallen@aero.tamu.edu*
[2]*Sandia National Labs, Albuquerque, NM 88101*

Received 28 December 1998; accepted in revised form 31 December 1998

**Abstract.** A model is developed herein for predicting the mechanical response of inelastic crystalline solids. Particular emphasis is given to the development of microstructural damage along grain boundaries, and the interaction of this damage with intragranular inelasticity caused by dislocation dissipation mechanisms. The model is developed within the concepts of continuum mechanics, with special emphasis on the development of internal boundaries in the continuum by utilizing a cohesive zone model based on fracture mechanics. In addition, the crystalline grains are assumed to be characterized by nonlinear viscoplastic mechanical material behavior in order to account for dislocation generation and migration. Due to the nonlinearities introduced by the crack growth and viscoplastic constitution, a numerical algorithm is utilized to solve representative problems. Implementation of the model to a finite element computational algorithm is therefore briefly described. Finally, sample calculations are presented for a polycrystalline titanium alloy with particular focus on effects of scale on the predicted response.

**Keywords:** Polycrystalline solids, viscoplasticity, damage, cohesive zone, crack propagation, grain boundary fracture, titanium, Timetal-21S, scale effects.

## 1. Introduction

Polycrystalline solids such as metals experience a variety of thermodynamically dissipative processes during inelastic deformation. These dissipative processes may include several types of dislocation rearrangement mechanisms within grains (Ashby, 1983), as well as fracture and/or sliding along grain boundaries. When these processes occur at temperatures that are generally in excess of about three-tenths of the melting point of the solid, they are inherently rate dependent. Due to the complexity of these events, and the fact that they occur in multiple grains simultaneously, analytic models of these phenomena have not been completely developed at this time.

One part of this problem that is quite difficult deals with the dissipative processes that occur along grain boundaries. Due to the migration of dislocations to grain boundaries, vacancies can open up, thereby assisting sliding and fracture between grains. In some environmental circumstances, this process can produce the majority of energy dissipation during loading. Thus, models that can physically account for this mechanism are needed.

In the current paper, the authors will discuss the deployment of a rate-dependent cohesive zone model to account for the process of grain boundary fracture and/or sliding. Inelasticity within the grains will be accounted for by employing a unified viscoplasticity model. The resulting boundary value problem will be formulated within the framework of continuum mechanics. In order to solve the resulting highly nonlinear set of integral-partial differential equations, the finite element method will be employed.

### 1.1. Recent Research

Considerable research has focused on developing models for predicting the behavior of inelastic materials that undergo evolving damage. For purposes of the present paper, these efforts are distinguished herein as either continuum damage models, or discrete fracture models. In the continuum damage class of models, an internal variable is formulated within the macroscopic constitutive model for the multi-granular material that accounts for the locally averaged dissipation due to microfracture (Voyiadjis and Kattan, 1993). In the discrete fracture class of models, a micromechanical analysis of a multi-granular material is performed, with each microcrack modeled explicitly as an evolving internal boundary in a representative volume element (RVE) of the material. The former of these approaches has been studied in much greater detail, and due to the rather large quantity of research by this approach, no attempt will be made herein to review the works that utilize continuum damage mechanics approaches. In the current paper, the latter approach will be considered. We detail below some recent research related to this approach.

Experimental evidence indicates that numerous polycrystalline solids undergo significant grain boundary cracking, such as metals (Hertzberg, 1976; Suresh, 1991; Majumdar and Newaz, 1991), argillaceous salt (Chan et al., 1997a; Chan et al., 1997b), and sea ice (Schapery, 1997). In the current paper we have chosen to model the multi-granular behavior of Timetal-21S, a beta titanium alloy with microstructure as shown in Figure 1. Introduced in 1989, Timetal-21S was developed as a strip producible, oxidation resistant beta titanium alloy (Larsen et al., 1993). Timetal-21S is cold formable, age hardenable, and weldable. In addition to being used as a matrix material in metal matrix composites, Timetal-21S has also found limited applications in neat form. Because of many promising qualities, Timetal-21S has become the subject of much research (Foulk et al., 1998).

Numerous models have been proposed to predict the undamaged inelastic material behavior of Timetal-21S. At elevated temperatures, titanium is highly rate dependent. In fact, some viscoplastic theories were developed specifically for titanium-based matrices (Arnold et al., 1996). However, because the Bodner-Partom (B-P) model was used herein, the discussion has been limited to the B-P viscoplastic constitutive model (Chan et al., 1988). Neu (1993) first determined the material parameters for the B-P model in the temperature range of 23° to 815°C. Although these parameters were extremely accurate for monotonic simulations, cyclic fatigue yielded negative hardening values. The initial parameters emphasized directional hardening and upon stress reversal (compression), the directional hardening term $Z_D$, and consequently the total hardening term $Z$, became negative. Zuiker and Sanders (1995) proposed a new set of constants stressing the effects of isotropic hardening. After further review of more test data, Neu and Bodner (1995) presented a revised set of constants which also emphasized the effects of isotropic hardening. In addition, exponential evolution laws were proposed to govern the hardening rates and thermal recovery terms. A revised set of constants was developed for two specific temperatures, 482°C and 650°C.

In order to model the progression of cracks in a multi-granular inelastic solid, certain pragmatic assumptions must be made in order to construct a model that is both reasonably accurate and computationally solvable. Although singular finite elements can be employed to simulate crack propagation in a general viscoplastic crystalline material, highly refined

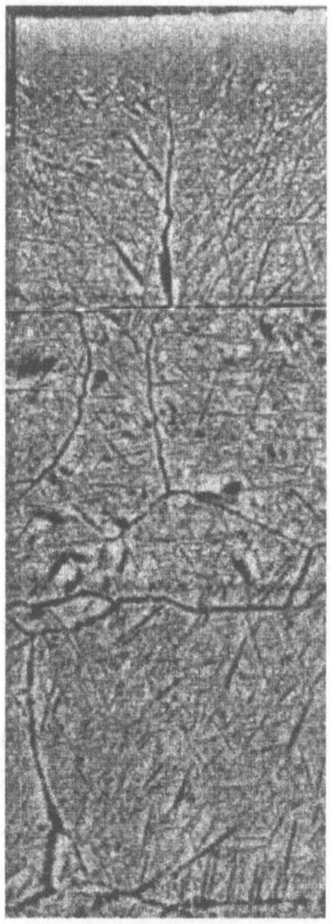

*Figure 1.* Multi-grain microstructure of Timetal 21S.

finite element meshes are required to accurately predict even the growth of a single crack. In addition, the critical energy release rate is generally not a material constant in inelastic solids. Furthermore, the geometric scale of intergranular zones is such that it is feasible to approximate them by planes of zero thickness in the undeformed state.

Cohesive zone models, first introduced by Dugdale (1960) and Barenblatt (1962), provide a computationally efficient and thermodynamically consistent method for the transition from internal surfaces to traction-free boundaries (Yoon and Allen, 1998; Allen and Searcy, 1998). Additionally, a cohesive zone formulation can account for a non-stationary critical energy release rate, and it does not require monitoring of the critical energy release rate in order to predict crack extension. For the above reasons, and due to the fact that numerous evolving microcracks are to be modeled simultaneously, we have chosen to employ a cohesive zone model.

Recent cohesive zone models that have influenced the current research have been proposed by Needleman (1987) and Tvergaard (1990). Needleman provided a unified framework to describe the evolution from initial debonding to complete separation in a cohesive zone using a cubic traction-displacement relation that was intended to simulate the microphysics of molecular bond stretching. Needleman has employed cohesive zones to model debonding of spherical (Needleman, 1987) and cylindrical (Nutt and Needleman, 1987) inclusions. Other works followed that studied the behavior of ceramic inclusions in metal matrices (Christman et al., 1989; Llorca et al., 1991), as well as modeling the decohesion of a viscoplastic block from a rigid substrate (Needleman, 1990a; Needleman, 1990b).

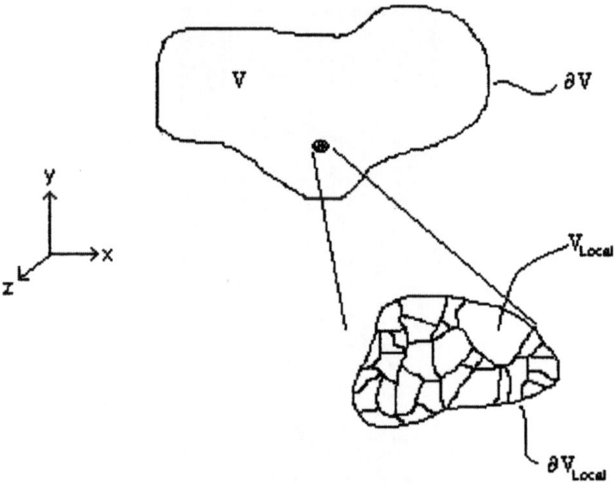


*Figure 2.* Continuum representation of a polycrystalline material.

Tvergaard (1990) developed a model similar to Needleman's model. The formulations were identical for purely normal separation, but Tvergaard (1990) coupled normal and tangential displacements to predict other modes of separation. Tvergaard (1990) modeled the decohesion of transversely staggered SiC whiskers in an aluminum matrix and subsequent void formation by normal and tangential separation. Tvergaard (1995) also investigated fiber fracture and decohesion for a similar system with transversely aligned fibers. In addition to modeling metal matrix composites, Tvergaard also employed similar methods to predict crack bridging in ceramics (Tvergaard, 1992).

Several efforts have used Tvergaard's model to predict crack growth. Allen et al. (1994) predicted interface cracking in continuous fiber metal matrix composites. Lissenden et al. (1996) also used cohesive zone elements to predict interface cracking. Allen et al. (1996) modeled the effect of surface cracking on fiber/matrix debonding.

The cohesive zone models proposed by both Needleman and Tvergaard are elasto-plastic in the sense that they both account for dissipation in the cohesive zone that is both rate and history independent (for the case of monotonic loading). Recently the second author and coworkers have developed a cohesive zone model that can account for both history and rate dependent features of crack growth (Allen and Searcy, 1998; Yoon and Allen, 1998). It is somewhat similar to models previously developed by Schapery (1986) and Knauss and Losi (1993). This model will be utilized in the current approach.

## 2. Model development

This section details the development of a model for predicting grain boundary cracking in polycrystalline viscoplastic solids. Towards this end, Figure 2 shows a representation of the material of interest along with a magnified view of the heterogeneous microstructure of the polycrystalline material. Both views are assumed to be sufficiently large that the concepts of continuum mechanics are applicable. We consider herein an analysis of the heterogeneous microscale. Using local continuum mechanics, we formulate the governing field equations in the following sections.

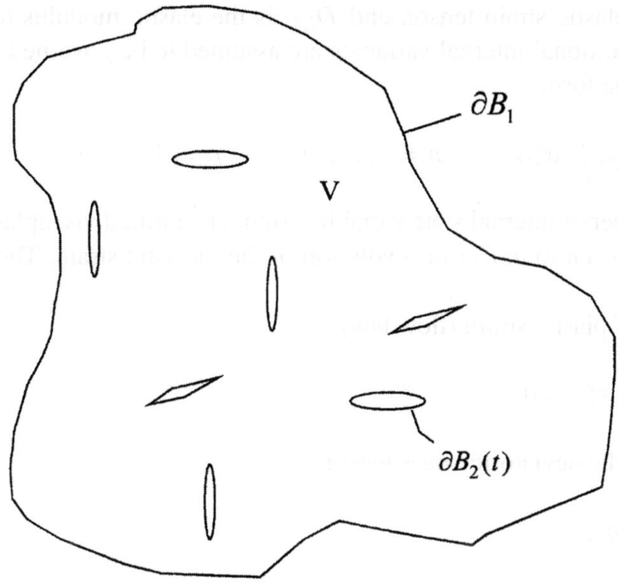


*Figure 3.* A general body $B$, with initial boundary $\partial B_1$, and time dependent internal boundaries, $\partial B_2(t)$.

### 2.1. Conservation laws

Consider a general continuum $B$, with initial boundary $\partial B_1$, and time dependent internal boundaries, $\partial B_2(t)$, as depicted in Figure 3. The primary observable state variables to be modeled are the displacement $u_i$, the Cauchy stress tensor $\sigma_{ij}$, and the infinitesimal strain tensor $\varepsilon_{ij}$. The conservation of mass is assumed to be trivially satisfied, assuming small displacements. The conservation of energy can be shown to be satisfied by the constitutive equations used herein, together with the assumption of constant temperature throughout the analysis (Allen, 1991). The conservation of angular momentum is satisfied by the assumption that there are no body moments when the stress tensor is symmetric. Finally, the conservation of linear momentum is satisfied by the following

$$\sigma_{ji,j} = 0, \tag{1}$$

where it has been assumed that body forces and inertial effects are negligible.

### 2.2. Kinematics

Assuming small displacement gradients, the strains are defined as follows

$$\varepsilon_{ij} \equiv \tfrac{1}{2}(u_{i,j} + u_{j,i}). \tag{2}$$

The above description produces strains that are an energy conjugate to the stress tensor.

### 2.3. Viscoplastic constitutive model

The mechanical constitutive behavior of each of the grains is assumed to be accurately modeled by a unified viscoplastic constitutive model of ordinary differential equation type. This formulation is given by (Chan, et al., 1988)

$$\sigma_{ij} = D_{ijkl}(\varepsilon_{kl} - \varepsilon^{\mathrm{I}}_{kl}), \tag{3}$$

where $\varepsilon_{kl}^{\mathrm{I}}$ is the inelastic strain tensor, and $D_{ijkl}$ is the elastic modulus tensor. The inelastic strain tensor and additional internal variables are assumed to be governed by internal variable evolution laws of the form

$$\dot{\alpha}_{ij}^{\eta} = \Omega_{ij}^{\eta}(\varepsilon_{kl}, \varepsilon_{kl}^{\mathrm{I}}, T, \alpha_{kl}^{\mu}); \qquad \eta = 1, \ldots, n; \qquad \mu = 1, \ldots, n, \tag{4}$$

where $n$ is the number of internal state variables. Bodner's unified viscoplasticity model (Chan et al., 1988) was chosen to model the evolution of the inelastic strain. The evolution laws are as follows.

The evolution of plastic strain (flow law)

$$\dot{\varepsilon}_{ij}^{P} = \lambda' S_{ij}, \qquad \dot{\varepsilon}_{kk}^{P} = 0, \tag{5}$$

which is related to the deviatoric stress tensor

$$S_{ij} = \sigma_{ij} - \tfrac{1}{3}\delta_{ij}\sigma_{kk} \tag{6}$$

and $\lambda'$ through the kinetic equation

$$D_2^P = D_0^P \left\{ - \left[ \frac{Z^2}{3J_2} \right]^n \right\}, \tag{7}$$

with

$$Z = Z_{\mathrm{I}} + Z_D, \tag{8}$$

$$D_2^P = \tfrac{1}{2}\dot{\varepsilon}_{ij}^P \dot{\varepsilon}_{ij}^P, \tag{9}$$

$$J_2 = \tfrac{1}{2} S_{ij} S_{ij}, \tag{10}$$

$$\lambda'^2 = \frac{D_2^P}{J_2}, \tag{11}$$

$Z$ is the summation of isotropic and directional hardening while $D_0$ represents the maximum shear strain rate. $J_2$ is a deviatoric stress invariant. The exponential behavior is governed by the kinetic parameter, $n$.

Evolution equations for the remaining internal state variables are expressed in terms of isotropic and directional hardening. The evolution law for isotropic hardening is

$$\dot{Z}_{\mathrm{I}} = m_1[Z_1 - Z_{\mathrm{I}}]\dot{W}_p - A_1 Z_1 \left[ \frac{Z_{\mathrm{I}} - Z_2}{Z_1} \right]^{r_1}, \tag{12}$$

where the plastic work rate and the initial value of isotropic hardening are

$$\dot{W}_p = \sigma_{ij}\dot{\varepsilon}_{ij}^P, \tag{13}$$

$$Z_{\mathrm{I}}(0) = Z_0. \tag{14}$$

The first term in (12) involves the isotropic hardening rate term $m_1$, and the upper limit of isotropic hardening $Z_1$. The second term contains the isotropic thermal recovery rate term $A_1$, the lower limit of isotropic hardening $Z_2$, and the exponential term dictating isotropic thermal recovery $r_1$. Directional hardening is governed by

$$\dot{\beta}_{ij} = m_2(Z_3 u_{ij} - \beta_{ij})\dot{W}_p - A_2 Z_1 \left[\frac{(\beta_{kl}\beta_{kl})^{1/2}}{Z_1}\right]^{r_2} V_{ij}, \tag{15}$$

where $V_{ij}$ and $u_{ij}$ are the norms of $\beta_{ij}$ and $\sigma_{ij}$, respectively, given by

$$V_{ij} = \frac{\beta_{ij}}{(\beta_{kl}\beta_{kl})^{1/2}}, \tag{16}$$

$$u_{ij} = \frac{\sigma_{ij}}{(\sigma_{kl}\sigma_{kl})^{1/2}}. \tag{17}$$

Coupled with the following initial conditions

$$Z_D(0) = 0, \tag{18}$$

$$\beta_{ij}(0) = 0, \tag{19}$$

the evolution law yields the expression for directional hardening

$$Z_D = \beta_{ij} u_{ij}. \tag{20}$$

The first term in (15) involves the directional hardening rate term $m_2$, and the upper limit of isotropic hardening $Z_3$. The second term contains the directional thermal recovery rate term $A_2$, and the exponential term governing directional thermal recovery $r_2$. In order to more accurately capture the behavior of Timetal-21S, Neu and Bodner (1995) wrote isotropic and directional hardening rate terms as exponential functions of the current hardening levels, $Z_{\mathrm{I}}$ and $Z_D$

$$m_1 = m_{1b} + (m_{1a} - m_{1b}) \exp[-m_{1c}(Z_{\mathrm{I}} - Z_o)], \tag{21}$$

$$m_2 = m_{2b} + (m_{2a} - m_{2b}) \exp[-m_{2c}(Z_D], \tag{22}$$

where $m_{1a}$ and $m_{2a}$ represent initial values while $m_{1b}$ and $m_{2b}$ represent asymptotic values. In addition, the isotropic thermal recovery rate term was also rewritten as an exponential function

$$A_1 = A_{1b} + (A_{1a} - A_{1b}) \exp[-A_{1c}(Z_{\mathrm{I}} - Z_2)], \tag{23}$$

where $A_{1a}$ and $A_{1b}$ represent the initial and asymptotic values, respectively.

Implementation of Bodner's model into the finite element algorithm is most easily accomplished in the following alternate form

$$\dot{\sigma}_{ij} = C_{ijkl}\dot{\varepsilon}_{kl}, \tag{24}$$

where $C_{ijkl}$ is the tangent modulus tensor and $\dot{\varepsilon}_{kl}$ is the total strain rate tensor. See Allen, et al. (1994) for the derivation of the tangent modulus tensor.

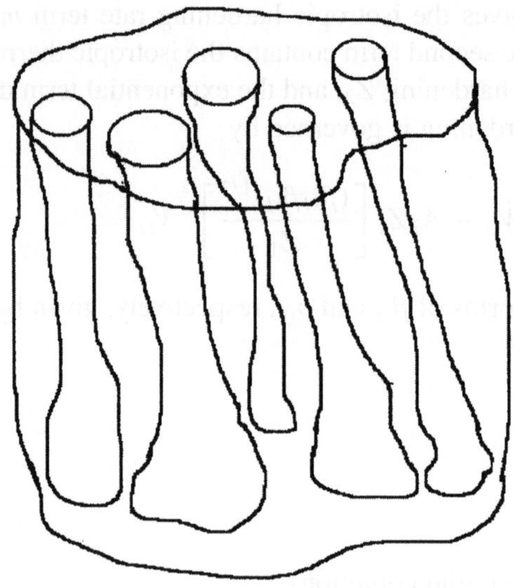

*Figure 4.* Depiction of assemblage of parallel fibrils bridging across grain boundaries.

## 2.4. Cohesive zone constitutive model

The second author and coworkers have recently developed a micoromechanically based cohesive zone model that is based on the assumption that the intergranular material can be characterized roughly as an assemblage of parallel fibrils bridging across grains, as shown in Figure 4. The constitutive equations for this cohesive zone model are given by (Allen and Searcy, 1998; Yoon and Allen, 1998)

$$T_i(t) = \frac{u_i(t)}{\delta_i \lambda(t)}(1-\alpha(t))\left[\sigma_i^f + \int_0^t E^c(t-\tau)\frac{\partial \lambda}{\partial \tau}\,\mathrm{d}\tau\right], \tag{25}$$

where the index $i$ specifies the coordinates normal and tangential to the crack tip plane; $T_i$ are the corresponding components of the cohesive zone traction; $u_i$ are the corresponding components of the displacements; $\sigma_i^f$ is the associated stress component at the crack tip upon the initiation of damage; $\alpha$ is a damage evolution parameter describing successive fracture of the fibrils due to elongation; $E^C$ is the relaxation modulus of the cohesive zone; and $\lambda$ is the boundary imposed longitudinal strain of fibrils in the cohesive zone such that

$$\lambda \equiv \sqrt{\left(\frac{u_n}{\delta_n}\right)^2 + \left(\frac{u_t}{\delta_t}\right)^2 + \left(\frac{u_s}{\delta_s}\right)^2}, \tag{26}$$

where the subscripts $n, t$, and $s$ denote local coordinates that are normal and parallel to the cohesive zones. The empirical material length parameters, $\delta_n$, $\delta_s$, and $\delta_t$ are additional material constants for the cohesive zones. Specifically $\delta_n$, is arbitrary while $\delta_s$ and $\delta_t$ are determined empirically by mixed-mode fracture experiments.

The following power law representation is used for the damage evolution

$$\dot{\alpha} = \alpha_1 \lambda^m; \qquad \dot{\lambda} \geqslant 0 \quad \text{and} \quad \alpha \leqslant 1, \tag{27}$$

$$\dot{\alpha} = 0; \qquad \dot{\lambda} \leqslant 0 \quad \text{and} \quad \alpha = 1, \tag{28}$$

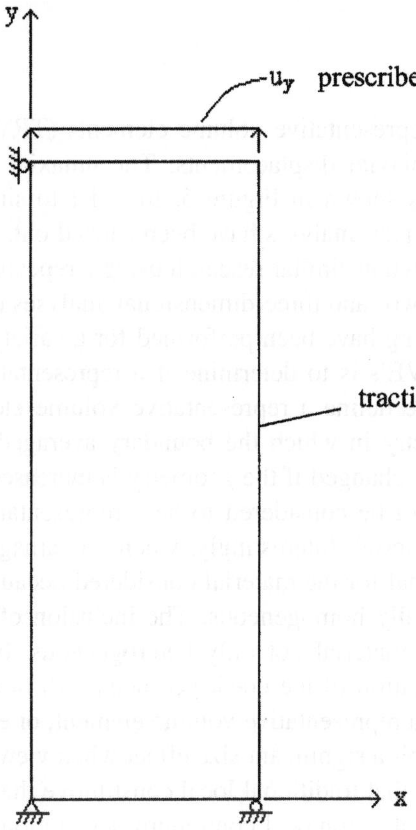


*Figure 5.* Boundary conditions imposed on representative volume element.

where $\alpha_1$ and $m$ are material parameters. When $\alpha$ reaches the value of unity, the tractions go to zero and the crack propagates.

The mechanics and thermodynamics of the fracture process as predicted by the cohesive zone model are described in Yoon and Allen (1998). Briefly, it has been shown that the above cohesive zone model is consistent with the concepts of fracture mechanics in that there exists an available and critical energy release rate necessary for crack extension. However, in the current model, the critical energy release rate is a material property that is history and rate dependent.

### 2.5. Initial and boundary conditions

Both tractions and displacements are specified on the boundary of the body. Figure 5 illustrates the displacement boundary conditions imposed on the RVE. Traction boundary conditions are imposed along the cohesive zones in accordance with (25) through (28).

## 3. Finite element implementation

The above equations have been implemented to the three-dimensional finite element code SADISTIC (Structural Analysis of Damage Induced Stresses in ThermoInelastic Composites) (Allen et al., 1994). The details of this inplementation are discussed in (Foulk et al., 1996).

## 4. Model results

Results are presented for several trial representative volume elements (TRVE's of Timetal-21S under monotonically increasing uniaxial displacements. The uniaxial displacement is prescribed vertically on the free end, as shown in Figure 5. In order to simplify the computations, only two-dimensional plane stress analyses have been carried out. Although this is clearly an approximation, Rodin has shown in similar research using a repeating cell structure that the trends are comparable for both two- and three-dimensional analyses (1995).

Simulations of grain boundary cracking have been performed for a variety of geometries. The objective of these analyses on TRVE's is to determine if a representative volume element exists for this material. Herein we define a representative volume element (RVE) to be a solution for a given size of geometry in which the boundary averaged traction versus displacement relation is not significantly changed if the geometry is increased in size. Below this size, the homogenized results cannot be considered to be a representation of the macroscopic constitutive behavior of the material. Interestingly, when no damage is present, the size of this RVE is essentially infinitesimal for the material considered because without grain boundary material, the medium is spatially homogeneous. The inclusion of cohesive zones along the grain boundaries renders the material not only heterogeneous, but also, at least potentially, anisotropic due to the orientation of the crack geometry. All of this leads to the possibility that there may indeed not be a representative volume element, or equivalently, that the material considered herein may exhibit a significant size effect when viewed on any scale. The ramifications of this result would be that traditional local constitutive theory would at the very least have to be discarded in favor of a nonlocal continuum description. Therefore, this is an important issue to be considered.

In order to address the above issues, solutions have been obtained for several geometries. The solution may in fact be sensitive to any or all of the following:

(1) the number of grains for fixed specimen size (grain size dependence);

(2) the fineness of the finite element mesh (mesh refinement);

(3) the size of the representative volume element with fixed grain size (scale effect); and

(4) the geometry of the grains obtained from the Voronoi tessellation model (orientation effect).

Each of these effects has been studied in some detail herein. Results have been obtained for the following geometries: 5 grain model (two geometries); 10 grain model (five geometries); 30 grain model (three geometries, one with two mesh refinements); and 100 grain model (one geometry). A Voronoi tessellation approach has been used to randomly establish grain boundaries. As it is well understood that the grain structure in granular materials can be represented by Voronoi tessellation, the details of the technique have not been included here. The interested reader is directed to other previous studies for more details (Van der Burg and Van der Giessen, 1994). The various geometries and meshes used in this study are shown in Figures 6 and Figure 7. Statistical information about each geometry and mesh has been collected into tabular form as shown in Table 1 and Table 2, respectively.

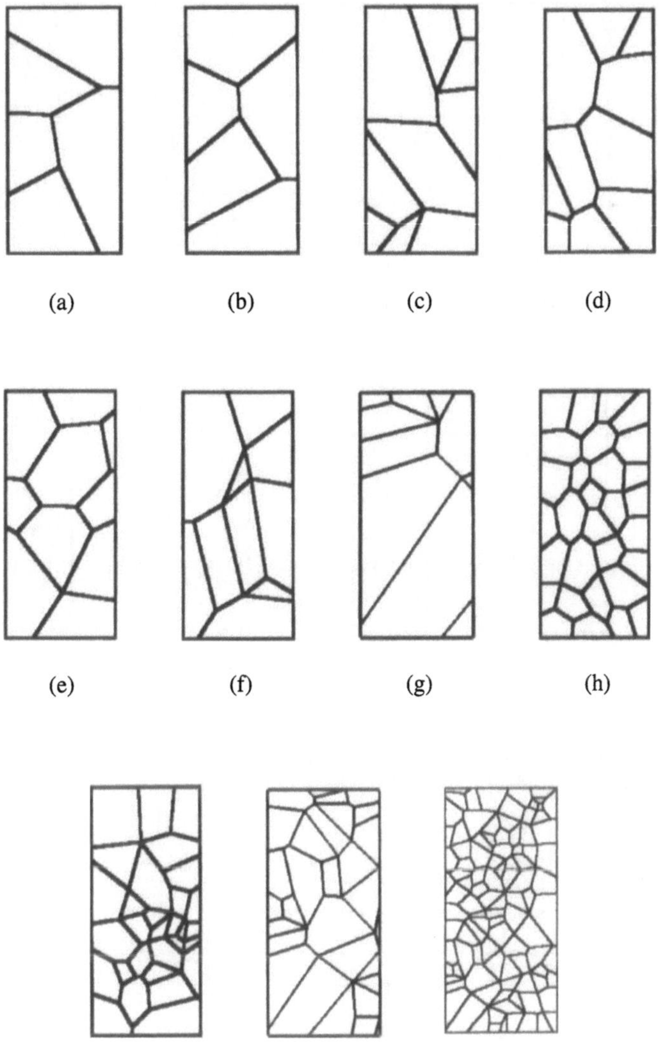

(a) (b) (c) (d)

(e) (f) (g) (h)

(i) (j) (k)

*Figure 6.* (a) Geometry 1: five grain Voronoi Tessellation 20 $\mu$m × 40 $\mu$m trial representative volume element; (b) Geometry 2: five grain Voronoi Tessellation 20 $\mu$m × 40 $\mu$m trial representative volume element; (c) Geometry 3: ten grain Voronoi Tessellation 20 $\mu$m × 40 $\mu$m trial representative volume element; (d) Geometry 4: ten grain Voronoi Tessellation 20 $\mu$m × 40 $\mu$m trial representative volume element; (e) Geometry 5: ten grain Voronoi Tessellation 20 $\mu$m × 40 $\mu$m trial representative volume element; (f) Geometry 6: ten grain Voronoi Tessellation 20 $\mu$m × 40 $\mu$m trial representative volume element; (g) Geometry 7: ten grain Voronoi Tessellation (part of Geometry 11) 5 $\mu$m × 10 $\mu$m trial representative volume element; (h) Geometry 8: thirty grain Voronoi Tessellation 20 $\mu$m × 40 $\mu$m trial representative volume element; (i) Geometry 9: thirty grain Voronoi Tessellation 20 $\mu$m × 40 $\mu$m trial representative volume element; (j) Geometry 10: thirty grain Voronoi Tessellation (part of Geometry 11) 10 $\mu$m × 20 $\mu$m trial representative volume element; (k) Geometry 11: one hundred grain Voronoi Tessellation 20 $\mu$m × 40 $\mu$m trial representative volume element.

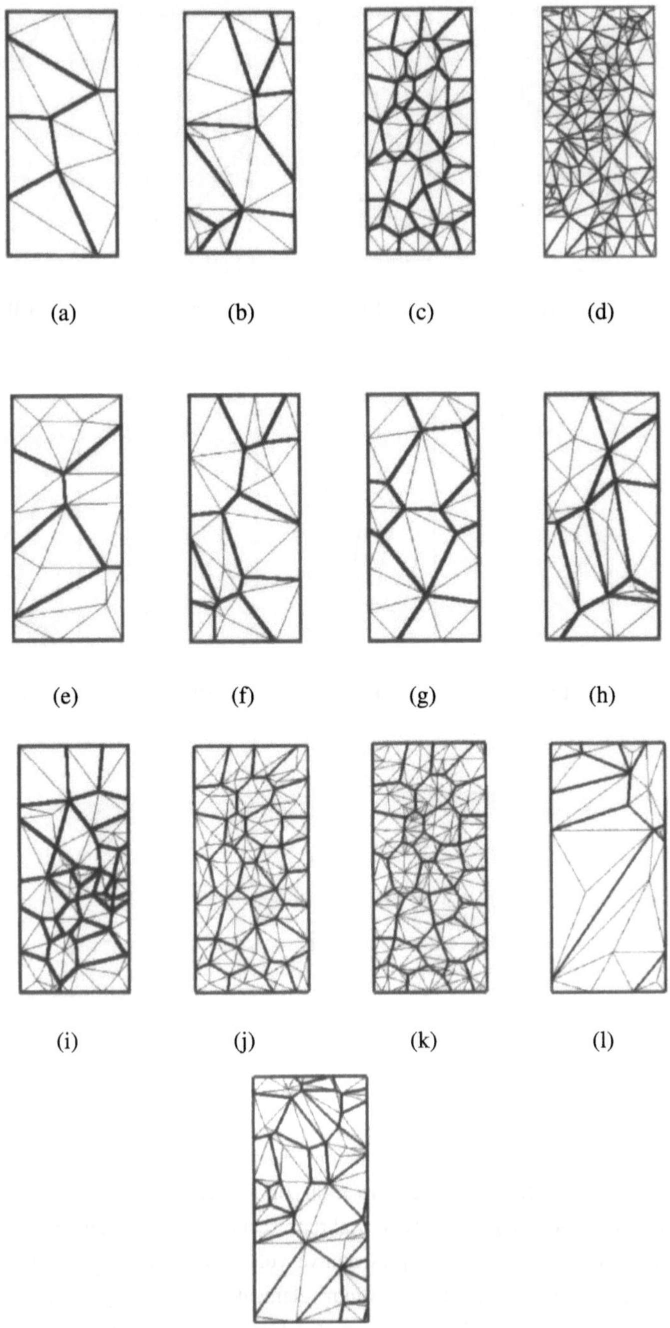


*Figure 7.* (a) Mesh 1: Geometry 1 (five grain) with 16 elements; (b) Mesh 2: Geometry 3 (ten grain) with 28 elements; (c) Mesh 3: Geometry 8 (thirty grain) with 97 elements; (d) Mesh 4: Geometry 11 (one hundred grain) with 292 elements; (e) Mesh 5: Geometry 2 (five grain) with 26 elements; (f) Mesh 6: Geometry 4 (ten grain) with 37 elements; (g) Mesh 7: Geometry 5 (ten grain) with 37 elements; (h) Mesh 8: Geometry 6 (ten grain) with 44 elements; (i) Mesh 9: Geometry 9 (thrity grain) with 109 elements; (j) Mesh 10: Geometry 8 (thirty grain) with 185 elements; (k) Mesh 11: Geometry 8 (thirty grain) with 290 elements; (l) Mesh 12: Geometry 7 (ten grain) with 37 elements; (m) Mesh 13: Geometry 10 (five grain) with 88 elements.

*Table 1.* Geometries considered.

| # | Geometry based on | TRVE size | # of grains | Average 2-D Grain Size |
|---|---|---|---|---|
| 1 | Voronoi tessellation | 20×40 | 5 | 160 |
| 2 | Voronoi tessellation | 20×40 | 5 | 160 |
| 3 | Voronoi tessellation | 20×40 | 10 | 80 |
| 4 | Voronoi tessellation | 20×40 | 10 | 80 |
| 5 | Voronoi tessellation | 20×40 | 10 | 80 |
| 6 | Voronoi tessellation | 20×40 | 10 | 80 |
| 7 | Part of Geometry 11 | 5×10 | 10 | 8 |
| 8 | Voronoi tessellation | 20×40 | 30 | 26.67 |
| 9 | Voronoi tessellation | 20×40 | 30 | 26.67 |
| 10 | Part of Geometry 11 | 10×20 | 30 | 8 |
| 11 | Voronoi tessellation | 20×40 | 100 | 8 |

*Table 2.* Meshes considered.

| # | Geometry | # of grains | # of elements | Average # elements/grain |
|---|---|---|---|---|
| 1 | 1 | 5 | 16 | 3.2 |
| 2 | 3 | 10 | 28 | 2.8 |
| 3 | 8 | 30 | 97 | 3.233 |
| 4 | 11 | 100 | 292 | 2.92 |
| 5 | 2 | 5 | 26 | 5.2 |
| 6 | 4 | 10 | 37 | 3.7 |
| 7 | 5 | 10 | 37 | 3.7 |
| 8 | 6 | 10 | 44 | 4.4 |
| 9 | 9 | 30 | 109 | 3.633 |
| 10 | 8 | 30 | 185 | 6.167 |
| 11 | 8 | 30 | 290 | 9.67 |
| 12 | 7 | 10 | 37 | 3.7 |
| 13 | 10 | 30 | 88 | 2.93 |

All results herein have been obtained using the material properties shown in Table 3. Note that in order to perform these sensitivity studies the constitutive behavior in the grains has been initially assumed to be linear elastic. Although not shown, model cases have been solved for each grain geometry wherein no damage is allowed. As expected, the stresses and strains resulting from these analyses are spatially homogeneous, thus lending credence to the correctness of the geometries tested.

*Table 3.* Material properties for Timetal-21S at 482°C.

| | |
|---|---|
| Young's modulus | 98100 MPa |
| Poisson's ratio | 0.34 |
| Elastic yield stress | 90000 MPa |

### 4.1. Effect of grain size

The effect of grain size has been simulated by 5, 10, 30, and 100 grain models. Each model has varying geometry and assumes the same TRVE size of 20 $\mu$m $\times$ 40 $\mu$m with a different number of grains thus yielding a different average grain size in each model. In the first level of analysis, one 5 grain, one 10 grain, one 30 grain, and one 100 grain model were considered. Both the length and the geometry of grain boundary cracks were observed to vary significantly in each of the models as is evidenced by the representations of crack growth for the 5 grain and 100 grain models as shown in Figures 8 and 9, respectively. A comparison of all four models resulted in predictions of force vs. prescribed displacement behavior as shown in Figure 10. From this figure, one could possibly conclude that the behavior is appearing to converge as the number of grains considered approaches 100. To confirm this conclusion, other geometries of 5 grain, 10 grain, and 30 grain models were considered as shown in Figure 11. From these simulations, one could possibly conclude that the behavior is converging with only 30 grains. This conclusion conflicts with the conclusion drawn from Figure 10. Consequently, we were unable to determine the effect of an average grain size from the limited cases considered.

### 4.2. Effect of finite element mesh refinement

In the geometries considered above, only coarse meshes have been used. As a finer mesh could impact convergence, two levels of mesh refinement were considered for one of the 30 grain geometries considered above. Each level of refinement approximately doubled the number of elements used. The finest mesh also incorporated multiple cohesive zone elements along the grain boundaries allowing for better ability to capture stress concentration behavior. The comparison of the predicted force vs. prescribed displacement for the three different meshes is shown in Figure 12. Although each of the finer meshes more smoothly captured the damage accumulation, there was no marked difference in overall response thus leading to the conclusion that mesh refinement alone may not have a significant effect on convergence.

### 4.3. Scale effect

As all of the TRVE's considered above were the same size, two different size TRVE's were extracted from the 100 grain geometry to allow for study of scale effect on convergence. The first TRVE extracted contained a 10 grain 5 $\mu$m $\times$ 10 $\mu$m portion from the lower left corner of the 100 grain geometry. The second TRVE extracted was larger containing a 30 grain 10 $\mu$m $\times$ 20 $\mu$m portion which included all of the 5 $\mu$m $\times$ 10 $\mu$m TRVE. The response of these two simulations are compared to the original 10 $\mu$m $\times$ 20 $\mu$m 100 grain model in Figure 13. Although the predicted responses did exhibit a noticeable trend in strength, larger TRVE's must be considered to identify convergence.

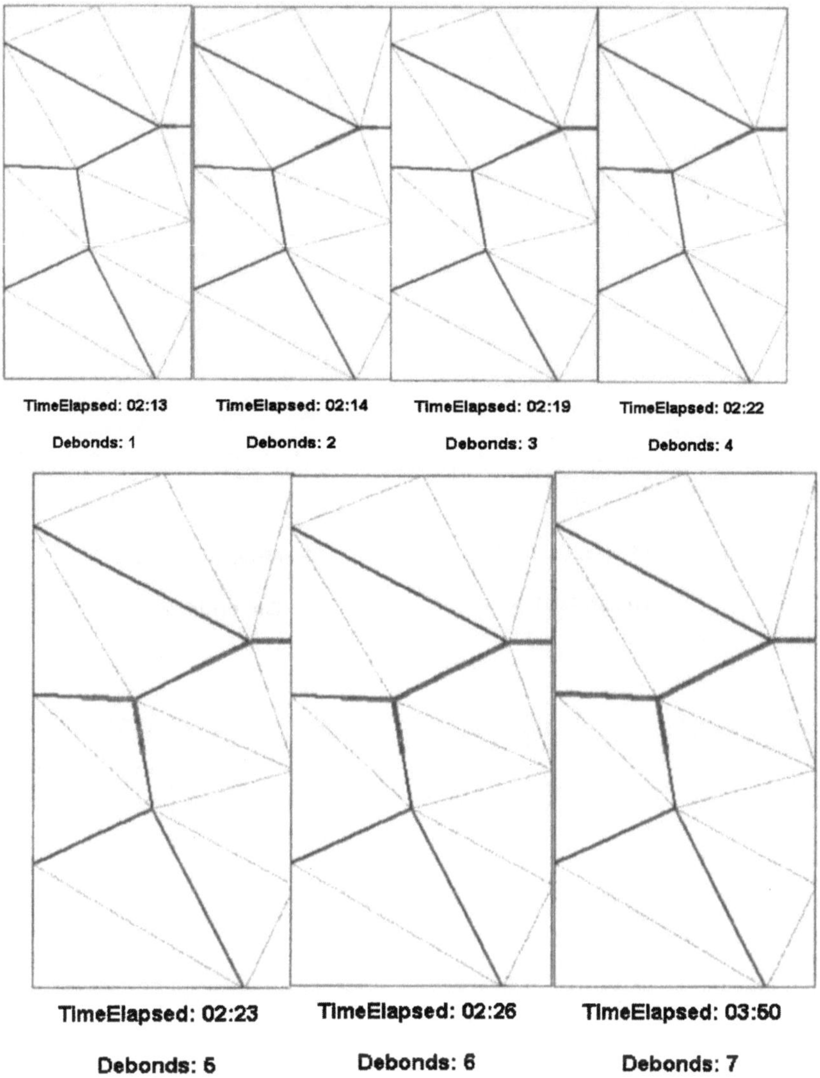


*Figure 8.* Predicted grain boundary crack growth for a 5 grain model (Mesh 1).

### 4.4. Orientation Effect

To simulate the effect of orientation of grains on predicted response, the grain size was kept constant while various geometries were considered. The response of four different geometries of a 20 $\mu$m × 40 $\mu$m TRVE each with 10 grains is shown in Figure 14. The responses show a great deal of variation leading to the conclusion that orientation has a significant effect on the predicted response of the TRVE.

## 5. Conclusion

A model to predict grain boundary cracking has been presented and applied to a general polycrystalline viscoplastic material, Timetal-21S, a titanium alloy. This model has incorporated both the effects of material inelasticity and damage. Material inelasticity has been predicted

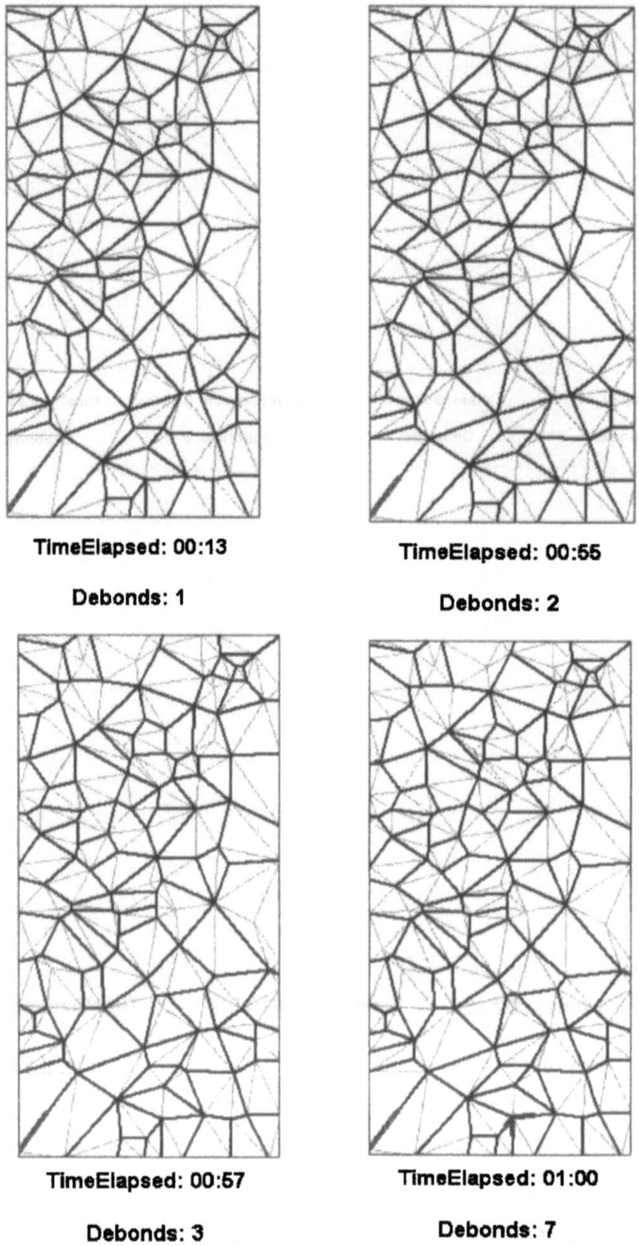


*Figure 9.* Predicted grain boundary crack growth for a 100 grain model (Mesh 4).

using Bodner's unified viscoplastic constitutive model. Damage has been included explicitly via cohesive zone elements.

Previous studies have shown that both inelastic material behavior which is rate dependent and damage effects must be incorporated into effective modeling of typical polycrystalline materials such as titanium, argillaceous salts, or sea ice. The current modeling efforts focused on determining a suitable representative volume element (if one exists) by considering the separate effects of (1) grain size, (2) mesh refinement, (3) scale, and (4) orientation. To this end, several geometries were considered to isolate these individual parameters. Average grain

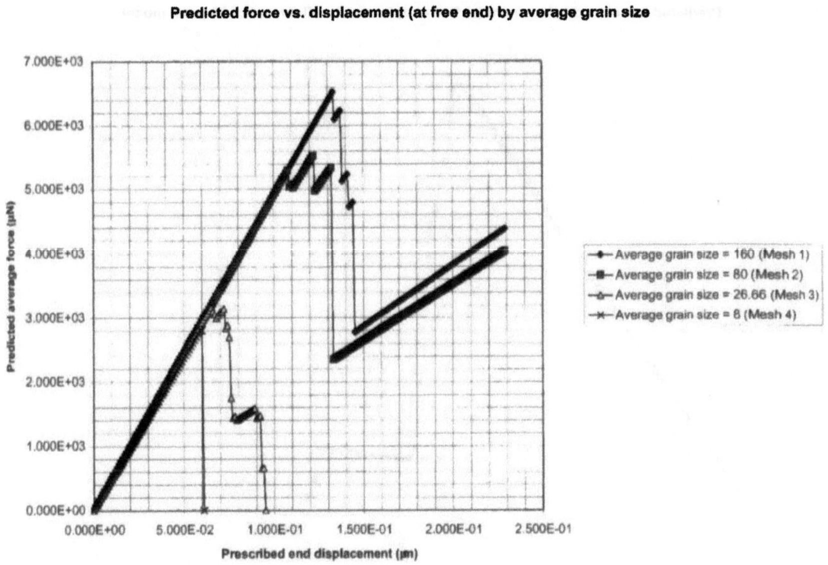


*Figure 10.* Predicted force vs. displacement by average grain size for 5, 10, 30, and 100 grain models (Mesh 1, Mesh 2, Mesh 3, and Mesh 4).

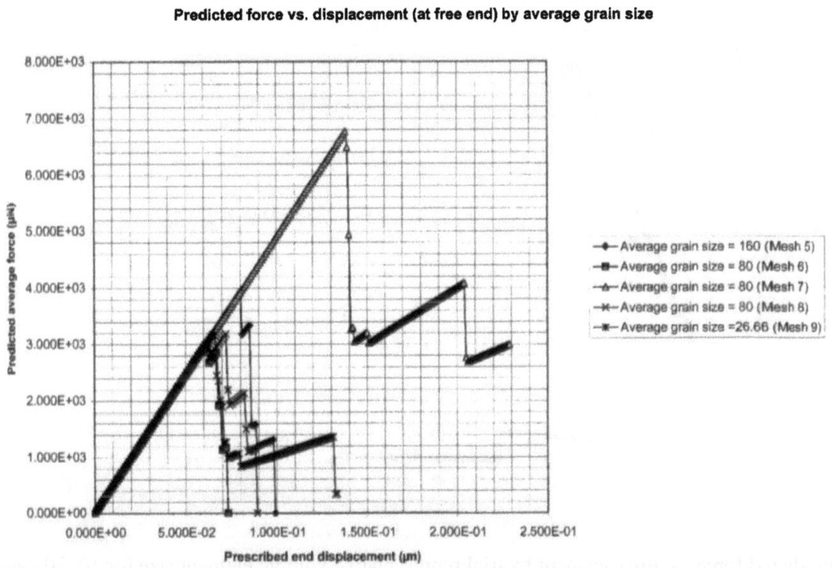


*Figure 11.* Predicted force vs. displacement by average grain size for 5, 10, and 30 grain models (Mesh 5, Mesh 6, Mesh 7, Mesh 8, and Mesh 4).

size alone was not found to lead towards convergent behavior. Using the one trial representative volume element considered, mesh refinement by itself was found to have only a small effect on predicted response. The scale parameter alone was found to affect predicted material response; however, for the three TRVE'S considered, no trend towards convergent behavior was observed. Varying only orientation was found to significantly affect material response, and further study is warranted to better describe this effect. No attempt in the current study, however, was made to fully vary the other three parameters simultaneously; consequently, a

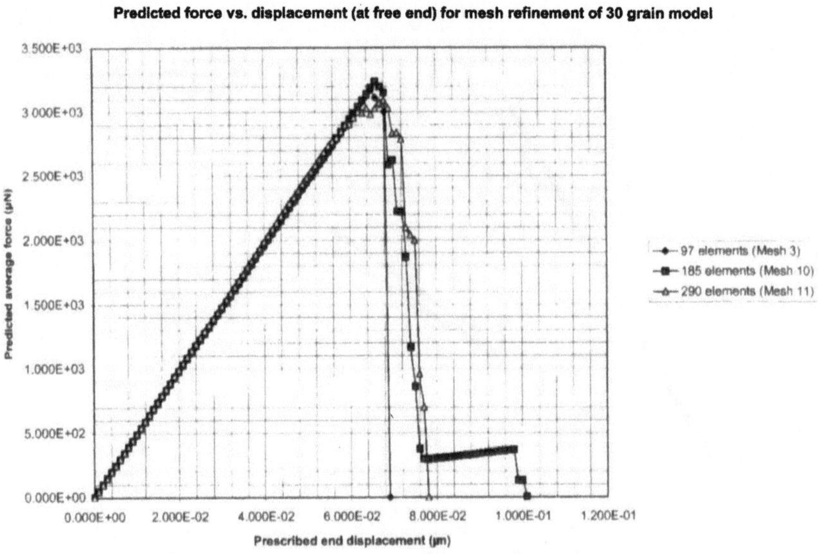


*Figure 12.* Predicted force vs. displacement for mesh refinement of a 30 grain model (Mesh 3, Mesh 10, and Mesh 11).

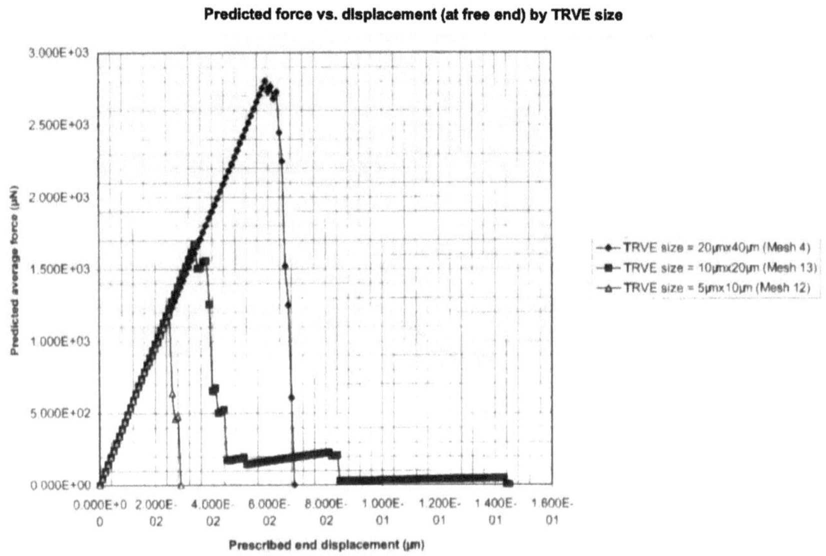


*Figure 13.* Predicted force vs. displacement by trial representative volume element size for 10, 30 , and 100 grain models (Mesh 4, Mesh 12, and Mesh 13).

definitive conclusion regarding convergence or lack of convergence to determine a RVE could not be drawn.

## Acknowledgments

The authors gratefully acknowledge the support provided for this research by Sandia National Laboratories under contract number AT-8781. Sandia is a multiprogram laboratory operated by Sandia Corporation, a Lockheed Martin Company, for the United States Department of En-

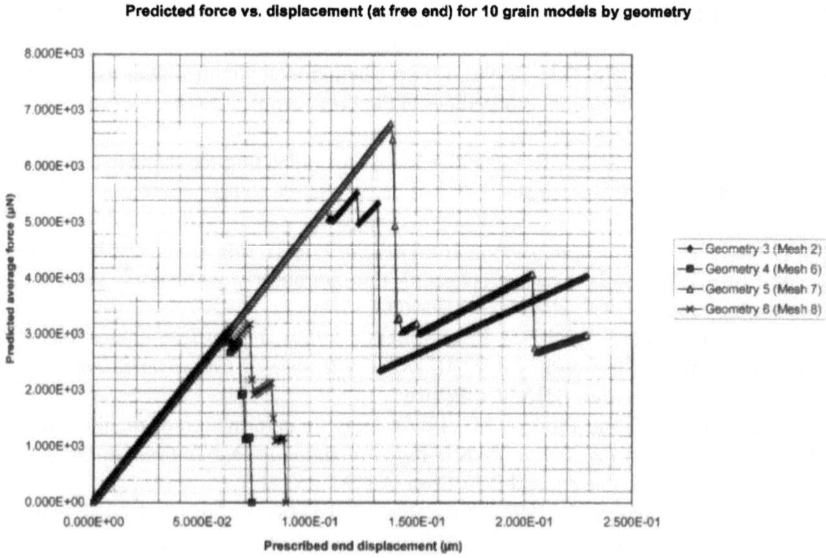


*Figure 14.* Predicted force vs. displacement for 10 grain models by for orientations (geometries) (Mesh 2, Mesh 6, Mesh 7, and Mesh 8).

ergy under Contract DE-AC04-94AL85000. We also wish to thank Jill Scheidt, Kyle Clayton, Chad Searcy, Ross Brown, Gary Seidel, Virginia Oxford, Lisa Biggs, and Annilee Travers for their assistance with mesh generation and Todd Rinaldo for assistance with crack animation software.

## References

Allen, D.H. (1991). Thermomechanical coupling in inelastic solids. *Applied Mechanics Review* **8**, 361–373.

Allen, D.H. and Searcy, C.R. (1998). A micromechanical model for viscoelastic cohesive zones, submitted to *International Journal of Fracture*.

Allen, D.H., Jones, R.H. and Boyd, J.G. (1994). Micromechanical analysis of a continuous fiber metal matrix composite including the effects of matrix viscoplasticity and evolving damage. *Journal of Mechanics and Physics of Solids* **42**(3), 502–529.

Allen, D.H., Foulk, J.W. and Helms, K.L.E. (1996). A model for predicting the effect of environmental degradation on damage evolution in metal matrix composites. *Applications of Continuum Damage Mechanics to Fatigue and Fracture*, Orlando, FL.

Arnold, S.M., Saleeb, A.F. and Castelli, M.G. (1996).A fully associative, nonlinear, kinematic, unified viscoplastic model for titanium-based matrices. *Life Prediction Methodology for Titanium Matrix Composites*, ASTM STP 1253 (Edited by W.S. Johnson, J.M. Larsen, and B.N. Cos), 231–256.

Ashby, M.F. (1983). Mechanisms of deformation and fracture. *Advances in Applied Mechanics* **23**, 117–177.

Barenblatt, G.I. (1962). The mathematical theory of equilibrium cracks in brittle fracture. *Advances in Applied Mechanics* **7**, 55–129.

Chan, K.S., Bodner, S.R., and Lindholm, U.S. (1988). Phenomenalogical modeling of hardening and thermal recovery in metals. *Journal of Engineering Materials and Technology*, 1–8.

Chan, K.S., Bodner, S.R., Fossum, A.F. and Munson, D.E. (1997). A damage mechanics treatment of creep failure in rock salt. *International Journal of Damage Mechanics* **6**, 121–152.

Chan, K.S., Brodsky, N.S., Fossum, A.F., Munson, D.E. and Bodner, S.R. (1997). Creep-induced cleavage fracture in WIPP salt under indirect tension. *Journal of Engineering Materials and Technology* **119**, 393–400.

Christman, T., Needleman, A. and Suresh, S. (1989). An experimental and numerical study of deformation in metal-ceramic composites. *Acta Metallurgica* **37**(11), 3029–3050.

Dugdale, D.S. (1960). Yielding of steel sheets containing slits. *Journal of the Mechanics and Physics of Solids* **8**, 100–104.

Foulk, J.W., Helms, K.L.E. and Allen, D.H. (1996). Life prediction in continuous fiber metal matrix composites subjected to environmental degradation. Proceedings of the 11th Technical Conference of the *American Society for Composites*, Atlanta, GA, October 7–9.

Foulk, J.W., Allen, D.H. and Helms, K.L.E. (1998). A model for predicting the damage and environmental degradation dependent life of SCS-6/Timetal 21S $[0]_4$ metal matrix composite, to appear in *Mechanics and Materials*.

Hertzberg, R.W. (1976). *Deformation and Fracture Mechanics of Engineering Materials*, John Wiley & Sons.

Knauss, W.G. and Losi, G.U. (1993). Crack propagation in a nonlinearly viscoelastic solid with relevance to adhesive bond failure. *Journal of Applied Mechanics* **60**, 793–801.

Larsen, J.M., Russ, S.M. and Jones, J.W. (1993). Possibilities of pitfalls in aerospace applications of titanium matrix composites, in the proceedings of the *NATO AGARD Conference on Characterization of Fibre Reinforced Titanium Metal Matrix Composites*, Bordeaux, France.

Lissenden, C.J., Herakovich, C.T. and Pindera, M. (1996). Inelastic deformation of titanium matrix composites under multiaxial loading. *Life Prediction Methodology for Titanium Matrix Composites*, ASTM STP 1253 (Edited by W.S. Johnson, J.M. Larsen, and B.N. Cox), ASTM, 257–277.

Llorca, L., Needleman, A. and Suresh, S. (1991). An analysis of the effects of matrix void growth on deformation and ductility in metal-ceramic composites. *Acta Metallurgica et Materialia* **39**(10), 2317–2335.

Majumdar, B.S. and Newaz, G.M. (1991). Thermomechanical fatigue of a quasi-isotropic metal matrix composite. *Composite Materials: Fatigue and Fracture (Third Volume), ASTM STP 11* (Edited by T.K. O'Brien), 732–752.

Needleman, A. (1987). A continuum model for void nucleation by inclusion debonding. *Journal of Applied Mechanics* **54**, 525–531.

Needleman, A. (1990). An analysis of tensile decohesion along an interface. *Journal of Mechanics and Physics of Solids* **38**(3), 289–324.

Needleman, A. (1990). An analysis of decohesion along an imperfect interface. *International Journal of Fracture* **42**, 21–40.

Neu, R.W. (1993). Nonisothermal material parameters for the Bodner-Parton model. In the proceedings of the *ASME Winter Annual Meeting*, New Orleans, LA.

Neu, R.W. and Bodner, S.R. (1993). Determination of the material constants of Timetal 21S for a constitutive model, internal report prepared for Wright Laboratory Materials Directorate, Wright–Patterson AFB, OH.

Nutt, S.R. and Needleman, A. (1987). Void nucleation at fiber ends in Al-SiC composites. *Scripta Metallurgica* **21**, 705–710.

Rodin, G.J. (1995). Stress transmission in polycrystals with frictionless grain boundaries. *Journal of Applied Mechanics* **62**, 1–6.

Schapery, R.A. (1986). A Micromechanical model for non-linear viscoelastic behavior of particle-reinforced rubber with distributed damage. *Engineering Fracture Mechanics* **25**, 845–867.

Schapery, R.A. (1997). Thermoviscoelastic constitutive equations for polycrystalline ice. *ASCE Journal of Cold Regions Engineering* **2**, 146–157.

Suresh, S. (1991). *Fatigue of Materials*, Cambridge University Press.

Tvergaard, V. (1990). Effect of fibre debonding in a whisker-reinforced metal. *Materials Science and Engineering* **A125**, 203–213.

Tvergaard, V. (1992). Effect of ductile particle debonding during crack bridging in ceramics. *International Journal of Mechanical Science* **34**(8), 635–649.

Tvergaard, V. (1995). Fibre debonding and breakage in a whisker-reinforced metal. *Materials Science and Engineering* **A190**, 215–222.

Van der Burg, M.W.D. and Van der Giessen, E. (1994). Simulation of microcrack propagation in creeping polycrystals due to diffusive grain boundary cavitation. *Applied Mechanics Review* **47**, 122–131.

Voyiadjis, G.Z. and Kattan, P.I. (1993). Local approach to damage in elasto-plastic metal matrix composites. *International Journal of Damage Mechanics* **2**, 96–114.

Yoon, C. and Allen, D.H. (1998). Damage dependent constitutive behavior and energy release rate for a cohesive zone in a thermoviscoelastic solid. submitted to *International Journal of Fracture*.

Zuiker, J.R. and Sanders, B.P. (1995). Analysis of pinned joints in viscoplastic monolithic plates. *Numerical Implementation and Application of Constitutive Models in the Finite Element Method.* (Edited by J.A. Sherwood), American Society of Mechanical Engineers – Applied Materials Division, New York, 37–48.

*International Journal of Fracture* **95:** 195–219, 1999.


# Effect of strain gradients on the size effect of concrete in uniaxial tension

MARCEL R.A. VAN VLIET* and JAN G.M. VAN MIER**
*Faculty of Civil Engineering and Geosciences, Delft University of Technology, P.O. Box 5048, 2600 GA Delft, The Netherlands*

Receiving 4 July 1998; accepted in revised form 18 December 1998

**Abstract.** A series of uniaxial tension experiments has been conducted to investigate the size effect on strength and fracture energy of quasi brittle materials like concrete and sandstone. This paper focuses on the results of the concrete tests, and specifically deals with the variation of the nominal strength for specimens of six different sizes in a scale range of 1:32. It was found that under given experimental conditions, the nominal strength strongly depended on the specimen size. More important however, is the fact that most of this size effect could be attributed to strain gradients which were present in the cross section of the specimens. These strain gradients were caused by the specimen shape, load eccentricity and material inhomogeneity. Through a combination of experimental data and a simple linear elastic analysis, the importance of strain gradients with respect to the ultimate load level could be visualized. This leads to the conclusion that studying a material size effect is not possible without taking into account structural stress/strain gradients.

**Key words:** Concrete, size effect, nominal strength, material inhomogeneity, material behaviour, structural effects, stress/strain gradients.

## 1. Introduction

In the field of material science, size differences between laboratory specimens and practical applications are often inevitable. This is especially so for applications concerning materials like concrete and rock. That this size difference can have an effect on, for instance, the nominal (or average) strength has been recognized for decades. Several researchers conducted various types of tests and developed theories trying to 'predict' the size effect for scale ranges which can not be tested under laboratory conditions. Best known in this respect is the work of Bažant (1997) and Carpinteri (1994). All these efforts have led to the knowledge that a size effect exists, and that it varies with the type of test in which it is observed. A true understanding of the real physical cause of the observed size effect is still lacking however. Many reasons and micro-mechanical mechanisms have been proposed, but a clear relation between the changes in the microstructure and the macroscopic size effect has never been established.

As a contribution to current and previous research dealing with size effect, a combined experimental and numerical investigation was started to study size effect of concrete and sandstone under uniaxial tension. These uniaxial tensile loading conditions were chosen because they constitute the most basic failure mechanism of quasi brittle materials, thereby offering the best possibility to study a size effect of the material itself. The experiments are carried out

* Address per 01/06/'99: TNO Building and Construction Research, Delft, The Netherlands
** Correspondence Address: Dept. of Civil Eng., Delft University of Technology, Stevin Laboratory, Stevinweg 1, 2628 CN Delft, The Netherlands. e-mail: j.vanmier@ct.tudelft.nl

under deformation control, so that a size effect on the nominal strength as well as the fracture energy can be studied. The main goal is to get insight in the physical mechanisms underlying the observed size effect. In this paper the authors focus on the results of the concrete tests and the size effect on the nominal strength in particular will be analyzed and discussed.

Experimental results always are a mixture of material effects and structural effects. Since these structural effects can strongly influence the outcome of an experiment, it is crucial to minimize them whenever possible. Knowledge and understanding of the remaining effects is then needed to separate the structural influences from the material effects.

In the paper first an outline is given of the experimental set-up. An overview is given of the obtained results, followed by a detailed analysis of two representative experiments on a small and a large concrete specimen respectively. Because of typical observations that were made during the uniaxial tension experiments, a series of tensile splitting tests was conducted. These verification tests focused on the variation of concrete properties within specimens of different size, and are briefly explained followed by a summary of the main findings. Finally, it will be shown that, by using the results of the verification tests, the observed size effect in the current uniaxial tension tests can be explained almost completely. Although the observed size effect must partly be attributed to the conditions which were present in these specific tests, the findings which are summarized in this paper have a general implication as well. In this respect a direction is proposed in which size effect research should focus.

## 2. Size effect tests

### 2.1. Specimens and concrete properties

When studying material size effect it would be most natural to use specimens scaled in three dimensions, or at least scaled in two dimensions but with a prismatic cross section. From experience however, it follows that the practical problems arising for these cases are huge. The dog-bone shaped specimens which were used in the present experiments therefore resulted from a careful consideration of theoretical demands and practical needs (Van Vliet and Van Mier, 1998). Inherent to the dog-bone shape is a nonuniform stress state within the specimens, similar to specimens in which notches are applied. The tests are still referred to as uniaxial tension tests however, since failure of the specimens takes place under mode I loading conditions. The effects of the nonuniform stress state are incorporated in the analysis of the experimental results.

By applying circular bays the cross section in the specimen middle is reduced with 40 percent compared to the specimen ends (Figure 1). Reducing the cross section is needed in order to avoid glue problems. The use of circular bays instead of notches is thereby preferred for several reasons. First of all because of stress concentrations: in case of circular bays the stress concentrations are not only smaller (Van Vliet and Van Mier, 1998), but also easier to reproduce for different specimen sizes. Scaling a notch in a material like concrete without changing its effect is less straightforward than it may seem, not to mention the damage which is introduced during the manufacturing of the notch. Another advantage of dog-bone shaped specimens is the fact that the area where failure can occur is larger compared to notched specimens. Failure is therefore more likely to occur at the weakest spot in the material, rather than being induced in a narrow zone with large stress concentrations whose influence on the ultimate load can only be guessed. As for the global dimensions of the specimens, the

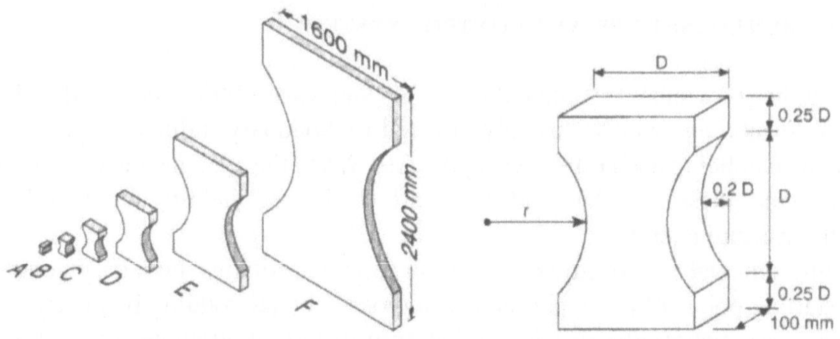


| Type | A | B | C | D | E | F |
|---|---|---|---|---|---|---|
| D (mm) | 50 | 100 | 200 | 400 | 800 | 1600 |
| r (mm) | 36.25 | 72.5 | 145 | 290 | 480 | 960 |

*Figure 1.* Full specimen range of 1:32 and specimen proportions.

height/width ratio was fixed at 1.5 to ensure that in the middle of the specimen boundary influences are absent and that the stress distribution has become uniform.

After the shape of the specimens was set, the next step was the choice of the concrete mixture because the dimensions of the smallest specimen are related to the maximum aggregate size $d_{a,\max}$. The only demands for the type of concrete concerned the workability and the stability of the mixture, since only mutual comparisons would be made between the different specimen sizes. Following these demands, a concrete mixture was chosen with an average cube compressive strength of 50 MPa and $d_{a,\max} = 8$ mm. The composition is shown in Table 1.

*Table 1.* Concrete mix properties.

| Unit content kg/m$^3$ | | | | | | | |
|---|---|---|---|---|---|---|---|
| 8-4$^a$ | 4-2$^a$ | 2-1$^a$ | 1-0.5$^a$ | 0.5-0.25$^a$ | 0.25-0.125$^a$ | PCB | w/c |
| 540 | 363 | 272 | 272 | 234 | 127 | 375 | 0.5 |

$^a$ aggregate size in mm, PCB: Portland Cement type B, w/c: water cement ratio

With $d_{a,\max}$ fixed at 8 mm the outer dimensions of the smallest specimen were taken 50×75 mm$^2$. Although the resulting minimal width of 30 mm is rather close to the common rule that the smallest structural dimension should be taken 3 to 5 times larger than the maximum aggregate size, it was decided that the results had to show whether 30 mm is too small. The thickness of the specimens was set at a constant value of 100 mm. Figure 1 illustrates the final test range of 1:32 that resulted after scaling the smallest specimen in two directions, as well as the final specimen dimensions.

### 2.2. Experimental set-ups and control system

Numerous studies previously carried out (see e.g. (Van Mier, 1995)) showed that the outcome of a uniaxial tension test will be strongly affected by boundary influences when keeping the specimen ends parallel to each other during the test. When the specimen ends can rotate freely on the other hand, these boundary influences are minimized yielding among others a lower bound for the fracture energy.

Performing uniaxial tension tests with these so called rotating boundary conditions, was one of the starting points of the experiments. In order to create rotating boundary conditions a special hinge construction was developed that ensures free rotations of the specimen ends in two orthogonal directions. Moreover, the point of rotation is situated exactly in the glue layer between specimen and loading platen (Van Vliet and Van Mier, 1998).

During the course of the research it became clear that conducting all tests in one single test set-up was not practically feasible. It was therefore decided to carry out the tests in three different test frames, each with its own hinge construction (Van Vliet and Van Mier, 1998). When using different frames however, the risk is present of stiffness differences between the frames influencing the experimental result. To take this into account, specimen type B will be tested in both the smallest and the middle-sized frame. By testing specimen type D in both the middle-sized frame and the large frame, differences between these two frames can be eliminated as well. Since the analysis of the stiffness influence of the frame had not been completed at the time of writing, the presented experimental results are the original results, i.e. they have not been corrected for possible frame influences. Generally, these stiffness influences originating from testing specimens in two different loading frames are neglected. For a full report on the results of the frame analysis the reader is referred to (Van Vliet, 1999).

Inherent to the specimen shape is the fact that the area of crack initiation scales with the specimen size. In order to control also the tests on the largest specimens, special care has to be taken. A possibility would be to scale the measuring length of the control LVDT with the specimen size. Hordijk however, clearly showed the increasing risk of snap-back behaviour at peak load with increasing measuring length (Hordijk, 1991), and also from the tension tests carried out by Ferro (1994) it can be concluded that the measuring length cannot be increased in an unlimited way. To eliminate the risk of running into snap-back problems caused by a too large measuring length, a small, constant measuring length was adopted in combination with a type of test control similar to that proposed by Li et al. (1993). On both the front and the back side of the specimen linear variable displacement transducers (LVDTs) with a measuring length of 75 mm are placed in line above each other. A specially developed control device checks up to sixteen LVDTs continuously for the LVDT measuring the largest deformation. This LVDT is used to control the test. When at some later instant another LVDT measures a larger deformation, test-control is switched automatically to this LVDT. In preliminary tests preceding the final test series also other types of test-control were studied, like taking the average signal of two or four LVDTs as control signal. It was found however, that controlling the test by means of the maximum deformation measured at a certain time and location, the best guarantee for stable post-peak specimen behaviour was obtained (Van Vliet, 1997b). For the largest specimen the total number of sixteen LVDTs is divided in two groups of eight, placed on the left front and the left rear side of the specimen. Consequently, it has to be ensured that the crack actually initiates at the left side of the specimen. This is achieved by applying a small eccentricity. The eccentricity, which is obtained by glueing the specimen eccentrically to the loading platens, is scaled according to the specimen size. In this way the (theoretical) linear elastic stress distribution in the middle of the specimen remains the same regardless of the size of the specimen. This is also reflected by the expression for the maximum stress $\sigma_{max}$ in the specimen middle

$$\sigma_{max} = \frac{F}{A}\left[1 + \frac{6e}{0.6D}\right], \quad e = 1, 2, 4, 8, 16, 32 \text{ mm}, \tag{1}$$

where $A$ is the cross-sectional area, $0.6D$ the width of the smallest cross section and $e$ the eccentricity of the external force $F$.

### 2.3. Specimen preparation

All specimens were cast horizontally, five specimens of different sizes at a time. Consequently, for casting all specimens several weeks were needed. Two days after casting the specimens were demoulded and stored under laboratory conditions (20°C , 50 percent RH) until the time of testing. All specimens have been tested at an age between 56 and 61 days, to ensure that the strength increase due to continuous hydration is negligible. Three to seven days before testing the loading platens were glued to the specimen ends using a two-component epoxy resin. Finally, the measuring points were glued on the specimen surface, which concluded the test preparations. During a test the load was transferred to the specimen by bolts connecting the loading platens with the hinge construction.

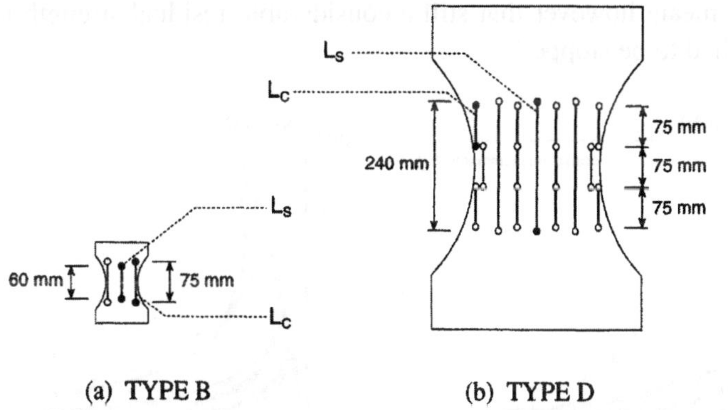


*Figure 2.* Placement of displacement transducers on specimen type B (a) and D (b). $L_c$ refers to constant control measuring length; $L_s$ to scaled measuring length.

### 2.4. Measurements

The size effect experiments that are found in the literature mainly focus on the peak load and failure strength, rather than the complete post-peak behaviour and (differences in) crack mechanisms between the various specimen sizes. In the present study it is attempted to fill this gap by closely monitoring the fracture behaviour of the concrete during the test. For this purpose a measurement scheme has been designed that was applied for all specimens. The scheme basically consists of 2 types of measurements, namely those needed for test control and additional measurements that will give insight in the process of crack propagation. Along the edges of a specimen, LVDTs with a constant measuring length of 75 mm are placed, which are used to control the test. The number of LVDTs varies with the size of the specimen. As is shown in Figure 2, one LVDT at each edge suffices to perform a test on a specimen type B (Figure 2a), whereas three LVDTs are placed along each edge of a specimen type D (Figure 2b). Furthermore, in the area covered by the control LVDTs, additional LVDTs with the same measuring length of 75 mm are placed over the specimen width. With these latter LVDTs the propagation of the crack through the specimen can be followed. Apart from these measurements with constant measuring length, also measurements with a measuring length scaled according to the specimen size are carried out (denoted by $L_s$ in Figures 2a, b). In this way the measurement covers the same part of the specimen independent of its size, thus allowing for mutual comparison of the measurements.

### 2.5. Results

#### 2.5.1. *Overview*

At least four successful tests were conducted for each specimen size. All force-deformation curves are shown in Figures 3a and b, where the deformation is the average deformation of the scaled measurements $L_s$ on the front and the back side of a specimen (Figure 2). The measured forces have been corrected for the dead weight of the upper loading platen and the part of the specimen above the crack. The weight of the specimen part below the crack, the lower loading platen and the lower hinge construction are incorporated in the measured force and can be seen as a constant, uncontrollable part of the external load. The tests were conducted with a deformation rate of 0.028 $\mu$m/s and continued until the first control LVDT exceeded its measuring range, which was at 500 $\mu$m. Because of the large width of the E and F specimens this meant however that still a considerable residual strength was present at the moment the test had to be stopped.

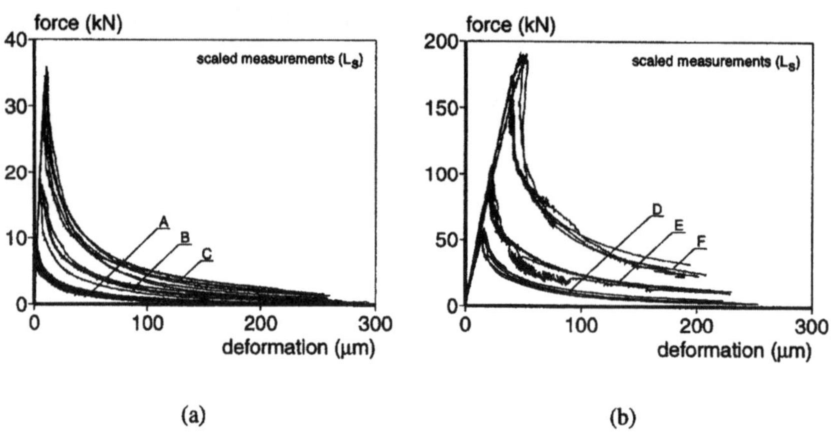


*Figure 3.* Overview of force-deformation curves for specimens A, B, C (a) and D, E, F (b).

Figure 3b shows the effect of using a constant, small control length $L_c$ (Figure 2). The softening curves of the type F specimens, which show a clear snap-back behaviour, could not have been retrieved if the scaled measuring length would have been used to control the test. Because of the large elastic deformations within $L_s$ the measured deformation *de*creases at peak load, which would have led to failure of the test if this signal would have been used for feed-back.

By dividing the ultimate force $F_u$ recorded during each test by the smallest cross section ($0.6bD$), the nominal stress $\sigma_N$ was computed. Figure 4a shows the values of $\sigma_N$ as a function of the characteristic specimen dimension $D$ in a log-log plot. Individual data for each specimen size have been replaced by mean values and standard deviations in Figure 4b. The corresponding values can be found in Table 2, where the number in brackets in the first row denotes the number of tests on which the average is based.

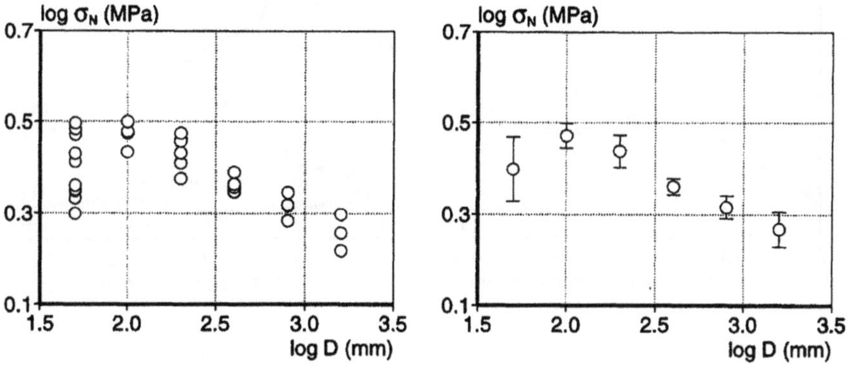


(a) (b)

*Figure 4.* Individual values (a) and mean values (b) of $\sigma_N$ vs. $D$.

Table 2 shows a decrease of the average nominal strength of 37 percent when comparing the values for the type B and the type F specimens. Let us however first take a closer look at the entire experimental outcome, rather than considering the values of the ultimate load alone. One of the questions that arises concerns the scatter in the results of the type A specimens. As can be seen in Figure 4a this scatter covers almost the entire band width in which all the other test results (i.e. for larger sizes) are lying. This matter will be addressed in Section 5.

2.5.2. *Uniaxial tension test on type B and type F specimen in detail*

To illustrate the failure process taking place in the uniaxial tension tests that were carried out, a type B and a type F specimen that are considered representative have been selected to be discussed in more detail. As was already pointed out in Section 2.4 multiple displacement transducers were placed on a specimen during a test. Figures 5a and 5b show the measured deformations in the failure zone of a type B and a type F specimen as a function of the maximum or control deformation (see Figures 6b, c for transducer numbers). As a reference also the external force is plotted against the maximum deformation, where it is mentioned that the magnitude of the force is not related to the values on the $y$ axis.

For both specimens the LVDT placed in the right front corner of the specimen measures the largest deformation and is therefore used to control the test. The deformations measured by the other LVDTs are smaller and roughly depend on their position on the specimen surface. When the crack, which propagates through the specimen from the right side towards the left side,

*Table 2.* Mean values for the nominal strength. The numbers between brackets indicate the number of tests and the standard deviation.

| Type (nr) | A (10) | B (4) | C (7) | D (5) | E (4) | F (4) |
|---|---|---|---|---|---|---|
| $\sigma_N$ (MPa) | 2.54<br>(0.41) | 2.97<br>(0.19) | 2.75<br>(0.21) | 2.30<br>(0.09) | 2.07<br>(0.12) | 1.86<br>(0.16) |

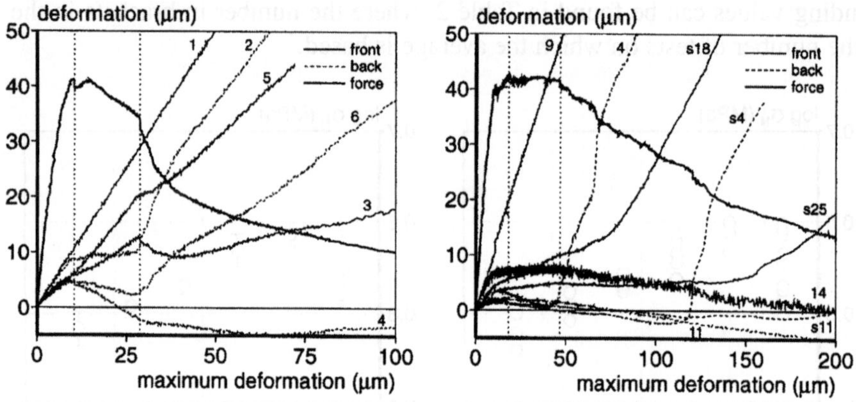


(a) (b)

*Figure 5.* Deformation readings in failure zone for B specimen (a) and F specimen (b), see also Figure 6b, c. The ultimate loads are 17.9 kN for the B specimen and 190.7 kN for the F specimen respectively.

passes a displacement transducer the measured signal strongly increases. Before this moment however, the measured deformations by that LVDT are mainly elastic.

Two typical stages in each test have been marked by the dotted vertical lines in Figures 5a and 5b. The first line indicates ultimate load. The second line marks the moment at which the deformation at the back side of the specimen, just beside the right corner, starts to increase rapidly. Whereas the load for the type B specimen increases almost linearly until peak load, this is not the case for the type F specimen. Here the load first increases linearly, followed by a much slower increase. Just before peak exactly the same phenomenon can be retrieved from the deformation readings, which for the large specimen strongly deviate around peak load. For the type B specimen on the other hand, hardly any deviation of the deformation readings on the front and the back side was found.

The second dotted line corresponds in case of the type B specimen with the moment in which a small load drop takes place. This is probably due to the fact that the length of the crack front suddenly decreases after having increased continuously. Although the same happens for the type F specimen, an equally abrupt load-drop was not observed in these tests. A possible reason might be the more gradual failure which is observed in the large specimens, like for instance at peak load. The moment the deformations at the back of the specimen start to increase, all deformations measured on the front side of both specimens are larger than those on the back side. This shows that crack propagation is not purely from the right to the left side, but an intermediate form between propagation in the in-plane and out-of-plane direction. This change of the preferential direction for crack growth can to some extent be attributed to the shape of the cross section, which is further elaborated in (Van Vliet, 1999).

Following these considerations, it would be interesting to verify whether a link exists between the ultimate load level and the deformation state of a specimen until peak load. In Figures 7 and 8 the deformations of the transducers covering the final crack are shown at ten loading stages before ultimate load (see also Figure 6a). For the type B specimen the measurements of the LVDTs in the middle of the specimen have not been incorporated since their measuring length is 60 mm instead of 75 mm like the other measurements. The measured deformations in the middle will therefore be about 20 percent lower than at the edges, purely

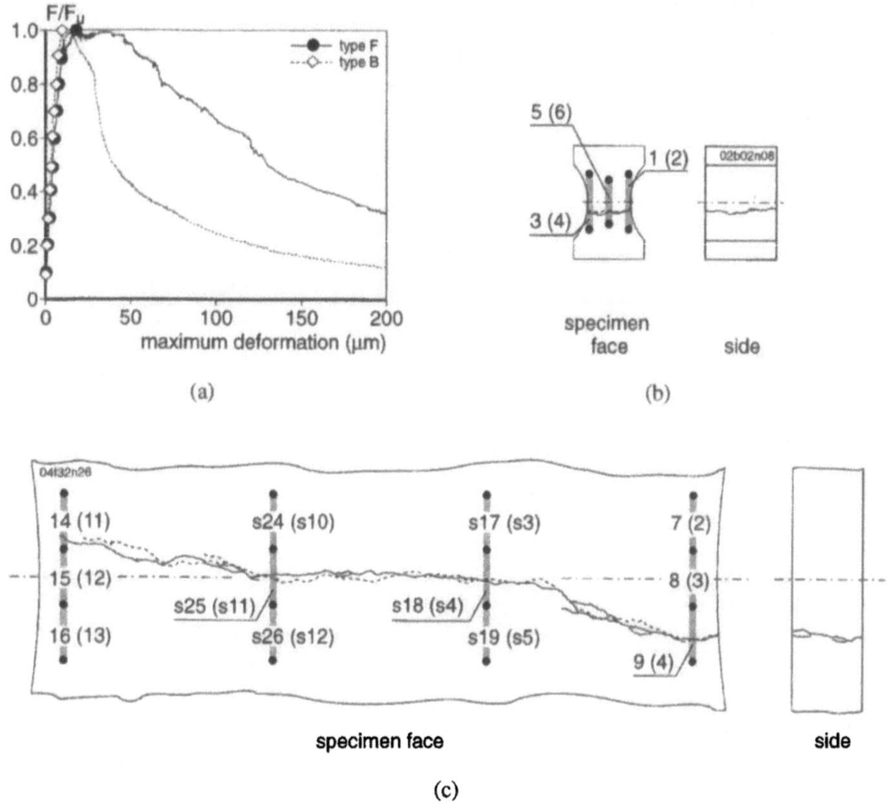


*Figure 6.* Different loading stages for the type B and type F specimen at which the deformations are shown in Figures 7. and 8. respectively (a), final crack pattern and transducer configuration for the type B specimen (b) and the type F specimen (c). The numbers between brackets indicate the transducers at the back side of the specimen.

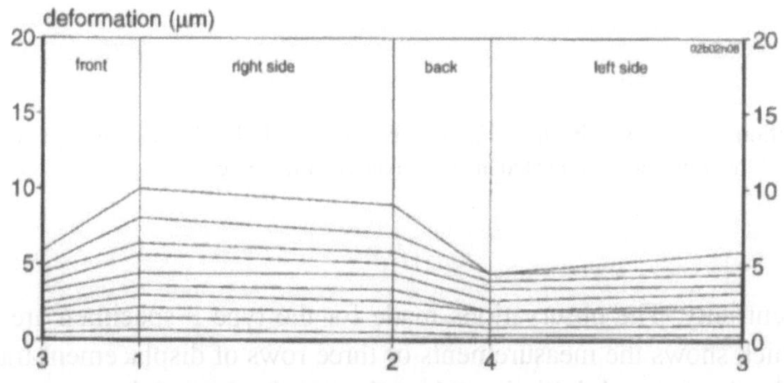


*Figure 7.* Deformation of type B specimen (of Figures 5a and 6a, b) during 10 loading stages before ultimate load. For the placement of the transducers indicated at the $x$-axis, see Figure 6b.

due to the different measuring lengths, which would give a wrong impression. The actual deformations on the front and back of the type B specimen are like the ones shown in Figure 8, rather than varying linearly.

In Figure 7 little difference is found between the deformation of the front and the back side of the type B specimen. Whereas a deformation gradient is absent over the specimen thickness (out-of-plane direction), a deformation gradient *is* found in the in-plane direction, mainly due

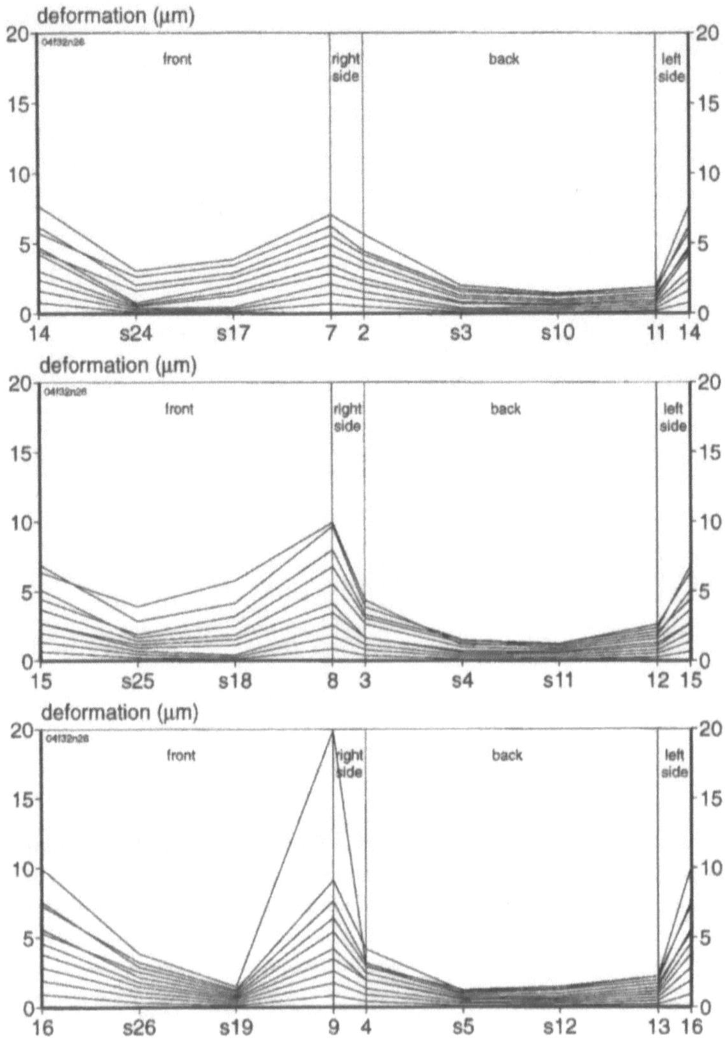


*Figure 8.* Deformation of type F specimen (of Figures 5b and 6a, c) during 10 loading stages before ultimate load. For the placement of the transducers indicated at the $x$-axis, see Figure 6c.

to the load eccentricity. The observations made for the type F specimen are quite different, see Figure 8 which shows the measurements of three rows of displacement transducers in the specimen middle. Again crack initiation takes place at the front right corner of the specimen. Contrary to the type B specimen however, a clear deformation gradient is present over the specimen thickness (out-of-plane direction). The variation of the deformations on the front and the back side of the specimen reflects the influence of the specimen shape.

From the foregoing it can be concluded that rather uniform, as well as highly nonuniform specimen deformations at peak load were observed in the out-of-plane direction. It is furthermore noteworthy that for the type B as well as the type F specimen the crack starts at the front side. For both specimens this was the casting side (upper side of the specimen during casting), whereas the back side was the mould or form side (lower side during casting). This observation and its implications will be discussed in Section 3.

## 3. Verification tests

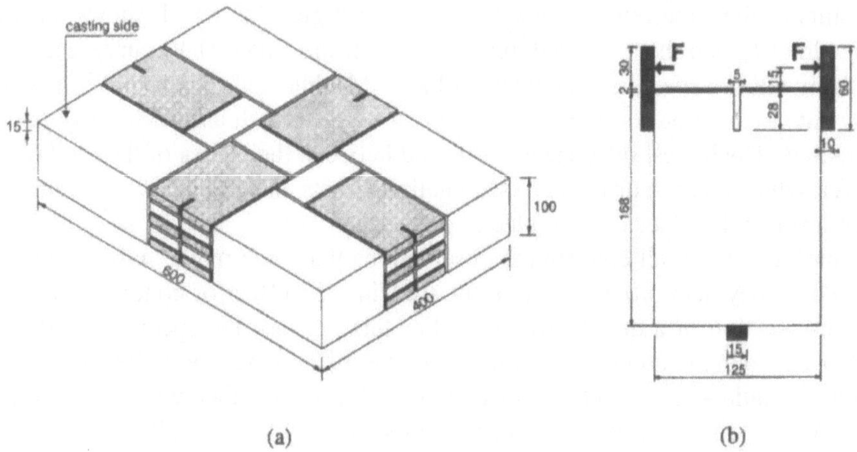


*Figure 9.* Method of specimen manufacturing (a) and adopted specimen geometry (b)

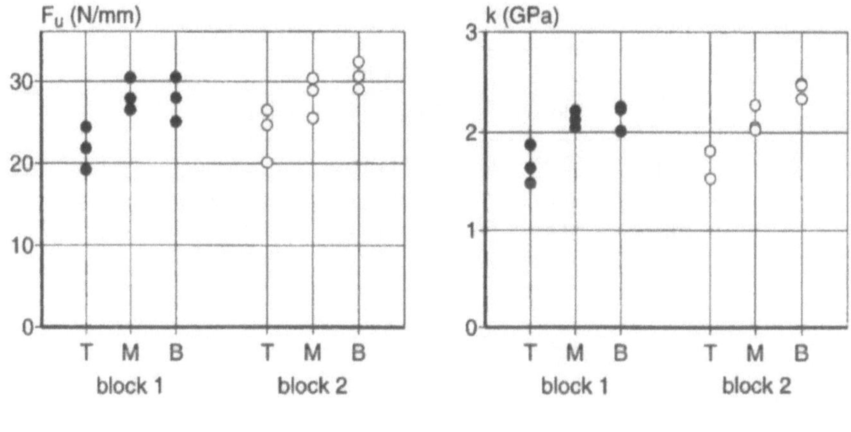


*Figure 10.* Values for the ultimate load $F_u$ (a) and the initial stiffness $k$ (b) of the splitting tests.

Throughout the experiments it was observed that both larger deformations and crack initiation frequently occurred at the casting side of the specimens. When considering the way in which the specimens were manufactured this finding could be explained however. Due to compaction problems during manufacturing of the specimens in the initial stage of the research, a casting procedure was developed in which all specimens were cast in three layers. Each layer was compacted with a vibration needle so that the first layer (mould side) is vibrated three times, the middle layer two times and the top layer (casting side) only once. Although this manufacturing method is likely to induce stiffness differences in the direction of casting, no better alternative was available. Therefore it was attempted to quantify the introduced effect in a series of verification tests (see (Van Vliet and Van Mier, 1997a) for details), which were carried out in the experimental set-up developed by Vervuurt (1997). These verification tests comprised tensile splitting tests on thin notched plates of $125 \times 168 \times 15$ mm$^3$ (Figure 9b) that were sawn from the top, middle and bottom layer of two concrete blocks with the dimensions of a type D specimen (Figure 9a). The blocks were cast and cured in the same way as the dog-bone shaped specimens.

The results of the experiments are summarized in Figure 10, which shows the values for the ultimate load $F_u$ and the measured initial slope $k$ of the load-CMOD curve as a function of the layer from which a specimen originated: (T)op, (M)iddle and (B)ottom. When going from the casting side to the mould side it was found that both strength and stiffness of the concrete block increased. The largest difference was found between the values of the top layer and the middle layer, which can be attributed to segregation effects and a higher water-cement ratio in the surface layer of the blocks due to bleeding.

These findings first of all confirm the observations that were made in the uniaxial tension tests. Secondly, they also provide a basis for studying the effect of concrete inhomogeneity over the cross section in a quantitative way. For this purpose the specimen stiffness $k$ was converted to the Young's modulus $E$ using an inverse method (Vervuurt, 1997). Table 3 shows the resulting variation of $E$ over the specimen thickness, together with the experimentally determined Young's modulus on three prisms (dimensions $400\times100\times100$ mm$^3$).

*Table 3.* Average values and standard deviations for calculated and experimentally determined Young's modulus ($E$).

| Layer | $E$ [GPa] | | |
|---|---|---|---|
| | Block 1 | Block 2 | Prism |
| Top | 20.93 (2.28) | 22.11 (2.51) | |
| Middle | 27.09 (1.19) | 26.91 (1.94) | 32.58 (1.08) |
| Bottom | 27.54 (1.87) | 30.99 (1.20) | |

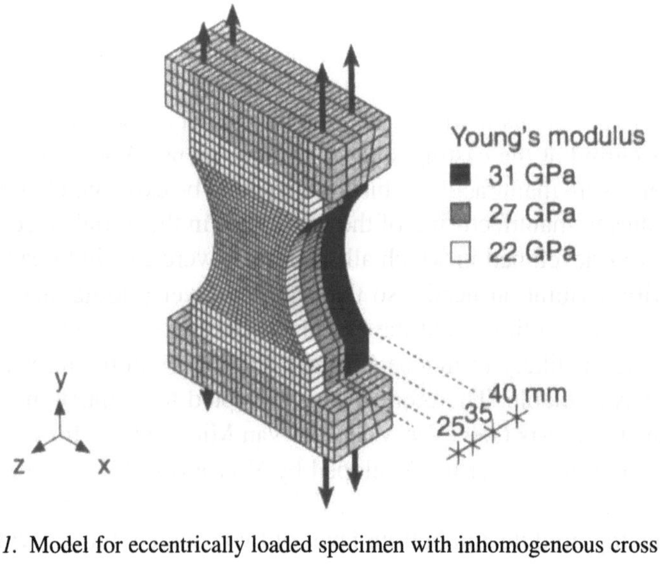


*Figure 11.* Model for eccentrically loaded specimen with inhomogeneous cross section.

## 4. Separation of structural and material size effects

### 4.1. Nonuniform strain distribution

The experimental results of the previous sections show that the behaviour before peak load hardly can be seen as that of a homogeneous specimen with uniform deformation distribution. It therefore may not be that straightforward to express the failure load in terms of the nominal stress $\sigma_N$, at least not without considering the actual pre-peak deformations in the critical cross section of the specimen. With respect to the ultimate load and a possible size effect this actual deformation distribution might be equally, if not more important than the properties of the material itself.

Let us therefore take a closer look at the nature of the specimen deformations, or rather strains when restricting to the situation until peak load. The strain and consequently the stress distribution within the specimens is affected by the specimen shape, the eccentricity of the external force and internal eccentricities. In the first loading stages these internal eccentricities can, among other causes, be due to the inhomogeneity of the cross section as shown in Section 3. As a consequence, certain areas of the specimen are more strained leading to more micro-cracking. These micro-cracks lead to a further decrease of the local stiffness and can therefore increase the internal eccentricity. To obtain more insight in the actual strain distribution in the specimens, a simple linear elastic calculation was carried out on the 3D finite element model shown in Figure 11. The model represents a type C specimen subjected to a load of 4 kN which is applied with an eccentricity $e_x = 4$ mm in $x$ direction. Inhomogeneity of the specimen is incorporated by subdividing the specimen into three layers with an arbitrarily chosen thickness (see Figure 11), and stiffness values according to Table 3.

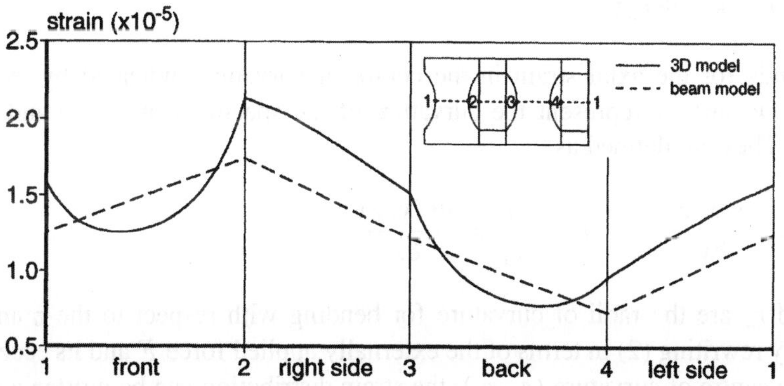


*Figure 12.* Strain distribution along the edges of the smallest cross section of an eccentrically loaded inhomogeneous specimen.

Figure 12 shows the strain distribution along the edges of the cross section in the middle of the specimen. Note the strain gradient over the specimen thickness which is caused by the material inhomogeneity. Because of this material inhomogeneity, the centre of curvature and the centroid of the cross section do no longer coincide. Hence the external load, which is applied centrically in $z$ direction, becomes eccentrical thereby causing a moment $M_z$ and thus bending in the out-of-plane direction (Figure 13). With the given load eccentricity $e_x$ and material inhomogeneity ($e_z$), the most strained part is the front corner on the right hand side of the specimen. Using Hooke's law the maximum stress in the smallest cross section of a specimen can be found from the strain distribution through the local Young's modulus $E$. This concept was used to estimate the maximum stress in the middle of a specimen for all experiments that were carried out, and will be outlined hereafter.

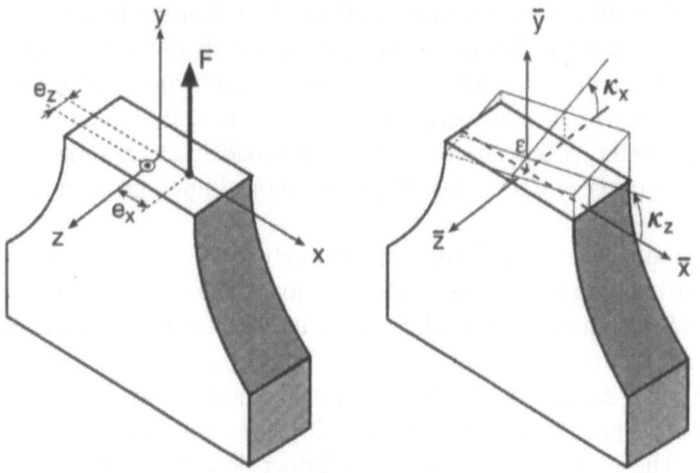


*Figure 13.* Schematization of the forces and deformations in the middle section of a specimen.

Consider the smallest cross section of the C specimen of Figure 11. Under the usual assumption of beam theory (i.e. cross sections which were originally plane and normal to the centre line remain so after bending), the strain distribution in the cross section can be obtained by superimposing its components (see also Figure 13)

$$\varepsilon(x, z) = \varepsilon + \kappa_z x + \kappa_x z. \tag{2}$$

In (2), $\varepsilon$ stands for the axial strain in the centre of curvature indicated by the $(\bar{x}, \bar{z})$ axes in Figure 13. $\kappa_z$ and $\kappa_x$ represent the curvature of the specimen in $x - y$ and $z - y$ plane respectively. They are defined as

$$\kappa_z = \frac{1}{r_z} = \frac{\partial \varepsilon(x, z)}{\partial x}, \qquad \kappa_x = \frac{1}{r_x} = \frac{\partial \varepsilon(x, z)}{\partial z}, \tag{3}$$

where $r_z$ and $r_x$ are the radii of curvature for bending with respect to the $z$ and $x$ axis respectively. By rewriting (2) in terms of the externally applied force $F$ and its eccentricity with respect to the centre of curvature $(e_x, e_z)$, the strain distribution can be written as

$$\varepsilon(x, z) = \frac{F}{EA} + \frac{Fe_x}{EI_z}x + \frac{Fe_z}{EI_x}z, \tag{4}$$

in which $EA$, $EI_z$ and $EI_x$ are the normal stiffness and bending stiffnesses of the composite cross section with respect to the $z$ and $x$ axis respectively. For the same values of $F = 4$ kN and $e_x = 4$ mm the resulting strains along the edges are presented by the dashed line in Figure 12. As can be seen, the bending effects are incorporated correctly and the only difference between the numerical and analytical computation of the strains along the edges is the influence of the specimen shape. When not the strain distribution in the entire cross section is of interest but only the maximum strain, the effect of the shape can be taken into account by introducing a theoretical stress concentration factor $K_t$, defined as $\sigma_{\max}/\sigma_N$. The maximum strain found in a corner point $(x_c, z_c)$, either on the front or the back side of the specimen, then follows from

$$\varepsilon_{\max} = K_t \frac{F}{EA} + \frac{Fe_x}{EI_z} x_c + \frac{Fe_z}{EI_x} |z_c| \,. \tag{5}$$

The value of $K_t$ for the present specimen geometry is taken $K_t = 1.28$, which is based on (Peterson, 1974) and on the results of some finite element calculations. Note that $K_t$ depends on the ratios $D/0.6D$ and $r/0.6D$ where $r$ is the radius of the circular edge. $D$ and $0.6D$ stand for the width of the specimen at the end and in the middle respectively. Since these two ratios are constant for all specimen sizes, the same is true for $K_t$, making it a constant purely geometrical factor.

### 4.2. Approximation of maximum stresses in critical cross section

Under the assumption of crack initiation in the middle of the specimen, and by using an expression like (5) together with experimental data, the stress in the most critical point of the smallest cross section can be estimated for each experiment that was carried out. Furthermore, because of the superposition principle, the different contributions to the maximum stress from the specimen shape and from bending effects can be separated. Starting from the measured values of $\sigma_N$, through correction for various strain gradients the following maximum stresses can be computed

$$\sigma_N = \frac{F_u}{A}, \tag{6a}$$

$$\sigma_{\max,1} = K_t \frac{F_u}{A}, \tag{6b}$$

$$\sigma_{\max,2} = K_t \frac{F_u}{A} + \frac{F_u e_x}{I_z} x_c, \tag{6c}$$

$$\sigma_{\max,3} = K_t \frac{F_u}{A} + E\kappa_{zu} x_c, \tag{6d}$$

$$\sigma_{\max,4} = K_t \frac{F_u}{A} + E\kappa_{zu} x_c + E\kappa_{xu} |z_c| \,, \tag{6e}$$

where $F_u$ is the ultimate load and $A$ the area of the cross section in the middle of the specimen. The values for $E$, $\kappa_{zu}$ and $\kappa_{xu}$ are computed from the deformations measured on each specimen.

The Young's modulus $E$ hereby represents the maximum slope of the average $\sigma - \varepsilon$ curve and is obtained using the least squares method. Once the maximum slope has been found, $E$ can be computed from the scaled measurements in the middle of the specimen through (see also Figures 2 and 14)

$$E = \frac{\sigma}{\varepsilon} = \frac{FL_s}{A_{\text{avg}} \overline{\delta_m}} = \frac{\overline{k_m} L_s}{A_{\text{avg}}}, \tag{7}$$

where $\overline{\delta_m}$ stands for the average deformation measured on the front and back side in the specimen middle (Figure 14), and $A_{\text{avg}}$ represents the average cross-sectional area within

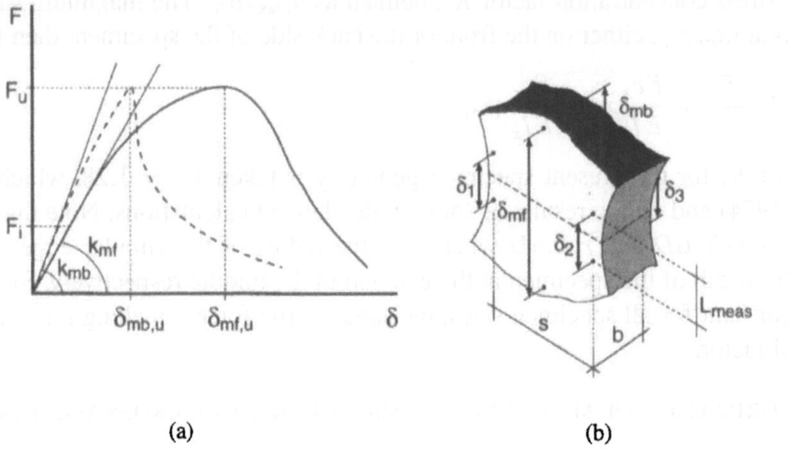


*Figure 14.* Determination of material parameters from measurements (a) and measurement configuration (b).

the scaled measuring length $L_s$. The curvatures $\kappa_{zu}$ and $\kappa_{xu}$ follow from the deformations measured at ultimate load $F_u$ through

$$\kappa_{zu} = \frac{(\delta_2 - \delta_1) + (\delta_3 - \delta_4)}{2sL_{\text{meas}}}, \qquad \kappa_{xu} = \frac{|\delta_{mf} - \delta_{mb}|}{bL_s} \quad \text{at} \quad F = F_u. \tag{8}$$

The subsequent $\delta$'s in (8) denote the deformations that were measured on the specimen surface at the positions indicated in Figure 14, within the measuring length $L_{\text{meas}}$. It is mentioned that applying $b$ in (8) implies that the deformations are measured right at the specimen surface. In reality however the LVDTs are situated at a small distance off the specimen surface, which has been neglected here. This was done because the influence on the computational results of taking a value slightly greater than $b$ is negligibly small. Moreover, due to possible rotations of the LVDTs and the measuring points (see (Van Mier, 1997) for details), other uncertainties are present in the measurements, reducing the relevance of a greater value of $b$.

Theoretically, (6c) and (6d) are identical but here they are explicitly taken into account. In that way the possibility exists to compare the $\sigma_{\max}$ values directly computed from the known eccentricity of the external load ($\sigma_{\max,2}$) and the measured bending deformations of the specimen due to the load eccentricity ($\sigma_{\max,3}$).

In Figure 15 the average maximum stresses and their standard deviations computed following (6b)-(6e) are shown for each specimen size. As a reference also the measured average $\sigma_N$ values are included in the graph.

For $\sigma_{\max,1}$ the maximum stress values have increased with a constant factor whereas the shape of the curve remains the same compared to $\sigma_N$. The same trivial result is found for $\sigma_{\max,2}$, which follows directly from (1). Interesting is however, that despite the fact that $\sigma_{\max,2}$ and $\sigma_{\max,3}$ were expected to be equal, this is not true. Especially for the smaller specimens, $\kappa_z$ computed from the measured deformations appears to be considerably larger than the curvature following from the load eccentricity $e_x$. The reason for the observed difference is that locally the deformations of the cross section can be considerably higher than expected, especially at ultimate load (see for instance Figure 8). When these local deformation peaks are not taken into account, like in (6c), the maximum stresses will be underestimated. Accounting for these nonlinearities by means of a first order approximation based on experimental data is a rather rough approach, which has quite some restrictions. Despite these drawbacks additional insight

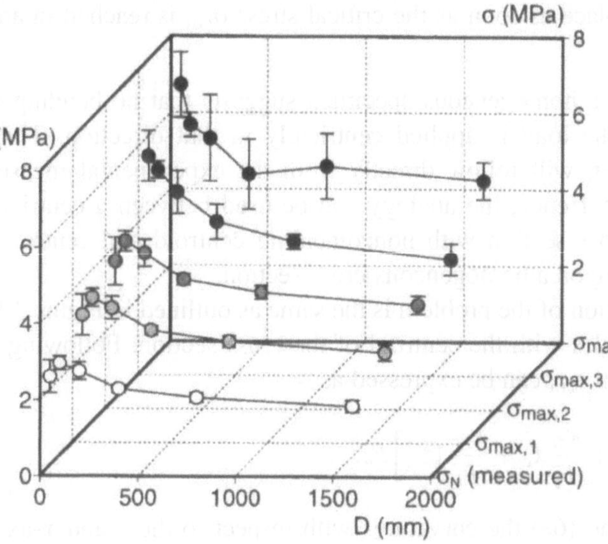


*Figure 15.* Approximation of maximum stresses in the smallest cross section.

is gained as well, which will be further elucidated in the next section. Regarding $\sigma_{max,4}$ Figure 15 shows a further increase of the maximum stress when also $\kappa_x$ is taken into account. This indicates that – as expected – a considerable material inhomogeneity in $z$ direction was present in the experiments.

The above considerations are made under the strong assumption of linear elasticity until the ultimate load is reached. Furthermore it can be questioned whether trying to predict the maximum stresses in the cross section can offer an explanation for the observed size effect in the experiments. These computations *do* show however, that a far from homogeneous strain distribution was present in the laboratory specimens. Since accounting for this fact is already a step ahead of the assumption of a *uniform* strain distribution, in the following section an attempt is made to predict the ultimate load under the presence of strain gradients.

### 4.3. Ultimate load predictions based on nonuniform deformations

Using a model to predict any kind of observation made in reality, usually implies that certain features will be absent due to assumptions that were made or mechanisms that were not incorporated. This however also means that the mismatch between the model predictions and the actual observations can be attributed to 'erroneous' assumptions or to neglected mechanisms. In the following it is attempted to show the direct influence of the presence of strain gradients in a specimen on the ultimate load. This will be done using a very simple schematization of the experiments that were carried out, similar to the one in the previous section. A crucial difference however, is that the ultimate load of each of the experiments carried out is *predicted* using the initial strain gradients and an arbitrarily chosen value for the concrete strength.

The following assumptions are made:

- the specimen follows beam theory, i.e. plane sections remain plane
- the stress-strain relation is defined by Hooke's law
- the specimen is homogeneous
- failure takes place in the smallest cross section

- failure takes place as soon as the critical stress $\sigma_{\text{crit}}$ is reached in any point of the cross section

The assumption of a homogeneous specimen suggests that no bending would take place in $z$ direction since the load is applied centrically in that direction. The strain gradients in $z$ direction however, will follow directly from the experimental measurements as will be explained hereafter. Hence, the analogy can be used between a centrical load acting on an inhomogeneous cross section with noncoinciding centroid and centre of curvature, and an eccentric load acting on a homogeneous cross section.

The schematization of the problem is the same as outlined in Figure 13, but now the centre of curvature coincides with the centroid of the cross section. Following expressions (2) and (4) the critical strain $\varepsilon_{\text{crit}}$ can be expressed as

$$\varepsilon_{\text{crit}} = \left[K_t \frac{1}{EA} + \frac{\kappa_{zi}}{F_i} x_c + \frac{\kappa_{xi}}{F_i} |z_c|\right] F_u. \tag{9}$$

Contrary to (6d) and (6e) the curvatures with respect to the $z$ and $x$-axis are not computed from the deformation measurements at peak load. Instead, the *initial* curvatures $\kappa_{zi}$ and $\kappa_{xi}$ were computed at a load level $F_i$ using the maximum slope of the $F - \delta_j$ curves of the subsequent displacement transducers (see Figure 14). The expressions for $\kappa_{zi}$ and $\kappa_{xi}$ are now

$$\kappa_{zi} = \frac{F_i\left[(k_2^{-1} - k_1^{-1}) + (k_3^{-1} - k_4^{-1})\right]}{2sL_{\text{meas}}}, \quad \kappa_{xi} = \frac{F_i\left|k_{mf}^{-1} - k_{mb}^{-1}\right|}{bL_s}, \tag{10}$$

where $k_j$ are the slopes of the various $F - \delta_j$ curves and where the used subscripts are according to Figures 2 and 14. The curvature at peak load can now be computed directly from (10) by linear extrapolation. This of course means that nonlinear deformations at peak load will not be taken into account. Whether this is of crucial importance or not must follow from the results.

Apart from $K_t$, which is equal to 1.28 as in the previous section, the only remaining unknown is $\varepsilon_{\text{crit}}$. Following Hooke's law $\varepsilon_{\text{crit}}$ can be computed from $\sigma_{\text{crit}}$. Since the Young's modulus computed for the different specimens varied quite a lot, especially for the smallest specimens, setting $\sigma_{\text{crit}}$ seems more straightforward than keeping $\varepsilon_{\text{crit}}$ constant. The value for the critical or local failure strength is, arbitrarily using the average strength of the type B specimens (Table 2), set as

$$\sigma_{\text{crit}} = \sigma_N\left[K_t + \frac{6e_x}{0.6D}\right] = 4.39 \text{ MPa} \tag{11}$$

with $e_x = 2$ mm and $D = 100$ mm. As a matter of fact any other value for $\sigma_{\text{crit}}$ could have been chosen, since this would merely affect the height and scatter of the computed $\sigma_N$ values. The tendency for different specimen sizes however, remains the same. With

$$\varepsilon_{\text{crit}} = \frac{\sigma_{\text{crit}}}{E} \tag{12}$$

and $E$ according to (7), $\sigma_N$ can now be predicted for each test. In the same fashion as (6b)-(6e), $\sigma_N$ can be obtained via the computed ultimate load $F_u$ for different cases

$$F_{u,1} = \frac{\varepsilon_{\text{crit}}}{K_t(EA)^{-1}}, \tag{13a}$$

$$F_{u,2} = \frac{\varepsilon_{\text{crit}}}{K_t(EA)^{-1} + e_x(EI_z)^{-1}x_c}, \tag{13b}$$

$$F_{u,3} = \frac{\varepsilon_{\text{crit}}}{K_t(EA)^{-1} + \kappa_{zi}F_i^{-1}x_c}, \tag{13c}$$

$$F_{u,4} = \frac{\varepsilon_{\text{crit}}}{K_t(EA)^{-1} + \kappa_{zi}F_i^{-1}x_c + \kappa_{xi}F_i^{-1}|z_c|}. \tag{13d}$$

The conclusions that can be drawn from this whole analysis follow directly from Figure 16. First of all, the difference between $\sigma_{N,2}$ and $\sigma_{N,3}$ clearly shows that the measured strain gradients in the plane of the specimen are larger than expected. Moreover, they are the onset for the variation of strength with specimen size. When the strain gradients in the out-of-plane direction are incorporated in the analysis as well ($\sigma_{N,4}$), the predicted values of the ultimate strength vary with the specimen size in a way which corresponds quite well with the experimental results (Figure 17). As for the comparison, the tendency of the predicted strength values is of importance in this analysis and not the real values, since the critical stress in the computation was arbitrarily set.

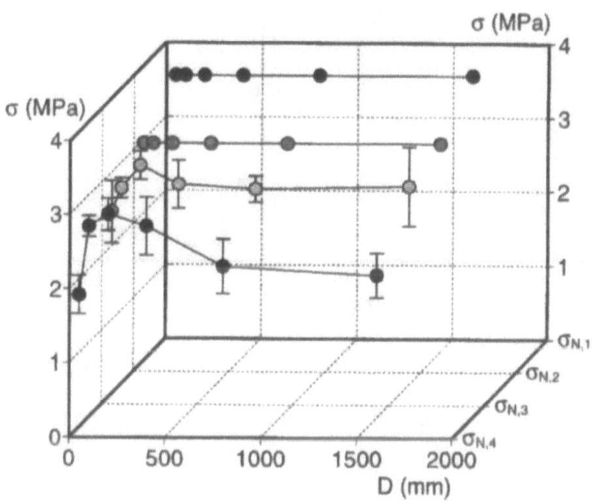


*Figure 16.* Calculation of the nominal stress in the smallest cross section. Assuming constant material strength, $\sigma_{N,1}$ through $\sigma_{N,4}$ are the corrections made for the specimen size (13a) and various eccentricities (13b-d).

A general observation in Figure 17 is that for the three largest specimen sizes the scatter of the predicted values is much larger than for the experimental values. This is not the case for the three smallest sizes but there the tendency of the values is not similar: contrary to the measured values, the predicted strength values decrease with decreasing specimen size. The explanation for this is two-fold, and is illustrated by Figures 18a and 18b. In Figure 18 the measured and predicted values for the normal strain $\varepsilon$ in the centroid of the cross section, as well as the total curvature $\kappa$ are compared. $\kappa$ is composed from $\kappa_x$ and $\kappa_z$ through

$$\kappa = \sqrt{\kappa_x^2 + \kappa_z^2}. \tag{14}$$

Starting with the three largest sizes, it is found that the predicted values for $\kappa$ are lower than the measured ones. Both predicted and measured values are likely to underestimate the real, local strain gradients which follows from Figure 8. Inherent to predicted $\kappa$ values which

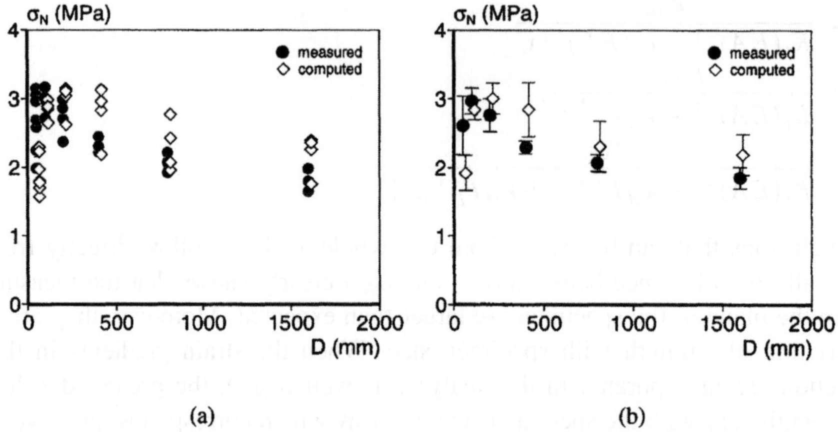


*Figure 17.* Individual values (a) and mean values (b) of measured and computed values for the nominal stress.

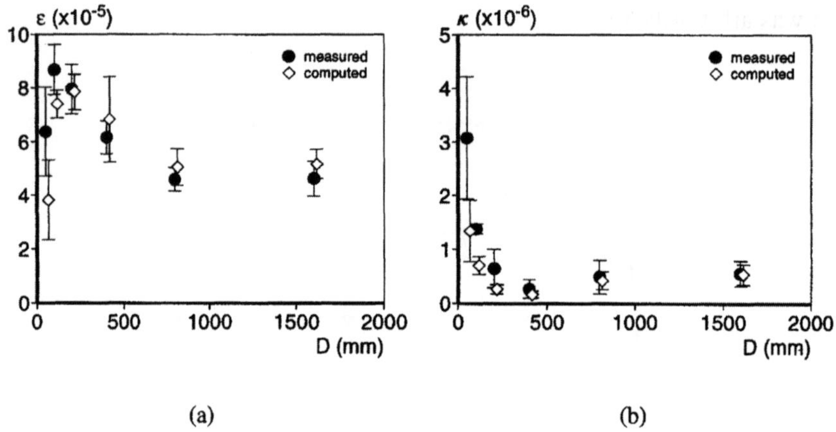


*Figure 18.* Measured and computed values of normal strain (a) and total curvature (b).

are too low, are predictions for $\varepsilon$ and $\sigma_N$ which are too high. When comparing each pair of predicted and experimental value for one of the larger specimen sizes it was found that the values deviate when the strain gradients are smaller. For this case the computed stress distribution tends to the homogeneous case resulting in a far too high prediction of the failure strength. This is due to the absence of a material structure in the model.

Figure 18b shows that the computed curvature strongly under-estimates the measured one for the three smallest specimen sizes. Although one would therefore expect too high values for the computed $\varepsilon$ and $\sigma_N$, the actual findings are completely opposite. The experimentally obtained values of $\sigma_N$ are higher, despite the presence of considerable strain gradients over the cross section. This could indicate that the material structure overrules the influence of the strain gradients. In this respect one has to consider the effect of a 8 mm grain which is situated in an area with a very high strain gradient, which was also mentioned by Heilmann (1976). Another explanation can be found in the statistical interpretation of the curvatures. A high value for the curvature implies that the part of the cross section in which the highest strains are present is small. The probability of a weak spot being present in this part of the cross section is therefore small as well. As a consequence, the ultimate loads which are found for small specimens in which high strain gradients are present will be high. In the experiments this phenomenon could very well occur, whereas it is absent in the computations leading to differences between the two sets of $\sigma_N$ values.

A point which has not been discussed until now, concerns the variation of the stress and/or strain gradients with size. Note that a direct relation exists between this variation and the observed size effect, since a constant gradient for all sizes implies that in the presented approach no size effect would be found (see (13a), (13b) and Figure 16). Several reasons can be found for the variation of the strain gradients. One of the most important must be attributed to the manufacturing process. Although all specimen sizes were manufactured with the greatest care, it is much easier to maintain a constant quality for large specimens than for small ones. Consequently stiffness differences will be present in the casting direction for specimens of different sizes, and thus variations of the strain gradients in the out-of-plane direction. Another reason can be found in different deformation nonlinearities in the in-plane direction. These nonlinearities can develop due to a nonconstant influence of the material heterogeneity (grain size) with respect to the specimen width. Finally, differences in the amount of shrinkage cracks and the magnitude of eigenstresses due to differential shrinkage can induce variations in the strain gradients. This will be further discussed in the next section.

In the present tests differences in stress and/or strain gradients were found to be the main reason for the observed size effect on the nominal strength. Although these gradients will always be present in concrete subjected to a certain load, the significance may vary depending on the presence of other size effects. In this respect one may consider e.g. statistical size effect, the amount of energy which can be released into a unit crack extension or the stress state in the specimen depending on the type of loading which has been applied.

## 5. Discussion

It is common knowledge that concrete exhibits an apparent size effect on strength. Also well known is that a dependency exists between the extent of the observed size effect, and the presence of stress gradients inherent to the type of test which is considered. Actually taking these stress gradients, or maybe rather *strain* gradients, into account in an attempt to explain the real nature of the size effect, is however less common practice.

In this paper the results are presented of a series of uniaxial tension tests on concrete, that were conducted to study the size effect on the uniaxial tensile strength and fracture energy caused by the material itself. Apart from the smallest specimens, the nominal strength values which resulted from the experiments showed a decrease with increasing specimen size. The scatter of the strength values of the smallest specimens was disproportionally large however, indicating that the aspect ratio of $D/d_{a,\max}=3.75$ was too small. The most important reason for the scatter is the presence or absence of large grains in the smallest cross section. Numerical simulations with a lattice model revealed that if such a large grain is present in the surface layer of the specimen, the ultimate load can decrease considerably (Van Vliet, 1999). The main reasons for this reduction are the consequent stress concentrations in combination with the few possibilities for stress redistributions in the cross section of such a small specimen. This scatter was not only found in the strength values, but also in the values of the Young's modulus which were derived from the measurements. In this respect, the findings could be related to the work of Müller (1964). He reported a direct relation between the standard deviation of the strains measured on the surface of a uniaxially loaded concrete prism on the one hand, and the ratio of $d_{a,\max}$ and the measuring length on the other hand.

In a majority of the experiments it was observed that the deformations that were measured on the casting side of a specimen, were larger than those measured on the mould side. This result is analogous to that of Scheidler (1987), who found a variation of the concrete stiffness parallel to the direction of casting. Another experimental observation was that crack initiation often occurred at the casting side. A series of verification tests on slices of concrete originating from different positions over the specimen cross section, confirmed the presumed strength and stiffness variations within a specimen. In the present research these variations can be attributed to the manufacturing process of the specimens. Of general importance is however, that due to boundary and segregation effects such variations are inherent to the material concrete. The resulting stiffness differences induce strain gradients and thus bending, also in case of a centrically applied external load. By explicitly taking into account the strain gradients due to the specimen shape, eccentricity of the external load and material inhomogeneity, most of the observed size effect on the nominal strength could be explained under the strong assumption of simple linear elastic beam theory. It was also found that for small specimen sizes, the strain gradients which are present in a cross section seem to be overruled by (local) gradients caused by the structure of the material. In case of larger sizes on the other hand, the effect of strain gradients (if present) can overrule any effects of the material structure.

Despite the fact that a size dependence of the peak load was found by only taking into account deformation gradients over a specimen's cross section, it is mentioned that at the same time other phenomena are implicitly incorporated as well. One of these phenomena is the relation between the specimen stiffness and the ultimate load which was found in the tensile splitting verification tests. When these two properties are directly related, the presence of strong stiffness differences over the specimen's cross section also implies variations of the concrete strength within the cross section. In that sense the most critical point in the cross section with respect to maximum strains then simultaneously is the point with the lowest strength. Other phenomena which have not been incorporated explicitly in the analysis but which are however likely to play a role, are for instance eigenstresses due to differential shrinkage and differential temperatures during hardening of the concrete. Both phenomena may lead to cracking in an early stage of the life cycle of the material, even before any external load is applied. Shrinkage cracks will be present in the outer layers of the specimens. The amount of cracks can vary, and will among others depend on the concrete composition in the outer specimen layers. Because of these shrinkage cracks the concrete stiffness will locally be lower, thus resulting in larger deformations. Eigenstresses on the other hand do not lead to deformation differences but to stress concentrations, again in the surface layers of the specimens. For both phenomena a size effect can be present which is not known beforehand. Such a size effect could in addition to casting effects be a reason for the observed variation of the deformation gradients in the uniaxial tension tests, and thus for the size effect on the nominal strength. The impact of shrinkage cracks and eigenstresses due to differential shrinkage has been investigated in an additional test series. For this purpose uniaxial tension tests have been conducted on dog-bone shaped specimens of the four smallest sizes. Contrary to the specimens of the main test series, these specimens were cured in a fog-room for two months, rather than stored in the laboratory. For the results of these tests, the reader is referred to (Van Vliet, 1999).

As has already been mentioned in Section 4.3, various size effects may occur at the same time. It has been shown that in the present tests nonuniformity of the deformations seems to have a clear influence on the observed size effect. Meanwhile however, it is well known that especially in uniaxial tension tests material size effects should play a dominant role due to the 'uniform' stress state. These material (or statistical) size effects can be described with Weibull's weakest link theory (Weibull, 1939), from which an inverse proportionality can be deduced between the nominal strength $\sigma_N$ and the characteristic dimension $D$ following (see also (Bažant et al., 1991))

$$\sigma_N \propto D^{-n/m}. \tag{15}$$

In (15) $n$ represents the number of dimensions (1, 2 or 3) and $m$ the Weibull modulus. For concrete $m$ has been found to be 12 (Zech and Wittmann, 1978) and together with $n = 2$ for the present tests this results in

$$\sigma_N \propto D^{-1/6}. \tag{16}$$

This relation between $\sigma_N$ and $D$ is shown graphically in Figures 19a and b. The result for the experimental values needs little discussion and could lead to the conclusion that a pure material size effect is observed in the tests. Interesting however, is the fact that also the computed values from Figure 17 approximate Weibull theory when plotted on a log-log scale. Despite this observation it is clear that no strength variations at all were used to compute the latter values. Consequently, size effects other than material size effects can lead to comparable observations, showing the need to distinguish the separate effects and to study their importance for a specific case.

To conclude, in the authors' opinion the general term 'size effect' should be better defined. When 'size effect' is interpreted as a micro-structural effect inherent to the material, then it needs to be separated from macro-structural effects which will always to some extent be present in experimental results. Macro-structural effects can be boundary condition effects and shape effects, but also effects originating from the material which become visible at a more global level of observation. In this respect one can think of material inhomogeneity due to stiffness differences but also moisture gradients or eigenstresses due to differential shrinkage (Sadouki and Van Mier, 1996) and differential temperatures during hardening of the concrete. All these effects have in common that they cause strain and/or stress gradients which can strongly blur the actual *material* size effect. For this reason they can not be excluded from any analysis of experimental size effect data.

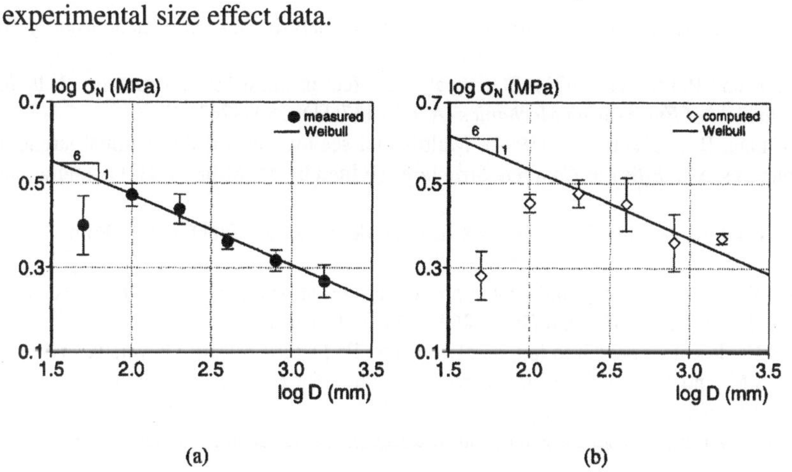


*Figure 19.* Measured (a) and computed values (b) for the nominal stress.

## 6. Conclusions

Based on the results presented in this paper, the following conclusions can be drawn.

- To come to a better understanding of the size dependency of material parameters, *material size effects* should be separated from *structural size effects*. Material size effects are micro-structural effects inherent to the material. Structural size effects on the other hand prevail at the macro-level of observation. They can be induced by boundary and shape effects, but also by the material itself as in the case of variation of the material properties within a specimen.
- Strain and/or stress gradients can not be excluded from any analysis of experimental size effect data. In the present investigation it was found that by taking these gradients into account, most of the observed size effect on the nominal strength could be explained. The gradients can be due to

  – the specimen shape
  – eccentricity of the external load
  – material inhomogeneity
  – eigen-stresses due to differential shrinkage

- When conducting experiments, it is crucial to maintain an aspect ratio of characteristic specimen dimension to maximum aggregate size $D/d_{a,\max}$ which is large enough. For an aspect ratio of 3.75 the experimental results showed a large scatter whereas the scatter was considerably smaller for larger aspect ratios. Also in the discussion about a strength asymptote for small sizes this should be carefully considered.


## Acknowledgements

The authors wish to thank Mr. Gerard Timmers for his expert assistance in conducting the experiments and solving numerous technical problems. The research was financially supported by the Dutch Technology Foundation (STW), which is gratefully acknowledged.


## References


Bažant, Z.P. (1997). Scaling of quasibrittle fracture: asymptotic analysis. *International Journal of Fracture* **83**(1), 19–40.

Bažant, Z.P., Xi, Y. and Reid, S.G. (1991). Statistical size effect in quasi brittle structures: I. Is Weibull theory applicable?. *Journal of Engineering Mechanics (ASCE)* **117**(11), 2609–2622.

Carpinteri, A., Chiaia, B. and Ferro, B. (1994). Multifractal scaling law for the nominal strength variation of concrete structures. *Size Effect in Concrete Structures* (Edited by M. Mihashi, H. Okamura, and Z. Bažant), 173–185.

Ferro, G. (1994). Effetti di scala sulla resistenza a trazione dei materiali. Ph.D. thesis, Politecnico di Torino (in Italian).

Heilmann, H.G. (1976). Zugspannung und dehnung in unbewehrten betonquerschnitten bei exzentrischer Belastung. In: *Deutscher Ausschuss für Stahlbeton*, **269**, 1–61 (in German).

Hordijk, D.A. (1991). Local Approach to fatigue of concrete. Ph.D. thesis, Delft University of Technology.

Li, Z., Kulkarni, S.M. and Shah, S.P. (1993). New test method for obtaining softening response of unnotched concete specimen under uniaxial tension. *Experimental Mechanics* (9), 181–188.

Müller, R.K. (1964). Der einfluss der messlänge auf die ergebnisse bei dehnmessungen an beton. *Beton* (5), 205–208 (in German).

Peterson, R.E. (1974). *Stress concentration factors*. John Wiley & Sons Ltd.

Sadouki, H. and Van Mier, J.G.M. (1996). Analysis of hygral induced crack growth in multiphase materials. *HERON* **41**(4), 267–286.

Scheidler, D. (1987). Experimentelle und analytische untersuchungen zur wirklichkeitsnahen bestimmung der bruchschnittgrössen unbewehrter betonbauteile unter zugbeanspruchung'. In: *Deutscher Ausschuss für Stahlbeton*, **379**, 1–94 (in German).

Van Mier, J.G.M. (1997). *Fracture processes of concrete*. CRC Press.

Van Mier, J.G.M., Schlangen, E. and Vervuurt, A. (1995). Lattice type fracture models for concrete. *Continuum Models for Materials with Microstructure* (Edited by H.-B. Mühlhaus), John Wiley & Sons Ltd., 341–377.

Van Vliet, M.R.A. (1999). Ph.D. thesis, Delft University of Technology (in preparation).

Van Vliet, M.R.A. and Van Mier, J.G.M. (1997a). Concrete behaviour under uniaxial tension: Effect of casting and drying conditions. *Brittle Matrix Composites 5* (Edited by A. Brandt, V. Li, and I. Marshall), 329–339.

Van Vliet, M.R.A. and Van Mier, J.G.M. (1997b). Size effect in tensile fracture of concrete: Effect of test-control. *Ninth International Conference on Fracture* (Edited by R. Ritchie, B. Karihaloo, and Y. Mai), 957–964.

Van Vliet, M.R.A. and Van Mier, J.G.M. (1998). Experimental and numerical investigation of size scale effects in concrete fracture. *Material Instabilities in Solids* (Edited by R. De Borst and E. Van Der Giessen), John Wiley & Sons Ltd., 185–206.

Vervuurt, A.H.J.M. (1997). Interface fracture in concrete. Ph.D. thesis, Delft University of Technology.

Weibull, W. (1939). A statistical theory of the strength of materials. *Royal Swedish Academy of Eng. Sci. Proc.* **151**, 1–45.

Zech, B. and Wittmann, F.H. (1978). A complex study on the reliability assessment of the containment of a PWR. Part II, Probabilistic approach to describe the behaviour of materials. *Nuclear Engineering and Design* **48**, 563–593.

*International Journal of Fracture* **95:** 221–238, 1999.


# Scaling phenomena due to fractal contact in concrete and rock fractures

M. BORRI-BRUNETTO, A. CARPINTERI and B. CHIAIA
*Department of Structural and Geotechnical Engineering, Politecnico di Torino, 10129 Torino, Italy*
*e-mail: chiaia@polito.it*

Received 10 July 1998; accepted in revised form 18 December 1998

**Abstract.** Concrete-to-concrete friction contributes in many cases to the stability of a structure. At different scales, the slope stability of rock joints is deeply influenced by the surface morphology and shows a marked size-dependence. In this paper, the closure and sliding-dilatant behaviour of cracks in concrete and rocks is investigated by means of a coupled numerical/experimental approach. These natural interfaces show self-affine properties in the relevant scale range. Attention has been focused on the stress transfer mechanism across the interfaces, showing that the sets of contact points possess the self-similar character of lacunar fractal sets. Scaling laws come into play and the size-effects on the shear strength of rough interfaces, and on their closure deformability, can be explained.

**Key words:** Size-effects, fractal geometry, fracture interface strength, contact mechanics.

## 1. Introduction: frictional behaviour of cracks in concrete and rocks

Concrete-to-concrete friction comes into play in many situations and contributes to the stability of a structure. However, the effects of friction are especially valuable in massive structures, because the frictional contribution to stability is size-dependent, if dead loads are considered. Within the classical hypotheses (Johnson, 1985) the frictional force $T$ is independent of the nominal contact area $A_n$, but depends on the true contact area $A_r$, which, in turn, is proportional to the normal load $F(T = \mu F)$. If perfectly similar structures are considered, the nominal contact area scales according to $L^2$, whereas the dead loads (and the frictional force) scale proportionally to the weight of the structure, that is, according to $L^3$. Therefore, if $\tau = T/A_n$ is the frictional specific resistance, one obtains $\tau = T/A_n \sim L^3/L^2 \sim L$. Hence, stability due to friction increases with the size scale of the structure. Note, however, that this is no longer true when the shear actions also depend on gravity (e.g. seismic loading). The evaluation of the interface shear transfer capacity $T_u$ is based on the traditional theory of friction, that is, $T_u$ is proportional to the normal load $F$ by means of a friction coefficient $\mu$. The value $\mu = 1.4$ is suggested by Park and Paulay (1975) for concrete cast monolithically, while $\mu = 1.0$ applies for concrete placed against hardened concrete.

In concrete dams, because of the joints generated by a stepwise construction of layered concrete blocks, and because of the strong thermal gradients, large cracks can be present from the beginning of their life. The assessment of safety under the presence of the cracks cannot be adequately addressed if the shear resistance of these joints is not taken into account. In many concrete structures, shear forces may cause a sliding type of failure along a pre-existing plane. Because of external tension, shrinkage, or other causes, a crack may form along such plane even before shear occurs. Thus, the possibility of shear transfer by friction arises.

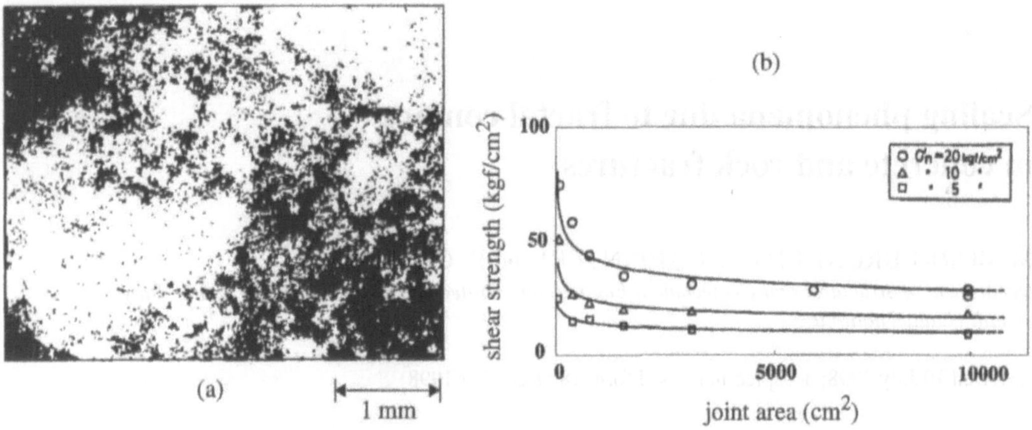


*Figure 1.* (a) Experimental contact domain in a natural rock joint at a mean apparent stress of 85 MPa (Cook, 1992). (b) Experimental decrease of the measured shear strength for increasing specimen size (Yoshinaka et al., 1993).

Earthquakes showed that construction joints in some members, particularly shear walls, form the weakest link in the load-resisting mechanism of the structure if large shear forces need to be transmitted.

Frictional effects in cracked flexural members can be critical only if the span to depth ratio is very small (deep beams). In these members, concrete friction is often accompanied by the dowel action of the longitudinal bars subjected to shear displacement. In fact, even a small shear displacement $\nu$ is accompanied by a normal displacement $\delta$ of the crack faces (*dilatancy*). To develop an acceptable friction capacity, dilatancy must be limited by normal loading, provided by external constraints or by the steel bars. Dilatancy may also reduce the durability of a joint, for example in seismic areas. If the bars reduce dilatancy, the interface shear transfer capacity increases. Thus, less shear displacement occurs and less deterioration of the shear capacity can be expected after a few cycles of high-intensity loading. Instead, if cracks are free to open, repeated loading will cause a deterioration of the interface roughness, with a corresponding reduction in the shear capacity.

At scales different from the usual scales of concrete structures, the engineering analysis of rock structures, like slopes, excavations or tunnels, needs a model of the mechanical response of the rock mass. Far from being a continuum, the rock mass is a complex system composed of blocks and fractures. In many cases, a detailed study of the morphology and of the behaviour of discontinuities subjected to normal and shear forces is necessary to evaluate, say, the factor of safety of a slope or a foundation. In this respect, it is worth mentioning the review paper by Cook (1992), where most of the basic mechanical issues are reported. In that paper, the crucial importance of the effective stress transfer at the interface is addressed (Figure 1a).

The measurement of the shear strength of natural discontinuities is usually obtained by means of laboratory testing on rock joints. The influence of the roughness of the contacting surfaces, as well as the importance of scaling in the assessment of the *in situ* resistance, is a well-known fact (Pinto da Cunha, 1993). Experimental tests showed a marked size-dependence of the shear strength of rock discontinuities (Figure 1b). A number of constitutive phenomenological models have been proposed with the purpose of estimating the most plausible full-scale strength parameters starting from data obtained on small scale specimens. The classical empirical approach of Barton and Choubey (1977), based on visual estimation of roughness of joint profiles, was extended in order to take into account the scale dependence of strength (Bandis et al., 1981), and is perhaps the most widely used shear strength criterion in rock engineering. Further attempts tried to get a more objective determination of the parameters through mathematical treatment of geometrical data obtained from surface analysis. A remarkable effort was due to Scholz and co-workers (Brown and Scholz, 1985; Yoshioka and Scholz, 1989; Wang and Scholz, 1993). In these works, the mechanical response of a rock interface is studied as a result of the interaction of asperities, paying attention to the real topography of the contacting surfaces.

## 2. Contact between rough surfaces: traditional models and fractal approach

A realistic characterization of the topography of rough surfaces represents a crucial point in the modelization of interface phenomena. Many tribologic phenomena, like friction, lubrication and wear of mechanical components, strongly depend on the surface morphology (Johnson, 1985). At the same time, the thermo-electric conductivity between two bodies in contact is intimately related to the interface characteristics.

The earliest attempts to extend the Hertzian theory of elastic contact between smooth bodies to real bodies with rough boundaries can be ascribed to Archard (1957) and to Greenwood and Williamson (1966). Archard was the first to solve the discrepancy between the classical Hertzian solution and the experimentally validated Amontons's law. This law is based on the hypothesis that the frictional force $T$ is proportional to the area of true contact $A_r$, which, in turn, is proportional to the normal load $F$. While, in fact, Hertz had shown that, in the case of smooth spheres, the real contact area $A_r$ is related to the normal load $F$ by a nonlinear relation ($A_r \sim F^{2/3}$), Amontons's law stated the direct proportionality between frictional force and normal load ($T = \mu F$). By means of a hierarchical model of hertzian spheres with progressively decreasing radius, Archard obtained the overall linear behaviour ($A_r \sim F$) requested by the experimental evidence.

Greenwood and Williamson (1966) considered one of the bodies in contact to be rough, with contact occurring at a number of discrete points. The asperities could undergo only elastic deformation, and at the top of each peak an Hertzian micro-sphere was supposed to exist. The asperities were modelled in a way for the peaks to follow a Gaussian distribution around a mean flat surface. The area of true contact was determined to be approximately proportional to the load, so that the basic hypothesis of the theory of friction could be supported also for purely elastic contact. The authors concluded that the behaviour of rough surfaces is determined primarily by the statistical distribution of asperity heights and secondarily by their mode of deformation. Thus, although plastic deformations cannot be excluded, at least for the highest and sharpest peaks, it was argued that no new features would be introduced by plasticity.

Because the interface phenomena involve all length scales, a scale-independent description of the surfaces should be pursued. Nevertheless, classical models as the one by Greenwood and Williamson (1966) make use of statistical parameters, like the mean and variance of the surface elevations, of the slopes and of the curvatures. Unfortunately, all these quantities, whose experimental determination is questionable, are strongly dependent on the resolution adopted for the description of the interface (Majumdar and Bhushan, 1990). On the other hand, the theory of fractals sheds new light on the Archard's model, which is, indeed, scale-independent. Archard's model can be thought of, in fact, as the first deterministic (pre-)fractal model applied to contact problems, as already pointed out by Ling (1989).

Recently, random fractal models of rough surfaces (Mandelbrot, 1982) have been shown to resemble very closely the morphology of natural and man-made surfaces. The surfaces of magnetic tapes and of metallic components, as well as natural boundaries like fractures and rock joints, show fractal properties over a wide scale range. This implies the morphological invariance (in the statistical sense) of the surfaces under a group of affine (anisotropic) scale transformations. Self-similarity is a particular case of the more general concept of *self-affinity*. Self-similarity, in fact, is characterized by isotropic dilatation symmetry and by the absence of an internal length scale. The scaling relation of a self-affine surface $z(x, y)$, if $r$ is the scaling (or magnification) factor, can be written as:

$$f(x, y, z) \rightarrow f(rx, ry, r^H z), \tag{1}$$

where $H < 1$ is called the *Hurst exponent*. Equation (1) implies that the vertical coordinates scale less than the horizontal ones. Thereby, sharper details appear in the surface as finer resolutions are adopted. The fractal dimension is obtained as $\Delta_{\mathcal{G}} = 3 - H$, and thus these 'surfaces' are invasive fractals ($\Delta_{\mathcal{G}} > 2.0$). According to the experimental evidence (Carpinteri, 1994) and to the theory of dissipative phenomena, the fractal dimension of a Brownian surface ($\Delta_{\mathcal{G}} = 2.5$) often represents the upper limit of disorder. This is the case, for example, of fracture surfaces. Therefore, for most natural surfaces, $0.5 \leqslant H \leqslant 1.0$.

Panagiotopoulos and Panagouli (1997) adopted an engineering-oriented fractal methodology to solve problems of unilateral and adhesive contact and problems of friction. Their numerical procedure exploited the approximation of fractal boundaries through classical surfaces of integer Hausdorff dimension (Iterated Function Systems). Borodich and Mosolov (1992) modelled polished fractal surfaces by means of Cantor punches. The problem of indentation was solved for this geometry, in the case of elastic as well as of plastic bodies. Using asymptotic methods, they obtained power-laws for the dependence of the load on the depth of the indentation. The exponents of the power-laws depended explicitly on the fractal dimension of the punching set. Warren and Krajcinovic (1995) developed a Cantorian fractal model, similar to the one by Borodich and Mosolov, and extended the discrete model to a continuous formulation of the elastic-perfectly plastic deformation model.

Regarding recent experimental investigations, it is worth mentioning Majumdar and Bhushan (1990), who adopted an optical profilometer to digitize magnetic tape surfaces. They used the Weierstrass-Mandelbrot fractal function to model the surface roughness, obtaining good correlations with experimental data. Majumdar and Tien (1990) carried out tests on various machined steel surfaces and textured magnetic thin-film disks. The power-law spectral behaviour of these surfaces unequivocally proved that statistically similar images of the surfaces appear when the surface is repeatedly magnified. However, although the self-similarity of the contact domain is exploited in their model, the mechanical quantities retain their usual physical dimensions. On the contrary, finite measures of fractal domains can be obtained only by means of noninteger dimensions. Accordingly, the physical quantities defined over these sets have to assume noninteger (anomalous) dimensions.

## 3. Anomalous scaling in the contact mechanics of natural interfaces

From a kinematical point of view, the linear elastic normal contact between two rough surfaces can be reduced, under some assumptions, to the problem of the so-called *composite topography* (given by the *sum* of the heights of two corresponding points in the undeformed condition) pushing against a flat plane. The mechanics is controlled by the following physical variables: the external normal force $F$ ([F]), the elastic modulus $E$ ([F][L]$^{-2}$), a linear representative size of the contact area $b$ ([L]), the interface closure displacement $w_{\text{int}}$ ([L]), and a morphological parameter of the interface, e.g. the standard deviation $\sigma_z$ ([L]) of the heights of the composite topography. According to the well-known Buckingham Theorem, there are three independent nondimensional groups, which can be determined in a straightforward manner. A scale-independent complete description of the mechanics of normal elastic contact can be obtained by simply finding the mutual relations among $\Phi = F/Eb^2$ and two other nondimensional quantities, for example $\xi = \sigma_z/b$ and $\omega = w_{\text{int}}/b$. A scale-independent governing relation $\Phi = \Phi(\xi, \omega)$ could be found. However, experimental investigations (e.g. Sage et al., 1990) provide data which do not fit in the nondimensional framework. This means that anomalous scaling occurs in the phenomenon.

Suppose, in fact, we are interested in extrapolating the nondimensional full-scale crack closure law $\Phi(\omega)$ from the laboratory tests. According to the nondimensional description, one should expect that $w$ scales proportionally to $b$, if a self-similar phenomenon were considered (i.e., if $\sigma_z$ were perfectly proportional to $b$). Keeping the elastic modulus $E$ constant, similarity would imply that, if $b \to 2b$ moving from the model to the prototype, then $F \to 4F$ and $\sigma_z \to 2\sigma_z$. In practical words, if the linear size of the full-scale structure is twice the linear size of the model, a double closure value $2w_{\text{int}}$, under the same nominal pressure $\sigma_n = F/b^2$, can be obtained only if the standard deviation $\sigma_z$ of the heights of the model's interface is doubled. Instead, when testing specimens of different size, *obtained from the same joint*, it is impossible to realize the condition $\sigma_z \to 2\sigma_z$ as $b \to 2b$. Schmittbuhl et al. (1995) affirmed that the standard deviation of heights, in natural rock discontinuities, far from being scale-independent scales with $b$ according to

$$\sigma_z \sim b^H, \tag{2}$$

where $H = 3 - \Delta_{\mathscr{g}}$ is the Hurst exponent of the self-affine composite topography. Statistical investigations yielded the value $H \approx 0.8$ for natural joints, whereas the value $H = 1$ (corresponding to $\Delta_{\mathscr{g}} = 2$, that is, to Euclidean surfaces) would be necessary to ensure perfect similitude.

## 4. Fractal analysis of real interfaces and numerical simulation of contact

In a previous paper by Borri-Brunetto et al. (1998), three mathematical self-affine surfaces were adopted for the contact simulation. In this paper, instead, real interfaces are considered, namely a rock joint and a concrete fracture surface. The rock joint has been obtained from a cylindrical specimen of fine-grained sandstone fractured at midspan in three-point bending. The diameter of the specimen was equal to 50 mm. The concrete fracture surface was obtained from a bone-shaped specimen broken under uniaxial tension.

The experimental acquisition technique, extensively described by Carpinteri et al. (1999), consists of a laser profilometer moving in the two orthogonal directions $X$ and $Y$ by means of two micrometric step-motors. At each point of a horizontal grid, with spacing equal to $s$, the laser reads the height of the corresponding surface point. Therefore, the digitized surfaces, as well as any computer-generated surface, are only images (*pre-fractals*, characterized by a precise value of the resolution $s$) of the fractal set, which is a limit concept for $s \to 0$.

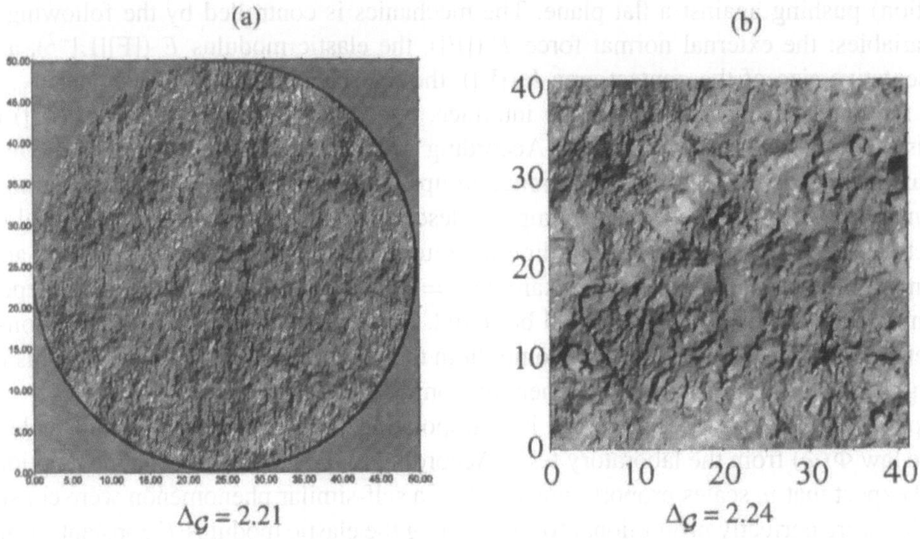


*Figure 2.* Shaded relief of the digitized fracture surfaces obtained from the sandstone specimen (a) and from the concrete one (b).

Shaded rendering of the rock and concrete surfaces are shown in Figures 2a and 2b, respectively. The scanned area measures $\pi(25)^2 = 1962\,\text{mm}^2$ in the case of the rock and $40 \times 40\,\text{mm}^2$ in the case of concrete. The digitisation interval $s$ was set equal to $100\mu$m and to $80\mu$m, respectively for rock and concrete. It is believed that further refinements (a limit precision of $2\mu$m can be attained by the laser) would not alter significantly the numerical simulation of contact, at least in the desired mesoscopic scale range. In both cases (concrete and rock), only one side of the fracture was digitized, because the hypothesis of initially perfectly matching surfaces at the opposite sides of the fracture is implicitly assumed in the following.

By applying both deterministic and statistic fractal tools the fractal dimension $\Delta_g$ of the rock surface results to be comprised in the range 2.16–2.25, with a mean value equal to 2.21. The fractal dimension of the concrete surface can be set in the range 2.15–2.29, with a mean value equal to 2.24.

The concrete and rock joints are modelled by two linear-elastic half-spaces exchanging forces through the rough fractal surfaces described before. The unilateral contact problem has been extensively approached by many authors as a nonlinear optimization problem (e.g. Panagiotopoulos, 1985). In the present scheme, instead, an *active set* strategy has been preferred, to reduce the storage requirements for the computation. The model we have chosen leads to a simple solution, still retaining some of the most important features of the real phenomenon.

The numerical procedure is designed in order to determine the geometrical properties of the contact domain $\mathcal{D}$ between the half-spaces. In particular, we focused on the consequences of varying the *resolution* in the representation of the interfaces undergoing contact. With the attempt to study the behaviour of the interface taking into account the shear displacement, different normal contact simulations have been carried out by shifting one side of the joint, with respect to the other, by a certain offset $\nu$. A discrete version of the problem is considered, by introducing in the reference plane $X - Y$ a square grid of points with spacing $s$. Of course, $s$ can be referred to as the resolution adopted in solving the problem and, in this

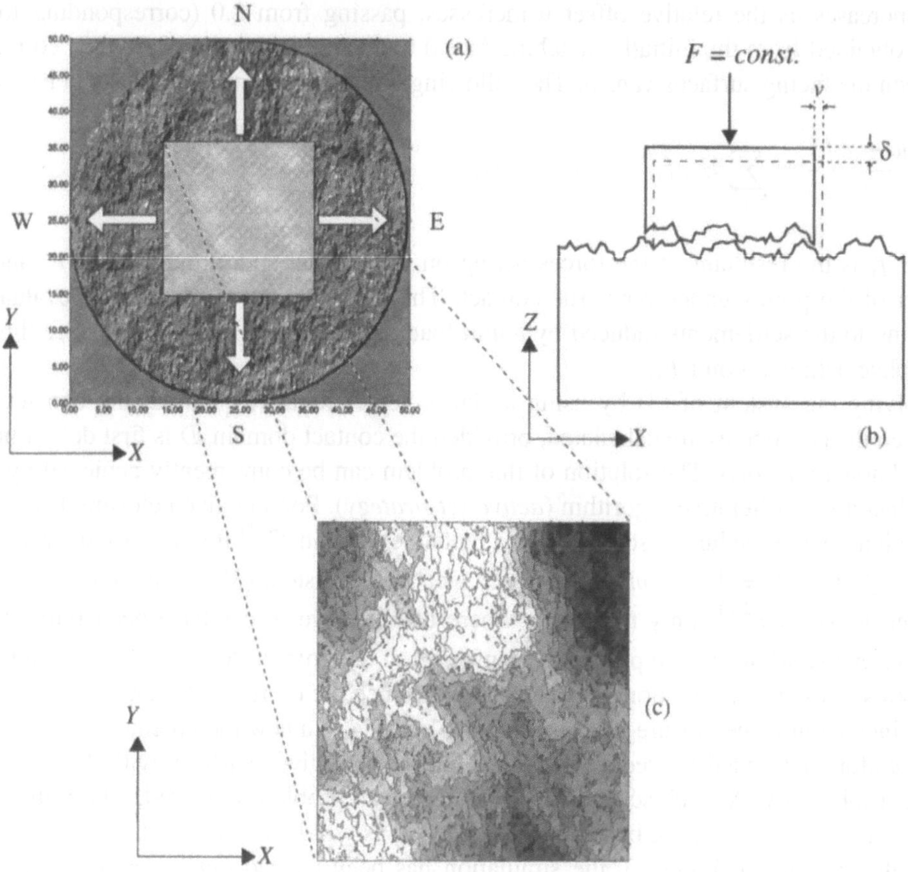


*Figure 3.* Scheme of the numerical simulation on the rock joint: dilatancy $\delta$, shear displacement $\nu$, and composite topography (c) obtained for $\nu = 10$ mm.

case, corresponds to the digitization resolution. At each node of the grid two facing points at the opposite sides of the crack can touch and transmit a force. We assume a small contact zone to be involved around the grid point, whose area is related to the grid spacing $s$.

Once the offset $\nu$ (in the $X - Y$ plane) between the initially perfectly matching surfaces is fixed, a relative closure displacement $w$ (in the $Z$ direction) is imposed to the half-spaces (Figure 3). The solution is sought in terms of pressure and surface displacements, paying attention to the unilateral condition at the interface. The approximations introduced as a first approach to the problem are:

(1) surface points displacements are perpendicular ($Z$ direction) to the boundary mean plane of both half-spaces;
(2) displacements are functions only of the normal components ($f$) of the surface forces;
(3) forces are related to displacements through influence functions.

Considering a reference plane where $z = 0$, let $w$ be the relative displacement between two points far from the interface, assumed as positive in the closure direction. The function $\Delta h(x, y)$ can be introduced, representing the *sum* of the heights of two corresponding points in the undeformed condition. The graph of this function represents a virtual invasive fractal surface (composite topography, or *delta-surface*, see Figure 3c) whose fractal dimension increases as the relative offset $\nu$ increases, passing from 2.0 (corresponding to a flat plane obtained from the initially matching faces) to a value larger than 2.0 when correlations between the facing surfaces vanish. The following linear system of equations can be written

$$\frac{-w + \Delta h_r}{2} = \sum_s H_{rs} f_s, \qquad \forall P_r \in \mathcal{D}, \tag{3}$$

where $f_s$ is the resultant of the forces acting on the contact spot around point $P_s$ and $\mathcal{D}$ is the set of the points undergoing true contact. The influence terms $H_{rs}$ can be evaluated by referring to the settlements induced by a unit load, applied to an elastic half space through a rigid plate acting at point $P_s$.

Solving the system of (3) by using a Gauss-Seidel iterative method, the contact forces between the two bodies are calculated, provided the contact domain $\mathcal{D}$ is first determined ($\mathcal{D}$ is not known *a priori*). The solution of this problem can be conveniently achieved by means of an incremental-iterative algorithm (*active set strategy*). For any given closure displacement $w_i$, an iterative procedure is started from a tentative domain $\mathcal{D}_i^{(1)}$ through a sequence of sets: $\mathcal{D}_i^{(k+1)} \subseteq \mathcal{D}_i^{(k)}$, $k = 1, \ldots, m-1$. At each step $k$, the system (3) is solved, retaining in the contact domain $\mathcal{D}_i^{(k+1)}$ only the points where compressive forces have been found (*unilateral contact condition*). The procedure converges to the correct domain $\mathcal{D}_i^{(m)}$ in a few steps, through successive elimination of tension points. Once the correct solution is reached for the given increment $i$, the closure $w_{i+1}$ is imposed, passing to a new increment.

The aforementioned procedure is carried out, for any given surface, with different values of the resolution $s$. At each scale, a classical euclidean problem is solved, and in the limit of the finest resolution a fractal behaviour is achieved, as will be explained in the following.

With reference to Figure 3, the simulation has been carried out by considering a $20 \times 20\,\text{mm}^2$ central portion of crack surface, initially perfectly matching the opposite side. The portion can slide over the facing surface along two orthogonal directions in the $X - Y$ plane. The minimum relative offset $\nu$ is equal to one pixel, (i.e., to $100\,\mu\text{m}$ for the rock and to $80\,\mu\text{m}$ for the concrete interface). The crack closure behaviour ($F - w$), for each relative horizontal position, can be obtained by applying an increasing closure displacement and computing the corresponding total normal force. By varying $\nu$ under constant normal force $F$, the crack opening $\delta$ can be computed, obtaining the crack dilatancy curves ($\nu - \delta$). In all cases, the simulations provide the true contact domain at each step, plus the forces and displacements at any contact point.

## 5. Mechanical response

The problem of joint closure under normal loads is of great importance in rock mechanics, and can be relevant also for concrete dams and reservoirs. In Figure 4a, the closure curves obtained, for the rock interface, in correspondence of five different values of the relative shear displacement $\nu$ (West direction in Figure 3a), are shown. The effective interface closure $w_{\text{int}}$ can be estimated by subtracting, from the total displacement $w$, the part due to the elasticity of the half-spaces ($w_{\text{int}} = w - w_b$). The bulk contribution $w_b$ depends only on $F$ and can be evaluated by considering uniform pressure on the nominal loaded area. By fitting the data with a hyperbolic function, the asymptotes of the curves can be estimated, which correspond to the complete closure of the joints. It can be noticed that, as the initial offset $\nu$ increases,

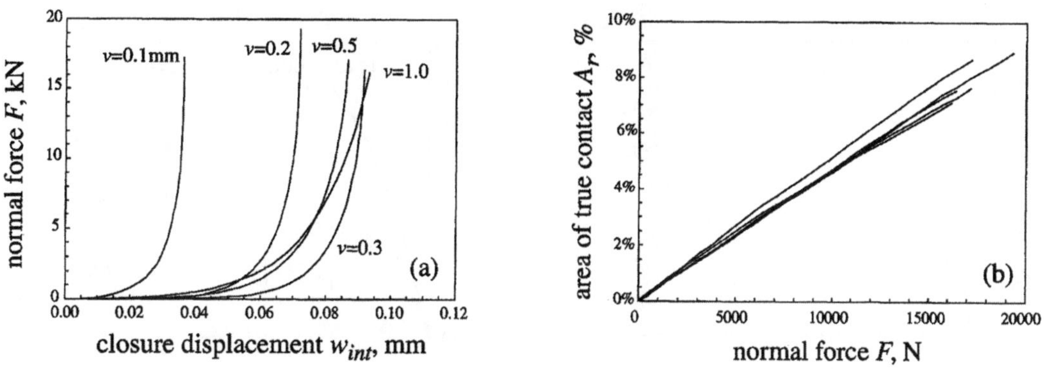


*Figure 4.* Crack closure laws (a) and plot of the real contact area vs. the applied force (b) for the rock interface (at fixed resolution).

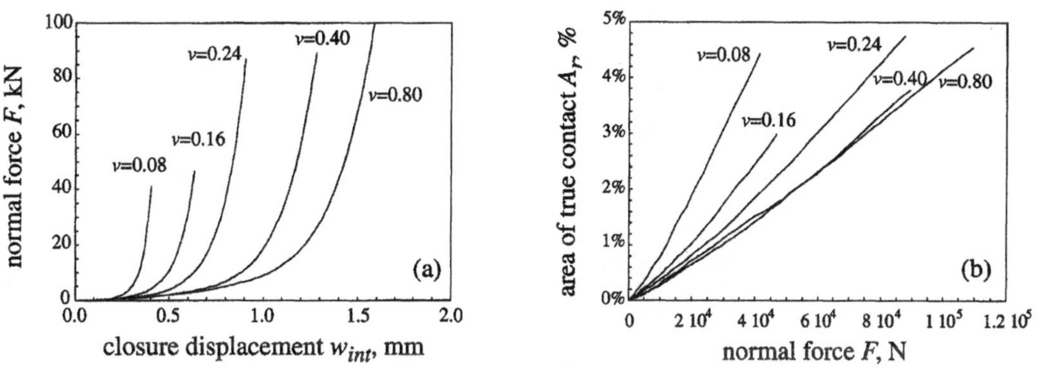


*Figure 5.* Crack closure laws (a) and plot of the real contact area vs. the applied force (b) for the concrete interface (at fixed resolution).

the extension of the closing stage increases, together with the interface compliance. This corresponds to the increase of the fractal dimension of the *delta-surface*, according to the results obtained with computer-generated surfaces by Borri-Brunetto et al. (1998). However, this is true only for the very initial stages of sliding. After a certain shear displacement $\nu$, the curves overlap and the joint deformability becomes constant. On the other hand, when $\nu$ becomes very large ($\nu \approx 10\,\mathrm{mm}$), the joint compliance increases again because no more correlations are present between the opposite pushing surfaces (see the last contact domain in Figure 6).

The crack closure curves for the concrete interface are shown in Figure 5a. The same observations put forward for the rock interface apply also here. Independently of the relative shear displacement, the closure behaviour of natural joints undergoes size-effects, i.e, the larger the samples, the larger the closure deformability. In a previous paper by Borri-Brunetto et al. (1998), in fact, the authors compared the numerical load-displacement diagrams obtained for the larger surfaces with the diagrams obtained for the smaller surfaces and concluded that the joint closure deformability is larger for larger specimens, in agreement with the experimental data reported by Sage et al. (1990).

If the area of true contact is plotted versus the applied load, a nearly linear behaviour ($A_r \sim F$) is observed, both for the rock (Figure 4b) and for the concrete (Figure 5b) interfaces. This is in agreement with early theoretical arguments and with many experimental results

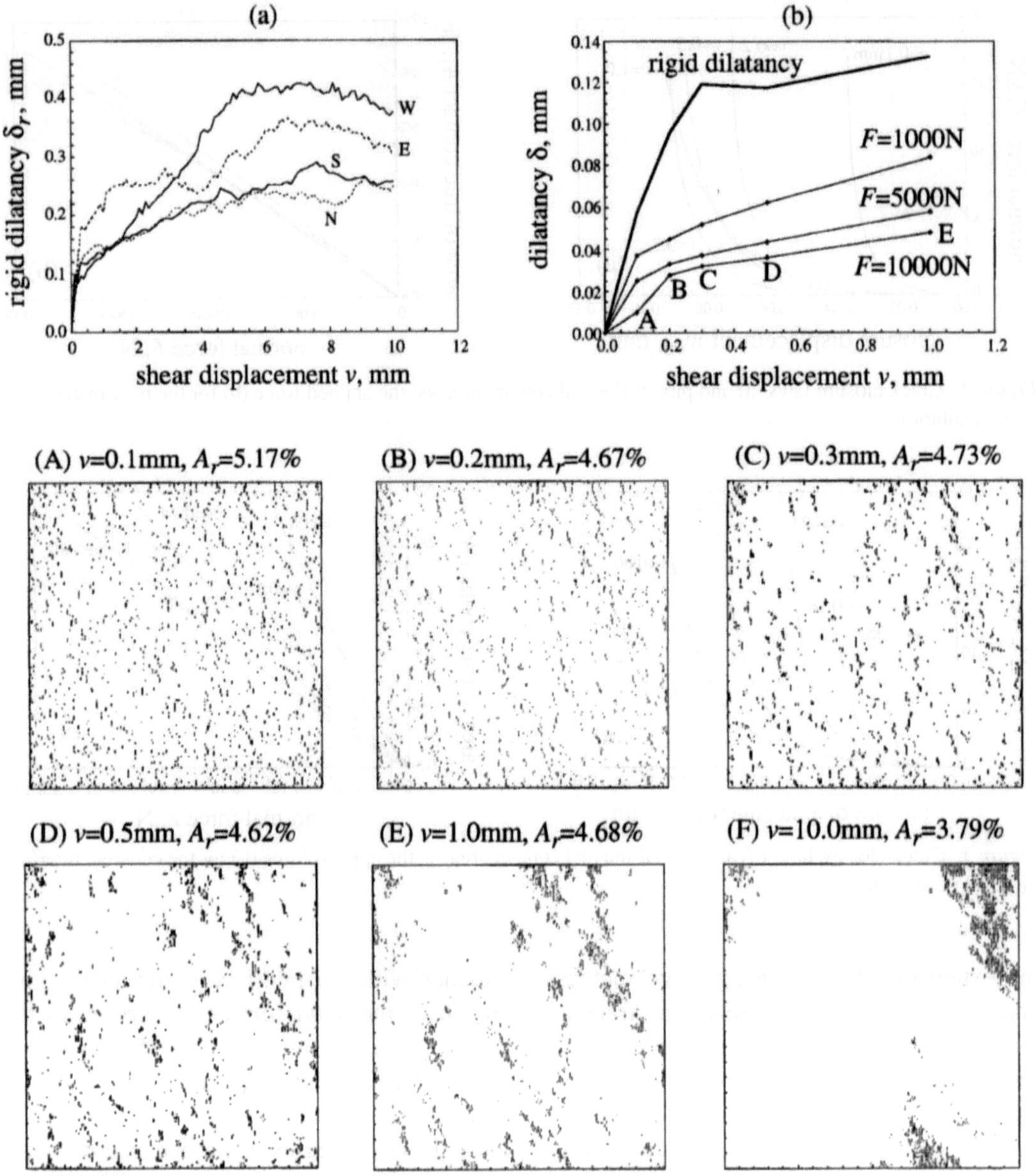


*Figure 6.* Rigid dilatancy curves (a), elastic curves (West direction) (b), and calculated contact domains corresponding to points $A - E$ on the $F = 10{,}000$N curve (rock interface).

(Johnson, 1985). However, as will be shown in the next section, because the euclidean measure of $A_r$ depends on the measurement precision, the classical contact theories do not provide predictive capabilities. Suffices to notice, for now, the decrease of the area of true contact for a fixed value of $F$ as the offset $v$ increases. Significantly, this is especially evident in the case of concrete (Figure 5b).

The rigid-body dilatancy curves (corresponding to $F = 0$) of the rock interface are shown in Figure 6a for the four orthogonal directions. The curves have been obtained after detrending the mean crack plane. In this way, a horizontal plateau is reached. The elastic dilatancy $\delta$, at a fixed value of $F$, is computed by subtracting, from the rigid dilatancy $\delta_r$, the joint closure displacement $w_{\rm int}$

$$\delta(F, \nu) = \delta_r(\nu) - w_{\rm int}(F, \nu) = \delta_r(\nu) - [w(F, \nu) - w_b(F)]. \quad (4)$$

The initial parts ($\nu \leqslant 1.0$ mm) of the elastic dilatancy curves obtained, in the West direction, for three values of $F$ (1000N, 5000N and 10,000N, corresponding, respectively, to the nominal pressures of 2.5MPa, 12.5MPa and 25MPa), are shown in Figure 6b. The analysis of the dilatancy curves could allow for an estimation of the shear strength developed during sliding of rock rough interfaces. In fact, as commonly assumed, the effective value of the friction angle can be calculated as the sum of a basic angle and the current dilatancy angle (e.g. Goodman, 1989).

Six contact domains, generated in the rock interface for fixed $F = 10{,}000$N, are depicted in Figure 6. As already observed in Figure 4b, the larger the displacement $\nu$ between the crack faces, the smaller the percentage of contact area, even if this trend is less evident than in the case of concrete. However, as $\nu$ increases, the domains become less homogeneous and the contact points, initially spread almost uniformly, tend to cluster only in some zones. The last domain (F), corresponding to the *delta-surface* in Figure 3c, is different from the others, due to the relatively large offset (10 mm). No more correlations between the crack sides are present, suggesting the existence of a threshold scale, related to a characteristic length of the fracture surface or to an internal length of the material.

In the case of the concrete interface, the rigid-body dilatancy curves (corresponding to $F = 0$) are shown in Figure 7a, while three elastic curves are shown in Figure 7b. A marked anisotropy is revealed in this case. For the West and North sliding directions, the rigid dilatancy $\delta_r$ immediately raises to a plateau value $\delta_r \approx 2$ mm (corresponding to $\nu \approx 0.8$ mm), and then remains approximately constant under shear displacement. Instead, in the East and South directions, after a steep increasing stage, $\delta_r$ continues to grow, attaining an oblique asymptote. Six contact domains for $F = 30{,}000$N are shown in the same figure. Again, the larger the shear displacement $\nu$ between crack faces, the smaller the percentage of true contact area. This is true up to a certain shear displacement, after which $A_r$ seems to remain approximately constant (see also Figure 5b, comparing the curves for $\nu = 0.40$ mm and $\nu = 0.80$ mm). Note also that the last domain (F) is totally different from the others, due to the large shear displacement (8 mm). Localization of the contact spots is even more pronounced than in the case of the rock interface, probably due to macro-heterogeneities.

## 6. Fractal lacunarity of the contact domain and related size effects

The results presented up to now have been obtained for a fixed value of the discretization $s$. Such an approach is particularly important because any real measurement is always performed at finite resolution. However, to consider the scale-dependence of the model response, a multi-resolution analysis has been performed. When the normal contact simulation is carried out, at a given offset, for a composite topography with given fractal dimension $\Delta_{\mathfrak{g}}$ but with *different discretizations* $s$, the resulting contact domains $\mathcal{D}_s$ can be compared (Figure 8). A visual comparison with the experimental domain shown in Figure 1a (where the black *islands* are contact points, floating in an *ocean* where contact does not occur) confirms the consistency of the numerical simulations and suggests a deeper insight into the geometric structure of the contact domain.

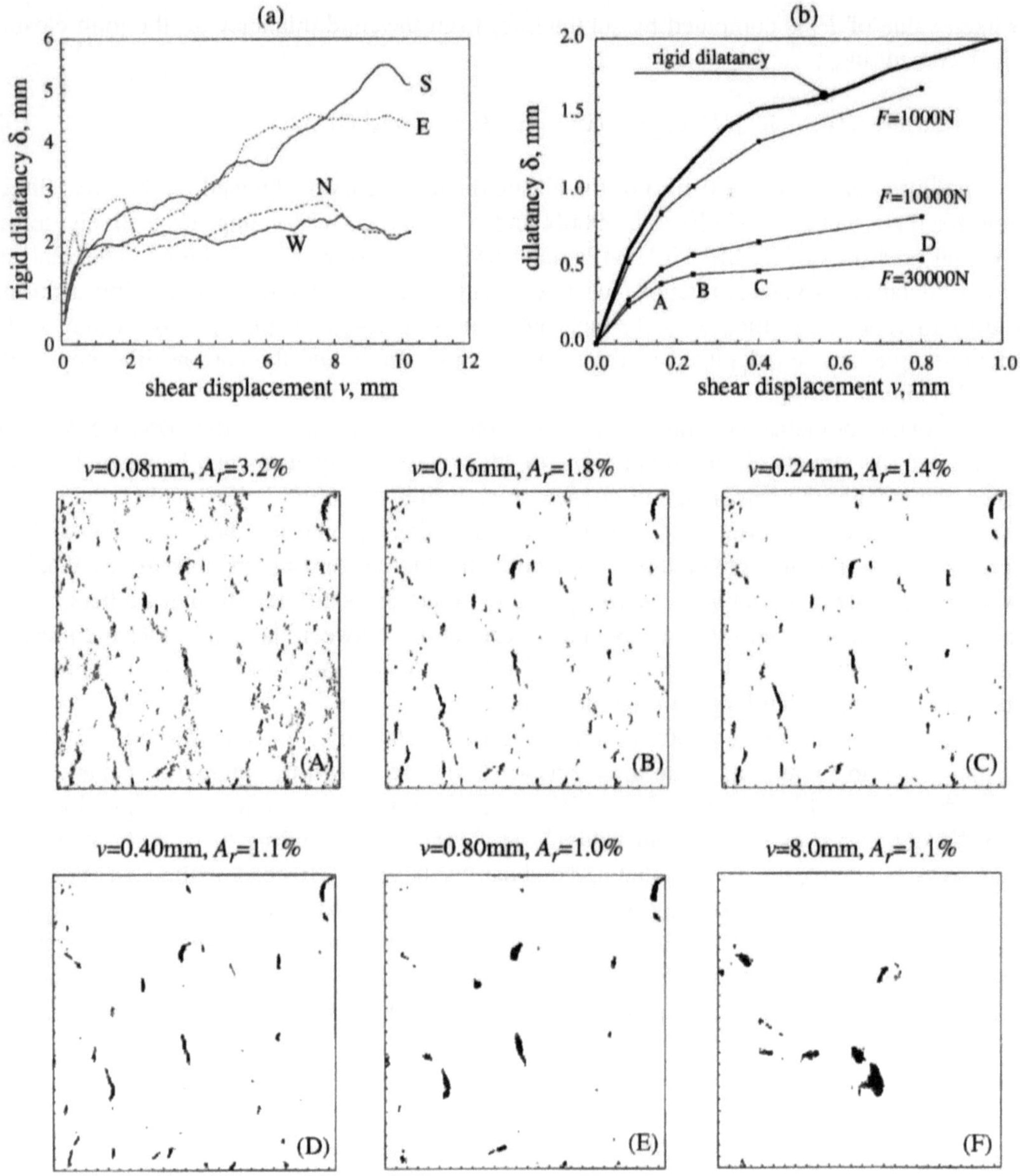


*Figure 7.* Rigid dilatancy curves (a), elastic curves (West direction) (b), and calculated contact domains corresponding to points $A - E$ on the $F = 30{,}000$N curve (concrete interface).

It can be easily realized that the concept of area of true contact, although representing a step forward with respect to the concept of apparent (nominally flat) area, is not able to describe consistently (that is, in a scale-independent manner) the interface interactions. As it clearly emerges from Figure 8 (which is relative to a mathematical joint), the real contact area $A_r$, at a fixed value of the normal pressure, progressively decreases with increasing the resolution of the discretization, ideally tending to zero in the theoretical limit of $s \to 0$. This power-law behaviour proves that the contact domain $\mathcal{D}$ is a *lacunar* (rarefying) fractal set in the plane, that is, a fractal set with null Lebesgue measure. This implies the necessity of abandoning

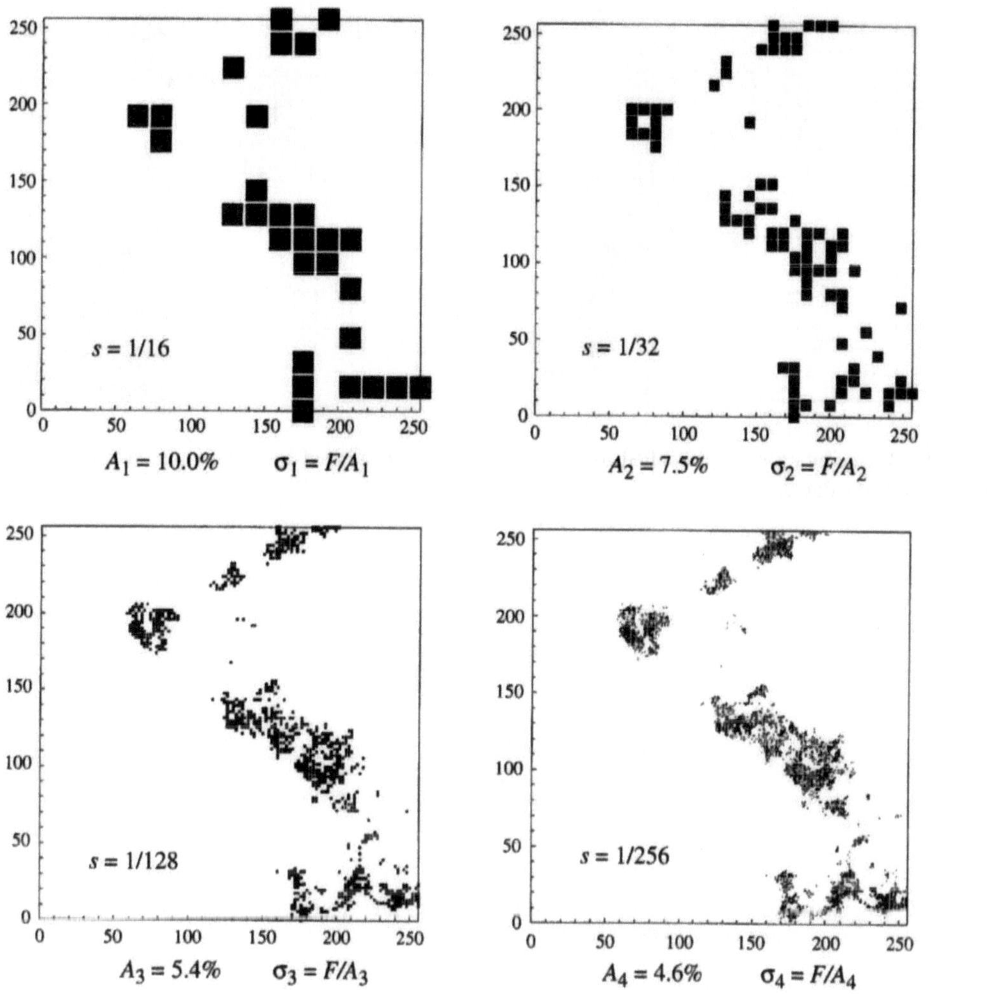


*Figure 8.* Decrease of the real contact area and increase of the corresponding real mean pressure (at fixed load) with increasing resolution (surface dimension $\Delta_{\mathcal{G}} = 2.1$, contact domain dimension $\Delta_\sigma = 1.39$).

the euclidean description and moving to the fractal model, characterized by the noninteger dimension $\Delta_\sigma\,(\Delta_\sigma \leqslant 2.0)$ of the domain $\mathcal{D}$.

Application of the *box-counting method* (Mandelbrot, 1982) to a large number of contact sets, obtained for different normal loads and various horizontal offsets, makes it possible to compute their fractal dimension $\Delta_\sigma$ (Figure 9). Note that the contact domains are, to a great extent, self-similar, that is, they show isotropic scaling. This is because they are related (but not coincident) to the so called *zero-sets* of the self-affine surfaces in contact, i.e., to the intersections of the surfaces with planes of equation $z = constant$ (Mandelbrot, 1985).

In perfect analogy with the theory of fractal scaling applied to nominal tensile strength (Carpinteri, 1994), let us consider the following renormalization group

$$F = \sigma_0 A_0 = \sigma_1 A_1 = \sigma_2 A_2 = \cdots = \sigma_n A_n = \cdots = \sigma^* \mathcal{H}_{\mathcal{D}}, \tag{5}$$

where $F$ is the applied normal load (scale-invariant quantity), $A_n$ and $\sigma_n$ are respectively the real contact area ($[L]^2$) and the real mean pressure ($[F][L]^{-2}$) measured at the pre-fractal

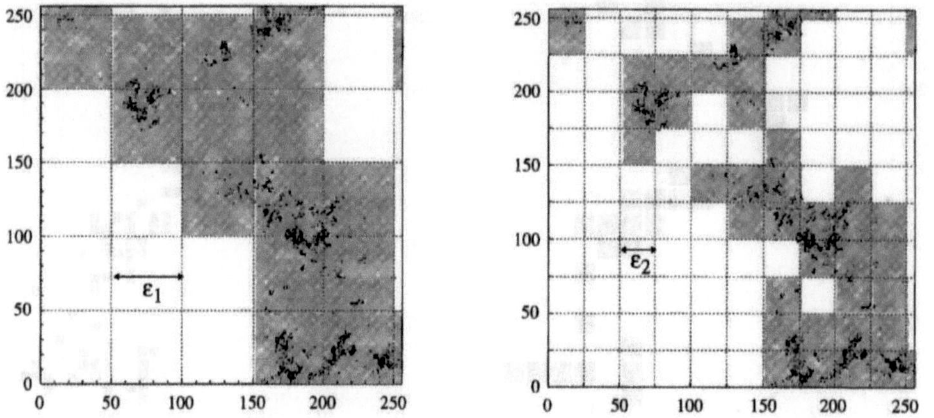


*Figure 9.* Box-counting method applied to the contact domain generated by a Brownian surface ($\Delta_{\mathcal{g}} = 2.5$) pushing against a flat plane.

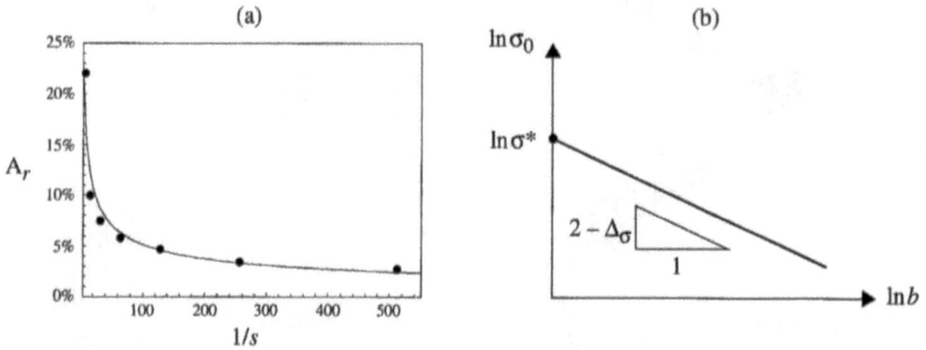


*Figure 10.* Computed power-law decrease of the real contact area with increasing resolution (a), and related scaling law for the nominal pressure (b).

resolution $n$, $A_0$ is the apparent (nominally flat) area and $\sigma_0$ is the apparent pressure. As one can realize from Figure 8, at any value of the normal load $F$, $\sigma_n$ increases with increasing resolution and tends to infinity as $s \to 0$, because the euclidean measure ($[L]^2$) of the contact domain $\mathcal{D}$ vanishes (see Figure 10a, relative to a mathematical joint). Hence, in the limit of the highest resolution ($s \to 0$), the euclidean description loses its significance and leaves place to the *fractal measure* $\mathcal{H}_{\mathcal{D}}$ of the domain $\mathcal{D}$ (which is univocally defined only by the noninteger dimensionality $[L]^{\Delta_\sigma}$). Correspondingly, the *fractal mean pressure* $\sigma^*$ can be defined as the anomalous flux of stress across the fractal interface. Owing to the dimensional homogeneity, this quantity holds the anomalous physical dimensions $[F][L]^{-\Delta_\sigma}$. By equating the second and the last term in (5) and taking the logarithm of both sides, if $b$ is a characteristic linear size of the interface, the following scaling law is provided

$$\log \sigma_0 = \log \sigma^* - (2 - \Delta_\sigma) \log b. \tag{6}$$

In this context, if $\sigma^*$ represents a scale-independent parameter governing the force flux across an interface, Equation (6) gives the dependence on the size of the specimen of the apparent normal pressure $\sigma_0$ required to activate that flux (Figure 10b). In other words, (6) allows to compute the size-dependent force $F$ necessary to produce $\sigma^*$ (and thus, possibly, to

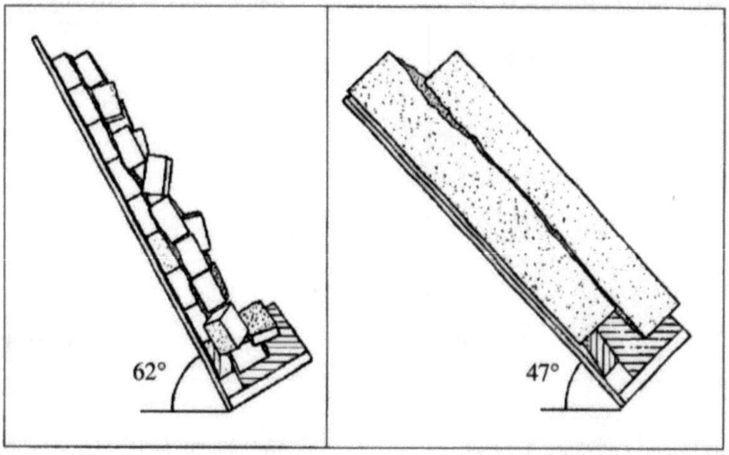


*Figure 11.* Experimentally detected size-scale effects on the limit slope angle (tilt tests after Bandis et al., 1981).

activate a certain friction) at the interface. In the case of euclidean interfaces, $\sigma_0$ would be the trivial governing parameter, and the classical $F \sim b^2$ scaling would be provided.

In close agreement with many experimental observations (Pinto da Cunha, 1993; Bandis et al., 1981), and assuming direct proportionality between the *fractal shear strength* $\tau^*$ and the fractal mean pressure $\sigma^*(\tau^* = \mu\sigma^*)$, Equation (6) affirms also that the apparent shear strength $\tau_0 = T/A_0$ decreases with increasing the size $b$ of the nominal contact area (see also Figure 1b). This conjecture implies that the friction coefficient $\mu$ is not constant for a given interface but varies with its size. Thus, the limit slope angle decreases with increasing the specimen size (Figure 11). This approach may shed light on many rock slope instabilities, which are not explicable when compared to the relatively high friction values measured on smaller specimens.

Another fundamental aspect to be highlighted is the dimensional evolution of the contact domain $\mathcal{D}$, which is initially very rarefied (like a Cantor dust) and progressively increases its topological density with increasing the applied load (Figure 12). As the apparent pressure $\sigma_0$ increases, the classical theories yield the increase of the real contact area $A_r$ (see the force vs. area laws in Figures 4b and 5b). Some relations can be found in the literature (Majumdar and Bhushan, 1990), connecting the area of true contact $A_r$ and the real mean pressure $\sigma_r = F/A_r$ to the apparent pressure $\sigma_0$. However, since the euclidean measure of $A_r$ depends on the measurement precision, the classical theories do not have unique predictive capabilities.

More significantly, the continuous variation of the box-counting dimension $\Delta_\sigma$ has to be considered. As the normal load $F$ increases, $\Delta_\sigma$ starts from the value 0.0 (pointwise non-structured contact), then takes values comprised between 0.0 and 1.0 (pointwise contact sets which are structured, or self-similar, in a well-defined scale range), and subsequently takes values larger than unity as soon as linear contact structures and rarefied contact *islands* are formed. The total saturation of the contact domain $\mathcal{D}$ (or, at least, of some *islands*) would imply $\Delta_\sigma = 2.0$. It may be argued that this value, in real materials, could be attained only under very high normal loads, and this would imply an extended plastic deformation of the material, which is not contemplated in the numerical model.

According to the topological evolution of $\mathcal{D}$, the slope of the scaling law progressively decreases with increasing normal pressure (Figure 13). Note that also the fractal governing parameter $\sigma^*$ is dependent on $F$, and therefore neither the slope, nor the intercept of the

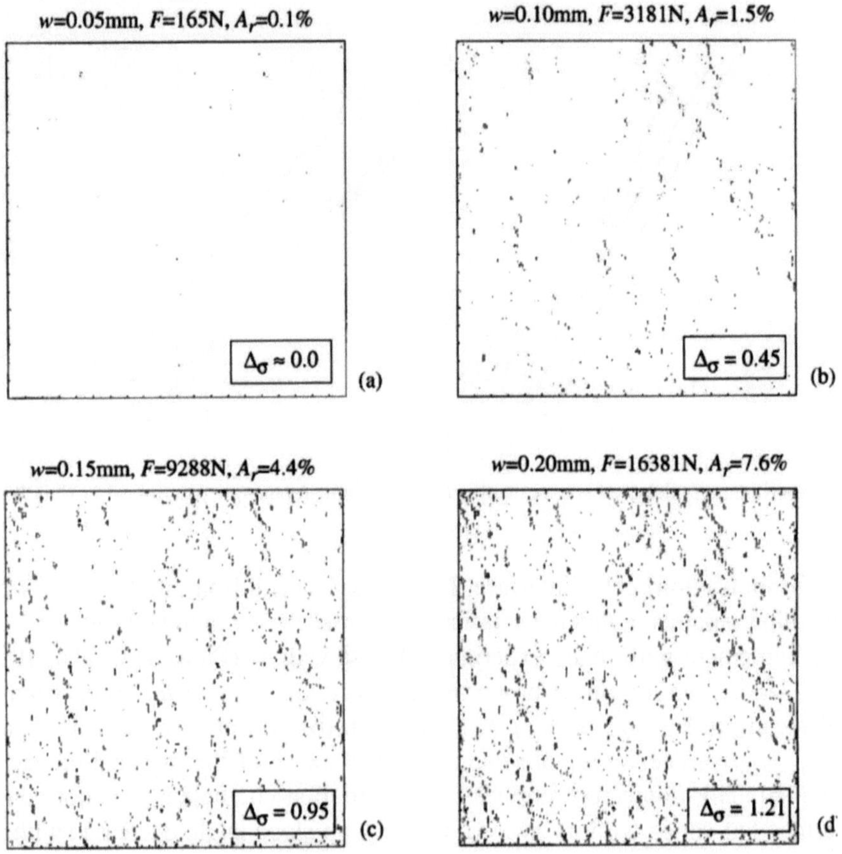


*Figure 12.* Evolution of the contact domain during closure of the rock joint (fixed offset $\nu = 0.3$ mm).

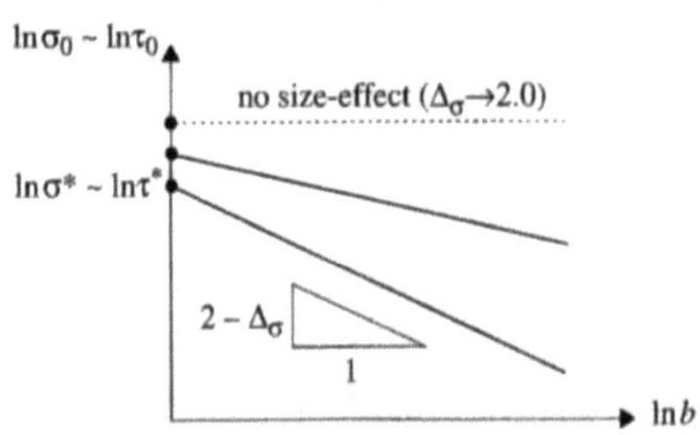


*Figure 13.* Size-scale effects on the nominal pressure (and on the nominal shear strength) for different lacunarities $\Delta_\sigma$ corresponding to different fractal mean pressures $\sigma^*$.

scaling law are unique, but vary with the load. In the limit case of $\Delta_\sigma = 2.0$, the euclidean description would be consistent. The physical quantities would retain their usual integer dimensions ($\sigma^* \rightarrow \sigma_0$) and a canonical scale-independent friction coefficient could be defined. Note that also in the model by Greenwood and Williamson (1966), the smooth behaviour is attained for very high loads.

## 7. Conclusions

In this paper, a numerical model has been presented for simulating the problem of normal contact between two elastic bodies with rough (fractal) boundaries. The model, although simple, captures some fundamental features of the mechanical problem, which can be easily extended to a wide class of interface phenomena, such as thermal and electric conductivity. The results of the simulations for a rock and a concrete interface are described, and the influence of fractality on the mechanical behaviour is discussed. The following conclusions may be traced.

- The higher the fractal dimension $\Delta_{\mathcal{G}}$ of the composite topography (related to the offset between the two facing sides of the crack), the higher the global closure deformability of the joint.
- Although the concept of area of true contact is ambiguous, the simulations, carried out at a pre-fractal level, yield a linear response in the normal load versus real area of contact diagram ($A_r \sim F$). Thus, the local self-affine complexity of the surfaces provides macroscopic linearity.

A completely new approach to the geometric structure of the contact domain (the horizontal projection of the set of points where contact truly occurs) has been put forward in the paper. A topologic analysis has shown that the real contact domain $\mathcal{D}$ is a *lacunar* fractal set in the two-dimensional plane, whose fractal (box-counting) dimension $\Delta_\sigma (\Delta_\sigma < 2.0)$ is progressively increasing with the load. As a major consequence, the physical quantities (mechanical stress, electricity or heat conductance) defined over this fractal set assume noninteger (anomalous) dimensions.

- The lacunarity of the contact domain $\mathcal{D}$ is proven numerically by the power-law scale-dependence of the real contact area $A_r$ and of the real mean pressure $\sigma_r = F/A_r$.
- Assuming direct proportionality between shear strength and normal pressure, the decrease of the apparent shear strength $\tau_0 = T/A_0$ with increasing the size of the nominal contact area $A_0$ is obtained (size-scale effect). Hence, the friction coefficient $\mu$ is not constant for a given interface, but varies with its size.
- However, this size-dependence, which is more pronounced for larger fractal dimensions $\Delta_{\mathcal{G}}$ of the composite topography, progressively decreases as the nominal pressure increases.
- For a fixed value of $\Delta_\sigma$, lacunarity implies that the contact domains of larger joints are less dense in the euclidean sense. This means that the probability of presence of large zones where contact does not occur increases with the size of the interface. This represents another way to justify the size effects on shear strength.

These preliminary conclusions suggest the possibility of re-interpreting the phenomenological friction laws on the basis of a fractal approach. However, further theoretical and experimental investigations are required to validate a scale-independent friction criterion based on a fractal mesomechanical framework.

## Acknowledgements

Support by the Italian Ministry of University and Scientific Research and by the EC-TMR Contract N. ERBFMRXCT 960062 is gratefully acknowledged.

## References


Archard, J.F. (1957). Elastic deformation and the laws of friction. *Proceedings of the Royal Society of London A* **243**, 190–205.

Bandis, S., Lumsden, A.C. and Barton, N.R. (1981). Experimental studies of scale effects on the shear behaviour of rock joints. *International Journal of Rock Mechanics, Mining Sciences and Geomechanics Abstracts* **18**, 1–21.

Barton, N. and Choubey, V. (1977). The shear strength of rock joints in theory and practice. *Rock Mechanics* **10**, 1–54.

Brown, S.R. and Scholz, C.H. (1985). Broad bandwidth study of the topography of natural rock surfaces. *Journal of Geophysical Research* **90**, 12575–12582.

Borodich, F.M. and Mosolov, A.B. (1992). Fractal roughness in contact problems. *Journal of Applied Mathematics and Mechanics* **56**, 681–690.

Borri-Brunetto, M., Carpinteri, A. and Chiaia, B. (1998). Lacunarity of the contact domain between elastic bodies with rough boundaries. *Probabilities and Materials. Tests, Models and Applications*, NATO ASI Series 3 (Applied Sciences), Vol. 46, Kluwer Academic Publishers, Dordrecht, 45–64.

Carpinteri, A. (1994). Fractal nature of material microstructure and size effects on apparent mechanical properties. *Mechanics of Materials* **18**, 89–101.

Carpinteri, A., Chiaia, B. and Invernizzi, S. (1999). Three-dimensional fractal analysis of concrete fracture at the meso-level. *Theoretical and Applied Fracture Mechanics*, in print.

Cook, N.G.W. (1992). Natural joints in rock: mechanical, hydraulic and seismic behaviour and properties under normal stress. *International Journal of Rock Mechanics, Mining Sciences and Geomechanics Abstracts* **29**, 198–223.

Goodman, R.E. (1989). *Introduction to Rock Mechanics*, John Wiley & Sons, New York.

Greenwood, J.A. and Williamson, J.B.P. (1966). The contact of nominally flat surfaces. *Proceedings of the Royal Society of London A* **295**, 300–319.

Johnson, K.L. (1985). *Contact Mechanics*, Cambridge University Press, Cambridge.

Ling, F.F. (1989). The possible role of fractal geometry in tribology. *Tribology Transactions* **32**, 497–505.

Majumdar, A. and Bhushan, B. (1990). Role of fractal geometry in roughness characterization and contact mechanics of surfaces. *Journal of Tribology (ASME)* **112**, 205–216.

Majumdar, A. and Tien, C.L. (1990). Fractal characterization and simulation of rough surfaces. *Wear* **136**, 313–327.

Mandelbrot, B.B. (1982). *The Fractal Geometry of Nature*, W.H. Freemann & Company, New York.

Mandelbrot, B.B. (1985). Self-affine fractals and fractal dimension. *Physica Scripta* **32**, 257–260.

Panagiotopoulos, P.D. (1985). *Inequality Problems in Mechanics and Applications*, Birkhäuser Verlag, Basel.

Panagiotopoulos, P.D. and Panagouli, O.K. (1997). Fractal geometry in contact mechanics and numerical applications. *Fractals and Fractional Calculus in Continuum Mechanics* (Edited by A. Carpinteri and F. Mainardi), Springer-Verlag, Wien, 109–171.

Park, R. and Paulay, T. (1975). *Reinforced Concrete Structures*, Wiley, New York.

Pinto da Cunha, A. (Ed.) (1993). *Scale Effects in Rock Masses 93*, Balkema, Rotterdam.

Sage, J.D., Aziz, A.A. and Danek, E.R. (1990). Aspects of scale effects on rock closure. *Scale Effects in Rock Masses 90* (Edited by A. Pinto da Cunha), Balkema, Rotterdam, 175–181.

Schmittbuhl, J., Schmitt, F. and Scholz, C.H. (1995). Scaling invariance of crack surfaces. *Journal of Geophysical Research* **100**, 5953–5973.

Yoshinaka, R., Yoshida, J., Arai, H. and Arisaka, S. (1993). Scale effects on shear strength and deformability of rock joints. *Scale Effects in Rock Masses 93* (Edited by A. Pinto da Cunha), Balkema, Rotterdam, 143–149.

Yoshioka, N. and Scholz, C.H. (1989). Elastic properties of contacting surfaces under normal and shear loads. *Journal of Geophysical Research* **94**, 17681–17690.

Wang, W. and Scholz, C.H. (1993). Scaling of constitutive parameters of friction for fractal surfaces. *International Journal of Rock Mechanics, Mining Sciences and Geomechanics Abstracts* **30**, 1359–1365.

Warren, T.L. and Krajcinovic, D. (1995). Fractal models of elastic-perfectly plastic contact of rough surfaces based on the Cantor set. *International Journal of Solids and Structures* **19**, 2907–2922.

*International Journal of Fracture* **95:** 239–259, 1999.


# Fractals and fractal scaling in fracture mechanics

FEODOR M. BORODICH
*Department of Mathematics, Glasgow Caledonian University, Glasgow G4 0BA, UK*
*e-mail: F.M.Borodich@gcal.ac.uk*

Received 3 September 1998; accepted in revised form 31 December 1998

**Abstract.** A review of modern fractal models of fracture in brittle and quasibrittle materials is given. The difference between mathematical and physical fractal approaches is emphasized. The scaling for both a fractal solitary crack and a fractal pattern of microcracks surrounding the main fracture is considered. Some concepts appropriate for fractal description of fracture are discussed. It is shown that if the layer of inelastic deformations in quasibrittle materials has the same order of magnitude as the upper cutoff of the fractal scaling then fractal properties of the main crack surface do not correlate with fracture energy. This observation selects the cases when the concept of the universal roughness exponent may be valid. It is shown that the correlation between fractal properties of fractal pattern of microcracks and the fracture energy of polyphase materials is usually possible. The case of different fractal dimensions for different length scales is also discussed.

**Key words:** Ceramics, concrete, rock, energy, failure, size, crack, pattern.

## 1. Introduction

The fracture processes in polyphase materials such as concrete, rock, and ceramics and in macrohomogeneous materials, such as glass and metal are different. The difference is caused by appearance of the process zone in the former materials (Bažant, 1984; Gettu and Shah, 1994; Cotterell and Mai, 1996). The process zone or multiple fracture appears not only in polyphase materials but also in various natural objects at all scales, e.g., in polymer materials. The multiple fracture consists of a cascade of interacting defects of various length scales, hence direct application of linear elastic fracture mechanics to such objects, disordered by defects, is impossible (Barenblatt, 1993). To study such a fracture it is necessary to develop new models.

### 1.1. Fractal approaches to fracture

It is still unclear what is the best way to describe the growing pattern of fracture. However, experimental data support the hypothesis that the multiple fracture process possesses self-similar and fractal properties in a wide range of scales. For example, these properties were found on a large scale for segments of geological faults (Brady, 1974; Sornette et al., 1990). It was also shown that patterns of microcracks and crack size distributions possess self-similar features (Barenblatt and Botvina, 1986; Barenblatt, 1993). Note that the use of lattice simulations of cracks has also discovered some fractal features of fracture process (Louis et al., 1986; Louis and Guinea, 1989; Herrmann et al., 1989; Meisner and Frantziskonis, 1997).

The term *fractal* was coined by Mandelbrot (1977). Roughly speaking, fractals can be defined as sets with noninteger fractal dimension. Precise definitions of fractal dimension will be given below. The dimension ($D$) of a fractal curve is less than or equal to 2, i.e., $1 < D < 2$.

Similarly, the dimension of a fractal surface is less than or equal to 3, i.e., $2 < D < 3$. If the fractional part $D^*$ ($0 < D^* < 1$) of the fractal dimension is used then we can write $D = 1 + D^*$ in the case of a fractal curve and $D = 2 + D^*$ in the case of a fractal surface. Mandelbrot (1977; 1983) argues that the language of fractal geometry often describes physical phenomena better than the language of smooth classic objects. Now fractal geometry is a branch of mathematics dealing with highly irregular sets, while fractality is attributed to irregular physical objects. Fractality is in a close connection with ideas of self-similarity and scaling. However, it is inappropriate to say that they are identical. There are a number of papers devoted to the fractal approach to fracture. Unfortunately, the concepts of fractal geometry are incorrectly used in a considerable number of these papers. One of typical mistakes is the confusion between mathematical and physical fractals. In this paper we emphasize the difference between these types of fractals. To warn the researchers in the field some of other typical mistakes will be also mentioned.

The existence of fractal law in physical systems is reported so often that one can speak about 'fractals everywhere' (Barnsley, 1988). For example, following the pioneering studies of Mandelbrot et al. (1984), fractality of roughness of fracture surfaces have been found in metal as well as concrete, ceramics and rock (reviews in Dauskardt et al., 1990; Saouma et al., 1990; Issa et al., 1993; Milman et al., 1994; Saouma and Barton, 1994; Botvina et al., 1997).

There arises the following question: is there any connection between the fractal dimension of fracture surfaces and the fracture toughness of the material? Studies of ceramics suggest a positive answer (Mecholsky et al., 1989), but in recent experiments with glass and porcelains no quantitative relationship has been established (Baran et al., 1995). To explain these results, one has to perform a theoretical analysis.

To the best of our knowledge, the first attempt to perform such an analysis was done by Lung (1986). He considered a plane steel sample as a set of identical right hexagons. It was supposed that a crack propagated along the sides of the hexagons. The size of each hexagon was equal to the size of a grain of the material. This size was taken as a length unit, which was in a formal way substituted into the fractal relationship between the measuring step size and the length of the crack trajectory. Evidently, that the set of hexagons is not a fractal object and the formal application of fractal approach to nonfractal objects is not justified.

### 1.2. Mosolov's models of fractal fracture

One of the first attempts to construct a theoretical mechanical model of a fractal crack was made by Mosolov (1991a). He employed the Griffith specific surface energy $\gamma$, which gives a measure of the elastic energy used on forming a unit of a new solid surface, and obtained a nonclassic formula for the stress intensity factor $K_1^f$ of a fractal crack or the '*fractal stress intensity factor*'

$$K_1^f \sim \sigma l^{(2-D)/2},$$

where $\sigma$ is nominal stress, $l$ is the macro-dimension of the crack and $D$ is the fractal or self-similarity dimension of the crack surface. In another paper he argued that the $J$-integral cannot be invariant for a fractal crack and he wrote a scaling relation for the integral (Mosolov, 1991b). He has also applied his approach to the problem of fractal fracture of brittle bodies during compression along the nominal direction of a crack, using the author's observation that the so-called fractal crack-thickness, i.e., the distance between the nominal crack direction and the fractal crack trajectory, is nonzero (Mosolov and Borodich, 1992). The further development of Mosolov's approach was fulfilled by Goldshtein and Mosolov (1991; 1992) who considered the elastic energy generated during the motion of the tip of a fractal crack and suggested a hierarchical (renormalization group) description of the cascade energy transfer from one scale to another, namely they wrote the following chain of equations

$$\Delta U = G_0 \Delta\delta = N(\delta_1) G_1 \Delta\delta_1 = \cdots = N(\delta_n) G_n \Delta\delta_n = \cdots,$$

where $\Delta U$ is the elastic energy released when the crack grows, $\delta_n$ is the scale of the fractal crack consideration, $N(\delta_n)$ is the number of elements of the scale $\delta_n$ and $G_n$ is the elastic energy release rate.

Mosolov's ideas above (Mosolov, 1991a,b; 1993; Goldshtein and Mosolov, 1991; 1992) are quite popular in the literature. For example, all theoretical models of fractal cracks in solids, reviewed by Cherepanov et al. (1995), were based on Mosolov's approaches. Notice that mathematical and physical fractals were confused and the concept of the Hausdorff–Besicovitch dimension was incorrectly used in the review.

It is well known that in order to obtain the formulae for the stress intensity factor $K_{\mathrm{I}}$ in the framework of classic linear elastic fracture mechanics, one must formulate and solve a boundary value problem of mathematical physics for an elastic plane with a rectilinear cut. Evidently, it is necessary to solve the boundary value problem for an elastic plane with a fractal cut to obtain a strict formula for $K_{\mathrm{I}}^f$, either for the mathematical or for the physical fractal approximations of the cut. In the former case, even the strict mathematical formulation to the problem has not yet been given. In the latter case, when the crack is modelled as a hierarchical structure consisting of a large number of small smooth cracks of size $\delta$, nobody has yet solved the problem for an elastic plane with such a highly irregular cut.

Hence, from the point of view of mathematical physics, Mosolov's papers did not give a new approach to the problem of fractal fracture. Nevertheless, his papers can be interpreted as having a new physical approach to the problem and giving its new physical vision. Evidently, his results should be considered critically and used cautious.

Unfortunately, Mosolov's ideas were adopted uncritically rather often which led to various mistakes and, in addition, the ideas were sometimes used without referring to the author. For example, Carpintery (1994) and Balankin (1996) employed his ideas about the fractal stress intensity factor with nonclassic physical dimension, and also other ideas such as scaling of the fractal $J$-integral or renormalization of the energy transfer from one scale to another. However, no boundary value problem for either mathematical or physical approximations of the fractal cut was solved in the papers.

### 1.3. Contemporary approaches

Note if the classic approaches are applied to fractal fracture they should be treated in a special way. Indeed, the criteria of linear fracture mechanics are equivalent to each other (Willis, 1967) but not all of them could be applied directly to fractal fracture. Even if some of them could formally be applied in a direct way to a fractal crack then they could lead to paradoxical conclusions. For example, it was shown for a problem of a solitary fractal crack propagation, that in the framework of fractal geometry the usage of the Griffith specific surface energy $\gamma$ leads to the paradoxical conclusion that no fractal cracking is possible (Borodich, 1992; Borodich, 1994; 1997a). Indeed, the area of each piece of crack increases to infinity and it could absorb an infinite value of energy. To resolve the paradox some new concepts were introduced, in particular, the concept of the *energy absorbed in a unit of the fractal measure* of the surface ($\beta(D)$) when the surface is simulated as the so-called $D$-measurable self-similar mathematical fractal (Borodich, 1992; 1994a).

The idea to attribute physical quantities to the finite fractal measure $m_D$ of the considered mathematical model, rather than to the infinite area of the fractal surface goes back to Barenblatt and Monin (1983). This idea now has a strict mathematical treatment and was applied to fractal fracture (Borodich, 1992). Bažant (1995; 1997a,b) suggested the use of the *fractal fracture energy* term in application to the $\beta(D)$-concept. Other authors have also employed this concept and called it as *renormalized (fractal) fracture energy* (Carpinteri and Chiaia, 1996).

Modelling a crack as a solitary fractal line and using the $\beta(D)$ concept, the value of critical stress $\sigma_c$ was obtained, when the crack of initial nominal length $l$ begins to propagate (Borodich, 1994a)

$$\sigma_c = \sqrt{\frac{2\beta(D)m_D(\mathcal{L}_0)DE_1}{k_1\mathcal{L}_0 l(1+a/l)}} \cdot (a/\mathcal{L}_0)^{D^*/2},$$

where $\mathcal{L}_0$ is the nominal length of some measured interval, $a$ is the nominal crack advance during loading, $E_1$ is the elastic modulus, equal to the Young modulus $E$ for plane stress and to $E/(1-\nu^2)$ for plane strain ($\nu$ is the Poisson ratio), and $k_1$ is a constant. Thus, it was shown that the fractal crack propagated in a perfect brittle solid is stable. Indeed, $\sigma_c$ grows with $a$. Later Bažant (1995; 1997a) obtained a similar formula using the asymptotic expansion technique.

Considering the physical model of a solitary fractal crack of dimension $D$, Borodich (1992) noted that due to a thin surface layer of inelastic deformations near the fracture surface, even if the surface shows fractal properties down to the atomic scale, the fractal laws become important in the evaluation of the quasibrittle fracture energy only in some mesoscale range $\delta_* < x < \Delta_*$, where $\Delta_*$ is the upper cutoff for the fractal law and $\delta_*$ is the lower limit of the covering which could be taken as the width of the layer. Then the effective energy-absorbing capacity $G_f$ of an elementary segment $\delta_*$ of the crack was introduced. This leads to the following estimations for the fracture energy $G_F$ of a quasibrittle sample of unit thickness

$$G_F \sim \begin{cases} G_f D(x/\delta_*)^{D-1}, & \delta_* < x < \Delta_*, \\ G_f D(\Delta_*/\delta_*)^{D-1}, & \Delta_* < x, \end{cases} \tag{1}$$

where $x$ the nominal length of the crack. It can be seen that $G_F$ increases exponentially in the mesoscale range, while it is constant and equal to $G_f D(\Delta_*/\delta_*)^{D-1}$ on the macroscopic scale.

It was shown by Borodich (1997a,b) that fractal properties of the main crack surface are not usually essential for characterization of the fracture energy for quasibrittle polyphase materials, while it can be characterized by fractal properties of fractal patterns of microcracks. These estimation were obtained in models based on similarity principles for a cascade of defects. Depending on the considered scales, different kinds of self-similarity were used. When the size of the fracture region was less than some critical size $\Delta_*$, the microcrack pattern growth was considered using the continuous fractal scaling, i.e., it was assumed that the growth of microcrack pattern in intact material near a cut or a notch is statistically self-similar with respect to the continuous group of the homogeneous coordinate dilations, when $\mathbf{x} \to \lambda\mathbf{x}$ for positive $\lambda$, and this growth resulted in a continuous growth of the main crack. When the size of the fracture region was more than the critical size $\Delta_*$, another scaling hypothesis was used, namely that a fully developed fracture pattern picture is repeated statistically, i.e., the growing fracture is self-similar with respect to the continuous group of the coordinate translations. It was shown by estimating of the fracture energy of the pattern that the above models are in qualitative agreement with experimental results regarding the behaviour of fracture energy of concrete (Brameshuber and Hilsdorf, 1990).

Here both mathematical and physical fractal models will be considered. The paper is organised as follows: in Section 2 some relevant definitions and methods attributed to fractal geometry and fractal scaling are recalled. Then the properties of self-similarity and self-affinity of fracture surfaces are defined and both deterministic and random fractals are considered. Finally, scaling of fracture energy for mathematical fractals is studied. In Section 3 applications of physical fractals to fracture are discussed. Some practical methods of estimation of fractal dimensions of both fracture surfaces and patterns are described, and some experimental results concerning evaluation of their fractal dimensions are presented. Finally, possible sources of the apparent fractality are discussed. In Section 4 a development of fractal scaling approaches to multiple fracture is presented. A tip of a continuous crack is modelled when surrounded by a growing cloud of mesocracks whose active part is postulated to be a fractal cluster with multiple branches. The fracture energy of a physical fractal pattern is calculated. Also considered is the case of different fractal regimes for different length scales. Using these estimates, an explanation is obtained for those cases when a positive correlation between the fractal dimension of the fracture surface and the fracture energy is observed. It is shown that the reason for using the concept of a universal roughness exponent is that usually fractal properties of the main crack surface have no correlation with fracture energy. In Section 5 some rules are discussed which could help us to avoid some of the typical mistakes in the use of fractal approaches.

## 2. Mathematical fractals and scaling

### 2.1. Basic definitions

What is a mathematical fractal? The following definition given by Mandelbrot (1977) can be used. A set in a metric space is called a *fractal set* if the *Hausdorff dimension* or Hausdorff–Besicovitch dimension of the set is greater than its topological dimension.

Let $X$ be a compact metric space and $O$ be the totality of open balls in $X$. Let us consider $\delta$-covers of a subset $S \subset X$ (Falconer, 1990), i.e., finite or denumerable subsets $G$ of $O$ which cover $S$ and diameters diam $V$ of balls $V \in G$ are less than or equal to $\delta$. For $s \geqslant 0$ we consider the following objects

$$\mathcal{H}_\delta^s(S) = \inf_{G \in O} \left\{ \sum_{V \in G} (\operatorname{diam} V)^s : S \subseteq \bigcup_{V \in G} V, \operatorname{diam} V \leqslant \delta \right\}.$$

The *Hausdorff s-measure* of a subset $S \subset X$ is defined for $s \geqslant 0$ as the following limit

$$m_H(S, s) = \lim_{\delta \to 0+} \mathcal{H}_\delta^s(S). \tag{2}$$

It was proved that there exists a value $s_0$ such that

$$m_H(S, s) = \begin{cases} \infty, & \text{for } s < s_0, \\ 0, & \text{for } s > s_0. \end{cases} \tag{3}$$

The *Hausdorff dimension* of the set $S$, denoted by $\dim_H S$, is the number $s_0$ such that (3) holds.

The calculation of the Hausdorff dimension of mathematical objects is usually a rather difficult procedure. Even the finding of some estimations of the dimension demands a lot of effort. This stipulate the use of other definitions of fractal dimensions for problems of applied mathematics, for examples, Bouligand, Minkowski, self-similarity, Frostman and packing dimensions. To calculate these dimensions for a mathematical fractal, various methods are used (Falconer, 1990; Tricot, 1995). Sometimes, when using different definitions, different values of fractal dimensions can be obtained. At the same time, it was shown that some of the above mentioned definitions are mathematically equivalent to each other.

Another example of such alternative dimensions is the box dimension, whose definition is usually attributed to Pontrjagin–Schnirelman (1932) and Kolmogorov. The analytical calculations of box dimension are usually easier because they involve covering by spheres of equal radii.

Let $E$ be the Euclidean dimension of the space in which a set $S$ is embedded. For $\delta > 0$, let $N(\delta)$ be the smallest number of $E$-dimensional balls or cubes of diameter $\delta$ needed to cover the set $S$. The *box counting dimension* or *box dimension*, denoted by $\dim_B S$, can be defined if the following limit exists

$$\dim_B S = \lim_{\delta\to 0+} \frac{\log N(\delta)}{-\log \delta}. \tag{4}$$

It can be proved that $\dim_B S$ does not change if one takes $N(\delta)$ as: (i) the smallest number of $\delta$-cubes that cover $S$; (ii) the number of $\delta$-mesh cubes that intersect $S$; (iii) the smallest number of sets of diameter at most $\delta$ that cover $S$; (iv) the largest number of disjoint $\delta$-balls with centres in $S$. In general, one can define two numbers, called the upper and lower box dimensions. To calculate these numbers, one has to replace the expression lim in the above definition by lim sup and lim inf, respectively. If the upper and lower dimensions are equal to each other, then the true limit exists, and this value is defined as box dimension (Falconer, 1990).

The box dimension is not always equal to the Hausdorff dimension. It is only possible to prove in the general case that $\dim_H S \leqslant \dim_B S$. For example, discussing lacunar and nonlacunar fractals, Mandelbrot (1983) noted that the Hausdorff dimension of a nonlacunar fractal in the space $\mathbb{R}^E$ equals to some noninteger number $D$ while its similarity and box dimensions are equal to the dimension of the space $E$.

As a simple alternative to the Hausdorff measure, we could introduce the $s$-measure $m_s$ of a set as the following limit

$$m_s(S) = \lim_{\delta\to 0+} \inf N(\delta)\delta^s \tag{5}$$

and define box dimension as the value, $s = D$ such that $m_s(S)$ has a jump from 0 to $\infty$ similar to the behaviour of $m_H(S, s)$ in (3). However, the $m_s$ is not a $\sigma$-additive measure. This was one of the reasons to note that box dimensions have a number of unfortunate properties, and can be awkward to handle mathematically (Falconer 1990). On the other hand, the difficulties in calculations of Hausdorff dimension are the reason for Tricot's (1995) opinion that this dimension has no practical application in the study of curves originated in other sciences: physics, biology, or engineering.

The scaling properties of mathematical fractals are continuous and based on the following property of homogeneity for the Hausdorff $s$-measure $m_H(S, s)$ (Falconer, 1990)

$$m_H(\lambda S, s) = \lambda^s m_H(S, s) \quad \text{for every } \lambda > 0, \qquad (\mathbf{x} \in \lambda S) \Longleftrightarrow (\lambda^{-1}\mathbf{x} = \mathbf{x}_1 \in S). \tag{6}$$

Clearly, if a set $S$ is a subset of $\lambda S$ for every $\lambda > 0$ then $S$ is a nonfractal set. Thus, the statement $S \subset \lambda S, \forall\lambda > 0$, where $S$ is a fractal, should be understood only in a statistical sense. If there are some discrete values of $\lambda$ such that $S \subset \lambda S$ is realised exactly then $S$ is a fractal PH-set (Borodich, 1994b; 1997d; 1998b).

The Hausdorff dimension of a set does not change under the transformation of homogeneous dilation of coordinates because the fractal measures $m_H(\lambda S, s)$ and $m_H(S, s)$ have the jump from 0 to $\infty$ for the same $s$.

Finally, note that $m_H(S, \dim_H S)$ may be zero or infinite, or may satisfy

$$0 < m_H(S, \dim_H S) < \infty. \tag{7}$$

If a Borel set satisfies the condition (7) then it is called an *s-set* (Falconer, 1990). One says that a set is *D-measurable* if its $s$-measure has a finite positive value $m_D(S)$ for $s$ equal to the fractal dimension $D$ (Borodich, 1997b). It follows from (5) a scaling property similar to the property (6)

$$m_s(\lambda S) = \lambda^s m_s(S). \tag{8}$$

### 2.2. Self-similarity and self-affinity of fracture surfaces

Let us recall that a one-to-one mapping $M$ of a plane $\pi$ onto a plane $\pi'$ is called a *similarity mapping* with coefficient $\lambda > 0$, or simply a *similarity*, when the following property holds: if $A$ and $B$ are any two points of $\pi$, and $A'$, $B'$ are their images under $M$, then $|A'B'| = \lambda|AB|$ (Modenov and Parkhomenko, 1965).

It is known that any similarity transformation of a plane is a homogeneous (isotropic) dilation of coordinates $x' = \lambda x,\ z' = \lambda z$ up to rotation and translation.

A set $S$ is called *statistically self-similar* if under homogeneous scaling with the coefficient $\lambda$, $0 < \lambda < 1$, it is identical from the statistical point of view to the set $S' = \lambda S$. In practice, it is impossible to verify that all statistical moments of the two distributions are identical. Frequently, a set $S$ is said to be self-similar if only a few moments do not change under scaling.

A one-to-one mapping $M$ of a plane $\pi$ onto a plane $\pi'$ is called an *affine mapping*, if the images of any three collinear points are collinear in turn (Modenov and Parkhomenko, 1965). In general, an affine transformation of a plane may be given in any coordinate system as a nondegenerative linear transformation. In the practical study of fracture surfaces, one often considers a particular affine mapping, namely anisotropic scaling, that is given coordinatewise by

$$x' = \lambda x, \qquad z' = \lambda^H z.$$

Here $z$ is a graph of surface profile and $H$ is some scaling exponent.

One says that a fractal is self-affine if it is invariant from the statistical point of view under quasi-homogeneous (anisotropic) scaling.

It is possible to show that usually quasi-homogeneous transformation is a particular case of *Lipschitz homeomorphism* and the Hausdorff dimension of a set $S$ does not change under the action of the Lipschitz homeomorphism $L$ (Borodich, 1994b), i.e.,

$$\dim_H S = \dim_H L(S). \tag{9}$$

Thus, a self-similar curve can be transformed in a self-affine curve preserving its Hausdorff dimension.

The ideas of self-similarity and self-affinity are very popular in studying surface roughness because experimental investigations show that usually profiles of vertical sections of real surfaces are statistically similar to themselves under repeated magnifications; however, the profiles should be scaled differently in the direction of nominal surface plane and in the vertical direction. The self-affine fractals were used in a number of papers as a tool for description of fracture surfaces (Tricot et al., 1994; Schmittbuhl et al., 1995; Vandembroucq and Roux, 1997a). Note that not all fractal curves can be used for modelling of a fractal crack, for example, Bažant (1997a) pointed out that the classic von Koch curve does not allow kinematic separation of surfaces, because of the coupling of segments of fracture path. Therefore, such a curve is a physically impossible model of fractal fracture. Other examples of fractal curves both self-similar and self-affine which allow kinematic separation were considered by Borodich (1997a). Two standard examples of self-affine fractals are the trace of the Weierstrass function (see discussion by Borodich and Onishchenko 1999) and the fractional Brownian motion. The former is a deterministic fractal, while the latter is a statistical (random) fractal.

### 2.3. Brownian surfaces and random fractals

The fractional Brownian motion is a generalisation of the ordinary Brownian process. The fractional Brownian processes are widely used in creating computer-generated surfaces, in particular landscapes. For example, a profile can be constructed as a graph of one-dimensional fractional Brownian motion $V_H(x)$, i.e., we have

$$z = z(x), \qquad z(x) = V_H(x),$$

where $x$ is taken as the time and $z$ is the random variable of the single valued function $V_H(x)$.

It is known (Falconer, 1990) that with probability equal to 1

$$\dim_H V_H(x) = \dim_B V_H(x) = 2 - H.$$

The auto-correlation function $R(\delta)$ is one of the main tools for studying statistical models of rough surfaces. The auto-correlation function $R(\delta)$ of the profile is

$$R(\delta) = \lim_{T\to\infty} \frac{1}{2T} \int_{-T}^{T} [z(x+\delta) - \bar{z}][z(x) - \bar{z}]\, \mathrm{d}x = \langle [z(x+\delta) - \bar{z}][z(x) - \bar{z}] \rangle$$

or

$$R(\delta) = \lim_{T\to\infty} \frac{1}{2T} \int_{-T}^{T} z(x+\delta) z(x)\, \mathrm{d}x - (\bar{z})^2,$$

where $\bar{z}$ is the average value of the profile function $z(x)$

$$\bar{z} = \lim_{T\to\infty} \frac{1}{2T} \int_{-T}^{T} z(x)\, dx.$$

Another tool for the characterization of surfaces is the spectral density function $G(\omega)$, which is the Fourier transform of $R(\delta)$, i.e.,

$$G(\omega) = \frac{2}{\pi} \int_0^{\infty} R(\delta) \cos \omega\delta \, d\delta \quad \text{and} \quad R(\delta) = \int_0^{\infty} G(\omega) \cos \omega\delta \, d\omega.$$

The following point of view is accepted (Falconer, 1990)

(i) if the auto-correlation function $R(\delta)$ of the profile $z(x)$ satisfies

$$R(0) - R(\delta) \sim \delta^{2(2-s)}$$

then it is reasonable to expect that $\dim_B z(x) = s$;

(ii) if the profile $z(x)$ has spectral density

$$G(\omega) \sim 1/\omega^H \tag{10}$$

then it is reasonable to expect the following formula for the box dimension of the graph $z(x)$

$$\dim_B z(x) = (5 - H)/2. \tag{11}$$

### 2.4. Fractal scaling and fracture energy

As it has been mentioned previously, the usage of the Griffith specific surface energy $\gamma$ leads to the paradoxical conclusion that no fractal cracking is possible. The paradox was resolved by the use of a new concepts of specific energy absorbing capacity of a fractal surface $\beta(D)$ (Borodich, 1992; 1997a). We have agreed to consider the fracture pattern as a fractal set. Similarly to the case of a solitary crack, one can use the concept of *specific energy absorbing capacity of a fractal measure of the microcrack pattern* $\beta(C^*)$, where $C^*$ is the fractional part of $C$. The value $\beta(C^*)$ has the physical dimension

$$[\beta(C^*)] = FL/L^{2+C^*} \tag{12}$$

and gives the amount of elastic energy spent on forming a unit of the fractal measure $m_C(\mathcal{R})$. Here $F$ and $L$ denote the dimensions of force and length respectively, and $\mathcal{R}$ is the scale of consideration of the pattern.

Let us consider a fractal pattern of microcracks developed from a cut or a notch in a sample of the thickness $t$. Then the fracture energy ($\Pi$) absorbed by the pattern in some region of a size $\mathcal{R}_0$ is

$$\Pi \sim t\beta(C^*)m_C(\mathcal{R}_0).$$

Using the property of homogeneity (scaling law) for the fractal measure (4)

$$m_C(\mathcal{R}) = (\mathcal{R}/\mathcal{R}_0)^C m_C(\mathcal{R}_0),$$

we can estimate the total fracture energy for the fractal pattern of microcracks of a scale $\mathcal{R}$ when $\mathcal{R} < \Delta_*$

$$\Pi \sim t\beta(C^*)m_C(\mathcal{R}_0)(\mathcal{R}/\mathcal{R}_0)^C. \tag{13}$$

There is no mathematical difference whether one considers the fractal fracture as a self-similar solitary crack or as a self-similar pattern. Therefore, (13) is also valid for a solitary fractal crack when we replace $C^*$ by $D^*$ and $\mathcal{R}_0$ by $\mathcal{L}_0$.

## 3. Physical concept of fractals

Evidently, it is impossible to carry out the scaling down procedure for any real physical object to infinitely small scales. Hence, the mathematical concept of the Hausdorff measure can be applied only to mathematical models of objects and not to these objects themselves. Evidently, the Hausdorff dimension cannot be obtained by experimental or numerical procedures. In this sense there is no fractal objects in nature.

Similarly, box dimension cannot be calculated analytically for physical objects but it can be estimated by experimental or numerical calculations. Various errors can arise during such numerical calculations.

There is no canonical definition of physical fractals and there are numerous methods for the practical estimation of the fractal dimension of an object (see the review by Borodich and Onishchenko 1999).

### 3.1. Practical estimations of fractal dimension

The cluster fractal dimension is taken as the first example of a physical fractal dimension definition.

Let a whole cluster be imagined as consisting of elementary parts of the size $\delta_*$. An object can be modelled as a fractal cluster with dimension $D$ when the model considers scales $\mathcal{R}$ such that $\delta_* < \mathcal{R} < \Delta_*$, where $\delta_*$ and $\Delta_*$ are upper and lower cutoff for fractal law.

To get the value $D$ of the dimension, the region under consideration is divided into discrete cubes with side length $\delta_*$. Then the smallest number of $E$-dimensional cubes needed to cover the cluster ($N(\delta_*)$) is counted. One says that the cluster is fractal if the numbers $N(\delta_*)$ satisfy the so-called *number-radius* relation for different sizes of the considered region of the cluster $\mathcal{R}$

$$N(\delta_*) \approx (\mathcal{R}/\delta_*)^D, \qquad \delta_* < \mathcal{R} < \Delta_*. \tag{14}$$

The value of $D$ is estimated as the slope of linear growth of $\ln(N(\delta_*))$ against $\ln(\mathcal{R})$. The power $D$ is usually called the *cluster dimension* or *mass dimension*.

The name of the latter term can be explained in the following way. Let some 'mass' $M(S_*)$ be assigned to the elementary particle $S_*$ of the size $\delta_*$. Then instead of the $s$-measure $m_s$ of the cluster $S$ used in the definition of box dimension one has the 'mass' of the whole cluster $M(S) = N(\delta_*)M(S_*)$.

As an example of the use of the definition, let us consider a profile which is imaged on a computer screen as a union of points (pixels) of the size $\delta_*$. Then we can obtain a computerised estimation of the number of pixels $N(\delta_*)$ forming the line and lying inside a circle or a square box of size $\mathcal{R}$ centered at a point $\mathbf{x}$. If the profile has the fractal properties then both the relation (14) and a corresponding scaling property for the the cluster mass can be obtained

$$M(S) = \lambda^D M(S_*), \quad \lambda = (\mathcal{R}/\delta_*) \tag{15}$$

repeating the procedure of estimation for different values of $\mathcal{R}$. Note the similarity between the scaling properties (8) and (15).

Another definition of the physical fractal dimension is based on the Richardson method. The method uses dividers which are set to a prescribed opening $\delta$ (Mandelbrot, 1983). Moving these dividers along the contour so that each new step starts where the previous step leaves off, one obtains the number of steps $N(\delta)$. The contour is said to be of a fractal nature if by repeating this procedure for different values of $\delta$ the relation

$$N(\delta) \sim (\delta)^{-D} \tag{16}$$

is obtained in some interval $\delta_* < \delta < \Delta_*$ of sizes $\delta$. The power $D$ is usually called the *Richardson dimension* $D_R$.

The last method which we would like to mention is the following. We cover a physical object $S$ on a plane by a square grid of a size $\delta_i$. Then successively dividing each initial square of the grid into four subsquares of the size $\delta_{i+1} = \delta_i/2$ and calculating the number of subsquares which contain points of the $S$, we may obtain the relation (16) holding in some interval of sizes $\delta$. In this case, the power $D$ is usually called the *physical box dimension* $D_B$.

We see that there are various methods to calculate fractal dimension of physical objects. However, if the mathematical box dimension $\dim_B S$ is the same for various specific schemes of covering (Falconer, 1990), the estimations obtained for physical fractals can depend on the used technique.

### 3.2. Spectral density of fracture surfaces

The previously mentioned conclusions concerning Brownian surfaces and random fractals are valid for mathematical models of the fracture surfaces. An effective method for determining the physical fractal dimension of a random structure is based on the idea that the conclusions obtained for mathematical fractals are also valid for physical fractals and

$$D_B(S) \approx \dim_B(S). \tag{17}$$

The moments $m_n$ of the spectral density $G(\omega)$ provide a useful description of the surface roughness

$$m_n = \int_{\omega_0}^{\infty} \omega^n \, G(\omega) \, \mathrm{d}\omega,$$

where $\omega_0 = 2\pi/\lambda_0$ is the wavenumber corresponding to the profile length $\lambda_0$. It is possible to show (Brown, 1995) that $m_0$ is the variance of heights (rms height) of the surface, $m_2$ is thc variancc of slopcs (rms slopc) and $m_4$ is thc variancc of curvaturcs (rms curvaturc). This gives us the connection between the spectral density and the standard engineering parameters of roughness.

It is shown that real surfaces approximately satisfy the property (10) in wide range of scales (Sayles and Thomas, 1978; Brown, 1995). The exponent $H$ varies typically between 1 and 3.

It follows from this that all wavelengths are equally represented in the profile and that there exists no characteristic scale; in other words, after magnification roughness looks like before.

### 3.3. EXPERIMENTAL STUDIES OF FRACTAL DIMENSIONS OF FRACTURE

There are various experimental techniques which are used to estimate fractal dimensions (Saouma and Barton, 1994; Tricot et al., 1994; Xie et al., 1998). It was noted by Tricot et al. (1994) that when a surface is digitized, it is necessary to avoid smoothing or filtering processes because they can affect the estimate of fractal dimension. In particular, they conclude that the 'slit-island' method, which is commonly used in fractography for determining fractal dimension, is not appropriate.

The above described technique of square grids is usually used for analysis of patterns of multiple fracture (Zhao et al., 1993; Chelidze et al., 1994). While a technique of slices, which are taken through a surface, is usually used for analysis of a fracture surface (Saouma and Barton, 1994; Botvina et al., 1997). This technique suggests to replace the study of a three-dimensional problem by the study of two-dimensional problems for profiles taken at different places of the surface and use then a well-known result obtained for some mathematical fractals. This result says that the dimension of the surface differs from the dimension of profiles by unity, i.e., the fractional part $D^*$ of the dimensions is the same for the surface and the profiles. There arises the following question: can we use this result obtained for mathematical fractals in application to physical fractals?

The question is still open (see, e.g., recent studies by Xie et al., 1998). Using their extensive studies, Tricot et al. (1994) argued that slicing techniques should be used in experiments, at least for isotropic surfaces, since they are much more economical than three-dimensional techniques.

When Bouchaud et al. (1990) found that for samples of aluminium alloys the fractal dimension of fracture surface lies in a quite narrow region interval $0.11 \leqslant D^* \leqslant 0.30$, they concluded that the fractal dimension of fractured surfaces can have a universal value. Moreover, the analysis fulfilled by Borodich (1997a) of published experimental data concerning fractal dimensions ($D$) of profiles of fracture surfaces for quasibrittle materials shows that the values belong mainly to the interval $0.04 < D^* < 0.33$. The fractal dimensions of patterns of fracture ($C$) belong mainly to $0.47 < C^* < 0.73$.

### 3.4. SOURCES OF FRACTALITY

The question 'Why are fractals so common in Nature?' was recently studied by Avnir et al. (1997) and Malcai et al. (1997). Their extended analysis of data published, in particular concerning rough surfaces, shows that the overwhelming majority of reported physical fractals span only around 1.5 orders of magnitude. It seems to the author that this is a truly physical limit for the majority of physical objects encountered as the object structure moves from one paradigm to another.

To give a possible interpretation for this observation, Avnir et al. (1997) and Malcai et al. (1997) considered various numerical models which were pure random. For example, they considered a model in which rods of some specific length are randomly placed on the unit interval. It has been shown that the characteristic relation (16) holds for the structures generated in their random models approximately over the above physically meaningful range. They point out that such properties of random structures can be attributed to apparent fractality. Finding that randomness obeys such a dilation similarity, they conclude that 'fractals everywhere' may be caused by 'randomness everywhere'.

Thus, the physical fractal approach can be considered as a tool to describe some random objects having dilation similarity without the use of the terminology of probability methods.

## 4. Modelling of fractal fracture and fracture energy

Here the physical fractal approach is used to construct a model of multiple fracture in quasi-brittle materials which develops the models of our previous papers (Borodich, 1997a,b). The concept of quasibrittle materials assumes that there is a narrow layer of inelastic deformations near the fracture surfaces. During the crack propagation this layer absorbs energy as well as the new fracture surface.

### 4.1. Basic features of multiple fracture

It is believed that the fracture mechanisms in rock, concrete and ceramics are similar (Bažant, 1984; Bažant et al., 1990). Hence, the similarity between fracture process in concrete and some rock can be employed, where the process zone can be visualised (see, e.g., experiments on marble Nolen–Hoeksema and Gordon, 1987; Huang et al., 1993).

It is possible to identify the following stages of fracture on loading a specimen of a polyphase material with a blunt notch. At the beginning of the fracture process, microdefects grow and become microcracks. On further loading the microcracks accumulate within the volume of a highly stressed intact material, coalesce and a few isolated mesocracks are formed in the area near the notch tip. Then the intensity of mesocracking increases. The process of the coalescence of the mesocracks in turn causes a crack pattern (the process zone). By some critical value of external load, an abruptly propagation of main fracture is observed.

Thus, the process of fracture in polyphase quasibrittle materials should be considered on several scales: micro, meso and macro.

### 4.2. Basic hypotheses of fractal fracture

We study the fracture process within the intermediate scale and suppose that the mesocrack pattern can be described as a fractal cluster, i.e., it is imagined as consisting of elementary mesocracks of the size $\delta_*$, with dimension $C$ when the size $\mathcal{R}$ of the fracture pattern is less than some critical $\Delta_*$, which is the upper cutoff for the fractal law.

We suppose that (i) the width $h$ of the layer of inelastic deformations near any fracture surface is constant and it is the same for mesocracks and the macrocrack; (ii) the material of every cube of size $h$ centered in a point of fracture surface absorbs the same quantity of energy $G_f$. These hypotheses concerning the layer of inelastic deformations are the same as the basic hypotheses in our previous models (Borodich, 1992; 1994a; 1997a).

To distinguish micro- and mesolevels, all cracks of the length less than $h$ are considered as microcracks.

#### 4.2.1. *Energy absorbed by a fractal pattern*

Let us calculate the average amount $\langle W \rangle$ of the energy absorbed by a fractal pattern. This energy differs from the surface energy $\Pi$ because the former concept takes into account the energy spent to create inelastic deformations.

Let us cover all cracks of the pattern by cubes of the size $h$ centered in points of fracture surfaces. Then the average amount $\langle W \rangle$ of absorbed energy can be calculated as

$$\langle W \rangle = G_f N(h), \tag{18}$$

where $N(h)$ is the minimal number of the cubes in the cover. Evidently, this approach can be used if the width of the plastic layer is small with respect to the average distance between mesocracks.

The pattern of mesocracks is a self-similar physical fractal of dimension $C$ (scales $\delta_* = h < \mathcal{R} < \Delta_*$). Hence, the numbers $N(h)$ satisfy the number-radius relation or the fractal law (14) by change of sizes of the considered region of the cluster $\mathcal{R}$.

Substituting (14) into (18), one can calculate the average amount of energy $\langle W \rangle$ absorbed by mesocracks situated within an area of the size $\mathcal{R}$

$$\langle W \rangle \simeq G_f (\mathcal{R}/h)^C. \tag{19}$$

4.2.2. *Change of fractal dimension with the scale*

We cannot consider intermediate scales of an object by applying the mathematical concept of fractal dimensions because this concept uses the limit values only (see (2) and (4)). However, using the concept of physical fractals, we have a possibility to take into account intermediate scales in the case when different fractal dimensions are observed on different length scales. This case is quite common in study of physical objects (Mandelbrot et al., 1984; Dauskardt et al., 1990; Milman et al., 1994; Botvina et al., 1997).

Let the fracture pattern have two intervals with different fractal scales, i.e., it has the fractal dimension $C_1$ for scales $\delta_* = h < \mathcal{R} < \Delta_1$ and the fractal dimension $C_2$ for scales $\Delta_1 < \mathcal{R} < \Delta_*$. Can we calculate the average amount of energy $\langle W \rangle$ absorbed by mesocracks situated within an area of the size $\mathcal{R}$ when $\Delta_1 < \mathcal{R}$?

It follows from (19) that the average amount of energy $\langle W_1 \rangle$ absorbed by mesocracks situated within an area of the size $\Delta_1$ can be calculated by the formula

$$\langle W_1 \rangle = \langle W(\Delta_1) \rangle \simeq G_f (\Delta_1/h)^{C_1}. \tag{20}$$

Using (20) and $\Delta_1$ as a new measuring size, we can calculate the average amount of energy $\langle W \rangle$ absorbed by the pattern of a larger size

$$\langle W(\mathcal{R}) \rangle \simeq \langle W_1 \rangle (\mathcal{R}/\Delta_1)^{C_2}, \quad \Delta_1 < \mathcal{R} < \Delta_*. \tag{21}$$

It is easy to generalize (21) to the case of any number of intervals with different fractal scales.

### 4.3. Basic formulae of the model

A two-dimensional problem for a crack can be formulated as follows. A continuous sample in the plane $0x_1x_2$ with a cut $[-l_0, l_0]$ along the $x_1$-axis is subjected to stretching forces which are applied symmetrically with respect to the $x_1$-axis. The process zone $S$ is formed in the neighbourhood of the cut tips. The considered problem is symmetric with regard to the $x_2$-axis. Therefore, the problem can be reduced to the case $x_1 > 0$ only.

The process zone develops during loading and its size (its width and length) depends on both the fine structure of material and the stress field. When we estimated the energy released in an infinite elastic solid under a uniaxial tensile stress after creating a straight-through planar crack perpendicular to the stress direction, we followed the idea that there exist a domain where the stress field relaxes (Borodich, 1994a; 1997a). We will suppose the same in our model. The propagation of the main crack tip leads to a relaxation of the stress field in some domain behind the tip, and the micro- and mesocracks situated in the domain will not grow further. Thus, the process zone can be separated on an active $S_A$ and a passive part $S_P$. The stress concentration regions near the mesocracks of the periphery of the active domain are sources for growth of new microcracks. We suppose that the active part only is essential for propagation of the main crack.

Thus, it is assumed in our model that (a) the growth of the process zone results in a continuous growth of the main crack; (b) the fractal scaling is applicable to describe the beginning of the self-similar growth of the process zone; (c) while a fracture pattern is already fully developed, i.e., when the width of the process zone reaches some critical size $w_*$, the pattern picture is repeated. This means that the active part of the process zone is invariant with respect to continuous group of coordinate translations.

Let the main crack have an advance $x$. The process zone of the crack with the tip at a point $x_1 = l_0 + x$ is wedge-shaped with some angle $2\alpha$ at the wedge vertex for $l_0 < x_1 < x$ and it is described by a segment of a circle of some radius $w_c$, $w_c(x) = x \sin\alpha$ for $l_0 + x < x_1$.

Now let us estimate the energy absorbed by the active part of the process zone. We cover all cracks of the active part of the process zone by cubes of the size $h$ centered in points of fracture surfaces of the pattern. Similarly to (19), we obtain

$$\langle W \rangle = G_f N(h) + \text{const.} \tag{22}$$

Substituting (14) into (22), one can calculate the average amount of energy $\langle W \rangle$ absorbed by mesocracks of the fractal process zone

$$\langle W \rangle \sim G_f (w_c/h)^C + \text{const.} \tag{23}$$

The average value of fracture energy $G_F$ per unit of thickness is defined by the formula

$$G_F = \frac{d\langle W \rangle}{\mathrm{d}x}. \tag{24}$$

Substituting (23) into (24), one obtains

$$G_F \sim G_f(1 + C^*)(w_c/h)^{C^*} \cdot w_c'(x)/h = G_f(1 + C^*)(w_c/h)^{C^*} \sin\alpha/h, \tag{25}$$

for the growing pattern of mesocracks and

$$G_F \sim G_f(1 + C^*)(w_*/h)^{C^*} \sin\alpha/h = \text{const} \tag{26}$$

for the the repeated pattern of mesocracks.

Evidently, formulae (25) and (26) are similar to the formulae (1) obtained in the case of a single fractal crack. However, (25) and (26) give a more detailed description of the fracture proccss.

### 4.4. SOME APPLICATIONS OF THE MODEL

#### 4.4.1. *Correlation between D, C and the fracture toughness of the material*

It follows from the basic hypotheses of the model concerning the layer of inelastic deformations that fractal dimension of a fracture surface can correlate with fracture energy only in materials with extremely narrow layer, i.e., $h \ll w_* \sim \Delta_*$. Then we can expect that (1) is valid. If fracture surface exhibits fractal features, however $h \sim \Delta_*$, then fracture energy is mainly related to the work done within the nonfractal zone of inelastic deformations. Hence, there is no correlation between the fractal dimension of the fracture surface and the fracture toughness of the material (Borodich, 1992; 1994; 1997a).

Experimental studies show (Dauskard et al., 1990; Bouchaud et al., 1990; Botvina et al., 1997) that for metals $\Delta_* \simeq 0.1$ mm, while $h \simeq 0.4$ mm (Botvina et al., 1997), i.e., $\Delta_* \leqslant h$. Thus, the fractal properties of fracture surface are not essential for fracture of ductile materials (Borodich, 1997a). This theoretical prediction was supported by recent experimental data (Ikeshoji and Shioya, 1997).

The experimental studies show (Nolen-Hoeksema and Gordon, 1987; Chelidze et al., 1994; Zhao et al., 1993) that for polyphase materials the upper cutoff for fractal law for fracture patterns is $\Delta_* \sim 1$ cm, i.e., $\Delta_* \gg h$. Therefore, the fractal properties of the pattern of mesocracks could characterize the fracture energy of the materials and formulae (25) and (26) are valid even if the fractal properties of the main crack surface are not essential. Thus, the experimental evaluation of the dimension $C$ of the fracture pattern in quasibrittle materials is more important for fracture mechanics than the evaluation of the dimension $D$ of the fracture surface. Fractal properties of developing patterns of microfractures and pores have already been studied in some papers (Zhao et al., 1993; Chelidze et al., 1994; Xie et al., 1996). It is necessary to continue such studies.

#### 4.4.2. *Universal exponents and fracture*

In addition to the above mentioned observations by Bouchaud et al. (1990), it was recently suggested that scaling of the roughness of brittle cracks is universal (Hansen et al., 1991; Måløy et al., 1992). This means that one observes the following scaling

$$w_r \sim \mathcal{L}^{\zeta}$$

of roughness $w_r$, defined as the width of the profile

$$w_r = (\langle z^2 \rangle - \langle z \rangle^2)^{1/2}, \qquad \langle z \rangle = \frac{1}{\mathcal{L}} \int_0^{\mathcal{L}} z(x)\,\mathrm{d}x, \qquad \langle z^2 \rangle = \frac{1}{\mathcal{L}} \int_0^{\mathcal{L}} z^2(x)\,\mathrm{d}x,$$

with the length $\mathcal{L}$ of a slice profile. Here $\zeta$ is treated as a *universal* exponent independent of the material.

To determine the roughness exponent $\zeta$, two different methods were used. One of the methods was the above described spectral density function method. It was found that $G(\omega) \sim 1/\omega^{1+2\zeta}$. Hence, it follows from (11) and (17) that the profile dimension is $D_B = 2 - \zeta$ or $D^* = 1 - \zeta$. For six different brittle materials including graphite, porcelain and steel, the average value of the obtained roughness exponent was $0.86 \pm 0.06$, i.e., $0.08 < D^* < 0.20$. Note this value is in a good agreement which date obtained for profiles of concrete fracture surfaces $0.071 < D^* < 0.165$ (Saouma and Barton, 1994).

A discussion followed the above observations. Milman et al. (1993) noted that it is not uncommon to find different fractal dimensions of fracture surfaces and respectively different values of $\zeta$ on different length scales (Mandelbrot et al., 1984; Dauskardt et al., 1990). In addition, some studies find correlations between $\zeta$ and mechanical properties of materials, an observation that contradicts the very notation of universality.

Måløy et al. (1993) replied that it would be too much to expect a universal $\zeta$ for all materials on any length scale. In particular, they obtained the value of $\zeta$ from measurements in the micrometre range. This value differs from the values obtained by Milman et al. (1993) from measurements in the nanometre range. With regards to correlation between $\zeta$ and mechanical properties of materials, they noted that no systematic dependence of the $\zeta$ on the impact energy may be found. For example, Mandelbrot et al. (1984) reported $\zeta$ for steel in the range 0.7 to 0.9, increasing with increasing impact energy. Other groups, which repeated this experiment, found the same range of $\zeta$, but with conflicting dependence of this exponent on the impact energy (Tzschichholz and Pfuff, 1991).

Let us try to explain these contradictions using the theoretical model presented. The above arguments show that studies of the size $h$ of the layer of inelastic deformations near the fracture surface are very important for the evaluation of the validity of fractal models. For example, such studies were fulfilled by Botvina et al. (1997). It seems to us that concept of the universal roughness exponent may be valid on some mesoscale and some interval of temperatures for ductile and quasibrittle materials when $h \sim \Delta_*$ and there is no correlation between the fractal dimension of the fracture surface and the fracture toughness of the material. The above mentioned analysis of published data concerning fractal dimensions of fracture surfaces (Borodich, 1997a) supports this hypothesis. If $h \ll \Delta_*$ then it is unlikely that the hypothesis is valid.

## 5. Discussion

The fractal approaches need a rather delicate treatment. Indeed, there is a number of mistakes and misunderstandings, which one can meet even in the best papers and books concerning applications of fractals. It has been mentioned above that the main mistake is the confusion between mathematical and physical fractals. The wrong use of the Hausdorff dimension is also typical.

There is also a negative tendency in the literature to use results obtained for mathematical fractals directly in applications to physical fractals. For example, results obtained for fractional Brownian motion are considered as valid for all self-affine fractals, both mathematical and physical. By the way, authors should define accurately what they mean by the term 'self-affine fractal' when a physical fractal is considered.

One can meet wrong statements that curves are either smooth or fractal, and nowhere differentiable functions are always fractal. An example was given by Borodich (1998b) when the graph of a continuous, nowhere differentiable function $U_M$ contains an infinite number of reduced copies of the graph of the whole $U_M$ function and $\dim_H$ graph $U_M = 1$, i.e., the curve is nonfractal in the spirit of the original definition (Mandelbrot, 1977).

It is worth to mention that not all sets are D-measurable (see (7)), i.e., sometimes a fractal measure of a set can be either 0 or $\infty$ only.

Let us formulate some rules which could help to avoid some of the typical mistakes:

(i) realize the difference between mathematical and physical fractals;

(ii) realize the difference between the Hausdorff dimension and the box counting dimension;
(iii) do not forget that the Hausdorff dimension may be established for pure mathematical objects only and it cannot be determined experimentally, e.g., using 'probe molecules', or numerically;
(iv) do not forget to say about the upper cutoff for the fractal law;
(v) do not confuse self-similarity and fractality;
(vi) do not forget that scaling properties of fractal objects are valid in a statistical sense only;
(vii) do not use new terminology, e.g., '*fractal space*' or '*stress intensity factor for a self-affine crack*', without appropriate definitions;
(viii) do not use well-known concepts in a nontraditional way without appropriate explanations.

## 6. Conclusion

The questions concerning the size effect of fracture have not been considered in the paper. One can find an extensive study of this problem in the papers by Bažant and his co-workers (Bažant, 1984; 1995; 1997a,b; Bažant et al., 1991). Although, our model presented above describes scaling of fracture energy in an infinite elastic solid under a uniaxial tensile stress, it can be applied for explanation of the size effect of fracture. In this case, it is necessary to introduce one scale more, namely the size of sample and consider the effects of the sample boundary. Indeed, when cracks ahead of the tip of the main crack reach the sample boundary, the formula (26) for the value of the fracture energy is not valid more. Such a model was recently developed by Borodich (1998a).

Here we have concentrated mainly on modern fractal models of fracture. We have seen above that if we consider fracture as having fractal features then we can use one of the following ways: (i) to model the phenomenon as a pure mathematical fractal and apply the strict mathematical approach; (ii) to consider physical fractal, i.e., an object obeys fractal laws when the scale varies in interval between upper and lower cutoff.

The use of the former way leads often to rather complex mathematical constructions and stipulates the introduction of new concepts which are appropriate for mathematical description of fractal fracture processes, for example $\beta(D)$-concept (Borodich, 1992; 1994; 1997a).

Of course, the mathematical problems arise in such studies are very interesting (Borodich and Volovikov, 1997; Vandembroucq and Roux, 1997a,b). However, it is rather difficult to expect practical applications in the near future. Indeed, we have seen that the employment of mathematical fractal approach in the study of fracture has several drawbacks. We list only several of them: it is often unclear how we can use results obtained for mathematical fractals in applications to physical fractals; it is impossible to use the classical formulations of boundary value problems for bodies with fractal boundaries; and it is still unclear which definition of fractal dimension is preferable for the study.

The use of physical fractal approach looks more encouraging. Using this approach, it is better to avoid the direct transfer of results obtained for mathematical fractals to physical fractals if it not justified by some discussion. Evidently, it is necessary to indicate the upper and lower cutoffs as well.

We have seen that the physical fractal approach allows us (i) to calculate the scaling of the fracture energy for quasibrittle materials, (ii) to explain the cases of correlation between the fractal dimension of fracture surfaces or fracture patterns and the fracture toughness of the material, (iii) to give some explanations concerning the concept of the universal roughness exponent.

We have also seen that the physical approach is more flexible than mathematical, for example, it can take into account the change of fractal dimension (similarity exponent) with changing the scale of consideration.

It seems that such a universal property of random structures as their dilation similarity on some intermediate scales is a crucial feature for the fracture process and it does not matter if one calls it as the power law exponent of the similarity, the physical fractal dimension or the similarity exponent. What is important is to involve this similarity in models and calculations.

## Acknowledgements

Thanks are due to the Centre de Recerca Matemàtica, Institut d'Estudis Catalans (Bellterra, Barcelona) for funding a visit during which the main part of this work was undertaken.

## References

Avnir, D., Biham, O., Lidar (Hamburger), D. and Malcai, O. (1997). On the abundance of fractals. In: *Fractal Frontiers* (Edited by M.M. Novak and T.G. Dewey), World Scientific, Singapore, 199–234.

Balankin, A.S. (1996). The effect of fracture surface morphology on the crack mechanics in a brittle material. *International Journal of Fracture* **76**, R63–R70.

Baran, G.R., Roques-Carmes, C., Wehbi, D. and Degrange, M. (1992). Fractal characteristics of fracture surfaces. *Journal of the American Ceramic Society* **75**, 2687–2691.

Barenblatt, G.I. (1993). Some general aspects of fracture mechanics. In: *Modeling of Defects and Fracture Mechanics* (Edited by G. Herrmann), Springer-Verlag, Wien, New York, 29–59.

Barenblatt, G.I. and Botvina, L.R. (1986). Similarity methods in mechanics and physics of fracture. *Fiz.-Khim. Mekh. Mater.* **1**, 57–62.

Barenblatt, G.I. and Monin, A.S. (1983). Similarity principles for the biology of pelagic animals. *Proceeding of the National Academic Science of USA* **80**, 3540–3542.

Barnsley, M. (1988). *Fractals Everywhere*, Academic Press, Boston.

Bažant, Z.P. (1984). Size effect in blunt fracture: concrete, rock, metal. *Journal of Engineering Mechanics* **110**, 518–535.

Bažant, Z.P. (1995). Scaling of quasi-brittle fracture and the fractal question. *ASME Journal of Materials and Technology* **117**, 361–367.

Bažant, Z.P. (1997a). Scaling in nonlinear fracture mechanics. In: *IUTAM Symposium on Nonlinear Analysis of Fracture, University of Cambridge, 3–7 September 1995* (Edited by J.R. Willis), Kluwer Academic Publishers, Dordrecht, 1–12.

Bažant, Z.P. (1997b). Scaling of quasibrittle fracture: hypotheses of invasive and lacunar fractality, their critique and Weibull connection. *International Journal of Fracture* **29**, 1699–1709.

Bažant, Z.P., Gettu, R. and Kazemi, M.T. (1991). Identification of nonlinear fracture from size effect tests and structural analysis based on geometry-dependent $R$-curves. *International Journal of Rock Mechanics and Mining Sciences* **28**, 43–51.

Borodich, F.M. (1992). Fracture energy in a fractal crack propagating in concrete or rock. *Doklady Akademii Nauk (Russia)* **325**, 1138–1141 (English transl. in: *Transactions (Doklady) of the Russian Akademy of Sciences: Earth Science Sections* **327**(8), 36–40).

Borodich, F.M. (1994a). Fracture energy of brittle and quasi-brittle fractal cracks. *IFIP Transactions: A. Computer Science and Technology* **41**, 61–68.

Borodich, F.M. (1994b). Some applications of the fractal parametric-homogeneous functions. *Fractals* **2**, 311–314.

Borodich, F.M. (1997a). Some fractal models of fracture. *Journal of Mechanics and Physics of Solids* **45**, 239–259.

Borodich, F.M. (1997b). Scaling of microcrack patterns and crack propagation. In: *Fractal Frontiers* (Edited by M.M. Novak and T.G. Dewey), World Scientific, Singapore, 235–243.

Borodich, F.M. (1997c). Non-classical scaling of microcrack patterns and crack propagation. In: *Multiple Scale Analyses and Coupled Physical Systems (Saint-Venant Symposium, August 28th-29th, 1997)*, Presses de l'école nationale des Ponts et Chaussées, Paris, 493–500.

Borodich, F.M. (1997d). Parametric-homogeneity and self-similar phenomena. *Nonlinear Analysis* **30**, 409–418.

Borodich, F.M. (1998a). Self-similar models of multiple fracture and size effect. Centre de Recerca Matemàtica Bellaterra, Barcelona, Preprint núm. 380, 1–31.

Borodich, F.M. (1998b). Parametric-homogeneity and non-classical self-similarity. I. Mathematical background. *Acta Mechanica* **131**, 27–45.

Borodich, F.M. and Onishchenko, D.A. (1999). Similarity and fractality in the modelling of roughness by multilevel profile with hierarchical structure. *International Journal of Solids and Structures* **36**, 2585–2612.

Borodich, F.M. and Volovikov, A.Yu. (1997). *Continuity of Surface Integrals for Domains with Fractal Boundaries.* Technical Report TR/MAT/FMB-AYV/97-80, Glasgow Caledonian University, Glasgow, 1–18.

Botvina, L.R., Ioffe, A.V. and Tetyueva, T.V. (1997). Effect of the zone of plastic deformation on the fractal properties of a fracture surface. *Metal Science of Heat Treatment* **39**, 296–300.

Bouchaud, E., Lapasset, G. and Planès, J. (1990). Fractal dimension of fractured surfaces: a universal value? *Europhys. Lett.* **13**, 73–79.

Brady, B.T. (1974). Theory of earthquakes. I. A scale independent theory of rock failure. *Pageoph* **112**, 701–725.

Brameshuber, W. and Hilsdorf, H.K. (1990). Influence of ligament length and stress state on fracture energy of concrete. *Engineering Fracture Mechanics* **35**, 95–106.

Brown, S.R. (1995). Simple mathematical model of rough fracture. *Journal of Geophysical Research* **100**(B4), 5941–5952.

Carpinteri, A. (1994). Scaling laws and renormalization groups for strength and toughness of disordered materials. *International Journal of Solids and Structures* **31**, 291–302.

Carpinteri, A. and Chiaia, B. (1996). Crack-resistance behavior as a consequence of self-similar fracture topologies. *International Journal of Fracture* **76**, 327–340.

Chelidze, T., Reuschlé, T. and Guéguen, Y. (1994). A theoretical investigation of the fracture energy of heterogeneous brittle materials. *Journal of Physics: Condens. Matter.* **6**, 1857–1868.

Cherepanov, G.P., Balankin, A.S. and Ivanova, V.S. (1995). Fractal fracture mechanics – a review. *Engineering Fracture Mechanics* **51**, 997–1033.

Cotterell, B. and Mai, Y.-W. (1996). *Fracture Mechanics of Cementitious Materials.* Blackie Academic and Professional, London.

Dauskardt, R.H., Haubensak, F. and Ritchie, R.O. (1990). On the interpretation of the fractal character of fracture surfaces. *Acta Metallurgica et Materialia* **38**, 143–159.

Falconer, K.J. (1990). *Fractal Geometry: Mathematical Foundations and Applications.* John Wiley, Chichester.

Goldshtein, R.V. and Mosolov, A.B. (1991). Cracks with fractal surface. *Doklady Akademii Nauk (Russia)* **319**, 840–844. (English transl. in: *Sov. Phys. Dokl.* **36**(8), 603–605).

Goldshtein, R.V. and Mosolov, A.B. (1992). Fractal cracks. *Journal of Applied Mathematics and Mechanics (PMM)* **56**, 563–571.

Gettu, R. and Shah, S.P. (1994). Fracture mechanics. In: *High Performance Concretes and Applications* (Edited by S.P. Shah and S.H. Ahmad), Edward Arnold, London, 161–212.

Hansen, A., Hinrichsen, E.L. and Roux, S. (1991). Roughness of crack interfaces. *Phys. Rev. Lett.* **66**, 2476–2479.

Herrmann, H.J., Hansen, A. and Roux S. (1989). Fracture of disordered, elastic lattices in two dimensions. *Phys. Rev.* **39B**, 637–648.

Huang, J., Wang, Z. and Zhao, Y. (1993). The development of rock fracture from microfracturing to main fracture formation. *International Journal of Rock Mechanics and Mining Sciences* **30**, 925–928.

Ikeshoji, T. and Shioya, T. (1997). Fractal dimension of fracture surfaces in ductile-brittle transition regime. In: *Fractal Frontiers* (Edited by M.M. Novak and T.G. Dewey), World Scientific, Singapore, 255–263.

Issa, M.A., Hammad, A.M., Chudnovsky, A. (1993). Correlation between crack tortuosity and fracture-toughness in cementitious material. *International Journal of Fracture* **60**, 97–105.

Louis, E., Guinea, F. and Flores, F. (1986). The fractal nature of fracture. In: *Fractals in Physics* (Edited by L. Pietronero and E. Tosatti), Elsevier North-Holland, Amsterdam, 177–180.

Louis, E. and Guinea, F. (1989). Fracture as a growth-process. *Physica D* **38**, 235–241.

Lung, C.W. (1986). Fractals and the fracture of cracked metals. In: *Fractals in Physics* (Edited by L. Pietronero and E. Tosatti), Elsevier North-Holland, Amsterdam, 189–192.

Nolen-Hoeksema, R.C. and Gordon, R.B. (1987). Optical detection of crack patterns in the opening-mode fracture of marble. *International Journal of Rock Mechanics and Mining Sciences* **24**, 135–144.

Malcai, O., Lidar, D.A., Biham, O. and Avnir, D. (1997). Scaling range and cutoffs in empirical fractals. *Phys. Rev. E* **56**, 2817–2828.

Måløy, K.J., Hansen, A., Hinrichsen, E.L. and Roux, S. (1992). Experimental measurements of the roughness of brittle cracks. *Phys. Rev. Lett.* **68**, 2266–2269.

Måløy, K.J., Hansen, A., Hinrichsen, E.L. and Roux, S. (1993). Reply. *Phys. Rev. Lett.* **71**, 205.

Mandelbrot, B.B. (1977). *Fractals: Form, Chance, and Dimension.* W.H. Freeman, San Francisco.

Mandelbrot, B.B. (1983). *The Fractal Geometry of Nature.* New York, W.H. Freeman.

Mandelbrot, B.B., Passoja, D.E. and Paullay, A.J. (1984). Fractal character of fracture surfaces of metals. *Nature* **308**, 721–722.

Mecholsky, J.J., Passoja, D.E. and Feinberg-Ringel, K.S. (1989). Quantitative analysis of brittle fracture profiles using fractal geometry. *J. Amer. Ceram. Soc.* **72**, 60–65.

Meisner, M.J. and Frantziskonis, G.N. (1997). Heterogeneous materials – scaling phenomena relevant to fracture and to fracture toughness. *Chaos, Solitons and Fractals* **8**, 151–170.

Milman, V.Y., Stelmashenko, N.A. and Blumenfeld, R. (1994). Fracture surfaces: a critical-review of fractal studies and novel morphological analysis of scanning tunneling microscopy measurements. *Progress in Mater. Sci.* **38**, 425–474.

Milman, V.Y., Blumenfeld, R., Stelmashenko, N.A. and Ball, R.C. (1993). Comment on Experimental measurements of the roughness of brittle cracks. *Phys. Rev. Lett.* **71**, 204.

Modenov, P.S. and Parkhomenko, A.S. (1965). *Geometric Transformations. Vol. 1. Euclidean and Affine Transformations.* Academic Press, New York.

Mosolov, A.B. (1991a). Fractal Griffith fracture. *Zhurn. Tekhn. Fiziki* **61**, 57–60 (English transl. in: *Sov. Phys.–Tech. Phys.*).

Mosolov, A.B. (1991b). Fractal $J$-integral under the destruction. *Pis'ma v Zhurn. Tekhn. Fiziki* **17**, 45–50 (English transl. in: *Tech. Phys. Lett.*).

Mosolov, A.B. (1993). Mechanics of fractal cracks in brittle solids. *Europhys. Lett.* **24**, 673–678.

Mosolov, A.B. and Borodich, F.M. (1992). Fractal fracture of brittle bodies during compression. *Dokl. Akad. Nauk (Russia)* **324**, 546–549 (English transl. in: *Soviet Phys. Dokl.* **37**, 263–265).

Pontrjagin, L. and Schnirelmann, L. (1932). Sur une propriété métrique de la dimension. *Annals of Mathematics* **33**, 156–162.

Saouma, V.E. and Barton, C.C. (1994). Fractals, fractures, and size effects in concrete. *Journal of Engineering Mechanics* **120**(4), 835–854.

Saouma, V.E., Barton, C.C. and Gamaleldin, N.A. (1990). Fractal characterization of fracture surfaces in concrete. *Engineering Fracture Mechanics* **35**, 47–53.

Sayles, R.S. and Thomas, T.R. (1978). Surface topography as a nonstationary random process. *Nature* **271**, 431–434.

Schmittbuhl, J., Schmitt, F. and Scholz, C. (1995). Scaling invariance of crack surfaces. *J. Geophys. Res.* **100**(B4), 5953–5973.

Sornette, A., Davy, P. and Sornette, D. (1990). Growth of fractal fault patterns. *Phys. Rev. Lett.* **65**, 2266–2269.

Tzschichholz, F. and Pfuff, M. (1991). Influence of crackpath-roughness on crack resistance in brittle materials. In: *Fracture processes in concrete, rock and ceramics. Vol. 1-2, Ch. 85* (Edited by J.G.M. Van Mier, J.G. Roth and A. Bakker), Cambridge University Press, Cambridge, 251–260.

Tricot, C. (1995). Curves and Fractal Dimension. Springer-Verlag, Berlin.

Tricot, C., Ferland P. and Baran, G. (1994). Fractal analysis of worn surfaces. *Wear* **172**, 127–133.

Vandembroucq, D and Roux, S. (1997a). Conformal mapping on rough boundaries. 1. Applications to harmonic problems. *Phys. Rev. E* **55**(5B), 6171–6185.

Vandembroucq, D and Roux, S. (1997b). Mode III stress intensity factor ahead of a rough crack. *Journal of Mechanics and Physics of Solids* **45**, 853–872.

Willis, J.R. (1967). A comparison of the fracture criteria of Griffith and Barenblatt. *Journal of Mechanics and Physics of Solids* **15**, 151–162.

Xie, H., Wang, J. and Qan, P. (1996). Fractal characteristics of micropore evolution in marbles. *Phys. Lett. A* **218**, 275–280.

Xie, H., Wang, J. and Stein, E. (1998). Direct fractal measurement and multifractal properties of fracture surfaces. *Phys. Lett. A* **242**, 41–50.

Zhao, Y., Huang, J. and Wang, R. (1993). Fractal characteristics of mesofractures in compressed rock specimens. *International Journal of Rock Mechanics and Minerals of Sciences* **30**, 877–882.

*International Journal of Fracture* **95:** 261–277, 1999.


# A unified framework for concrete damage and fracture models including size effects

RENÉ DE BORST and MIGUEL A. GUTIÉRREZ
*Delft University of Technology, Koiter Institute Delft, NL-2600 GB Delft, The Netherlands*

Received 13 November 1998; accepted in revised form 4 December 1998

**Abstract.** A unified approach is given for isotropic and anisotropic damage formulations for concrete fracture. The formulation encompasses the classical fixed and rotating smeared crack models, but also more refined approaches based on the microplane concept. Higher-order strain gradients are introduced to avoid the boundary value problem from becoming ill-posed at the onset of softening. Again, this has been accomplished in a unified setting. By analysing a geometrically identical anchor bolt of three different sizes it is shown that higher-order strain gradients also introduce a proper size effect in the model.

**Key words:** Concrete fracture, damage, smeared cracking, microplane model, localisation, regularisation, higher-order continua, strain gradients, size effect, anchor bolt.

## 1. Introduction

Computational modelling of concrete structures started in the late 1960s with the landmark papers of Ngo and Scordelis (1967) and Rashid (1968), in which the discrete and smeared crack approaches were introduced. The following three decades have seen many refinements, both for the discrete crack models and for the smeared crack models. The two major deficiencies of the discrete crack approach, namely that crack propagation was restricted to interelement boundaries and that it caused a continuous change in the topology of the mesh, were remedied by the introduction of automatic remeshing at the crack tip (Ingraffea and Saouma, 1985), and by a priori inserting interface elements in the finite element discretisation (Rots, 1991). However, neither of these solutions removes both disadvantages. Smeared crack models have proven to be more flexible in the sense that, in principle, arbitrary crack propagation, including curved cracks, can be simulated, since no topological constraints exists. Nevertheless, experience has shown that this advantage is less rigorous than it would seem, since it appeared difficult to simulate curved crack paths by smeared representations (de Borst, 1986; Rots, 1991; Feenstra, 1993; Feenstra and de Borst, 1995). Indeed, discrete approaches have proven to be more successful, either when using lattice type approaches (Herrmann et al., 1989; van Mier et al., 1994), or when adopting finite element methods with remeshing, e.g. Ingraffea and Saouma (1985), who used pure linear elastic fracture mechanics, or van Vroonhoven and de Borst (1997), who employed a hybrid approach, incorporating concepts of fracture mechanics as well as damage mechanics.

In the early 1990s it became apparent that the failure of the smeared approach to properly predict curved crack paths is rooted in the fact that a smeared concept inevitably introduces *strain* softening into the constitutive model, which at a certain level of loading causes a loss of well-posedness of the incremental boundary value problem. This ill-posedness creates an infinite number of solutions (Benallal et al., 1988; de Borst et al., 1993), from which a numerical method selects the solution with the smallest energy dissipation that is available in the finite dimensional solution space. In the limit of an infinitely dense mesh, solutions are computed which predict failure without energy dissipation, thus rendering the solution physically meaningless. Early solutions as fracture-energy models in their various forms (Pietruszczak and Mróz, 1981; Bažant and Oh, 1983; Willam, 1984), provide a partial solution for the mesh densification problem, but fail to repair the mesh bias issue, i.e. they still predict crack propagation along the direction of the grid lines.

A rigorous solution is the introduction of higher-order continuum models. The first models that were applied to fracture in concrete were nonlocal damage models (Bažant, 1984; Pijaudier-Cabot and Bažant, 1987; Bažant and Pijaudier-Cabot, 1988) and gradient plasticity models (de Borst and Mühlhaus, 1992; de Borst and Pamin, 1996). Fully nonlocal approaches, in which spatially averaged quantities are employed in the constitutive models, are computationally unwieldy and are not believed to have potential for large-scale computations of concrete structures. The gradient approaches are more promising, but the gradient plasticity model suffers from the drawback that there is an internal boundary between the elastic and plastic domain, which necessitates a smooth interpolation of the plastic strain field. The required $C^1$-continuity is believed to reduce the ability of the gradient plasticity model to simulate curved crack propagation accurately, although globally proper directions of crack propagation were computed, independent of the discretisation. Gradient damage approaches, first introduced in a computationally feasible format by Peerlings et al., (1996), do not necessarily require a higher-order continuity of the interpolants of the strain field or the damage field and, as was recently shown by Peerlings et al. (1998), are capable of simulating curved cracks.

We shall start this article by local, isotropic damage formulations, and then extend the formulation to anisotropic models, including various forms of smeared crack concepts, such as the fixed crack model and the rotating crack model (Cope et al., 1980), and the microplane models (Bažant and Gambarova, 1984). Then, starting from fully nonlocal damage models, we will develop isotropic and anisotropic gradient damage models. Specifically, we will derive a gradient smeared crack model, and we will indicate how gradient microplane models can be developed (Kuhl et al., 1998). The contribution will be concluded by the example of an anchor bolt, which, using the isotropic gradient damage model of Peerlings et al. (1996), has been analysed for three different dimensions. The results exhibit a clear size effect, which compares favourably with experimental results (Van Mier, 1996).

## 2. Standard damage models

The basic structure of constitutive models that are set up in the spirit of damage mechanics is simple. We have a total stress-strain relation,

$$\sigma_{ij} = D^s_{ijkl}(q_m, \omega_{mn}, \Omega_{mnop}, \ldots)\varepsilon_{kl}, \tag{1}$$

where $\sigma_{ij}$ is the stress tensor, $\varepsilon_{kl}$ is the strain tensor and $D^s_{ijkl}$ is a secant stiffness tensor, which can depend on a number of internal variables, like scalar-valued internal variables $q_m$, and tensor-valued internal variables like $\omega_{mn}$, $\Omega_{mnop}$ and so on. Relation (1) distinguishes itself from classical nonlinear elasticity by a history dependence, which is incorporated via a loading-unloading function, $f$, which vanishes upon loading, and is negative otherwise. For damage growth, $f$ must remain zero for an infinitesimal period, so that we have the additional requirement that $\dot{f} = 0$. The theory is completed by specifying the appropriate (material-dependent) evolution equations for the internal variables.

### 2.1. Isotropic damage models

For the case of isotropic damage evolution, the total stress-strain relation (1) specialises as

$$\sigma_{ij} = [(1-\omega_1)G(\delta_{ik}\delta_{jl} + \delta_{il}\delta_{jk} - 2/3\delta_{ij}\delta_{kl}) + (1-\omega_2)K\delta_{ij}\delta_{kl}]\varepsilon_{kl}, \tag{2}$$

with $G$ the virgin shear modulus and $K$ the virgin bulk modulus, which are degraded by the scalar damage variables $\omega_1$ and $\omega_2$, respectively (cf. Ladeveze, 1983). A further simplification can be achieved if it is assumed that the secant shear stiffness and bulk moduli, $(1-\omega_1)G$ and $(1-\omega_2)K$, degrade in the same manner during damage growth. Essentially, this means that Poisson's ratio $\nu$ remains constant throughout the damage process and we have

$$\sigma_{ij} = (1-\omega)D^{\mathrm{e}}_{ijkl}\varepsilon_{kl}, \tag{3}$$

with $\omega$ the damage variable which grows from zero to one (at complete loss of integrity) and $D^{\mathrm{e}}_{ijkl}$ the fourth-order elastic stiffness tensor. The total stress-strain relation (3) is again complemented by a damage loading function $f$, which now reads: $f = f(\tilde{\varepsilon}, \tilde{\sigma}, \kappa)$, with $\tilde{\varepsilon}$ and $\tilde{\sigma}$ scalar-valued functions of the strain and stress tensors, respectively, and $\kappa$ the only remaining scalar history variable. The damage loading function $f$ and the rate of the history variable, $\dot{\kappa}$, have to satisfy the discrete Kuhn-Tucker loading-unloading conditions

$$f \leqslant 0, \qquad \dot{\kappa} \geqslant 0, \qquad f\dot{\kappa} = 0. \tag{4}$$

In this contribution we consider the case that the damage loading function does not depend on $\tilde{\sigma}$. Then,

$$f(\tilde{\varepsilon}, \kappa) = \tilde{\varepsilon} - \kappa. \tag{5}$$

For metals a commonly accepted choice for $\tilde{\varepsilon}$ is (Lemaitre and Chaboche, 1990)

$$\tilde{\varepsilon} = \tfrac{1}{2}\varepsilon_{ij}D^{\mathrm{e}}_{ijkl}\varepsilon_{kl}. \tag{6}$$

Equation (6) is less convenient from the point of view that it does not reduce to the uniaxial strain for uniaxial stressing. For this reason it is sometimes replaced by the modified expression

$$\tilde{\varepsilon} = \sqrt{\frac{1}{E}\varepsilon_{ij}D^{\mathrm{e}}_{ijkl}\varepsilon_{kl}}, \tag{7}$$

with $E$ the virgin Young's modulus. Expression (7) is represented graphically in the principal strain space for plane-stress conditions in Figure 1(a). In this figure, a scaling has been applied such that $\tilde{\varepsilon} = 1$, while $\nu = 0.2$. The dashed lines are uniaxial stress paths. The above energy release rate definition for $\tilde{\varepsilon}$ gives equal weights to tensile and compressive strain components, which makes it unsuitable to describe the mechanical behaviour of quasi-brittle materials like

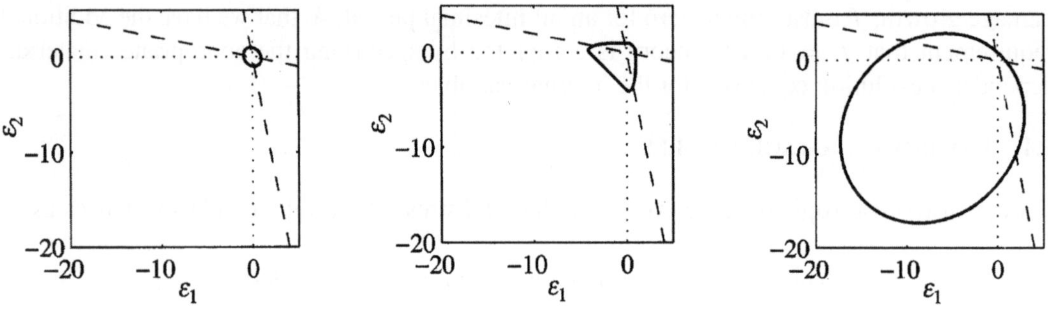


*Figure 1.* Contour plots for $\tilde{\varepsilon}$ for the local energy release concept (left), the Mazars definition (centre) and the Modified von Mises definition (right), after Peerlings et al. (1998).

concrete and rock. To remedy this deficiency, Mazars (1984) and Mazars and Pijaudier-Cabot (1989) have suggested the definition

$$\tilde{\varepsilon} = \sqrt{\sum_{i=1}^{3} (\langle \varepsilon_i \rangle)^2}, \tag{8}$$

with $\varepsilon_i$ the principal strains, and $\langle \varepsilon_i \rangle = \varepsilon_i$ if $\varepsilon_i > 0$ and $\langle \varepsilon_i \rangle = 0$ otherwise. A contour plot for $\tilde{\varepsilon} = 1$ is given in Figure 1(b). A third definition for the equivalent strain has been proposed by de Vree et al. (1995). This proposition, which can be named a Modified von Mises definition, is given by

$$\tilde{\varepsilon} = \frac{k-1}{2k(1-\nu)} \mathrm{I}_1 + \frac{1}{2k} \sqrt{\frac{(k-1)^2}{(1-2\nu)^2} \mathrm{I}_1^2 + \frac{6k}{(1+\nu)^2} J_2}, \tag{9}$$

with $\mathrm{I}_1 = \varepsilon_{kk}$ the first invariant of the strain tensor and $J_2 = e_{ij} e_{ij}$ the second invariant of the deviatoric strain tensor $e_{ij} = \varepsilon_{ij} - \frac{1}{3}\varepsilon_{kk}$. The parameter $k$ governs the sensitivity to the compressive strain components relative to the tensile strain components. The definition of $\tilde{\varepsilon}$ is such that a compressive uniaxial stress $k\sigma$ has the same effect as a uniaxial tensile stress $\sigma$. The parameter $k$ is therefore typically set equal to the ratio of the compressive uniaxial strength and the tensile uniaxial strength: $k = f_c / f_t$. A graphical representation of the Modified von Mises definition is given in Figure 1(c) for $k = 10$.

The history parameter $\kappa$ starts at a damage threshold level $\kappa_i$ and is updated by the requirement that during damage growth $f = 0$. Damage growth occurs according to an evolution law such that $\omega = \omega(\kappa)$, which can be determined from a uniaxial test. For instance, in the size effect simulations of the anchor bolt in the final part of this contribution a relation has been used which shows an elastic behaviour up to the peak tensile strength $f_t = E\kappa_i$, followed by an exponential decay

$$\omega(\kappa) = 1 - \frac{\kappa_i}{\kappa}(1 - a + a\,\mathrm{e}^{-b(\kappa - \kappa_i)}) \tag{10}$$

### 2.2. ANISOTROPIC DAMAGE MODELS

While isotropic damage models have been used successfully for describing progressive crack propagation, their disadvantage is that possible compressive strut action is eliminated. This is a drawback especially for the analysis of reinforced concrete members. Directional dependence of damage evolution can be incorporated by degrading the Young's modulus $E$ in a preferential direction. When, for plane-stress conditions, distinction is made between the global $x$, $y$-coordinate system and a local $n$, $s$-coordinate system a simple loading function in the local coordinate system would be

$$f(\varepsilon_{nn}, \kappa) = \varepsilon_{nn} - \kappa, \tag{11}$$

with $\varepsilon_{nn}$ the normal strain in the local $n$, $s$-coordinate system, subject to the standard Kuhn-Tucker loading-unloading conditions. The secant stiffness relation now reads

$$\boldsymbol{\sigma}_{ns} = \mathbf{D}^s_{ns}\boldsymbol{\varepsilon}_{ns}, \tag{12}$$

with $\boldsymbol{\sigma}_{ns} = [\sigma_{nn}, \sigma_{ss}, \sigma_{ns}]^T$, $\boldsymbol{\varepsilon}_{ns} = [\varepsilon_{nn}, \varepsilon_{ss}, \gamma_{ns}]^T$ and $\mathbf{D}^s_{ns}$ defined as

$$\mathbf{D}^s_{ns} = \begin{bmatrix} \frac{(1-\omega_1)E}{1-(1-\omega_1)\nu^2} & \frac{(1-\omega_1)\nu E}{1-(1-\omega_1)\nu^2} & 0 \\ \frac{(1-\omega_1)\nu E}{1-(1-\omega_1)\nu^2} & \frac{E}{1-(1-\omega_1)\nu^2} & 0 \\ 0 & 0 & (1-\omega_2)G \end{bmatrix}, \tag{13}$$

with $\omega_1 = \omega_1(\kappa)$ and $\omega_2 = \omega_2(\kappa)$. The factor $1 - \omega_2$ represents the degradation of the shear stiffness and can be identified with the traditional shear retention factor $\beta$ (Suidan and Schnobrich, 1973). It is emphasised that because of the choice of a preferential direction in which damage takes place, the damage variables $\omega_1$ and $\omega_2$ have an entirely different meaning than those that were introduced in the basic isotropic formulation of (2).

If we introduce $\phi$ as the angle from the $x$-axis to the $n$-axis, we can relate the components of $\boldsymbol{\varepsilon}_{ns}$ and $\boldsymbol{\sigma}_{ns}$ to those in the global $x$, $y$-coordinate system via the standard tranformation matrices $\mathbf{T}_\varepsilon$ and $\mathbf{T}_\sigma$

$$\boldsymbol{\varepsilon}_{ns} = \mathbf{T}_\varepsilon(\phi)\boldsymbol{\varepsilon}_{xy} \tag{14}$$

and

$$\boldsymbol{\sigma}_{ns} = \mathbf{T}_\sigma(\phi)\boldsymbol{\sigma}_{xy}. \tag{15}$$

With aid of (14) and (15) the damage loading function (11) can be written in terms of the strain components $\varepsilon_{xx}$, $\varepsilon_{yy}$ and $\gamma_{xy}$ of the global $x$, $y$-coordinate system

$$f = \varepsilon_{xx}\cos^2\phi + \varepsilon_{yy}\sin^2\phi + \gamma_{xy}\sin\phi\cos\phi - \kappa. \tag{16}$$

Similarly, we obtain for the secant stress-strain relation instead of (12)

$$\boldsymbol{\sigma}_{xy} = \mathbf{T}_\sigma^{-1}(\phi)\mathbf{D}^s_{ns}\mathbf{T}_\varepsilon(\phi)\boldsymbol{\varepsilon}_{xy}. \tag{17}$$

Equations (16) and (17) incorporate the traditional fixed crack model and the rotating crack model. The only difference is that in the fixed crack model the inclination angle $\phi$ is fixed when the major principal stress first reaches the tensile strength ($\phi = \phi_0$), while in the rotating crack concept $\phi$ changes such that the $n$-axis continues to coincide with the major principal strain direction, and, through the coaxiality requirement, also with the major principal stress direction. In fact, the enforcement of coaxiality has profound consequences when deriving the tangential stiffness, especially with regard to the shear term.

The above framework also allows for incorporation of constitutive models that are based on the microplane concept. As an example we shall consider a microplane model based on the so-called kinematic constraint, which implies that the normal and tangential strains on a microplane that is labelled $\alpha$, can be derived by a simple projection of the global strains $\boldsymbol{\varepsilon}_{xy}$ similar to (14)

$$\boldsymbol{\varepsilon}^{\alpha}_{ns} = \mathbf{T}_{\varepsilon}(\phi^{\alpha})\boldsymbol{\varepsilon}_{xy}. \tag{18}$$

The stresses on this microplane can be derived in a fashion similar to (12)

$$\boldsymbol{\sigma}^{\alpha}_{ns} = \mathbf{D}^{\alpha}_{ns}\boldsymbol{\varepsilon}^{\alpha}_{ns}, \tag{19}$$

with $\mathbf{D}^{\alpha}_{ns}$ given by

$$\mathbf{D}^{\alpha}_{ns} = \begin{bmatrix} (1-\omega^{\alpha}_{N})E_N & 0 & 0 \\ 0 & 0 & 0 \\ 0 & 0 & (1-\omega^{\alpha}_{T})E_T \end{bmatrix}, \tag{20}$$

where the initial stiffness moduli $E_N$ and $E_T$ are functions of the Young's modulus, the Poisson's ratio and a weight parameter, see e.g. Bažant and Prat (1988). The damage parameters $\omega^{\alpha}_{N}$ and $\omega^{\alpha}_{T}$ for the normal stiffness and the shear stiffness are functions of the history parameters $\kappa^{\alpha}_{N}$ and $\kappa^{\alpha}_{T}$ in a standard fashion: $\omega^{\alpha}_{N} = \omega^{\alpha}_{N}(\kappa^{\alpha}_{N})$ and $\omega^{\alpha}_{T} = \omega^{\alpha}_{T}(\kappa^{\alpha}_{T})$. The main departure from the fixed crack model as outlined above is the fact that we need two damage loading functions on microplane $\alpha$

$$f^{\alpha}_{N} = \varepsilon^{\alpha}_{nn} - \kappa^{\alpha}_{N} \tag{21}$$

and

$$f^{\alpha}_{T} = \gamma^{\alpha}_{ns} - \kappa^{\alpha}_{T}, \tag{22}$$

each subject to the standard Kuhn-Tucker loading-unloading conditions. Finally, the stresses in the global $x$, $y$ coordinate system are recovered by summing over all the microplanes and by transforming them in a standard fashion according to (15). With (18) and (19) we finally arrive at

$$\boldsymbol{\sigma}_{xy} = \sum_{\alpha=1}^{n} w^{\alpha}\mathbf{T}^{-1}_{\sigma}(\phi^{\alpha})\mathbf{D}^{\alpha}_{ns}\mathbf{T}_{\varepsilon}(\phi^{\alpha})\boldsymbol{\varepsilon}_{xy}, \tag{23}$$

with $n$ the chosen number of microplanes and $w^{\alpha}$ the weighting factors.

Attention is drawn to the fact that the second row of $\mathbf{D}^{\alpha}_{ns}$ consists of zeros. This is because in the microplane concept only the normal stress and the shear stress are resolved on each microplane. The normal stress parallel to this plane therefore becomes irrelevant. Furthermore, attention is drawn to the fact that here we use a relatively simple version of the microplane model, namely one in which no splitting in volumetric and deviatoric components is considered (Bažant and Gambarova, 1984). Nevertheless, more sophisticated microplane models, e.g. that by Bažant and Prat (1988), which incorporate such a split, can be captured by the same formalism (Kuhl and Ramm, 1998). It is finally noted that the microplane is very similar to the multiple fixed-crack model (de Borst and Nauta, 1985; de Borst, 1987).

For the fixed crack model differentiation of (17) yields the tangential stress-strain relation needed in an incremental-iterative procedure which utilises the Newton–Raphson method

$$\dot{\boldsymbol{\sigma}}_{xy} = \mathbf{T}_{\sigma}^{-1}(\phi_0)(\mathbf{D}^{s}_{ns} - \Delta\mathbf{D}_{ns})\mathbf{T}_{\varepsilon}(\phi_0)\dot{\boldsymbol{\varepsilon}}_{xy}, \tag{24}$$

with $\mathbf{D}^{s}_{ns}$ given by (13) and

$$\Delta\mathbf{D}_{ns} = \begin{bmatrix} d_{11} & 0 & 0 \\ \nu d_{11} & 0 & 0 \\ d_{31} & 0 & 0 \end{bmatrix}, \tag{25}$$

with

$$d_{11} = \frac{\partial\omega_1}{\partial\kappa}\frac{\partial\kappa}{\partial\varepsilon_{nn}}\frac{E(\varepsilon_{nn}+\nu\varepsilon_{ss})}{(1-(1-\omega_1)\nu^2)^2}$$

and

$$d_{31} = \frac{\partial\omega_2}{\partial\kappa}\frac{\partial\kappa}{\partial\varepsilon_{nn}}G\gamma_{ns},$$

$\partial\kappa/\partial\varepsilon_{nn} = 1$ upon loading and zero otherwise. We observe that the local material tangential stiffness matrix $\mathbf{D}^{s}_{ns} - \Delta\mathbf{D}_{ns}$ generally becomes non-symmetric.

The fact that in the rotating smeared crack model the local coordinate system of the crack and the principal axes of stress and strain coincide throughout the entire deformation process implies that the secant stiffness matrix $\mathbf{D}^{s}_{ns}$ relates principal stresses to principal strains, and that a secant shear stiffness becomes superfluous. Consequently, there is only one remaining damage parameter, $\omega_1 = \omega$, and we have

$$\mathbf{D}^{s}_{ns} = \begin{bmatrix} \frac{(1-\omega)E}{1-(1-\omega)\nu^2} & \frac{(1-\omega)\nu E}{1-(1-\omega)\nu^2} & 0 \\ \frac{(1-\omega)\nu E}{1-(1-\omega)\nu^2} & \frac{E}{1-(1-\omega)\nu^2} & 0 \\ 0 & 0 & 0 \end{bmatrix} \tag{26}$$

instead of expression (13). With (26) differentiation of the secant stiffness relation in global coordinates yields (Bažant, 1983; Willam et al., 1986; Feenstra, 1993)

$$\dot{\boldsymbol{\sigma}}_{xy} = \mathbf{T}_{\sigma}^{-1}(\phi)(\mathbf{D}^{s}_{ns} - \Delta\mathbf{D}_{ns})\mathbf{T}_{\varepsilon}(\phi)\dot{\boldsymbol{\varepsilon}}_{xy}, \tag{27}$$

with $\Delta\mathbf{D}_{ns}$ now given by

$$\Delta\mathbf{D}_{ns} = \begin{bmatrix} d_{11} & 0 & 0 \\ \nu d_{11} & 0 & 0 \\ 0 & 0 & d_{33} \end{bmatrix}, \tag{28}$$

with

$$d_{11} = \frac{\partial\omega}{\partial\kappa}\frac{\partial\kappa}{\partial\varepsilon_{nn}}\frac{E(\varepsilon_{nn} + \nu\varepsilon_{ss})}{(1-(1-\omega)\nu^2)^2}$$

and

$$d_{33} = -\frac{\sigma_{nn} - \sigma_{ss}}{2(\varepsilon_{nn} - \varepsilon_{ss})}.$$

The tangential stress-strain relation of the microplane model can be cast in the same formalism as that of the rotating and the fixed crack models. Indeed, upon linearisation of (23) we obtain

$$\dot{\boldsymbol{\sigma}}_{xy} = \sum_{\alpha=1}^{n} w^\alpha \mathbf{T}_\sigma^{-1}(\phi^\alpha)(\mathbf{D}^\alpha_{ns} - \Delta\mathbf{D}^\alpha_{ns})\mathbf{T}_\varepsilon(\phi^\alpha)\dot{\boldsymbol{\varepsilon}}_{xy}, \tag{29}$$

with $\mathbf{D}^\alpha_{ns}$ given by (20) and $\Delta\mathbf{D}_{ns}$ defined as

$$\Delta\mathbf{D}^\alpha_{ns} = \begin{bmatrix} d^\alpha_{11} & 0 & 0 \\ 0 & 0 & 0 \\ 0 & 0 & d^\alpha_{33} \end{bmatrix}, \tag{30}$$

with

$$d^\alpha_{11} = \frac{\partial\omega^\alpha_N}{\partial\kappa^\alpha_N}\frac{\partial\kappa^\alpha_N}{\partial\varepsilon^\alpha_{nn}}E_N\varepsilon^\alpha_{nn},$$

where $\partial\kappa^\alpha_N/\partial\varepsilon^\alpha_{nn} = 1$ if $f^\alpha_N = 0$ and zero otherwise, and

$$d^\alpha_{33} = \frac{\partial\omega^\alpha_T}{\partial\kappa^\alpha_T}\frac{\partial\kappa^\alpha_T}{\partial\gamma^\alpha_{ns}}E_T\gamma^\alpha_{ns},$$

while $\partial\kappa^\alpha_T/\partial\gamma^\alpha_{ns} = 1$ if $f^\alpha_T = 0$ and zero otherwise.

## 3. Enhanced damage models

In a nonlocal generalisation the equivalent strain $\tilde{\varepsilon}$ is normally replaced by a spatially averaged quantity in the damage loading function (Pijaudier-Cabot and Bažant, 1987; Bažant and Pijaudier-Cabot, 1988)

$$f(\bar{\varepsilon}, \kappa) = \bar{\varepsilon} - \kappa, \tag{31}$$

where the nonlocal strain $\bar{\varepsilon}$ is computed from

$$\bar{\varepsilon}(\mathbf{x}) = \frac{1}{V_r(\mathbf{x})} \int_V \psi(\mathbf{s}) \tilde{\varepsilon}(\mathbf{x}+\mathbf{s})\,\mathrm{d}V, \qquad V_r(\mathbf{x}) = \int_V \psi(\mathbf{s})\,\mathrm{d}V, \tag{32}$$

with $\psi(\mathbf{s})$ a weight function, e.g., the error function, and $\mathbf{s}$ a relative position vector pointing to the infinitesimal volume $\mathrm{d}V$. In this formulation all the other relations remain local: the Kuhn-Tucker loading-unloading conditions (4), the local stress-strain relation (3) and the dependence of the damage variable $\omega$ on the history variable $\kappa$.

Alternatively, the locally defined history parameter $\kappa$ may be replaced in the damage loading function $f$ by a spatially averaged quantity

$$f(\tilde{\varepsilon}, \bar{\kappa}) = \tilde{\varepsilon} - \bar{\kappa}, \tag{33}$$

where the nonlocal history parameter $\bar{\kappa}$ follows from

$$\bar{\kappa}(\mathbf{x}) = \frac{1}{V_r(\mathbf{x})} \int_V \psi(\mathbf{s}) \kappa(\mathbf{x}+\mathbf{s})\,\mathrm{d}V, \tag{34}$$

where $\psi$ typically has a different form than the weight function in (32). This alternative formulation requires some slight modifications with respect to some of the other governing equations. The local stress-strain relation (3) and the damage evolution law $\omega = \omega(\kappa)$ remain unaltered, but the Kuhn-Tucker loading-unloading conditions must be rephrased as

$$f \leqslant 0, \qquad \dot{\bar{\kappa}} \geqslant 0, \qquad f\dot{\bar{\kappa}} = 0. \tag{35}$$

### 3.1. Isotropic gradient damage models

Nonlocal constitutive relations can be considered as a point of departure for constructing gradient models, although we wish to emphasise that the latter class of models can also be defined directly by supplying higher-order gradients in the damage loading function. Yet, we will follow the first-mentioned route to underline the connection between integral and differential type nonlocal models. Again, this can either be done by expanding the kernel $\tilde{\varepsilon}$ of the integral in (32) in a Taylor series, or by expanding the history parameter $\kappa$ in (34) in a Taylor series. We will first consider the expansion of $\tilde{\varepsilon}$ and then we will do the same for $\kappa$. If we truncate after the second-order terms and carry out the integration implied in (32) under the assumption of isotropy, the following relation ensues

$$\bar{\varepsilon} = \tilde{\varepsilon} + g\nabla^2\tilde{\varepsilon}, \tag{36}$$

where $g$ is a material parameter of the dimension length squared. It can be related to the averaging volume and then becomes dependent on the precise form of the weight function $\psi$. For instance, for a one-dimensional continuum and taking

$$\psi(s) = \frac{1}{\sqrt{2\pi}\,l} \mathrm{e}^{-s^2/2l^2}, \tag{37}$$

we obtain $g = \frac{1}{2}l^2$. Here, we adopt the phenomenological view that $\sqrt{g}$ reflects the length scale of the failure process which we wish to describe macroscopically.

Formulation (36) has a severe disadvantage when applied in a finite element context, namely that it requires computation of second-order gradients of the local equivalent strain $\tilde{\varepsilon}$. Since this quantity is a function of the strain tensor, and since the strain tensor involves first-order derivatives of the displacements, third-order derivatives of the displacements have to be computed, which would necessitate $C^1$-continuity of the shape functions. To obviate this problem, (36) is differentiated twice and the result is substituted again into (36). Again neglecting fourth-order terms then leads to

$$\bar{\varepsilon} - g\nabla^2\bar{\varepsilon} = \tilde{\varepsilon}. \tag{38}$$

When $\bar{\varepsilon}$ is discretised independently and use is made of the divergence theorem, a $C^0$-interpolation for $\bar{\varepsilon}$ suffices (Peerlings et al., 1996).

Higher-order continua require additional boundary conditions. With (38) governing the damage process, either the averaged equivalent strain $\bar{\varepsilon}$ itself or its normal derivative must be specified on the boundary $S$ of the body

$$\bar{\varepsilon} = \bar{\varepsilon}_s \quad \text{or} \quad \mathbf{n}^T\nabla\bar{\varepsilon} = \bar{\varepsilon}_{ns}, \tag{39}$$

with $\mathbf{n}$ the outward normal vector to the boundary of the body $S$. Typically, the natural boundary condition $\bar{\varepsilon}_{ns} = 0$ is adopted in numerical calculations (Peerlings et al., 1996; 1999).

In a fashion similar to the derivation of the gradient damage models based on the averaging of the equivalent strain $\tilde{\varepsilon}$, we can elaborate a gradient approximation of (34), i.e., by developing $\kappa$ into a Taylor series. For an isotropic, infinite medium and truncating after the second term we then have (de Borst et al., 1996)

$$\bar{\kappa} = \kappa + g\nabla^2\kappa. \tag{40}$$

Since the weight functions for the different gradient formulations may be quite different, also the gradient parameter $g$ may have different values for the various formulations. The additional boundary conditions now apply to $\kappa$. Although formally quite similar to those of (39), namely

$$\kappa = \kappa_s \quad \text{or} \quad \mathbf{n}^T\nabla\kappa = \kappa_{ns}, \tag{41}$$

they have a different character, since they apply to an internal variable instead of to a kinematic quantity, which seems somewhat suspect. On the other hand, the physical interpretation that can be given to the boundary condition ($41_2$) is rather clear. Since the damage variable $\omega$ is a function of the internal variable $\kappa$, and therefore, the differential equation (40) and the boundary conditions (41) can be replaced by (e.g., de Borst et al., 1996; Comi, 1999)

$$\bar{\omega} = \omega + g\nabla^2\omega, \tag{42}$$

where $\bar{\omega}$ is a spatially averaged damage field, similar to $\bar{\varepsilon}$ or $\bar{\kappa}$, and the corresponding boundary conditions

$$\omega = \omega_s \quad \text{or} \quad \mathbf{n}^T\nabla\omega = \omega_{ns}. \tag{43}$$

Equation ($43_2$) with $\omega_{ns} = 0$ can then be identified as a condition of no damage flux through the boundary $S$ of the body, see also Mühlhaus et al. (1994).

### 3.2. Anisotropic gradient damage models

Anisotropic gradient damage models can be developed in a manner similar to isotropic gradient damage models. We take as point of departure the damage loading function (11) in the local $n, s$-coordinate system, where we replace the normal strain $\varepsilon_{nn}$ by its nonlocal equivalent

$$f(\bar{\varepsilon}_{nn}, \kappa) = \bar{\varepsilon}_{nn} - \kappa, \tag{44}$$

or, identically

$$f = \bar{\varepsilon}_{xx} \cos^2 \phi + \bar{\varepsilon}_{yy} \sin^2 \phi + \bar{\gamma}_{xy} \sin \phi \cos \phi - \kappa, \tag{45}$$

instead of (16). We can now apply the averaging process to each of the (nonlocal) strain components $\bar{\varepsilon}_{xx}$, $\bar{\varepsilon}_{yy}$ and $\bar{\gamma}_{xy}$. In the spirit of (38) they have to satisfy

$$\begin{aligned} \bar{\varepsilon}_{xx} - g\nabla^2\bar{\varepsilon}_{xx} &= \varepsilon_{xx}, \\ \bar{\varepsilon}_{yy} - g\nabla^2\bar{\varepsilon}_{yy} &= \varepsilon_{yy}, \\ \bar{\gamma}_{xy} - g\nabla^2\bar{\gamma}_{xy} &= \gamma_{xy}, \end{aligned} \tag{46}$$

or written in a vector format

$$\bar{\boldsymbol{\varepsilon}}_{xy} - g\nabla^2\bar{\boldsymbol{\varepsilon}}_{xy} = \boldsymbol{\varepsilon}_{xy}, \tag{47}$$

with the boundary conditions

$$\bar{\boldsymbol{\varepsilon}}_{xy} = \bar{\boldsymbol{\varepsilon}}_s \quad \text{or} \quad \mathbf{n}^T \nabla \bar{\boldsymbol{\varepsilon}}_{xy} = \bar{\boldsymbol{\varepsilon}}_{ns}. \tag{48}$$

After solution of the set of Helmholtz equations (47), $\bar{\boldsymbol{\varepsilon}}_{xy}$ is substituted into the damage loading function (45) to determine whether loading occurs and next into the secant stress-strain relation to solve for the stresses $\boldsymbol{\sigma}_{xy}$.

The tangential stress-strain relation attains a slightly different format due to the fact that the nonlocal strain components must be discretised independently. For a gradient-enhanced version of the fixed crack model, we obtain upon linearisation of (17)

$$\dot{\boldsymbol{\sigma}}_{xy} = \mathbf{T}_\sigma^{-1}(\phi_0)\mathbf{D}_{ns}^s\mathbf{T}_\varepsilon(\phi_0)\dot{\boldsymbol{\varepsilon}}_{xy} - \mathbf{T}_\sigma^{-1}(\phi_0)\Delta\mathbf{D}_{ns}\mathbf{T}_\varepsilon(\phi_0)\dot{\bar{\boldsymbol{\varepsilon}}}_{xy}, \tag{49}$$

with

$$\Delta\mathbf{D}_{ns} = \begin{bmatrix} d_{11} & 0 & 0 \\ \nu d_{11} & 0 & 0 \\ d_{31} & 0 & 0 \end{bmatrix}, \tag{50}$$

with

$$d_{11} = \frac{\partial\omega_1}{\partial\kappa}\frac{\partial\kappa}{\partial\bar{\varepsilon}_{nn}}\frac{E(\varepsilon_{nn} + \nu\varepsilon_{ss})}{(1 - (1 - \omega_1)\nu^2)^2}$$

and

$$d_{31} = \frac{\partial \omega_2}{\partial \kappa} \frac{\partial \kappa}{\partial \bar{\varepsilon}_{nn}} G \gamma_{ns},$$

$\partial \kappa / \partial \bar{\varepsilon}_{nn} = 1$ upon loading and zero otherwise.

For the gradient-enhanced rotating crack model, we obtain upon linearisation of (17)

$$\dot{\boldsymbol{\sigma}}_{xy} = \mathbf{T}_\sigma^{-1}(\phi) \mathbf{D}_{ns}^s \mathbf{T}_\varepsilon(\phi) \dot{\boldsymbol{\varepsilon}}_{xy} - \mathbf{T}_\sigma^{-1}(\phi) \Delta \mathbf{D}_{ns} \mathbf{T}_\varepsilon(\phi) \dot{\bar{\boldsymbol{\varepsilon}}}_{xy}, \tag{51}$$

but now $\Delta \mathbf{D}_{ns}$ is given by

$$\Delta \mathbf{D}_{ns} = \begin{bmatrix} d_{11} & 0 & 0 \\ \nu d_{11} & 0 & 0 \\ 0 & 0 & d_{33} \end{bmatrix}, \tag{52}$$

with

$$d_{11} = \frac{\partial \omega}{\partial \kappa} \frac{\partial \kappa}{\partial \bar{\varepsilon}_{nn}} \frac{E(\varepsilon_{nn} + \nu \varepsilon_{ss})}{(1 - (1 - \omega)\nu^2)^2}$$

and

$$d_{33} = -\frac{\sigma_{nn} - \sigma_{ss}}{2(\varepsilon_{nn} - \varepsilon_{ss})}.$$

In a gradient-enhanced microplane model (Kuhl et al., 1998) the single damage loading condition (44), or equivalently (45), must be replaced by two loading functions for each microplane

$$f_N = \bar{\varepsilon}_{nn}^\alpha - \kappa_N^\alpha, \tag{53}$$

$$f_T = \bar{\gamma}_{ns}^\alpha - \kappa_T^\alpha, \tag{54}$$

where $\bar{\varepsilon}_{nn}^\alpha$ is obtained by solving (47) for $\bar{\varepsilon}_{xx}$, $\bar{\varepsilon}_{yy}$ and $\bar{\varepsilon}_{xy}$ followed by a transformation to the coordinate system of microplane $\alpha$ via the matrix $\mathbf{T}_\varepsilon(\phi^\alpha)$. The tangential stiffness relation can then be derived to be of a form similar to the fixed and rotating crack models. Indeed, upon linearisation of (23) we obtain

$$\dot{\boldsymbol{\sigma}}_{xy} = \sum_{\alpha=1}^{n} w^\alpha [\mathbf{T}_\sigma^{-1}(\phi^\alpha) \mathbf{D}_{ns}^\alpha \mathbf{T}_\varepsilon(\phi^\alpha) \dot{\boldsymbol{\varepsilon}}_{xy} - \mathbf{T}_\sigma^{-1}(\phi^\alpha) \Delta \mathbf{D}_{ns}^\alpha \mathbf{T}_\varepsilon(\phi^\alpha) \dot{\bar{\boldsymbol{\varepsilon}}}_{xy}], \tag{55}$$

with $\mathbf{D}_{ns}^\alpha$ given by (20) and $\Delta \mathbf{D}_{ns}^\alpha$ defined as

$$\Delta \mathbf{D}_{ns}^\alpha = \begin{bmatrix} d_{11}^\alpha & 0 & 0 \\ 0 & 0 & 0 \\ 0 & 0 & d_{33}^\alpha \end{bmatrix}, \tag{56}$$

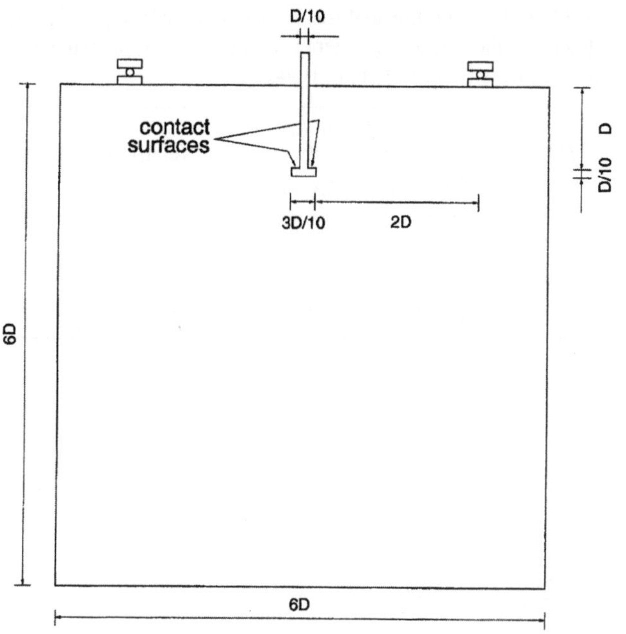


*Figure 2.* Geometry of pull-out test of anchor bolt.

with

$$d_{11}^{\alpha} = \frac{\partial \omega_N^{\alpha}}{\partial \kappa_N^{\alpha}} \frac{\partial \kappa_N^{\alpha}}{\partial \bar{\varepsilon}_{nn}^{\alpha}} E_N \varepsilon_{nn}^{\alpha},$$

with $\partial \kappa_N^{\alpha} / \partial \bar{\varepsilon}_{nn}^{\alpha} = 1$ if $f_N^{\alpha} = 0$ and zero otherwise, and

$$d_{33}^{\alpha} = \frac{\partial \omega_T^{\alpha}}{\partial \kappa_T^{\alpha}} \frac{\partial \kappa_T^{\alpha}}{\partial \bar{\gamma}_{ns}^{\alpha}} E_T \gamma_{ns}^{\alpha}, \tag{57}$$

$\partial \kappa_T^{\alpha} / \partial \bar{\gamma}_{ns}^{\alpha} = 1$ if $f_T^{\alpha} = 0$ and zero otherwise.

## 4. Size effects in gradient models: A numerical example

Anchor bolts embedded in concrete serve well to investigate the size effect phenomenon. The pull-out test that is considered here was proposed by RILEM committee 90FMA as a round-robin analysis to compare the predictive capabilities of the different models with respect to experiments, in particular size effects. The geometry considered in this study consists of a $T$-shaped steel anchor embedded in a concrete specimen of thickness $b = 100$ mm. Different embedment depths have been considered, which also define the dimensions of the specimen. The generic geometry is represented in Figure 2.

The material properties are prescribed by the RILEM proposal. The Young's modulus is $E = 30,000$ MPa, Poisson's ratio is $\nu = 0.2$, the tensile and compressive strengths are $f_t = 3$ MPa and $f_c = 30$ MPa, respectively and the fracture energy is $G_f = 100$ N/m. The results obtained by using different models have been summarised by van Mier (1997). A significant scatter of the computed peak loads is observed. The isotropic gradient damage model (38) is

*Table 1.* Round robin analysis of anchor bolts: Computed peak load with the isotropic gradient damage model compared to the experiments (Vervuurt et al., 1994).

| $D$ [mm] | $P/D$ [N/mm] | | Relative error [%] |
|---|---|---|---|
| | Simulation | Experiment | |
| 50 | 274.48 | 268 | +2.04 |
| 100 | 236.76 | 245 | −3.36 |
| 150 | 219.94 | 220 | −0.02 |

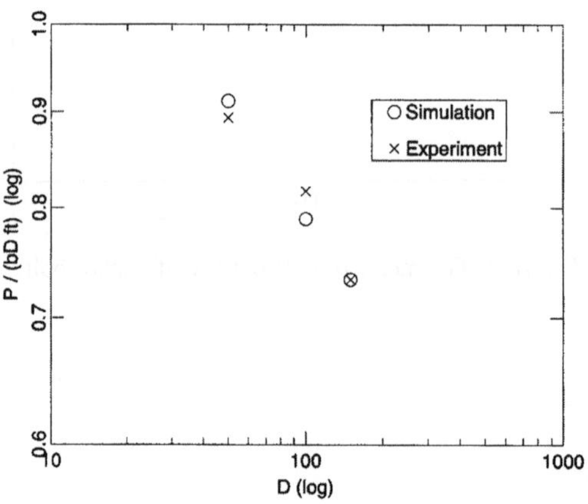


*Figure 3.* Results of size effect calculations of anchor bolt pull-out test.

used here to assess the ability of gradient-enhanced continuum models to capture size effects in concrete in a quantitative sense. The material properties are taken as $\kappa_i = 10^{-4}$, $k = 10$, $a = 0.96$, $b = 350$ (see 9 and 10) and the gradient parameter $g = 2\,\text{mm}^2$.

The specimen is discretised into eight-noded, plane-stress finite elements with a $2 \times 2$ Gauss-Legendre integration quadrature. The loading is controlled by a selective path-following technique (de Borst 1986) such that the vertical displacement of the interface anchor-concrete is equal to the path parameter. Following the analyses by Vervuurt et al. (1994) three embedment depths have been studied, namely $D = 50$ mm, 100 mm and 150 mm. The finite element mesh has been scaled to match the three considered depths. The results, which are summarised in Table 1, show that the simulations follow the experiments closely.

Because of the use of a mere scaling of the finite element mesh to simulate different embedment depths the peak load can be overestimated for larger values of $D$. If a finer discretisation had been used for the medium and large specimens, the computed peak load would probably have been slightly lower. Then, the size effect, which is represented in Figure 3 for the simulations together with the experimental results, might have been captured even more closely.

The results suggest that gradient-enhanced damage models are capable of providing accurate estimates of the peak load in pull-out tests as well as of reproducing the deterministic size effect inherent in quasi-brittle materials.

## 5. Concluding remarks

A unified framework has been proposed which encompasses most of the existing smeared crack formulations, the damage approaches and the microplane concepts and captures size effects in a proper manner, as has been shown by a numerical simulation and comparison with the experimental evidence of a anchor bolt test. Although many of these conceptions have arisen from completely different backgrounds, the resemblances are remarkable. A striking example is the similarity between the multiple fixed-crack model (de Borst and Nauta, 1984) and the microplane models (e.g., Bažant and Gambarova, 1984; Bažant and Prat, 1988). The multiple-fixed crack model is a purely phenomenological approach, while it is often suggested that the microplane models are based on micromechanical ideas. Nevertheless, the resulting equations are very similar, and, indeed, both models can be cast within the same formalism.

## Acknowledgements

Illuminating discussions with Ellen Kuhl (University of Stuttgart, Germany), Marc Geers (Royal Military School, Belgium) and Ron Peerlings (Eindhoven University of Technology) are gratefully acknowledged.

## References

Bažant, Z.P. (1983). Comment on orthotropic models for concrete and geomaterials. *ASCE Journal of Engineering Mechanics*. **109**, 849–865.

Bažant, Z.P. and Oh, B. (1983). Crack band theory for fracture of concrete. *RILEM Materials and Structures* **16**, 155–177.

Bažant, Z.P. (1984). Imbricate continuum and its variational derivation. *ASCE Journal of Engineering Mechanics* **110**, 1693–1712.

Bažant, Z.P. and Gambarova, P. (1984). Crack shear in concrete: Crack band microplane model. *ASCE Journal of Structural Engineering* **110**, 2015–2036.

Bažant, Z.P. and Pijaudier-Cabot, G. (1988). Nonlocal continuum damage, localization instability and convergence. *ASME Journal of Applied Mechanics* **55**, 287–293.

Bažant, Z.P. and Prat, P.C. (1988). Microplane model for brittle plastic material. I. Theory and II. Verification. *ASCE Journal of Engineering Mechanics* **114**, 1672–1702.

Benallal, A., Billardon, R. and Geymonat, G. (1988). Some mathematical aspects of the damage softening problem. *Cracking and Damage* (Edited by J. Mazars and Z.P. Bažant), Elsevier, Amsterdam and London, 247–258.

de Borst, R. and Nauta, P. (1985). Non-orthogonal cracks in a smeared finite element model. *Engineering Computations* **2**, 35–46.

de Borst, R. (1986). *Non-linear Analysis of Frictional Materials*. Dissertation. Delft University of Technology, Delft.

de Borst, R. (1987). Smeared cracking, plasticity, creep and thermal loading – a unified approach. *Computer Methods in Applied Mechanics and Engineering* **62**, 89–110.

de Borst, R. and Mühlhaus, H.-B. (1992). Gradient-dependent plasticity: Formulation and algorithmic aspects. *International Journal for Numerical Methods in Engineering* **35**, 521–539.

de Borst, R., Sluys, L.J., Mühlhaus, H.-B. and Pamin J. (1993). Fundamental issues in finite element analysis of localisation of deformation. *Engineering Computations* **10**, 99–122.

de Borst, R., Benallal, A. and Heeres, O.M. (1996). A gradient-enhanced damage approach to fracture. *Journal de Physique IV* **C6**, 491–502.

de Borst, R. and Pamin, J. (1996). Gradient plasticity in numerical simulation of concrete cracking. *European Journal of Mechanics: A/Solids* **15**, 295–320.

de Vree, J.H.P., Brekelmans, W.A.M. and van Gils, M.A.J. (1995). Comparison of nonlocal approaches in continuum damage mechanics. *Computers and Structures* **55**, 581–588.

Comi, C. (1999). Computational modelling of gradient-enhanced damage in quasi-brittle materials. *Mechanics of Cohesive-frictional Materials* **4**, 17–36.

Cope, R.J., Rao, P.V., Clark, L.A. and Norris, P. (1980). Modelling of reinforced concrete behaviour for finite element analysis of bridge slabs. *Numerical Methods for Non-Linear Problems* (Edited by C. Taylor, E. Hinton and D.R.J. Owen), Pineridge Press, Swansea, Vol. 1, 457–470.

Feenstra, P.H. (1993). Computational aspects of biaxial stress in plain and reinforced concrete. Dissertation. Delft University of Technology, Delft.

Feenstra, P.H. and de Borst, R. (1995). A plasticity model for mode-I cracking in concrete. *International Journal for Numerical Methods in Engineering* **38**, 2509–2529.

Herrmann, H.J., Hansen, H. and Roux, S. (1989). Fracture of disordered, elastic lattices in two dimensions. *Physical Review B* **39**, 637–648.

Ingraffea, A.R. and Saouma, V. (1985). Numerical modelling of discrete crack propagation in reinforced and plain concrete. *Fracture Mechanics of Concrete*, Martinus Nijhoff Publishers, Dordrecht, 171–225.

Kuhl, E. and Ramm, E. (1998). On the linearization of the microplane model. *Mechanics of Cohesive-frictional Materials* **3**, 343–364.

Kuhl, E., Ramm, E. and de Borst, R. (1998). Anisotropic gradient damage with the microplane model. *Computational Modelling of Concrete Structures* (Edited by R. de Borst, N. Bićanić, H.A. Mang and G. Meschke), Balkema, Rotterdam-Boston, 103–112.

Ladeveze, P. (1983). Sur une théorie de l'endommagement anisotrope. Report No. 34. Laboratoire de Mécanique et Technologi, Cachan.

Lemaitre, J. and Chaboche, J.L. (1990). *Mechanics of Solid Materials*, Cambridge University Press, Cambridge.

Mazars, J. (1984). Application de la mécanique de l'endommagement au comportement non linéare et à la rupture du béton de structure. These d'Etat. Université Paris VI, Paris.

Mazars, J. and Pijaudier-Cabot, G. (1989). Continuum damage theory – application to concrete. *ASCE Journal of Engineering Mechanics* **115**, 345–365.

Mühlhaus, H.-B., de Borst, R., Sluys, L.J. and Pamin, J. (1994). A thermodynamic theory for inhomogeneous damage evolution. *Computer Methods and Advances in Geomechanics* (Edited by H.J. Siriwardane and M.M. Zaman), Balkema, Rotterdam and Boston, 635–640.

Ngo, D. and Scordelis, A.C. (1967). Finite element analysis of reinforced concrete beams. *Journal of the American Concrete Institute* **64**, 152–163.

Peerlings, R.H.J., de Borst, R., Brekelmans, W.A.M. and de Vree, J.H.P. (1996). Gradient-enhanced damage for quasi-brittle materials. *International Journal for Numerical Methods in Engineering* **39**, 3391–3403.

Peerlings, R.H.J., de Borst, R., Brekelmans, W.A.M. and Geers, M.G.D. (1998). Gradient-enhanced modelling of concrete fracture. *Mechanics of Cohesive-frictional Materials* **3**, 323–342.

Pietruszczak, S. and Mróz, Z. (1981). Finite element analysis of deformation of strain softening materials. *International Journal for Numerical Methods in Engineering* **17**, 327–334.

Pijaudier-Cabot, G. and Bažant, Z.P. (1987). Nonlocal damage theory. *ASCE Journal of Engineering Mechanics* **113**, 1512–1533.

Rashid, Y.R. (1968). Analysis of prestressed concrete pressure vessels. *Nuclear Engineering and Design* **7**, 334–344.

Rots, J.G. (1991). Smeared and discrete representations of localized fracture. *International Journal of Fracture* **51**, 45–59.

Suidan, M. and Schnobrich, W.C. (1973). Finite element analysis of reinforced concrete. *ASCE Journal of Structural Div.* **99**, 2109–2122.

van Mier, J.G.M., Vervuurt, A. and Schlangen, E. (1994). Boundary and size effects in uniaxial tensile tests: a numerical and experimental study. *Fracture and Damage in Quasibrittle Structures* (Edited by G. Baker and B.L. Karihaloo), E & FN Spon, London, 289–302.

Mier, J.G.M. van (1997). *Fracture Processes of Concrete*. CRC Press, Boca Raton

van Vroonhoven, J.C.W. and de Borst, R. (1997). Uncoupled numerical method for fracture analysis. *International Journal of Fracture* **84**, 175–190.

Vervuurt, A., Mier, J.G.M. van and Schlangen, E. (1994). Analysis of anchor pull-out in concrete. *RILEM Materials and Structures* **27**, 251–259.

Willam, K.J. (1984). Experimental and computational aspects of concrete fracture. *Proceedings of the International Conference of Computer Aided Analysis and Design of Concrete Structures* (Edited by F. Damjanić et al.), Pineridge Press, Swansea, Vol. 1, 33–70.

Willam, K., Pramono, E. and Sture, S. (1986). Fundamental issues of smeared crack models. *Proceedings SEM/RILEM International Conference of Fracture of Concrete and Rock* (Edited by S.P. Shah and S.E. Swartz), Springer-Verlag, Heidelberg, Berlin, New York, 142–157.

*International Journal of Fracture* **95:** 279–297, 1999.


# Finite element simulation of ring expansion and fragmentation: The capturing of length and time scales through cohesive models of fracture

A. PANDOLFI[1], P. KRYSL[2] and M. ORTIZ[2]*
[1]*Dipartimento di Ingegneria Strutturale, Politecnico di Milano, 20133 Milano, Italy*
[2]*Graduate Aeronautical Laboratories, California Institute of Technology, Pasadena, CA 91125, U.S.A.*
*e-mail: ortiz@madrid.caltech.edu*

Received 11 September 1998; accepted in revised form 18 December 1998

**Abstract.** The expanding ring test of Grady and Benson (1983) is taken as a convenient yet challenging validation problem for assessing the fidelity of cohesive models in situations involving ductile dynamical fracture. Attention has been restricted to 1100-0 aluminum samples. Fracture has been modelled by recourse to an irreversible cohesive law embedded into cohesive elements. The finite element model is three-dimensional and fully Lagrangian. In order to limit the extent of deformation-induced distortion, we resort to continuous adaptive remeshing. The cohesive behavior of the material is assumed to be rate independent and, consequently, all rate effects predicted by the calculations are due to inertia and the rate dependency in plastic deformation. The numerical simulations are revealed to be highly predictive of a number of observed features, including: the number of dominant and arrested necks; the fragmentation patterns; the dependence of the number of fragments and the fracture strain on the expansion speed; and the distribution of fragment sizes at fixed expansion speed.

## 1. Introduction

Thin rings expanded using large, transient magnetic fields have been used to study the effect of deformation velocity on strains to failure of ductile metals (Grady and Benson, 1983; Wesenberg and Sagartz, 1977; Gourdin, 1989; Gourdin and Lassila, 1991; Altynova et al., 1996; Hu and Daehn, 1996; Juanicotena and Llorca, 1997) and to understand the dynamic fragmentation characteristics of many materials (Niordson, 1965, Rajendran and Fyfe, 1982; Grady and Benson, 1983; Altynova et al., 1996; Tamhane, 1996). Within the normal range of expansion velocities attained in these tests (50 to 300 m/s), the experimental results show that ductility of materials such as Al 6061 and OFHC Cu increases monotonically with increasing expansion velocity. Thus, the sample strain at failure in these materials is almost twice as great at 300 m/s as in the static condition (Altynova et al., 1996). The failure modes are observed to depend sensitively on the expansion velocity, leading to a varying numbers of fragments and arrested necks (Grady and Benson, 1983; Altynova et al., 1996; Tamhane et al., 1996). Expanding ring experiments have revealed other valuable insights into material behavior at high strain rates. For instance, Gourdin and Lassila (1991) found from ring tests that the work hardening rate varies significantly with strain rate but, unlike the flow stress, is independent of grain size. Gourdin and Lassila's observations also underscore the importance of dynamic recovery as regards material behavior at high temperatures and high strain rates.

* Correspondence author.

In the present work, we take the expanding-ring test as a convenient basis for assessing the predictive ability of cohesive models of fracture and fragmentation in applications involving ductile materials. These models–pioneered by Dugdale (1960), Barrenblatt (1962), Rice (1968) and others–regard fracture as a gradual phenomenon in which separation takes place across an extended crack 'tip', or cohesive zone, and is resisted by cohesive tractions. Cohesive models enable the incorporation into the analysis of *bona fide* fracture parameters such as the spall strength–the peak cohesive traction–and the fracture energy–the area under the cohesive law–of the material. In particular, the existence of a well-defined fracture energy endows the solid with characteristic, or 'intrinsic', spatial and temporal lengthscales (Camacho and Ortiz, 1996). The particular class of cohesive law contemplated in the present work has been proposed by Ortiz and Pandolfi (1999), and accounts for finite kinematics and irreversible behavior. Other alternative formulations may be found elsewhere (Rose et al., 1981; Needleman, 1987; Ortiz, 1988; Beltz, 1991; Rice, 1992; Ortiz and Suresh, 1993).

Cohesive laws have been built into finite element analyses as mixed boundary conditions (Hillerborg et al., 1976; Carpinteri, 1986; Needleman, 1987; 1990; 1992; Xu and Needleman, 1993; Planas et al., 1994; Tvergaard and Hutchinson, 1993; 1996a; 1996b) or have been embedded into *cohesive finite elements* (Willam, 1989; Ortiz and Suresh, 1993; Camacho and Ortiz, 1996; Xu and Needleman, 1994; 1995; 1996; De-Andrés et al., 1998; Ortiz and Pandolfi, 1999). These elements are surface-like and are compatible with general bulk finite element discretizations of the solid, including those which account for plasticity and large deformations. Cohesive elements bridge nascent surfaces and govern their separation in accordance with a cohesive law. In two-dimensional fragmentation simulations, both plane strain and axisymmetric, Camacho and Ortiz (1996), Ortiz (1996) have established the feasibility of using cohesive elements to account explicitly for individual cracks as they nucleate, propagate, branch and possibly link up to form fragments, as well as of simulating explicitly the granular flow which ensues following widespread fragmentation. Camacho and Ortiz (1996) have also shown that mesh-size independent results are obtained when the mesh adequately resolves the cohesive zone.

Several features of the expanding-ring test combine to make it particularly appealing as a benchmark problem. Firstly, owing to the linear geometry of the specimen, the computational models, while fully three dimensional, contain a modest number of elements, specially for thin rings. Secondly, large plastic deformations, frequently leading to necking instabilities, often precede fracture, which effectively tests theories of fracture in situations of high ductility. Finally, the wealth of experimental data available at present, including critical strains for necking and fracture, number of fragments as a function of radial expansion, and their dependence on such parameters as expansion velocity, provide a stringent test of fracture and fragmentation theories. Detailed comparisons between full three-dimensional finite-element simulations and the available experimental data presented in Section 4 demonstrate that cohesive theories accurately predict sensitive aspects of the fracture of the rings such as the strain to failure, the evolution of the number of fragments and the fragment size.

## 2. Dynamic necking instabilities in bars and rings

The development of necking stabilities in rapidly stretching sheets, round bars, jets and rings has been the focus of a number of studies (Fressengeas and Molinari, 1985a; 1985b; 1989; 1994; Knoche and Needleman, 1993; Nemes and Eftis, 1993; Han and Tvergaard, 1995; Hu and Daehn, 1996; Jeanclaude and Fressengeas, 1997). Thus, the dynamic necking instability of rapidly stretching sheets and jets of a non-linear viscoplastic material was investigated by Fressengeas and Molinari (1985b; 1989; 1994). Their analysis accounts for axial and lateral inertia effects, as well as for the three-dimensionality of the stress field. They found that the growth of small disturbances depends sensitively on the interplay between the destabilizing effect of geometrical softening and the stabilizing effect of inertia and rate sensitivity. In particular, inertia slows down the growth of long wavelengths, while viscosity predominantly damps the short wavelengths. By evaluating the most unstable wavelengths, which are time-dependent, Fressengeas and Molinari (1989; 1994) were able to predict breakup times and the fragment sizes in the jets. In a similar way, the dynamic necking instability in a viscoplastic rod was analyzed by Jeanclaude and Fressengeas (1997). They showed that, contrary to the static case, there is no single dominant unstable mode which operates throughout the necking process. Instead, the critical wavelength continuously shifts in time towards larger values. The delayed unstable modes are particularly important as it is such modes which mediate fragmentation and determine the fragment sizes.

The influence of inertia on failure initiation in the round-bar tensile test was analyzed numerically by Knoche and Needleman (1993). The material behavior was described in terms of a constitutive relation that models progressive micro-rupture by void nucleation and growth. Material strain-rate sensitivity and thermal softening due to adiabatic heating were incorporated into the analysis. Full dynamic transient analyses were carried out for differently sized, geometrically similar specimens, each with a geometric imperfection in the form of a constriction at the center, and subject to the same nominal strain rate. A critical void volume fraction criterion was used to identify the initiation of failure. For sufficiently small specimens, failure was found to occur by the development of a central neck, and the strain to failure was found to increase with specimen size. For a certain specimen size the location of the failure point moves toward the loaded end of the bar and the strain to failure subsequently decreases with increasing specimen size.

A similar numerical analysis of the round-bar tensile test was carried out by Nemes and Eftis (1993). They also found that the inhomogeneous fields of stress and deformation caused by wave propagation and wave reflection induce necking at different locations along the gauge section, depending upon the strain-rate imposed. This necking occurred without the introduction of geometrical or material imperfections. The imposed rate of strain was shown to affect the magnitude of the strain at which necking initiates, as well as the strain required for fracture.

A number of theoretical and numerical studies of ring expansion experiments have been put forth (Rajendran and Fyfe, 1982; Johnson, 1983; Kipp and Grady, 1985). Rajendran and Fyfe (1982) developed a model of thin ring expansion in which porous plasticity is identified as the principal failure mechanism. The predictions of the theory compare well with experimental data for statically and dynamically loaded rings. Johnson (1983) derived ordinary and partial differential equations governing inhomogeneous plastic flow and dynamic ductile fracture in expanding rings, which he proceeded to solve numerically. Imperfections in cross sectional area and porosity were introduced as a means of focusing the instability. Plastic deformations were predicted to be initially homogeneous and to rapidly localize to the weakest sections, leading to necking and eventually fracture. The imperfection wavelength, work-hardening rate, thermal softening and rate-dependency were accounted for. Kipp and Grady (1985) based their analysis on concepts of fracture energy and dynamic fracture, and extended Mott's theory (Mott, 1947) by explicitly accounting for the energy dissipation during the fracture process. Kipp and Grady's analysis predicts the fragment size, fracture time and dynamic fracture strain. Kipp and Grady also performed one-dimensional calculations with a view to examining the interaction between ductile fractures and ascertaining the conditions leading to the completion–or the arrest–of growing fractures. The results of the calculations are in close agreement with the experimental observations of Grady and Benson (1983).

More recently, one-dimensional calculations of expanding ring tests have been carried out by Hu and Daehn (1996) and Altynova et al. (1996). These simulations suggest that inertial effects are mainly responsible for enhanced ductility over a wide range of velocities. Numerical simulations by Han and Tvergaard (1995) of rapid radial expansion of rings containing small defects show that imperfections have a significant influence on the number of dynamic necks and the positions where they occur. Their analysis accounts for inertia and finite strain, but neglects thermal softening and material strain-rate sensitivity. For perfect rings under axisymmetric loading, Han and Tvergaard (1995) found no wave propagation around the circumference. When small initial imperfections are present, circumferential waves do arise and many necks appear almost simultaneously. The neck in the imperfect section develops more rapidly than other necks, which tend to be uniformly spaced. With larger initial imperfections, the necks concentrate around the initially imperfect cross section. The loading velocity strongly influences the strain to necking, with slow loading speeds resulting in earlier neck development. Numerical simulations carried out by Altynova et al. (1996) reveal that the rings stay almost perfectly circular up to the instant of fracture, at which point they break through multiple sections simultaneously. Numerous necks around the ring circumference are also observed to develop. The stress state during expansion is close to uniaxial, and the critical sections tend to shear off at about 30° to the ring cross section.

## 3. Finite element model

We begin by considering the general case of a deformable body occupying an initial configuration $B_0 \subset R^3$. The boundary $\partial B_0$ of the body is partitioned into a displacement boundary $\partial B_{0,1}$ and a traction boundary $\partial B_{0,2}$. The body undergoes a motion described by a deformation mapping $\boldsymbol{\varphi}$: $B_0 \times [0, T] \to R^3$, where $[0, T]$ is the duration of the motion, under the action of body forces $\rho_0 \boldsymbol{b}$ and prescribed boundary tractions $\bar{\boldsymbol{t}}$ applied over $\partial B_{0,2}$. Let $\boldsymbol{F}$ be the attendant deformation gradients and $\boldsymbol{P}$ the first Piola-Kirchhoff stress tensor (cf, e.g., Marsden and Hughes (1983)). In addition, the solid contains a collection of cohesive cracks. The locus of these cracks on the undeformed configuration is denoted $S_0$, Figure 1.

Under these conditions, the weak form of linear momentum balance, or virtual work expression, takes the form:

$$\int_{B_0} [\rho_0(\boldsymbol{b} - \ddot{\boldsymbol{\varphi}}) \cdot \boldsymbol{\eta} - \boldsymbol{P} \cdot \nabla_0 \boldsymbol{\eta}] \, \mathrm{d}V_0 - \int_{S_0} \boldsymbol{t} \cdot [\![\boldsymbol{\eta}]\!] \, \mathrm{d}S_0 + \int_{\partial B_{0,2}} \bar{\boldsymbol{t}} \cdot \boldsymbol{\eta} \, \mathrm{d}S_0 = 0, \tag{1}$$

where a superposed dot denotes the material time derivative, $\nabla_0$ is the material gradient, $\boldsymbol{\eta}$ is an arbitrary virtual displacement satisfying homogeneous boundary conditions on $\partial B_{0,1}$, $\boldsymbol{t}$ are the cohesive tractions over $S_0$, and $[\![\cdot]\!]$ denotes the jump across an oriented surface.

As is evident from (1), the presence of a cohesive surface results in the addition of a new term to the virtual work expression. Evidently, in order to complete the definition of the problem, a set of constitutive relations for the cohesive tractions $\boldsymbol{t}$ must to be provided. These constitutive relations are in addition to – and are independent of – the conventional

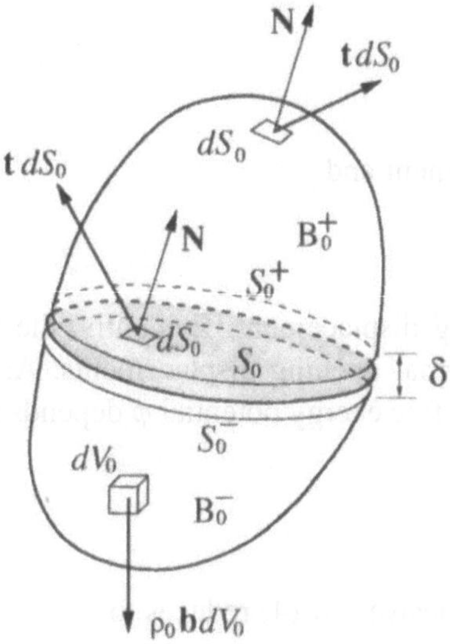


*Figure 1.* Cohesive surface traversing a 3D body.

constitutive relations describing the bulk behavior of the material. To this end, we postulate the existence of a free energy density per unit undeformed area over $S_0$ of the general form

$$\phi = \phi(\boldsymbol{\delta}, \theta, \boldsymbol{q}; \boldsymbol{n}), \tag{2}$$

where

$$\boldsymbol{\delta} = [\![\boldsymbol{\varphi}]\!] \tag{3}$$

are the opening displacements over the cohesive surface, $\theta$ is the local temperature, $\boldsymbol{q}$ is some suitable collection of internal variables which describe the inelastic processes attendant to decohesion, and $\boldsymbol{n}$ is the unit normal to the cohesive surface in the deformed configuration. The explicit dependence of $\phi$ on $\boldsymbol{n}$ is required to allow for differences in cohesive behavior for opening and sliding. By recourse to Coleman and Noll's method (e.g., Lubliner, 1972; 1973) it is possible to show that the cohesive law takes the form

$$\boldsymbol{t} = \frac{\partial \phi}{\partial \boldsymbol{\delta}}. \tag{4}$$

The potential structure of the cohesive law is a consequence of the first and second laws of thermodynamics. The evolution of the internal variables $\boldsymbol{q}$ is governed by a set of kinetic relations of the general form

$$\dot{\boldsymbol{q}} = \mathbf{f}(\boldsymbol{\delta}, \theta, \boldsymbol{q}). \tag{5}$$

A more general class of free energies which allows for surface anisotropy and finite opening displacements have been considered by Ortiz and Pandolfi (1999).

To further simplify the formulation of mixed-mode cohesive laws, we follow Camacho and Ortiz (1996) and introduce an effective opening displacement

$$\delta = \sqrt{\beta^2 \delta_S^2 + \delta_n^2}, \tag{6}$$

where

$$\delta_n = \boldsymbol{\delta} \cdot \boldsymbol{n} \tag{7}$$

is the normal opening displacement and

$$\delta_S = |\boldsymbol{\delta}_S| = |\boldsymbol{\delta} - \delta_n \boldsymbol{n}| \tag{8}$$

is the magnitude of the sliding displacement. Evidently, the parameter $\beta$ assigns different weights to the sliding and normal opening displacements. As simple model of cohesion is then obtained by assuming that free energy potential $\phi$ depends on $\boldsymbol{\delta}$ only through the effective opening displacement $\delta$, i.e.,

$$\phi = \phi(\delta, \theta, \boldsymbol{q}). \tag{9}$$

Under these conditions, the cohesive law (4) reduces to

$$\boldsymbol{t} = \frac{t}{\delta}(\beta^2 \boldsymbol{\delta}_S + \delta_n \boldsymbol{n}), \tag{10}$$

where

$$t = \frac{\partial \phi}{\partial \delta}(\delta, \theta, \boldsymbol{q}) \tag{11}$$

is a scalar effective traction. It follows from (6) and (10) that the effective traction is

$$t = \sqrt{\beta^{-2}|\boldsymbol{t}_S|^2 + t_n^2}, \tag{12}$$

where $|\boldsymbol{t}_S|$ and $t_n$ are the magnitudes of shear and the normal traction respectively. This relation shows that $\beta$ defines the ratio between the shear and the normal critical tractions. In brittle materials, this ratio may be estimated by imposing lateral confinement on specimens subjected to high-strain-rate axial compression (Chen and Ravichandran, 1994; 1996).

Upon closure, the cohesive surfaces are subject to the contact unilateral constraint, including friction. We regard contact and friction as independent phenomena to be modelled outside the cohesive law. Friction may significantly increase the sliding resistance in closed cohesive surfaces. In particular, the presence of friction may result in a steady – or even increasing – frictional resistance while the normal cohesive strength simultaneously weakens.

Figure 2 depicts the particular type of irreversible cohesive laws envisioned here. Irreversibility manifests itself upon unloading. Therefore, an appropriate choice of internal variable is the maximum attained effective opening displacement $\delta_{\max}$. Loading is then characterized by the conditions: $\delta = \delta_{\max}$ and $\dot{\delta} \geqslant 0$. Conversely, we shall say that the cohesive surface undergoes unloading when it does not undergo loading. We assume the existence of a loading envelope defining a relation between $t$ and $\delta$ under conditions of loading. A simple and convenient relation is furnished by Smith and Ferrante's universal binding law, Figure 2a, or by the linearly decreasing envelope shown in Figure 2b. Following Camacho and Ortiz (1996) we shall assume unloading to the origin, Figure 2, giving

$$t = \frac{t_{\max}}{\delta_{\max}}\delta, \quad \text{if} \quad \delta < \delta_{\max} \quad \text{or} \quad \dot{\delta} < 0. \tag{13}$$

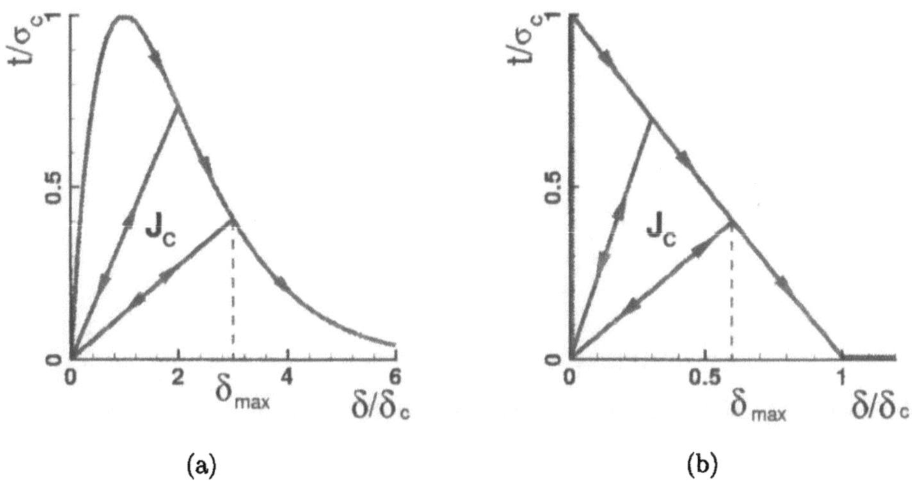


*Figure 2.* Two simple choices of cohesive law, expressed in terms of an effective opening displacement $\delta$ and traction $t$: (a) Loading-unloading rule from Smith-Ferrante envelope; (b) Loading-unloading rule from linearly decreasing loading envelope.

For the present model, the kinetic relations (5) reduce to a straightforward computation of $\delta_{\max}$.

A particularly appealing aspect of cohesive laws as models of fracture is that they fit naturally within the conventional framework of finite element analysis. One possible approach is to implement the cohesive law as a mixed boundary condition, relating tractions to displacements at boundaries or interfaces (Hillerborg et al., 1976; Carpinteri, 1986; Needleman, 1987; 1990; 1992; Planas et al., 1994; Tvergaard and Hutchinson, 1993; 1996a; 1996b). Here, by contradistinction, we follow Willam (1989), Ortiz and Suresh (1993), and Xu and Needleman (1994), and directly embed the cohesive law into surface-like finite elements, leading to the formulation of so-called 'cohesive' elements. In addition, we follow Camacho and Ortiz (1996) and adaptively create new surface as required by the cohesive model by duplicating nodes along previously coherent element boundaries. The introduction of cohesive surfaces may result in drastic changes in the topology of the model (Pandolfi and Ortiz, 1999). The nodes are subsequently released in accordance with a tension-shear cohesive law.

The geometry of the cohesive elements considered here is shown in Figure 3. The element consists of two six-node triangles endowed with quadratic displacement interpolation. Figure 4 also demonstrates the compatibility between the cohesive elements and ten-node volume elements. Inserting the displacement interpolation into the virtual work expression (1) leads to a system of semi-discrete equations of motion of the form

$$\boldsymbol{M}\ddot{\boldsymbol{x}} + \boldsymbol{f}^{\text{int}}(\boldsymbol{x}) = \boldsymbol{f}^{\text{ext}}(t), \tag{14}$$

where $\boldsymbol{x}$ is the array of nodal coordinates, $\boldsymbol{M}$ is the mass matrix, $\boldsymbol{f}^{\text{ext}}$ is the external force array, and $\boldsymbol{f}^{\text{int}}$ is the internal force array. In calculations we use the second-order accurate central difference algorithm to discretize (14) in time (Belytschko, 1983; Hughes, 1983; 1987). Despite the fact that the time step is bounded by stability (Hughes, 1983), explicit integration is particularly attractive in three-dimensional calculations, where implicit schemes lead to system matrices which often exceed the available in-core storage capacity. Yet another advantage of explicit algorithms is that they are ideally suited for concurrent computing (Mathur, 1996).

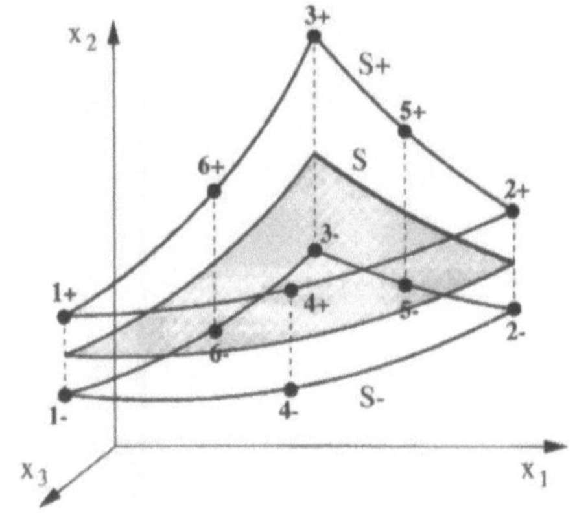


*Figure 3.* Geometry of cohesive element. The surfaces $S^-$ and $S^+$ coincide in the reference configuration of the solid.

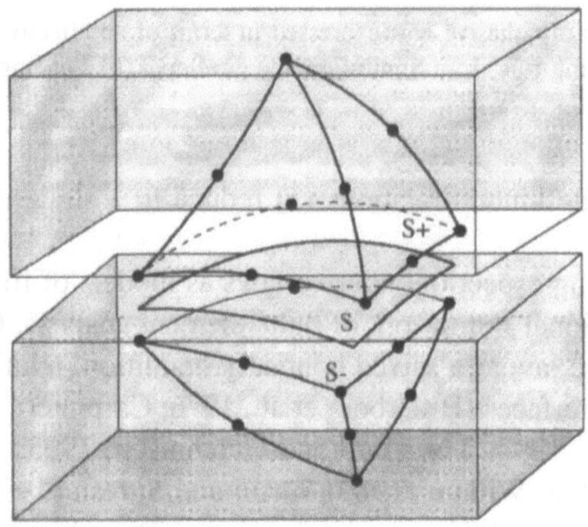


*Figure 4.* Assembly of 12-node triangular cohesive element and two 10-node tetrahedral elements.

## 4. Numerical tests

Next we turn to the expanding-ring test as a convenient yet exacting validation problem for assessing the fidelity of cohesive models in situations involving ductile fracture. Indeed the fragmentation of a body due to impulsive loads is a complex phenomenon in which stress waves, the effect of high strain rates on material properties and the statistics of fracture nucleation and growth interact strongly. For definiteness, we confine our attention to the experiments of Grady and Benson (1983) on aluminum rings. Grady and Benson subjected aluminum and copper rings to electromagnetic loading, by which device they imparted upon the rings a range of radial expansion speeds. Magnetic loading has several distinct advantages such as the fact that the motion of the sample results from continuous body forces instead of shock loading, and the ability to control loading rates through the driving-current pulse.

Grady and Benson proceeded to report a wealth of data including time-resolved strain and strain-rate measurements, the number of necks, recovered fragments, and others. A first set of twelve tests focused on the characteristics of fragmentation at a fixed strain rate. In these tests, fracture was observed to be preceded by the nucleation and growth of numerous necks and to result in the formation of multiple fragments. The fragment sizes exhibit a characteristic

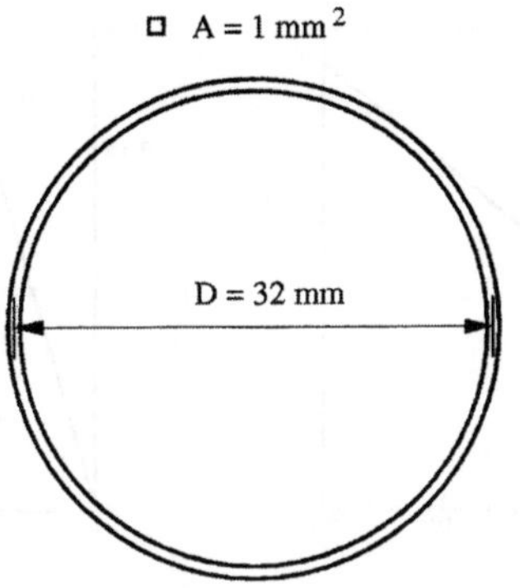


*Figure 5.* Sample geometry.

| **Bulk material** | | **Cohesive Model** | |
|---|---|---|---|
| ***J2 Plasticity with power hardening, rate dependency, thermal softening*** | | ***Loading - Unloading rule from linear decreasing envelope*** | |
| Mass density | $\rho$ = 2719 kg/m$^3$ | Surface energy | $G_c$ = 0.0165 N/mm |
| Young's modulus | $E$ = 69 GPa | Maximum cohesive stress | $\sigma_c$ = 110 MPa |
| Poisson's ratio | $\nu$ = 0.33 | Critical displacement | $\delta_c$ = 0.0003 mm |
| Initial yield stress | $\sigma_0$ = 90 MPa | Weighting coefficient | $\beta$ = 0.866 |
| Reference plastic strain | $\varepsilon_0^p$ = 1 x 10$^{-3}$ | | |
| Reference plastic strain rate | $\dot{\varepsilon}_0^p$ = 10$^4$/s | | |
| Rate sensitivity exponent | $m$ = 100 | | |
| Hardening exponent | $n$ = 25 | | |
| Heat capacity | $c_V$ = 904 J/kg K | | |
| Melting temperature | $T_m$ = 700°C | | |
| Thermal exponent | $\alpha$ = 1 | | |
| Taylor-Quinney coefficient | $\beta$ = 0.9 | | |

*Figure 6.* Material properties used in the numerical analysis.

distribution, which is best understood by the application of the statistical methods pioneered by Mott (1947). In a second set of sixteen experiments, the expansion velocity was varied from about 20 m/s to 200 m/s, corresponding to strain rates in the range of $10^3$ to $10^4\ s^{-1}$, with a view to ascertaining the dependence of the failure strain on strain rate. Static tensile tests were performed on dog-bone specimens of the same materials by way of reference. Grady and Benson measured the strain to failure based on the total length of the recovered fragments. They observed that the failure strain is an increasing function of strain rate and may attain values considerably in excess of the static ones, in keeping with previous observations (Rajendran, 1982; Winter, 1979). Grady and Benson also observed that the number of necks and fragments is an increasing function of the expansion speed and, therefore, of the strain rate.

In calculations we consider rings of 1100-0 aluminum with a diameter of 32 mm and a 1 mm$^2$ square cross section (cf., Grady and Benson (1983)). The geometry of the rings is shown in Figure 5. The material is assumed to obey $J_2$-plasticity with power-law hardening and rate dependency, and linear thermal softening. The details of the constitutive model may be found elsewhere (e.g., Cuitiño and Ortiz, 1992; Marusich and Ortiz, 1995; Camacho and Ortiz, 1997). The properties of 1100-0 aluminum assumed in the calculations are collected in

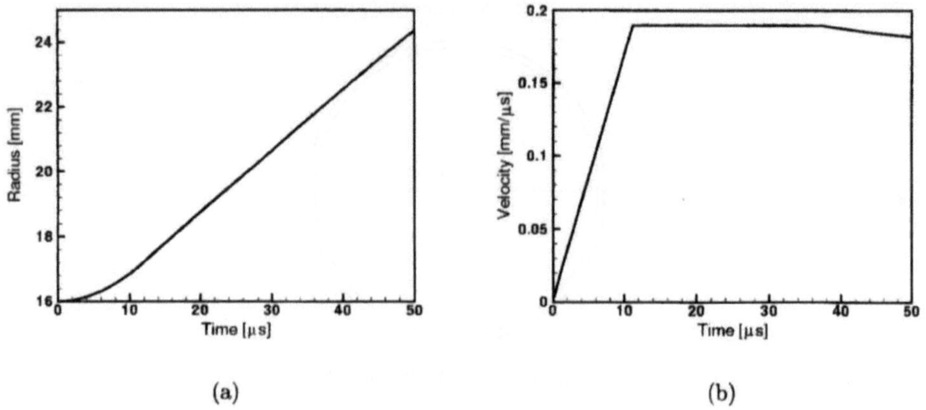


*Figure 7.* Evolution of the average ring radius for the 190 m/s test. a) trajectory; b) velocity.

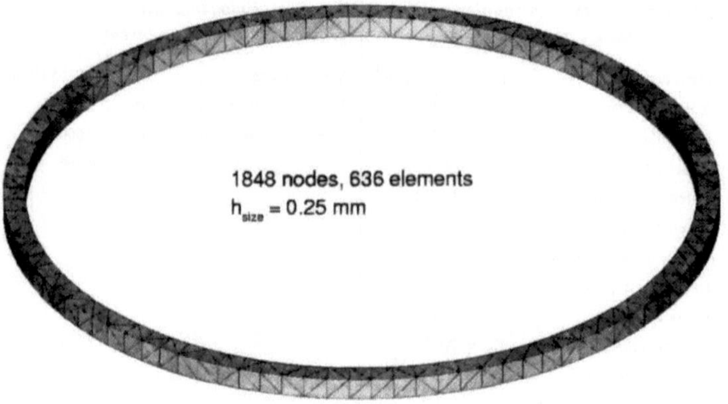


*Figure 8.* Initial mesh comprising 1,848 nodes and 636 10-node tetrahedra.

Figure 6. The state variables are updated by recourse to the method of extension of Cuitiño and Ortiz (1992). The temperature field is calculated locally assuming adiabatic conditions.

The finite element model consists of quadratic ten-noded tetrahedra. The calculations are fully lagrangian. The initial mesh is generated automatically by an advancing-front Delaunay method (Radovitzky and Ortiz, 1999b), Figure 8, and consists of 1,848 nodes and 636 elements. It should be carefully noted that the triangulation is somewhat irregular, and that this irregularity effectively breaks the rotational symmetry of the model and introduces initial imperfections which contribute to focusing the neck nuclei once the deformation becomes unstable. In order to limit the extent of deformation-induced distortion, we resort to continuous and adaptive remeshing. The mesh-adaption criterion is based on error indicators for nonlinear dynamic problems proposed by Radovitzky and Ortiz (1999a). The adaption procedure tends to refine the mesh at the necks, thus keeping the quality of the mesh uniformly good throughout the calculations. Upon adaption, meshes are constructed automatically by a variational Delaunay triangulation approach (Krysl and Ortiz, 1998). The topology of the solid model is described by a conventional boundary representation (see, e.g., Radovitzky and Ortiz (1999b)) and is continuously updated to account for fragmentation (Pandolfi and Ortiz, 1999).

Fracture is modelled by recourse to the irreversible cohesive law shown in Figure 2b. When the critical cohesive stress, or spall strength, is attained at the interface between two tetrahedral elements, a cohesive element of the type shown in Figure 3 is inserted between the elements.

The topological changes resulting from this operation are recorded and the solid model is updated to insure the generation of valid meshes upon adaption (Pandolfi and Ortiz, 1999). In calculations we have employed a critical cohesive stress of 110 MPa, which corresponds to the flow force determined by Grady and Benson's static experiments on aluminum (Grady and Benson, 1983). Subsequent to their insertion, the cohesive elements govern all aspects of the separation and closure of the incipient cracks. In particular, full fracture, resulting in the formation of free surface, occurs upon the attainment of a critical opening displacement. It bears emphasis that the cohesive behavior of the material is assumed to be rate independent, i.e., independent of the opening velocity. Therefore, all rate effects predicted by the calculations are due to inertia and the rate dependency in the constitutive behavior of the material.

The loading of the rings requires careful attention. The experimental data of (Grady and Benson, 1983) shows that, after a 10 $\mu$s transient, the average radial force imparted upon the ring by the magnetic field remains nearly uniform for as long as the ring remains continuous, and then drops to zero following the occurrence of the first full fracture. In keeping with these observations, we initially apply a constant radial force to the ring and then set the force to zero when the first through-crack occurs. All calculations are carried out by explicit dynamics, with the stable time step estimated from the Courant condition based on the elastic moduli. The time dependence of the average radius and radial velocity for representative tests is shown in Figure 7 for purposes of illustration.

In order to calibrate and validate the model as regards the remeshing and the development of necks, a first calculation was carried suppressing fracture. The radial velocity of the ring is of the order of 190 m/s and the ring expansion is taken to the experimental time of fracture. A distribution of necks is observed to nucleate throughout the ring early on in the calculation, signalling the onset of instability of the deformation. It bears emphasis that the material is assumed to undergo mechanical hardening at all times, and that the temperature rise at failure is of the order of 33 K, which is a modest rise indeed. Therefore, the necking instability is essentially geometrical in nature. Since, in addition, the material is assumed to have low rate sensitivity, the onset of necking may be expected to be reasonably well-characterized by Considere's criterion, at least for low expansion velocities. For a power-law hardening material this criterion gives the critical strain for the onset of necking as being equal to the hardening exponent $1/n$ (see, e.g., McClintock and Argon (1966)), which evaluates to a strain of 4 percent for the assumed hardening exponent $n = 25$. Indeed, incipient necks are clearly observed in the simulation at about that critical strain. Figure 9 shows the initial and three subsequent adapted meshes at intervals of 12 $\mu$s. The presence of well-developed necks is clearly visible in the third mesh. The good quality of all meshes despite plastic deformations in excess of 100 percent is also noteworthy.

The evolution of the necks is evident from Figures 10 and 11 which depict level contours of equivalent plastic strain at intervals of 9 $\mu$s. A first striking aspect revealed by these figures is the growing inhomogeneity of the deformation and the sheer number of necks. The plastic deformations attain high values in excess of 100 percent within the necks. Another striking feature is the competition between the necks, some of which thrive while other arrest. The arrested necks result from the arrival of relief waves emitted by the most active necks. These relief waves cause the driving force for the extension of the weak necks to drop, with the result that the necks eventually arrest. Since it takes time for waves to travel around the ring, the failure mode may be expected to be sensitive to the expansion speed. This expectation is born out by the variable-speed simulations described subsequently. It should also be carefully noted that the abundance of competing unstable modes renders the solution of the perfect

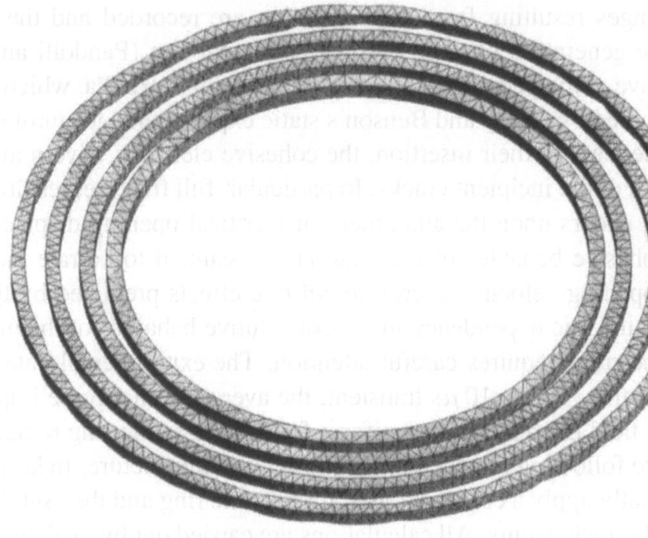

*Figure 9.* Initial and subsequent meshes produced by the remeshing procedure in the no-fracture reference analysis. The number of nodes of the three adapted meshes is 2680, 3239 and 3307 respectively. The number of tetrahedra is 902, 1098 and 1121 respectively.

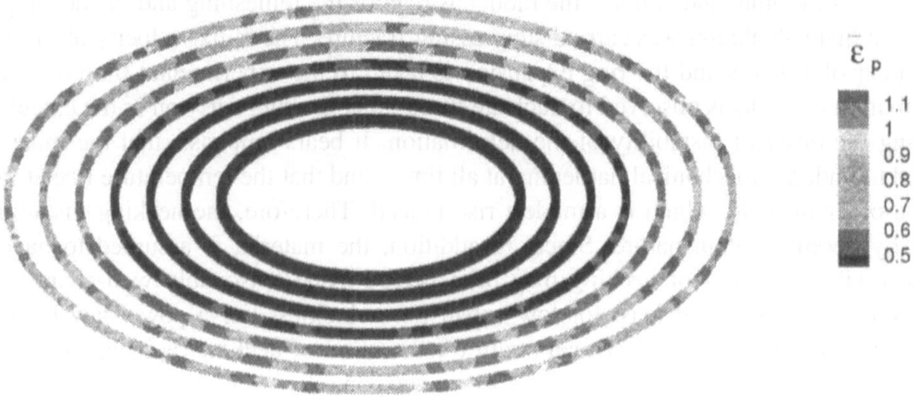


*Figure 10.* Evolution of the total equivalent plastic deformations. $\Delta t = 9\,\mu$s.

system massively nonunique and that, as is common in bifurcation problems, the particular choice of solution depends sensitively on initial imperfections such as effectively introduced by the mesh. Correspondingly, this imperfection sensitivity explains the variability in the experimental measurements, e.g., specially as regards the number of arrested necks. By contrast, the number of fragments appears to be far more reproducible (see Grady and Benson (1983), Table 1). With these provisos in mind, it is striking that the calculation predicts 13 dominant necks and 16 arrested necks at $t = 37\,\mu$s, Figure 11, which is in remarkable agreement with the 13 fragments and 17 arrested necks reported by Grady and Benson (1983) under the same test conditions.

In addition to the calibration test just described, we have carried out a number of simulations taking fracture into account. Fractures in both aluminum and copper are observed to occur across the work-hardened material in the dominant necks after a fair amount of extension and constriction of the cross section (Grady and Benson, 1983). As in the experimental

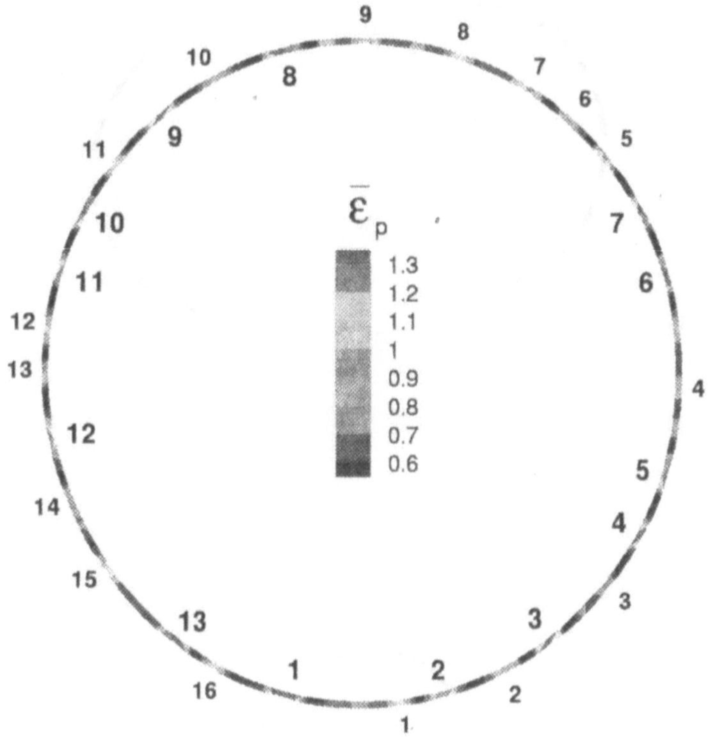


*Figure 11.* Distribution of the necks at $t = 37\,\mu s$. Interior numbers mark and count the 13 dominant necks. Exterior numbers mark and count weak or arrested necks.

work of Grady and Benson (1983), we have conducted a first set of simulations by varying the expansion speed through the experimental range of roughly 20 m/s to 200 m/s. The predicted fragmentation patterns at expansions speeds for 18 and 208 m/s are shown in Figure 12a and b, respectively. The proliferation of fragments at the higher expansion speed which is evident in these figures is particularly striking. The predicted dependence of the number of fragments on the expansion speed is compared in Figure 13 with the experimental data of Grady and Benson (1983). The overall agreement between simulation and observation, specially as regards the rate of increase in the number of fragments with expansion speed, is quite satisfactory.

Figure 14 compares the computed fracture strains with the experimental values reported by Grady and Benson (1983). As noted earlier, Grady and Benson calculate the fracture strain on the basis of the total length of the recovered fragments. This measure may be expected to overestimate somewhat the actual strain at fracture owing to the plastic deformation which inevitably takes places after the occurrence of the first fracture in the ring. Simulation affords the luxury of pinpointing the precise time at which the first fracture through the ring takes place, whence the fracture strain follows, e.g., from a computation of the average radius of the ring. Once these allowances are made, the agreement between the experimental and computed fracture strain as a function of expansion speed is satisfactory, Figure 14. In particular, the model captures remarkably well the rate of increase of the fracture strain with the expansion speed. This attests to the ability of the theory to account for loading-rate effects attendant to ductile fracture. Conversely, the predictive character of the theory appears to suggest that the cohesive response of aluminum, and presumably of other materials as well is, as assumed

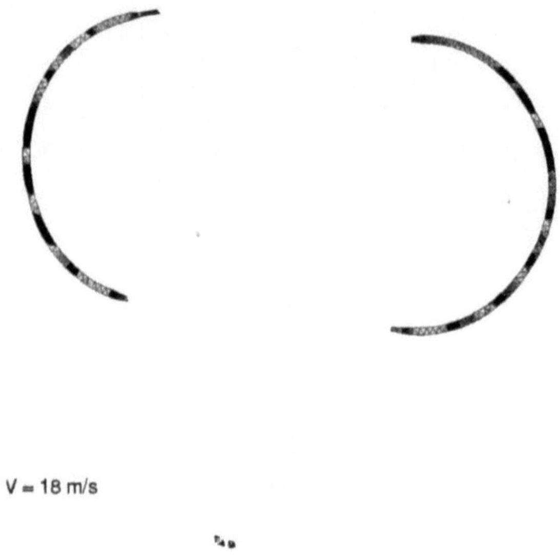


(a)

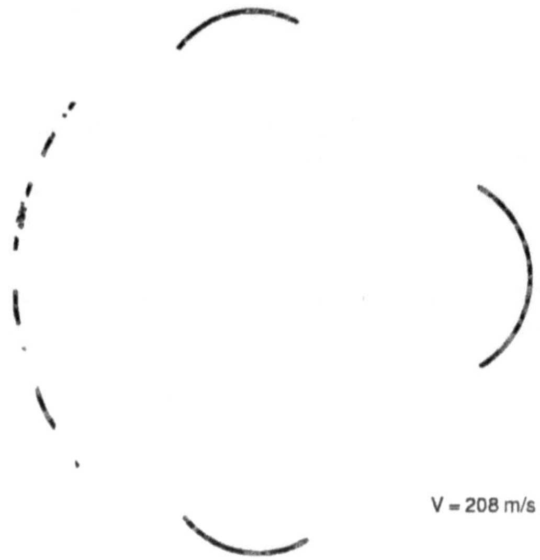


(b)

*Figure 12.* Fractured rings at expansion speeds: (a) 18 m/s; and (b) 208 m/s.

in the theory, rate-independent, and that all rate effects emanate from inertia and the rate-dependency of plastic deformation.

Another remarkable feature which is evident from Figure 12b is the variability in the fragment sizes at the higher speed. This variability justifies the use of statistical methods such as pioneered by Mott (1947). To this end, Grady and Benson (1983) carried out a set of six tests on aluminum corresponding to expansion velocities in the narrow range of 180 m/s to 220 m/s. Fragments from all tests were collated, counted and each was weighted separately. From these measurements, a statistical size and mass distribution of the fragments was determined. A

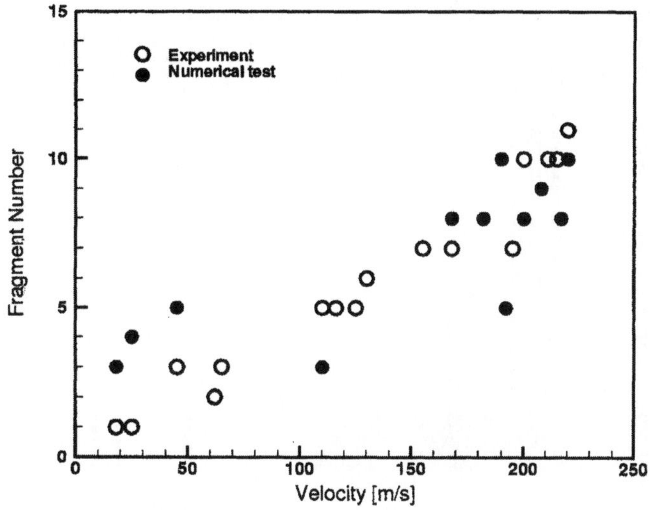


*Figure 13.* Experimental (Grady and Benson, 1983) and predicted dependence of the number of fragments on the expansion speed of 1100-0 aluminum rings.

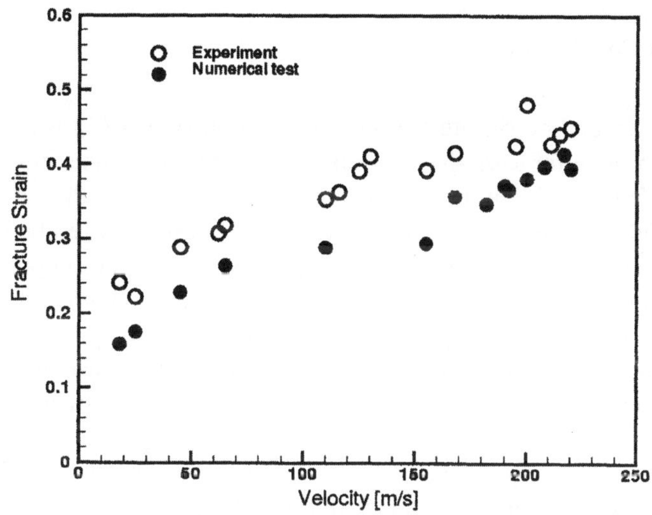


*Figure 14.* Experimental (Grady and Benson, 1983) and predicted dependence of the fracture strain on the expansion speed of 1100-0 aluminum rings.

comparison between the experimental and numerical fragment-mass distribution is shown in Figure 15. Keeping in mind the small size of the statistical sample, the correspondence between simulation and experiment may be considered satisfactory. As may be seen from Figure 15, the simulations predict fragment masses which cluster around 18 mg, with lighter and heavier fragments becoming increasingly infrequent, in accordance with observation. Both the experimental and numerical distributions agree fairly well with the analytical distribution derived by Grady (1981) by incorporating the ideas of Mott into a statistical fragmentation framework. It should be emphasized that our results demonstrate a fair ability on the part of cohesive models to predict not only the predominant fragment size but the distribution of fragment sizes as well in situations of high ductility.

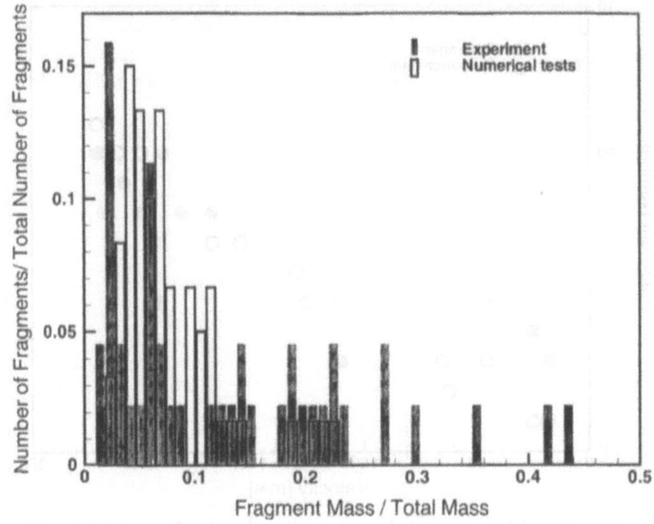


*Figure 15.* Experimental (Grady and Benson, 1983) and predicted fragment-mass distribution obtain from six tests on 1100-0 aluminum at expansions speed in the range of 180 m/s to 220 m/s.

## 5. Summary and Conclusions

We have taken the expanding ring test of Grady and Benson (1983) as a convenient yet exacting validation problem for assessing the fidelity of cohesive models in situations involving ductile dynamical fracture. Attention has been restricted to 1100-0 aluminum samples. Fracture has been modelled by recourse to an irreversible cohesive law embedded into cohesive elements. The cohesive elements are inserted adaptively into the mesh upon the attainment of a critical traction at tetrahedral element interfaces. Subsequent to their insertion, the cohesive elements govern all aspects of the separation and closure of the incipient cracks. The cohesive behavior of the material is assumed to be rate independent and, consequently, all rate effects predicted by the calculations are due to inertia and the rate dependency in plastic deformation. The finite element model is three dimensional and consists of quadratic ten-noded tetrahedra. The calculations are fully lagrangian. In order to limit the extent of deformation-induced distortion, we resort to continuous and adaptive remeshing.

Fragmentation of ductile solids subjected to impulsive loads furnishes a particularly exacting validation test of cohesive theories of fracture, in which stress waves, the effect of high strain rates on material properties and the statistics of fracture nucleation and growth interact strongly. In this challenging setting, the numerical simulations have proven highly predictive of a number of observed features, including: the number of dominant and arrested necks; the fragmentation patterns; the dependence of the number of fragments and the fracture strain on the expansion speed; and the distribution of fragment sizes at fixed expansion speed. The agreement between theory and experiment attests to the physical soundness of cohesive theories of fracture.

It should also be emphasized that cohesive models, and their finite-element embodiment, cohesive elements, greatly facilitate the solution of problems characterized by the development of very complex fracture and fragmentation patterns. Indeed, it is difficult to fathom how problems of this kind, in which cracks nucleate, branch and link up to form fragments in an entirely uncontrolled fashion, could possibly be analyzed numerically by any other means, specially by approaches which require a reckoning of fracture parameters such as stress-intensity factors at all crack tips. In this sense, cohesive theories may justifiably be regarded as an 'enabling technology'.

## Acknowledgements

AP and MO gratefully acknowledge support from the Office of Naval Research through grant N00014-95-1-0453. PK and MO are grateful for support from the Department of Energy through Caltech's ASCI Center of Excellence for Simulating Dynamic Response of Materials. MO also wishes to gratefully acknowledge the support of the Army Research Office through grant DAAH04-96-1-0056.

## References

Altynova, M., Hu, X.Y. and Daehn, G.S. (1996). Increased ductility in high-velocity electromagnetic ring expansion. *Metallurgical and Materials Transactions A* **27**(7), 1837–1844.

Barenblatt, G.I. (1962). The mathematical theory of equilibrium of cracks in brittle fracture. *Advances in Applied Mechanics* **7**, 55–129.

Beltz, G.E. and Rice, J.R. (1991). Book chapter: Dislocation nucleation versus cleavage decohesion at crack tips. *Modeling the Deformation of Crystalline Solids: Physical Theory, Application and Experimental Comparisons* (Edited by T.C. Lowe, A.D. Rollett, P.S. Follansbee and G.S. Daehn), Warrendale, PA, 457–480. TSM-AIME.

Belytschko, T. (1983). An overview of semidiscretization and time integration procedures. *Computational Methods for Transient Analysis* (Edited by T. Belytschko and T.J.R. Hughes), North Holland, 1–65.

Camacho, G.T. and Ortiz, M. (1996). Computational modelling of impact damage in brittle materials. *International Journal of Solids and Structures*, **33** (20–22), 2899–2938.

Camacho, G.T. and Ortiz, M. (1997). Adaptive Lagrangian modelling of ballistic penetration of metallic targets. *Computer Methods in Applied Mechanics and Engineering* **142**, 269–301.

Carpinteri, A. (1986). *Mechanical Damage and Crack Growth in Concrete*. Martinus Nijhoff, Dordrecht, The Netherlands.

Chen, G. and Ravichandran, W.N. (1994). Dynamic compressive behavior of ceramics under lateral confinement. *Journal de Physique IV* **4**, 177–182.

Chen, G. and Ravichandran, W.N. (1996). Static and dynamic compressive behavior of aluminum nitride under moderate confinement. *Journal of the American Ceramic Society* **79**, 579–584.

Cuitiño, A.M. and Ortiz, M. (1992). A material-independent method for extending stress update algorithms from small-strain plasticity to finite plasticity with multiplicative kinematics. *Engineering Computations* **9**, 437–451.

De-Andrés, A., Pérez, J.L. and Ortiz, M. (1999). Elastoplastic finite element analysis of three-dimensional fatigue crack growth in aluminum shafts to axial loading. *International Journal of Solids and Structures* **36**(15) 2231–2258.

Dugdale, D.S. (1960). Yielding of steel sheets containing slits. *Journal of the Mechanics and Physics of Solids* **8**, 100–104.

Fressengeas, C. and Molinari, A. (1985a). Inertia and thermal effects on the localization of plastic flow. *Acta Metallurgica* **33**, 387–396.

Fressengeas, C. and Molinari, A. (1985b). Localization criteria for plastic deformation of simple traction. *Journal de Physique, Colloque C5* **46**(C5), 121–125.

Fressengeas, C. and Molinari, A. (1989). Tensile instability at high strain rate-inertial and multiaxial effects. *Institute of Physics Conference Series* **102**, 57–64.

Fressengeas, C. and Molinari, A. (1994). Fragmentation of rapidly stretching sheets. *European Journal of Mechanics, A/Solids* **13**(2), 251–268.

Gourdin, W.H. (1989). Analysis and assessment of electromagnetic ring expansion as a high-strain rate test. *Journal of Applied Physics* **65**(2), 411–422.

Gourdin, W.H. and Lassila, D.H. (1991). Flow stress of OFE copper at strain rates from $10^{-4}$ to $10^4$ $s^{-1}$–grain-size effects and comparison to the mechanical threshold stress model. *Acta Metallurgica et Materialia* **39**(10), 2337–2348.

Grady, D.E. (1981). Fragmentation of solids under impulse stress loading. *Journal of geophysical Research* **86**(B2), 1047.

Grady, D.E. and Benson, D.A. (1983). Fragmentation of metal rings by electromagnetic loading. *Experimental Mechanics* **23**, 393–400.

Han, J.B. and Tvergaard, V. (1995). Effect of inertia on the necking behavior of ring specimens under rapid radial expansion. *European Journal of Mechanics–A Solids* **14**(2), 287–307.

Hillerborg, A., Modeer, M., Petersson, P.E. and Needleman, A. (1976). Analysis of crack formation and crack growth in concrete by means of fracture mechanics and finite elements. *Cement Concrete Research* **6**, 773–782.

Hu, X.Y. and Daehn, G.S. (1996). Effect of velocity on flow localization in tension. *Acta Materialia* **44**(3), 1021–1033.

Hughes, T.J.R. (1983). Analysis of transient algorithms with particular reference to stability behavior. *Computational Methods for Transient Analysis* (Edited by T. Belytschko and T.J.R. Hughes), North-Holland, 67–155.

Hughes, T.J.R. (1987). *The Finite Element Method: Linear Static and Dynamic Finite Element Analysis*. Prentice-Hall, Englewood Cliffs, N.J.

Jeanclaude, V. and Fressengeas, C. (1997). Dynamic necking of rods at high strain rates. *Journal de Physique IV, Colloque C3, Supplement au Journal del Physique III*, pages 699–704.

Johnson, J.N. (1983). Ductile fracture of rapidly expanding rings. *Journal of Applied Mechanics* **50**, 593–600.

Juanicotena, A. and Llorca, F. (1997). Utility of the ring expansion test in the analysis of different laws of elastoplastic behavior at high strain rates–application to copper and mars-190 steel. *Journal de Physique IV* **7**(C3), 541–546.

Kipp, M.E. and Grady, D.E. (1985). Dynamic fracture growth and interaction in one dimension. *Journal of the Mechanics and Physics of Solids* **33**(4), 399–415.

Knoche, P. and Needleman, A. (1993). The effect of size on the ductility of dynamically loaded tensile bars. *European Journal of Mechanics–A Solids* **12**(4), 585–601.

Krysl, P. and Ortiz, M. (1998). Generation of tetrahedral finite element meshes; variational Delaunay approach. In *7th International Meshing Roundtable*, Dearborn, Michigan.

Lubliner, J. (1972). On the thermodynamic foundations of non-linear solid mechanics. *International Journal of Non-linear Mechanics* **7**, 237–254.

Lubliner, J. (1973). On the structure of the rate equations of materials with internal variables. *Acta Mechanica* **17**, 109–119.

Marsden, J.E. and Hughes, T.J.R. (1983). *Mathematical foundations of elasticity*. Prentice-Hall, Englewood Cliffs, N.J.

Marusich, T.D. and Ortiz, M. (1995). Modelling and simulation of high-speed machining. *International Journal for Numerical Methods in Engineering* **38**, 3675–3694.

Mathur, K.K., Needleman, A. and Tvergaard, V. (1996). Three-dimensional analysis of dynamic ductile crack growth in a thin plate. *Journal of the Mechanics and Physics of Solids* **44**, 439–464.

McClintock, F.A. and Argon, A.S. (1966). *Mechanical Behavior of Materials*. Addison-Wesley, Reading, Mass.

Mott, N.F. (1947). Fragmentation of shell cases. *Proceedings of the Royal Society of London* **300**, 300–308.

Needleman, A. (1987). A continuum model for void nucleation by inclusion debonding. *Journal of Applied Mechanics* **54**, 525–531.

Needleman, A. (1990). An analysis of decohesion along an imperfect interface. *International Journal of Fracture* **42**, 21–40.

Needleman, A. (1992). Micromechanical modeling of interfacial decohesion. *Ultramicroscopy* **40**, 203–214.

Nemes, J.A. and Eftis, J. (1993). Constitutive modeling of the dynamic fracture of smooth tensile bars. *International Journal of Plasticity* **9**(2), 243–270.

Nordson, F.I. (1965). A unit for testing materials at high strain rates. *Experimental Mechanics* **5**, 29–32.

Ortiz, M. (1988). Microcrack coalescence and macroscopic crack growth initiation in brittle solids. *International Journal of Solids and Structures* **24**, 231–250.

Ortiz, M. (1996). Computational micromechanics. *Computational Mechanics* **18**, 321–338.

Ortiz, M. and Pandolfi, A. (1999). A class of cohesive elements for the simulation of three-dimensional crack propagation. *International Journal for Numerical Methods in Engineering* **44**(9), 1267–1282.

Ortiz, M. and Suresh, S. (1993). Statistical properties of residual stresses and intergranular fracture in ceramic materials. *Journal of Applied Mechanics* **60**, 77–84.

Pandolfi, A. and Ortiz, M. (1999). Solid modeling aspects of three-dimensional fragmentation. *Engineering with Computer* **14**(4) 287–308.

Planas, J., Elices, M. and Guinea, G.V. (1994). Book chapter: Cohesive cracks as a solution of a class of nonlocal problems. (Edited by Z.P. Bazant). *Fracture and Damage in Quasibrittle Structures. Experiment, modeling and Computer analysis* E & FN SPON.

Radovitzky, R. and Ortiz, M. (1999a). Error estimation and adaptive meshing in strongly nonlinear dynamic problems. *Computer Methods in Applied Mechanics and Engineering.*

Radovitzky, R. and Ortiz, M. (1999b). Tetrahedral mesh generation based on node insertion in crystal lattice arrangements and advancing-front-Delaunay triangulation. *Computer Methods in Applied Mechanics and Engineering.*

Rajendran, A.M. and Fyfe, I.M. (1982). Inertia effects on the ductile failure of thin rings. *Journal of Applied Mechanics* **49**, 31–36.

Rice, J.R. (1968). Mathematical analysis in the mechanics of fracture. *Fracture* (Edited by Liebowitz), Academic Press, 191–311.

Rice, J.R. (1992). Dislocation nucleation from a crack tip: an analysis based on the Peierls concept. *Journal of the Mechanics and Physics of Solids* **40**, 235–271.

Rose, J.H., Ferrante, J. and Smith, J.R. (1981). Universal binding energy curves for metals and bimetallic interfaces. *Physical Review Letters* **47**(9), 675–678.

Tamhane, A.A., Altynova, M.M. and Daehn, G.S. (1996). Effect of sample size on ductility in electromagnetic ring expansion. *Scripta Materialia* **34**(8), 1345–1350.

Tvergaard, V. and Hutchinson, J.W. (1993). The influence of plasticity on mixed-mode interface toughness. *Journal of the Mechanics and Physics of Solids* **41**, 1119–1135.

Tvergaard, V. and Hutchinson, J.W. (1996a). Effect of strain dependent cohesive zone model on predictions of interface crack growth. *Journal de Physique IV* **6**, 165–172.

Tvergaard, V. and Hutchinson, J.W. (1996b). Effect of strain dependent cohesive zone model on predictions of crack growth resistance. *International Journal of Solids and Structures* **33**, 3297–3308.

Wesenberg, D.L. and Sagartz, M.J. (1977). Dynamic fracture of 6061-T6 aluminum cylinders. *Journal Applied Mechanics* **44**(4), 643–646.

William, K. (1989). Simulation issues of distributed and localized failure computations. *Cracking and Damage* (Edited by J. Mazars and Z.P. Bazant), Elsevier Science, New York, 363–378.

Winter, R.E. (1979). Measurement of fracture strain at high strain rates. *Institute of Physics Conference Ser.* **47**, 81.

Xu, X.P. and Needleman, A. (1993). Void nucleation by inclusion debonding in a crystal matrix. *Modelling and Simulation in Materials Science and Engineering* **1**, 111–132.

Xu, X.P. and Needleman, A. (1994). Numerical simulations of fast crack growth in brittle solids. *Journal of the Mechanics and Physics of Solids* **42**, 1397–1434.

Xu, X.P. and Needleman, A. (1995). Numerical simulations of dynamic interfacial crack growth allowing for crack growth away from the bond line. *International Journal of Fracture* **74**, 253–275.

Xu, X.P. and Needleman, A. (1996). Numerical simulation of dynamic crack growth along an interface. *International Journal of Fracture* **74**, 289–324.

*International Journal of Fracture* **95:** 299–314, 1999.


# Strain gradient interpretation of size effects

E.C. AIFANTIS
*Laboratory of Mechanics and Materials, Aristotle University of Thessaloniki, GR 54006, Thessaloniki, Greece*
*e-mail: giannis@kelifos.physics.auth.gr*
*Center for Mechanics of Materials and Instabilities, Michigan Tech., Houghton, MI 49931, USA*

Received 10 July 1998; accepted in revised form 18 December 1998

**Abstract.** It is shown that the phenomenon of strength dependence on size, for otherwise geometrically similar specimens, may be interpreted on the basis of gradient elasticity and gradient plasticity arguments. This is illustrated by adopting a simple strength of materials approach for considering torsion and bending of solid bars and subsequent comparison with available experimental data. Solutions of boundary value problems based on gradient elasticity and gradient plasticity including those of fracture and shear banding can also be used to interpret size effects in more complex situations. This is not discussed here, however, in order to maintain simplicity and clarity when illustrating the theoretical predictions in comparison with the experimental trends.

**Key words:** Gradient elasticity, gradient plasticity, size effect.

## 1. Introduction

Size effects is a subject of increasing interest due to the fact that current applications in modern technology involve a variety of length scales ranging from a few centimeters (sheet metal forming) down to few nanometers (thin film technology). This range of scales and the corresponding necessity for modelling and experiment have revealed that there is a strong connection between the various length scales involved and that the response may depend on size for otherwise dimensionally similar specimens. Several authors are currently considering the problem of size effect by employing a variety of methods ranging from computer simulations and probabilistic mechanics to cohesive crack and fractal models. Among them one distinguishes the efforts of Bazant and co-workers (1976; 1986; 1991) who were motivated by the mechanical behavior of concrete and brittle geomaterials where size effects are more apparent. Scale effects have also been discussed for soils by Vardoulakis and Sulem (1995). Recent progress on size and scale effects issues for a variety of materials can be found in a recent book edited by Carpinteri (1996).

In the simplest case of scale- or size-dependent response, one may envision the interaction between the geometric length of the specimen/externally applied load and an internal length/internal force associated with the underlying microstructure. The interaction between macroscopic and microscopic length scales in the constitutive response and the corresponding interpretation of the associated size effects may be modelled through the introduction of higher order strain gradients in the respective constitutive equations. This suggestion has been taken up, among others, by Aifantis (1996) who considered the issue of size-dependent properties by building on his earlier proposals on gradient elasticity and plasticity. Specifically, he modified the usual constitutive equations of classical elasticity and plasticity by including the Laplacian of the elastic stress or the equivalent plastic shear strain into the standard expression of

Hooke's law or the yield condition respectively. In (Aifantis, 1996), it was shown, in particular, that the author's original 'symmetric stress' strain gradient theory (Aifantis, 1984; 1987; 1992; 1995) can successfully be used to interpret size effects exhibited by elastic bore holes and twisted wires. Later, it was shown by Zhu et al. (1997) that this theory can adequately model size effects exhibited by metal matrix composites. The results reported in both works of (Aifantis, 1996) and (Zhu et al., 1997) were obtained by solving boundary value problems based on the simplest possible forms of gradient elasticity and gradient plasticity theories.

The purpose of the present report is to adopt an even simpler scheme, i.e. a strength of materials approach, and use gradient elasticity and gradient plasticity arguments to interpret size effects in torsion and bending. In fact, our goal is not to consider a 'universal' gradient theory of elasticity or plasticity for interpreting the largest possible set of experimental data available on size effects. Our rather limited scope is to point out some interesting observations on size-dependent mechanical response which can easily be deduced by incorporating gradient terms into the standard strength of materials approach commonly used to consider torsion and bending configurations. A more rigorous consideration would require the formulation and solution of boundary value problems on the basis of three-dimensional equilibrium and constitutive equations of gradient elasticity or plasticity by incorporating nonstandard higher order boundary conditions and also dealing with nonstandard higher order stress components that may arise as a result of the strain gradient terms. Some typical problems in this direction are currently under consideration as briefly discussed in the next section.

In Section 2 we employ gradient elasticity arguments to interpret size effects in elastic torsion and elastic bending. The results are compared with relatively recent experimental data obtained by Lakes (1981; 1983; 1986) who also used a couple-stress elastic theory of Cosserat type to interpret his experiments. In contrast to this 'asymmetric stress' gradient elastic theory which uses at least two gradient coefficients, our 'symmetric stress' gradient elasticity approach employs only one single gradient coefficient whose size turns out to be of the order of the underlying microstructure. In Section 3 we employ gradient plasticity to interpret size effects in plastic torsion. The results are compared with recent experimental data obtained by Fleck et al. (1994) who also motivated the derivation of a couple-stress plastic theory of Cosserat type from their experiments. In contrast to this 'asymmetric stress' gradient plasticity theory which uses an internal couple with an extra constitutive equation for it, and only qualitatively captures the experimental trends, our 'symmetric stress' gradient plasticity is capable of employing a sufficient number of gradient coefficients to model the size-dependent response for the whole regime of specimen sizes and levels of strain tested. Finally, in Section 4 a simpler form of gradient plasticity is employed to interpret the size-dependent initial yielding response, in agreement with experimental results obtained several years ago by Morrison (1939) in torsion and Richards (1958) in bending.

## 2. Size effects in gradient elasticity

In this section, motivated by a 'symmetric stress' gradient elasticity theory discussed in (Aifantis, 1992), we employ a strength of materials approach in order to model the size-dependent elastic response noted by Lakes and co-workers (1981; 1983; 1986) for torsion and bending of bone and foam specimens. The simplest form of the aforementioned gradient elasticity theory reads (Aifantis, 1992)

$$\sigma = \lambda(\operatorname{tr}\boldsymbol{\varepsilon})\mathbf{1} + 2G\boldsymbol{\varepsilon} - c\nabla^2[\lambda(\operatorname{tr}\boldsymbol{\varepsilon})\mathbf{1} + 2G\boldsymbol{\varepsilon}], \tag{2.1}$$

where tr denotes trace, $\nabla^2$ is the Laplacian, $(\boldsymbol{\sigma}, \boldsymbol{\varepsilon})$ denote the stress and strain tensors respectively, $(\lambda, G)$ are the standard elastic constants and $c$ is the gradient coefficient. The theory given by (2.1) has been used to eliminate the strain singularity from crack tips and dislocation cores. It can be generalized to include additional gradient coefficients to measure different intensity of the gradient effect in shear and dilatation, as well as to include nonlinear effects.

Such generalizations in tensorial form are out of the scope of the present discussion which only aims into incorporating the reasoning leading into (2.1) within a strength of materials framework for elastic deformation in torsion and bending and produce corresponding strength of materials relations to interpret the observed size effects. Any required gradient generalization of (2.1) will be directly applied to the torsion and bending configurations considered below rather to the general tensorial form of (2.1). This way the presentation is kept as simple as possible and only the number and values of the required gradient coefficients for modelling the experimental data is of main concern. If the gradient elasticity embodied in (2.1) was to be rigorously applied to consider, for example, the torsion of circular shafts, then the theorem of Ru and Aifantis (1993) can be used to obtain the displacement field $\boldsymbol{u}$ corresponding to (2.1) in terms of the displacement field $\boldsymbol{u}_0$ of classical elasticity through the equation $\boldsymbol{u} - c\nabla^2\boldsymbol{u} = \boldsymbol{u}_0$. By using for $\boldsymbol{u}_0$ the displacement field corresponding to the torsion problem of classical elasticity, the displacement field $\boldsymbol{u}$ of the gradient elasticity can be found when higher order boundary conditions are properly assigned. Size effects can then be interpreted by using the obtained solution to express the applied torque in terms of the surface strain with the gradient coefficient $c$ and the diameter $d$ of the circular shaft entering explicitly in this relation (Tsagrakis et al., 1999).

Instead of the aforementioned procedure starting from a three-dimensional gradient elasticity, requiring higher order extra boundary conditions and based on the solution of a corresponding boundary value problem, we show below how size effects can be interpreted within a less rigorous but more robust method based on a direct gradient modification of the strength of materials approach for torsion and bending problems.

### 2.1. Torsion

We consider a solid cylinder of diameter $d$ subjected to a torsional deformation by an applied torque, such that

$$\gamma = \varphi r, \qquad M = \int_A \tau r \,\mathrm{d}A = 2\pi \int_0^{d/2} \tau r^2 \,\mathrm{d}r, \tag{2.2}$$

where $\gamma$ denotes shear strain, $\varphi$ is the angle of twist per unit length and $r$ is the radial coordinate of the point under consideration while $M$ denotes the applied torque and $\tau$ is the corresponding shear stress.

The appropriate gradient dependent shear stress-shear strain relation that we employ here reads

$$\tau = G[\gamma - c_1(\nabla\gamma \cdot \nabla\gamma)^{1/2} - c_2\nabla^2\gamma], \tag{2.3}$$

and involves two gradient coefficients $c_1$ and $c_2$, the nature and values of which will be discussed below, in conjunction with the fitting of the experimental results. It is pointed out that (2.3) should not be considered as a special case of (2.1) or its direct generalization to include first order gradient terms. In fact, it can readily be shown (Tsagrakis et al., 1999) that

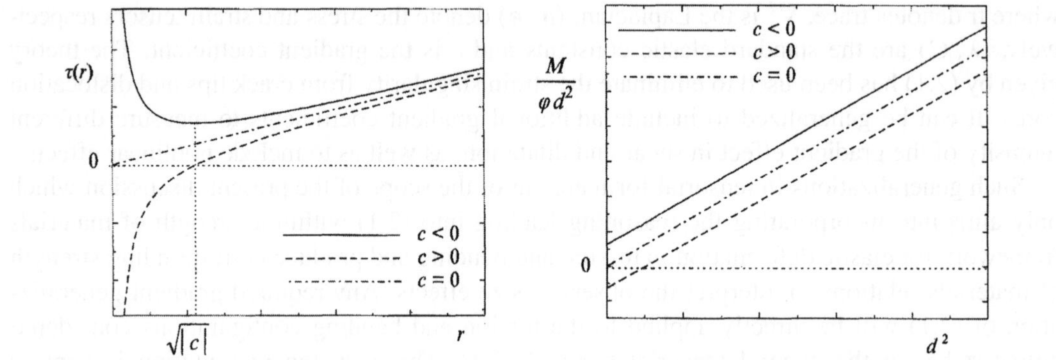


*Figure 1.* Qualitative shear stress and moment profiles.

a three-dimensional generalization of (2.3) reads $\boldsymbol{\sigma} = \lambda(\text{tr}\,\boldsymbol{\varepsilon})\mathbf{1} + 2G\boldsymbol{\varepsilon} - (\boldsymbol{\Sigma}_h + \boldsymbol{\Sigma}_d)$ where $\boldsymbol{\Sigma}_h = (\lambda+\frac{2}{3}G)\,[c_1\,\text{sign}(\text{tr}\,\boldsymbol{\varepsilon})|\nabla\text{tr}\,\boldsymbol{\varepsilon}|+c_2\nabla^2\,\text{tr}\,\boldsymbol{\varepsilon}]\mathbf{1}$ is the gradient contribution to the hydrostatic stress and $\boldsymbol{\Sigma}_d = (2G/\bar{\varepsilon})\,[c_1|\nabla\bar{\varepsilon}|+c_2\nabla^2\bar{\varepsilon}]\boldsymbol{\varepsilon}'$ is the gradient contribution to the deviatoric stress. The quantity $\bar{\varepsilon}$ denotes the equivalent strain $\bar{\varepsilon} = \sqrt{\frac{2}{3}\boldsymbol{\varepsilon}' \cdot \boldsymbol{\varepsilon}'}$ with $\boldsymbol{\varepsilon}' = \boldsymbol{\varepsilon} - (\frac{1}{3}\text{tr}\,\boldsymbol{\varepsilon})\mathbf{1}$ denoting the strain deviator.

On returning to (2.3) we consider its application to the interpretation of size effects for two different types of materials: bones and foams. In the first case it turns out that the contribution of the first gradient term is unimportant, while in the second case it turns out that this is the dominant term.

*Case 1*: $c_1 = 0, c_2 = c$. It follows that the corresponding gradient constitutive equation becomes

$$\tau = G[\gamma - c\nabla^2\gamma], \tag{2.4}$$

while the corresponding expression for the shear stress $\tau$ and the moment $M$ read

$$\tau = G\varphi\left(r - \frac{c}{r}\right), \qquad \frac{M}{\varphi d^2} = G\left(\frac{\pi d^2}{32} - c\frac{\pi}{4}\right). \tag{2.5}$$

The qualitative behavior of (2.5) is depicted in Figure 1, indicating that the obtained stress profile holds true only outside a core region determined by the gradient coefficient $c$ which, in turn, relates directly to the underlying microstructure as we shall see below.

The comparison of the theory to the experimental data obtained for human bone specimens, for which the dominant microstructure consists of osteons of 0.15–0.25 mm in diameter, is shown in Figure 2. Here and in the sequel, the values of the parameters, including both the standard material constants and the newly introduced gradient coefficients, are determined by using the least square method. Also, the dotted lines in the diagrams of this section correspond to the predictions of classical elasticity.

It is noted that the Cosserat theory employed by Yang and Lakes 1981 uses two gradient coefficients. It predicts the same behavior for the same $G$ and an internal length $l_t = \sqrt{|c|/6}$. It is also noted that if the gradient constitutive equation is written in the form $\tau = G\gamma - c'\nabla^2\gamma$ with $c' = Gc$, then it follows that in the above graphs $c' = -551\,\text{N}$ and $c' = -462\,\text{N}$ respectively.

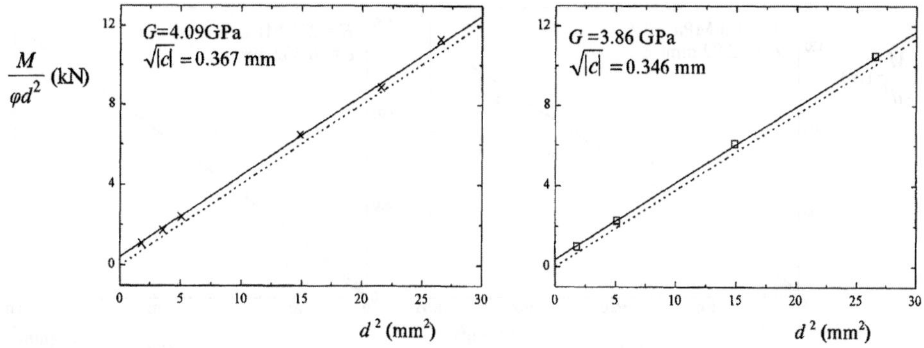


*Figure 2.* Qualitative comparison between theory and experiment for two types of human bones.

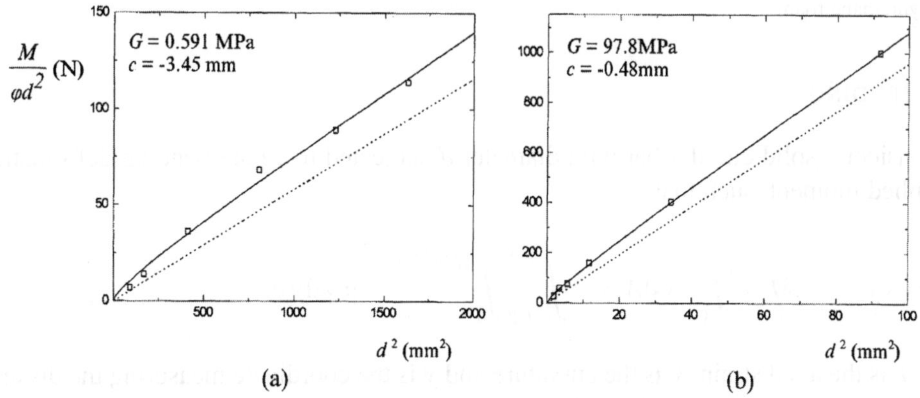


*Figure 3.* Qualitative comparison between theory and experiment for two types of foams: (a) polymeric foam and (b) polyurethane foam.

*Case 2* : $c_1 = c, c_2 = 0$. It follows that the corresponding gradient constitutive equation becomes

$$\tau = G(\gamma - c|\nabla\gamma|), \tag{2.6}$$

while the corresponding expression for the shear stress $\tau$ and the moment $M$ read

$$\tau = G\varphi(r - c), \qquad \frac{M}{\varphi d^2} = G\left(\frac{\pi d^2}{32} - c\frac{\pi\sqrt{d^2}}{12}\right). \tag{2.7}$$

The qualitative behavior of (2.7) does not present any peculiar feature as the one exhibited by (2.5). The comparison of the theory to experimental data obtained for foam specimens, for which the dominant microstructure consists of cells of 1 mm in diameter is shown in Figure 3 below for porous polymeric foam (Figure 3(a)) and polyurethane foam (Figure 3(b)) respectively.

It is noted that the Cosserat theory employed by Lakes (1983; 1986) uses four gradient coefficients. It predicts similar behavior for porous foam ($G = 0.60$ MPa, $l_t = 3.8$ mm) and polyurethane foam ($G = 104$ MPa, $l_t = 0.62$ mm). It is also noted that if the gradient constitutive equation is written in the form $\tau = G\gamma - c'|\nabla\gamma|$ with $c' = Gc$, then it follows that in the above graphs $c' = -2.04\ 10^3$ N/m and $c' = -4.7\ 10^4$ N/m respectively.

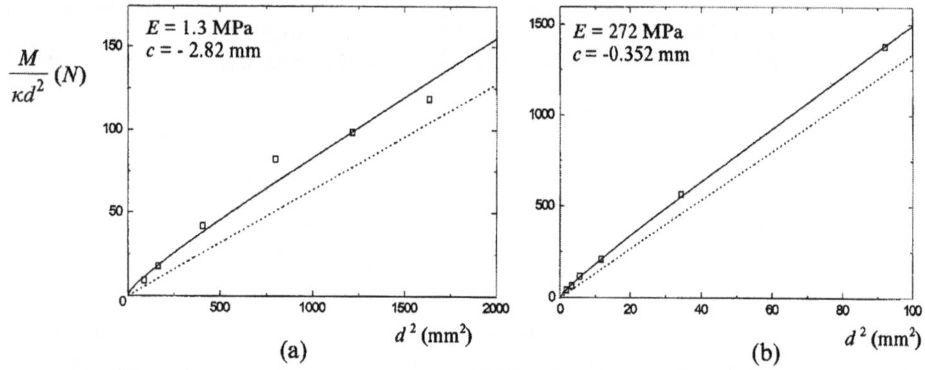


*Figure 4.* Quantitative comparison between theory and experiment for two types of foams:(a) polymeric foam and (b) polyurethane foam.

## 2.2. BENDING

We consider a solid circular bar with diameter $d$ subjected to a pure bending deformation by an applied moment, such that

$$\varepsilon = \kappa y, \qquad M = \int_A \sigma y \, \mathrm{d}A = 2 \int_{-d/2}^{d/2} \int_0^{\sqrt{(d^2/4)-z^2}} \sigma y \, \mathrm{d}y \, \mathrm{d}z, \tag{2.8}$$

where $\varepsilon$ is the axial strain, $\kappa$ is the curvature and $y$ is the coordinate measuring the distance of the point under consideration from the neutral axis.

The appropriate gradient dependent stress-strain relation which corresponds to (2.3) or (2.6) reads (note that $\nabla^2\varepsilon = 0$ for the pure bending configuration considered here)

$$\sigma = E(\varepsilon - c \, \mathrm{sign}(\varepsilon)|\nabla\varepsilon|). \tag{2.9}$$

This also results directly from the three-dimensional gradient generalization of (2.3) as discussed and explicitly written in the previous subsection. The corresponding expressions for the normal stress ($y > 0$) and the bending moment read

$$\sigma = E\kappa(y - c), \qquad \frac{M}{\kappa d^2} = E\left(\frac{\pi d^2}{64} - c\frac{d}{6}\right). \tag{2.10}$$

The comparison of the theory to experimental data obtained for the two types of foams discussed in the previous subsection is shown in Figure 4 below.

It is noted that the Cosserat theory employed by Lakes (1983; 1986) uses four gradient coefficients. It predicts similar behavior for polyurethane foam ($E = 299.5\,\mathrm{MPa}$, $l_b = 0.327\,\mathrm{mm}$), i.e. Figure 4(b), but this is not the case for polymeric foam, i.e. Figure 4(a). In fact, Lakes himself (1983) points out that his Cosserat type theory for the polymeric foam material used (with $E = 1.1\,\mathrm{MPa}$, $l_b = 5\,\mathrm{mm}$) does not predict satisfactorily the experimental trends for bending. In our case the agreement of the theoretical prediction to the experimental data is reasonably good, as shown in Figure 4(a). The values of the respective parameters were taken as $E = 1.3\,\mathrm{MPa}$ for Young's modulus (from Lakes (1983) reported experiments in tension; the corresponding measured value of Poisson's ratio is 0.07, while the one calculated from $E = 1.3\,\mathrm{MPa}$ and $G = 0.591\,\mathrm{MPa}$ listed in Figure 3(a) is 0.1) and $c = -2.82\,\mathrm{mm}$ for the gradient coefficient. It is also noted that if the gradient constitutive equation is written in the form $\sigma = E\varepsilon - c'|\nabla\varepsilon|$ with $c' = Ec$, then it follows that in the above graphs $c' = -3.67\ 10^3$ N/m and $c' = -9.57\ 10^4$ N/m respectively.

## 3. Size effects in gradient plasticity

In this section we employ a 'symmetric stress' gradient plasticity theory (Aifantis, 1984; 1987; 1992; 1995) to model the size-dependent plastic response noted by Fleck et al. (1994). Specifically, they report experimental results pertaining to the plastic flow behavior of high-purity (99.9 percent) copper wires of varying diameter twisted in the plastic range by controlling the magnitude of the surface shear strain to attain the same constant value. The experiments show that the flow stress increases with decreasing diameter and appropriate torque $M$ *vs.* surface strain $\gamma_s$ plots are obtained. They attributed the observed size effect to the radial gradient of the fixed surface shear strain that increases as the radius decreases. In fact, they built on this gradient dependent hardening to develop a three-dimensional first-order strain gradient couple stress theory of plasticity with asymmetric stress. However, they did not fully employ their theory to model their data for the whole range of the observed size-dependent response.

Specifically, they provided estimates of their internal length parameter, as well as of its average value, by utilizing only a single family of experimental isostrain points in the graphs of Figure 5 given below, without modelling the full range of the torque *vs.* surface strain curves as discussed in the remaining of this section. In presenting their experimental results Fleck et al. (1994) provide plots for the measured normalized torque $M/\alpha^3$ ($\alpha = d/2$ denotes in Fleck's notation wire radius) *vs.* surface strain $\gamma_s = \varphi\alpha$ ($\varphi$ denotes in Fleck's notation the angle of twist per unit length of the wire). They argued that if strain gradient hardening effects are unimportant, thc plots of normalized $M/\alpha^3$ *vs.* $\gamma_s$ would all lie on the same 'universal' curve instead of the family of curves obtained experimentally by varying $\alpha$. They claim that their first-order strain gradient theory can be used to interpret the observed size effect. Moreover, they comment that our second-order 'symmetric stress' strain gradient plasticity theory does not capture this effect. This comment, however, is not correct. It was apparently deduced by the implication that the (second-order) Laplacian term $\nabla^2\gamma$ which is central in the present author's gradient modification of flow stress, vanishes in their torsion experiments (if thc usual lincar relation between shear strain and radial distance is assumed). This is not the case, however, as in polar coordinates with $\gamma = \varphi r$ we have $\nabla^2\gamma = \partial^2\gamma/\partial r^2 + (1/r)(\partial\gamma/\partial r) = \varphi/r$ and, thus, the second gradient of strain contributes directly to the observed hardening.

It will be shown below that the second-order 'symmetric stress' gradient theory of plasticity can indeed interpret quantitatively the observed size effects in torsion. In fact, the experiments of Fleck et al. (1994) are used here to calibrate our gradient plasticity theory. This calibration suggests that in order to capture the observed behavior for the whole regime of the specimen sizes and strain ranges tested, additional gradient coefficients should be introduced and a special dependence of the gradient coefficients on the shear strain must be considered.

To illustrate this, we assume an effective strain measure $\bar{\gamma}$ in the form (Aifantis, 1995)

$$\bar{\gamma} = \gamma + \hat{c}(\gamma)\nabla^2\gamma. \tag{3.1}$$

While a formal argument based on a Taylor expansion of an integral expression for the average strain $\bar{\gamma} = (1/V)\int_V \gamma(x+s)\,\mathrm{d}V$ for a representative spherical volume $V$ and retaining the terms in the corresponding series up to second order (Mühlhaus and Aifantis (1991), Vardoulakis and Aifantis (1991)) leads to (3.1) with a constant gradient coefficient $\hat{c}$ proportional to the representative spherical volume radius $R$, it will be shown in the following that the generalization of a strain-dependent gradient coefficient $\hat{c}$ embodied in (3.1), is necessary for fitting the full set of experimental data for the whole range of wire diameters reported by Fleck et al. (1994) and also presented in Figure 5 below. In fact, if the aforementioned Taylor expansion for the average strain is combined with a self-consistent argument (Aifantis, 1995), it yields an explicit dependence of the gradient coefficient on the elastic constants and the plastic hardening modulus. This explicit dependence on the hardening modulus may be used as a motivation for the strain dependence of the gradient coefficient assumed in (3.1).

In passing, it is noted that the starting point of the integral representation for the average strain may be generalized to include a weighting function in the integrand (Bazant (1984), Bazant and Pijaudier-Cabot (1988)). In this connection, it is pointed out that expressions of the type (3.1) with $\hat{c}(\gamma)$ constant were proposed by Bazant (1984) for the strain tensor rather than the equivalent plastic strain in the context of elastic or brittle materials. In fact, such generalizations were motivated by polar and nonlocal theories of elasticity (e.g. Eringen (1966; 1971) and were also used in interpreting size effects (e.g. (Bazant and Ozbolt, 1990; Mazars et al., 1991)). The aforementioned generalizations on the entire set of tensorial stress-strain relations are often cumbersome and complicated to apply to the solution of specific problems and they exhibit excessive numerical difficulties. In contrast, it turns out that the gradient modification of the scalar equivalent strain $\gamma$ and of the corresponding equivalent stress $\tau$ overcome these technical problems resulting into a convenient structure of three-dimensional gradient theory (Aifantis, 1992) without loosing the essence of the physics we wish to model.

Next, we assume a general expression for the gradient dependent flow stress of the form

$$\tau = \tau_o + \kappa(\bar{\gamma}) - c_1|\nabla\bar{\gamma}| - c_2|\nabla\bar{\gamma}|^2, \tag{3.2}$$

with $\tau_0$ denoting an initial yield stress, $\kappa(\cdot)$ is the homogeneous-like part of the flow stress (apart of the fact that the effective strain measure $\bar{\gamma}$ instead of $\gamma$ enters now in this dependence), while $c_1$ and $c_2$ denote gradient coefficients. [The usual definitions

$$\nabla = \left(\frac{\partial}{\partial r}, \frac{1}{r}\frac{\partial}{\partial \theta}, \frac{\partial}{\partial z}\right) \quad \text{and} \quad \nabla^2 = \frac{\partial^2}{\partial r^2} + \frac{1}{r}\frac{\partial}{\partial r} + \frac{1}{r^2}\frac{\partial^2}{\partial \theta^2} + \frac{\partial^2}{\partial z^2}, \tag{3.3}$$

hold for the gradient and the Laplacian operators in polar coordinates].

Before we proceed further, we note that for the present case of torsion and for the proportional dependence of the gradient coefficient $\bar{c}$ on the strain $\gamma$ assumed in the sequel, the terms $|\nabla\bar{\gamma}|$ and $|\nabla\bar{\gamma}|^2$ entering (3.2) are identical to the terms $|\nabla\gamma|$ and $|\nabla\gamma|^2$. This results into a substantial simplification of the gradient dependence of the flow stress expression (3.2) and its subsequent application to interpret size effects in torsion, as only gradient terms up to second degree and order in the local strain $\gamma$ are considered. Such a simplification would hold true for general situations if for the gradient generalization of the flow stress we first replace $\kappa(\gamma)$ with $\kappa(\bar{\gamma})$ for its homogeneous-like part and then assume that any additional strain gradient dependence enters only through the local equivalent strain $\gamma$ rather than the average strain $\bar{\gamma}$. In passing, it is also noted that an initial stress $\tau_o$ has been included in (3.2) to represent the large differences in the initial portion of the experimental torque-surface strain graphs depicted in Figure 5 below. Such differences may be attributed to texture effects and this issue is considered in some detail in (Tsagrakis et al., 1999).

The strain field is given as usual by the relation

$$\gamma = r\varphi, \quad 0 \leqslant r \leqslant \alpha, \tag{3.4}$$

where $r$ is the radial coordinate, $\varphi$ denotes the angle of twist per unit length and $\alpha$ is the wire radius. By assuming that the entire cross-section of the circular bar is plastified, moment equilibrium implies the following relation for the applied torque $M$

$$M = 2\pi \int_0^\alpha \{\tau_o + \kappa(\bar{\gamma}) - c_1|\nabla\bar{\gamma}| - c_2|\nabla\bar{\gamma}|^2\} r^2\, dr, \tag{3.5}$$

where the flow stress $\kappa(\bar{\gamma})$ is assumed to obey a power law relation, while the gradient coefficient $\hat{c}(\gamma)$ is assumed to vary linearly with $\gamma$, i.e.

$$\kappa(\bar{\gamma}) = k\bar{\gamma}^n, \qquad \hat{c}(\gamma) = c + \bar{c}\gamma. \tag{3.6}$$

In the sequel, the value of $k$ is taken equal to 117 MPa and $n = 0.23$ as deduced by tension tests (Fleck et al., 1994). It is further assumed that $c \equiv 0$. It then follows that the final expression for the applied torque *vs.* the angle of twist $\varphi$ reads

$$M = 2\pi \int_0^\alpha \{\tau_o + k\varphi^n(r + \bar{c}\varphi)^n - c_1\varphi - c_2\varphi^2\} r^2\, dr. \tag{3.7}$$

This can be cast in the form used by Fleck et al. (1994) to represent their experimental measurements, i.e. $(M/\alpha^3)$ *vs.* $\gamma_s$ with $\gamma_s$ denoting the surface shear strain of the wire. It reads

$$\frac{M}{\alpha^3} = \frac{2\pi}{3}\left\{\tau_o - c_1\frac{\gamma_s}{\alpha} - c_2\left(\frac{\gamma_s}{\alpha}\right)^2\right\} + \frac{2\pi}{\alpha^3}\int_0^\alpha \left[k\left(r + \bar{c}\frac{\gamma_s}{\alpha}\right)^n\right] r^2\, dr. \tag{3.8}$$

where the analytical expression of the last integral is not given here because of its cumbersome form. The plot of (3.8) is depicted in Figure 5 below. The values of the parameters used to accomplish the fitting (by the least square method) are listed in Table 1 (the value of $G$ was taken to be 42 GPa). It is remarked that different values of the gradient coefficients are used for each curve which, however, are of the same order. While average-like values for the gradient coefficients could be used resulting into a less accurate fitting, we insist in using different values for each wire diameter as shown in Table 1, in order to reflect the possible dependence of the gradient coefficients on the substructure. Such substructure effects may be dominant in the present case where the grain size is comparable to the diameter of the wires.

It is also noted that the corresponding shear stress profile is given by the expression

$$\tau(r) = \tau_o + k\varphi^n(r + \bar{c}\varphi)^n - c_1\varphi - c_2\varphi^2, \tag{3.9}$$

which is plotted in Figure 6 for two values of $\varphi$ and average values of the gradient coefficients as given below.

In this connection, it is noted that at the wire axis ($r \to 0$) the shear stress $\tau(0)$ is given by the expression

$$\tau(0) = \tau_o + k\bar{c}^n\varphi^{2n} - c_1\varphi - c_2\varphi^2, \tag{3.10}$$

*Table 1.*

| $2\alpha$ ($\mu$m) | $\tau_o$ (MPa) | $\sqrt{\bar{c}}$ ($\mu$m) | $-c_1/k$ ($\mu$m) | $\sqrt{c_2/k}$ ($\mu$m) | $-c_1/G$ ($10^{-3}\,\mu$m) | $\sqrt{c_2/G}$ ($\mu$m) | $-c_1$ (N/m) | $c_2$ ($10^{-3}$ N) |
|---|---|---|---|---|---|---|---|---|
| 12 | 130 | 0.8 | 5.2 | 3.4 | 14.5 | 0.18 | 608.4 | 1.35 |
| 15 | 57.6 | 1.1 | 9.3 | 5.9 | 25.9 | 0.31 | 1088.1 | 4.07 |
| 20 | 15.3 | 0.8 | 4.6 | 4.1 | 12.8 | 0.22 | 538.2 | 1.97 |
| 30 | 5.5 | 0.5 | 7.5 | 7.6 | 20.9 | 0.40 | 877.5 | 6.76 |
| 170 | — | — | — | — | — | — | — | — |

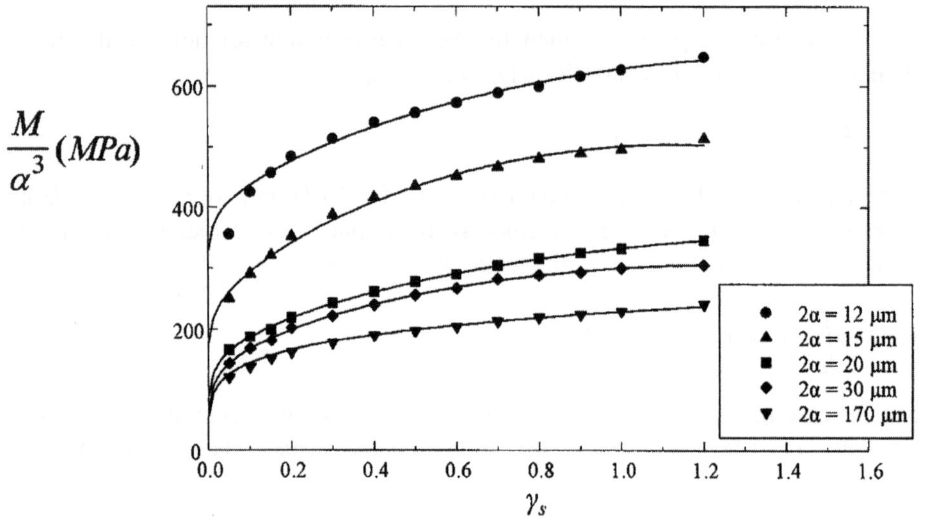


*Figure 5.* Quantitative comparison between theory and experiment for twisted Cu wires.

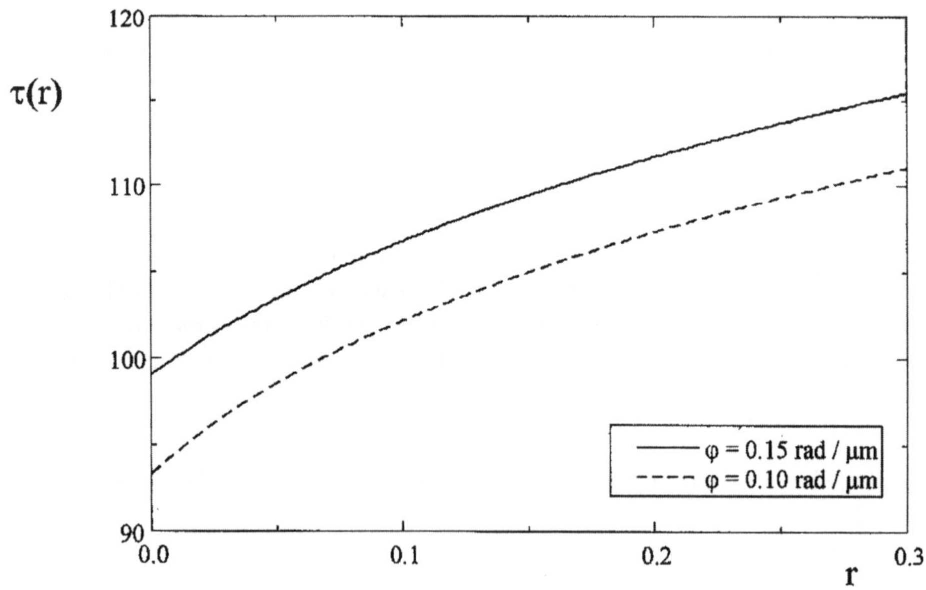


*Figure 6.* Shear stress profiles using $\tau_o = 5.5$ MPa and the average values of the gradient coefficients computed from Table 1, i.e. $\bar{c} = 0.64\,\mu\text{m}^2$, $c_1 = -778$ N/m and $c_2 = 3.54\ 10^{-3}$ N.

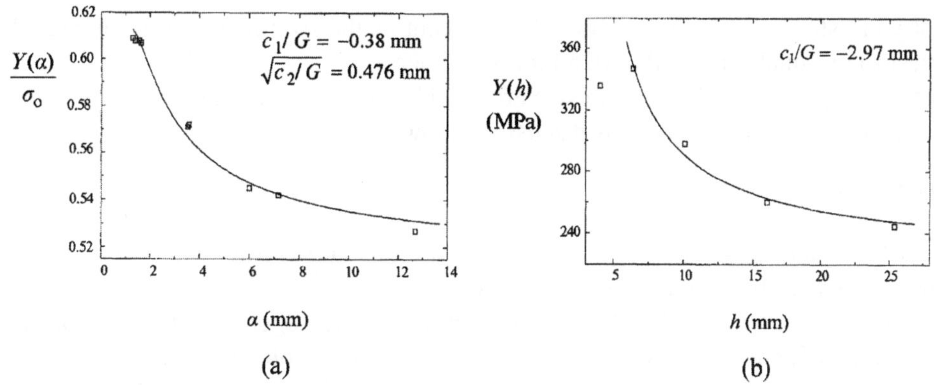


*Figure 7.* Quantitative comparison between theory and experiment in yielding behavior: (a) in torsion and (b) in pure bending.

indicating that a hardening occurs at the center, as twisting continually occurs. This type of hardening behavior at the center of twisted cylinders has been observed experimentally (Szekely, 00) and will be analyzed in detail in (Tsagrakis et al., 1999). Such hardening is not predicted by the Fleck et al. (1994) theory. A possible physical explanation of this may be sought to the fact that dislocations are pushed to the center of the shaft, thus contributing to the observed strain gradient hardening at this region. Of course, more detailed and careful experiments need to be done in this direction and the present conclusion should be viewed only as preliminary requiring further documentation.

## 4. Size effects in yield initiation

In this section, we provide an interpretation of the size effects observed by Morrison (1939) and Richards (1958) for yield initiation in torsion and bending. Within our simplified strength of materials approach we first assume classical gradient independent elastic behavior until yielding occurs at the outer surface. At this stage, we adopt a gradient dependent stress-strain relation of the form (Tsagrakis et al., 1998)

$$\sigma = k\varepsilon^n - c_1(\nabla\varepsilon \cdot \nabla\varepsilon)^{1/2} - c_2\nabla^2\varepsilon, \tag{4.1}$$

where $(\sigma, \varepsilon)$ denote the equivalent stress and strain, $(k, n)$ denote the characteristic parameters of the homogeneous plastic response and $(c_1, c_2)$ denote gradient coefficients. It is noted that a gradient elastic behavior could also be incorporated resulting into a similar behavior as the one depicted in Figure 7 below but with an increased number of gradient coefficients of different values than those listed there (Tsagrakis et al., 1999).

Morrison (1939) performed a series of careful torsion tests on the yield behavior of plain carbon steel cylindrical specimens of different size. The specimens were geometrically similar with varying diameter and a scale factor up to 8.67. The results indicated an increase in the apparent yield stress with decreasing specimen size. The value of yield stress increases by about 15.5 percent when the size is decreased from the largest to the smallest (diameter 2.586 mm) specimen.

Richards (1958) performed a series of pure bending tests for geometrically similar mild steel beam specimens of different size. In these tests a scale factor up to 6.31 was used and the value of yield stress increases by about 41.7 percent as the specimen size is decreased from the largest to the smallest (depth 4.026 mm) specimen.

For the interpretation of the above size effects, Hooke's law for the elastic region and (4.1) for the plastic region are adopted, as discussed below.

### 4.1. TORSION

In the elastic region the shear stress $\tau$ is given by the standard elastic relation $\tau = G\gamma$, where $\gamma$ denotes the shear strain and $G$ the shear modulus.

Carbon steel may be considered as perfectly plastic for homogeneous deformations beyond the elastic limit. Therefore, $n$ is zero in (4.1) where $k$ now denotes the tensile yield stress $\sigma_o$. For the present shear deformation problem, the effective axial stress $\sigma$ equals to $\tau/\Lambda$ and the effective axial strain $\varepsilon$ as $\gamma\Lambda$, where $\Lambda$ is the ratio of the shear yield stress $\tau_o$ over the tensile yield stress $\sigma_o$ ($\Lambda = 0.5$ for the Tresca yield criterion and $\Lambda = 1/\sqrt{3} = 0.577$ for the von Mises yield criterion). The appropriate gradient dependent stress-strain relation reads

$$\tau = \tau_o - \bar{c}_1|\nabla\gamma| - \bar{c}_2\nabla^2\gamma, \tag{4.2}$$

where $\bar{c}_1 = c_1\Lambda^2$ and $\bar{c}_2 = c_2\Lambda^2$.

By assuming that yielding first occurs when the stress at the outer surface of the specimen ($r = \alpha$) becomes equal to the yield stress value $Y$, we obtain

$$Y = G\varphi\alpha = \tau_o - \bar{c}_1\varphi - \bar{c}_2\varphi/\alpha, \tag{4.3}$$

since $\nabla\gamma = \varphi$ and $\nabla^2\gamma = \varphi/r$. On eliminating $\varphi$ from the system (4.3) we obtain the following relation for the dependence of the yield stress $Y$ on the specimen radius $\alpha$

$$Y(\alpha) = \tau_o\left(\frac{\alpha^2}{\alpha^2 + (\bar{c}_2/G) + (\bar{c}_1/G)\alpha}\right). \tag{4.4}$$

Using the least squares method, Morrison's torsion experimental results are fitted and the values of the internal lengths $l_1 = -\bar{c}_1/G$ and $l_2 = \sqrt{|\bar{c}_2|/G}$, are calculated as shown in Figure 7a. The ratio it $\Lambda$ of the shear yield stress $\tau_o$ to the tensile yield stress $\sigma_o$ is also calculated by the aforementioned fitting and is found to be equal to 0.516. This value is between 0.577 obtained by the von Mises yield criterion and 0.5 obtained by the Tresca yield criterion and it is comparable to the values found in the literature on carbon steel.

### 4.2. BENDING

In the elastic region the axial stress $\sigma$ in the beam axis is given by Hooke's law $\sigma = E\varepsilon$, where $E$ is the Young's modulus and $\varepsilon = \kappa y$ is the axial strain ($\kappa$ denotes the curvature of the beam and $y$ the coordinate along the depth of the beam, measured from the neutral axis). It turns out that the same relationship $\sigma = E\varepsilon$ holds for the equivalent stress $\sigma$ and equivalent strain $\varepsilon$ (which for convenience are denoted by the same symbols as the respective axial quantities) with $\varepsilon = \kappa|y|$.

Mild steel behaves as a perfectly plastic material for homogeneous deformations beyond the elastic limit. Therefore, $n$ is zero in (4.1) and $k$ is equal to the tensile yield stress $\sigma_o$. Then (4.1) takes the form

$$\sigma = \sigma_o - c_1|\nabla\varepsilon| - c_2\nabla^2\varepsilon. \tag{4.5}$$

We assume that yielding first occurs when the stress at the outer surface of the specimen ($y = \pm h/2$) becomes equal to the yield stress value $Y$. Therefore,

$$Y = E\kappa h/2 = \sigma_o - c_1\kappa, \tag{4.6}$$

since $\nabla\varepsilon = \kappa$ and $\nabla^2\varepsilon = 0$ for the case of pure bending. On eliminating $\kappa$ from the system (4.6) we obtain the following relation for the dependence of the yield stress $Y$ on the specimen depth $h$

$$Y(h) = \sigma_o\left(\frac{h/2}{(h/2) + (c_1/E)}\right). \tag{4.7}$$

Using the least squares method, Richards' bending experimental data are fitted with an internal length $l_1 = -c_1/G$, as shown in Figure 7(b), where the value $\nu = 0.32$ for the Poisson ratio was used. Because of the lack of data for the tension behavior of the material used, the tensile yield stress is also calculated by the aforementioned fitting and is found to be equal to 225.6 MPa. This value is comparable to the value 250 MPa found in the literature on mild steel. It is noted that the leftmost experimental point for the smallest specimen is not considered, as Richards himself have expressed doubts about its validity (Richards, 1958).

## 5. Concluding remarks

A gradient modification to the standard 'strength of materials' approach for elastic and plastic behavior was considered to interpret size effects in torsion and bending. The results compare well with the experimental trends and suggest that a more rigorous 'boundary-value' approach for interpreting size effects in deformation and fracture by using gradient elasticity and gradient plasticity models is in order. For example, various gradient elasticity models need to be employed for considering torsion and bending configurations in the sense that these problems are treated within a classical elasticity setting. The same should be done for gradient plasticity models.

In this connection, we refer to a recently published article by Vardoulakis et al. (1998) where a gradient elasticity model is employed to describe the mechanical behavior of marble and related size effects in three-point bending. While the relevant size effect is captured by the analysis, it seems that a minor inconsistency exists between the assumed gradient-dependent strain energy density, the corresponding equilibrium equations and the used boundary conditions. It turns out that the simple strain gradient elasticity model embodied in (2.1) with appropriate boundary conditions can be used to interpret the experimental results of Vardoulakis et al. (1998) as discussed in detail by Tsagrakis et al. (1999).

Similarly, reference is made to a recently published article by Stolken and Evans (1998) where the 'asymmetric stress' strain gradient plasticity model of Fleck and Hutchinson (1997) is used to interpret the observed size effect in bending of thin foils. This problem is also discussed in detail by Tsagrakis et al. (1999) by using 'symmetric stress' strain gradient models of the type of (3.2). In fact, by employing a gradient-dependent strain energy density formulation which allows a direct comparison of the aforementioned two approaches, one may consider the form $w = w(\varepsilon) + c_1|\nabla\varepsilon| + c_2|\nabla\varepsilon|^2$; $w(\varepsilon) = \varepsilon[3E_p\varepsilon + 4\sqrt{3}\Sigma_o]/8$ is the homogeneous part of the strain energy density with $\varepsilon = (2/\sqrt{3})\kappa|y|$ (plane strain conditions are assumed, instead of the pure bending configuration of Section 4.2) denoting the equivalent

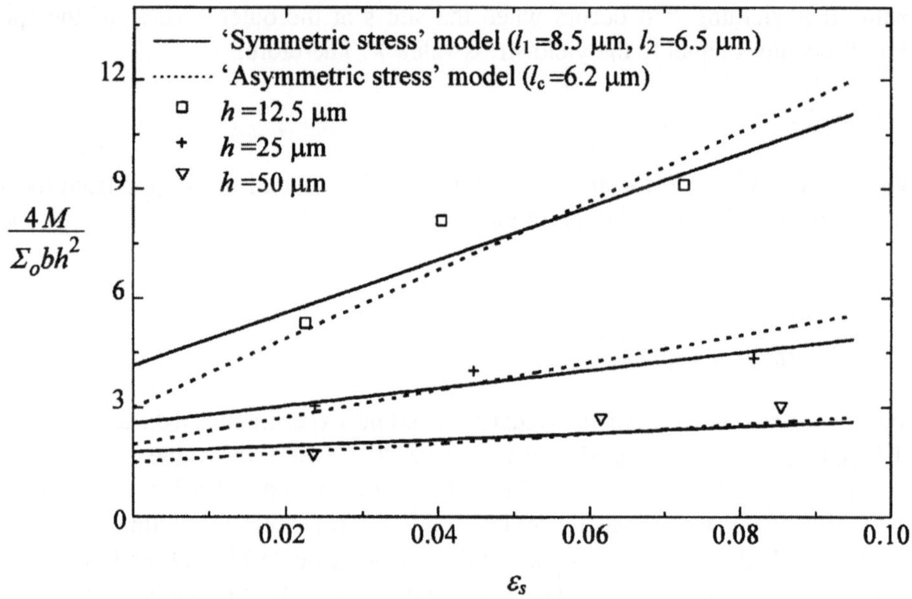


*Figure 8.* Fitting of Stolken and Evans (1998) thin foil bending experimental results with the 'symmetric stress' and 'asymmetric stress' strain gradient models.

strain and $(c, c_2)$ being gradient coefficients. The assumed expression for the homogeneous part of the strain energy density is the same as in Stolken and Evans (1998) with $E_p$ denoting plastic modulus (hardening coefficient) and $\Sigma_o$ effective yield strength. Finally, by using the relation $M = \mathrm{d}W/\mathrm{d}\kappa (W = \int_A w\,\mathrm{d}A)$ for the applied moment, we can obtain a moment $(M)$ vs. surface strain $(\varepsilon_s)$ relation depicted with solid lines in Figure 8. The dotted lines correspond to the prediction of the Fleck–Hutchinson theory employed by Stolken and Evans (1998) for a value of their internal length $l_c = 6.2\,\mu$m. [There is a departure of this prediction from the one reported in the aforementioned article due to a misprint in rescaling the moment and computing the corresponding theoretical values. This issue will be addressed in more detail in a future occasion]. The internal lengths $l_1$ and $l_2$ appearing in Figure 8 are defined by $l_1 = c_1/\Sigma_o$ and $l_2 = \sqrt{c_2/E_p}$.

**Acknowledgements**

Support of EU under the Reactor Vessel Integrity in Severe Accidents (REVISA) Project and Contract FI4S-CT96-0024 is acknowledged. The involvement of I. Tsagrakis (and his contribution throughout the preparation of the article) and I. Eleftheriadis and their help with the fitting of the experimental data is also acknowledged. Related discussions with T. Malmberg were very useful. Finally, the constructive comments of the reviewers on an earlier version of this report were helpful in revising it in its present form.

## References


Aifantis, E.C. (1984). On the microstructural origin of certain inelastic models. *Journal of Engineering Materials and Technology* **106**, 326–330.

Aifantis, E.C. (1987). The physics of plastic deformation. *International Journal of Plasticity* **3**, 211–247.

Aifantis, E.C. (1992). On the role of gradients in the localization of deformation and fracture. *International Journal of Engineering Science* **30**, 1279–1299.

Aifantis, E.C. (1995). Pattern formation in plasticity. *International Journal of Engineering Science* **33**, 2161–2178.

Aifantis, E.C. (1996). Higher order gradients and size effects. *Size-Scale Effects in the Failure Mechanisms of Materials and Structures* (Edited by A. Carpinteri), E & FN Spon, London, 231–242.

Bazant, Z.P. (1976). Instability, ductility and size effect in strain softening concrete. *Journal of Engineering Mechanics ASCE* **102**, 225–238.

Bazant, Z.P. (1984). The imbricate continuum and its variational variation. *Journal of Engineering Mechanics* **110**, 1693–1712.

Bazant, Z.P. (1986). Mechanics of distributed cracking. *Applied Mechanical Review* **39**, (1986) 675–705.

Bazant, Z.P. and Pijaudier-Cabot, G. (1988). Nonlocal continuum damage, localization instability and convergence. *Journal of Applied Mechanics* **5**, 287–293.

Bazant, Z.P. and Ozbolt, J. (1990). Nonlocal microplane model for fracture, damage and size effects in structures. *Journal of Engineering Mechanics ASCE* **116**, 2485–2505.

Bazant, Z.P. and Cedolin, L. (1991). *Stability of Structures: Elastic, Inelastic Fracture, and Damage Theories*, Oxford University Press, New York.

Carpinteri, A. (ed.) (1996). *Size-Scale Effects in the Failure Mechanisms of Materials and Structures*, E & FN Spon, London.

Eringen, A.C. (1966). Linear theory of micropolar elasticity. *Journal of Mathematical Mechanics* **15**, 909–923.

Eringen, A.C. (ed.) (1971, 1975, 1976). *Continuum Physics,* Vols I–IV, Academic Press, New York.

Fleck, N.A. and Hutchinson, J.W. (1997). Strain gradient plasticity. *Advances in Applied Mechanics* **33** (Edited by J.W. Hutchinson and T.Y. Wu), Academic Press, London, 295–361.

Fleck, N.A., Muller, G.M., Ashby, M.F. and Hutchinson, J.W. (1994). Strain gradient plasticity: Theory and experiment. *Acta Metallurgica et Materialia* **42**, 475–487.

Lakes, R.S. (1983). Size effects and micromechanics of a porous solid. *Journal of Materials Science* **18**, 2572–2580.

Lakes, R.S. (1986). Experimental microelasticity of two porous solids, *International Journal of Solids and Structures* **22**, 55–63.

Mazars, J., Pijaudier-Cabot, G. and Saouridis, C. (1991). Size effect and continuous damage in cementitious materials. *International Journal of Fracture* **51**, 159–173.

Morrison, J.L.M. (1939). The yield of mild steel with particular reference to the effect of size of specimen. *Proceedings of the Institute of Mechanical Engineers* **142**, 193–223.

Mühlhaus, H.B. and Aifantis, E.C. (1991). A variational principle for gradient plasticity. *International Journal of Solids and Structures* **28**, 845–857.

Richards, C.W. (1958). Effect of size on the yielding of mild steel beams. *Proceedings of the American Society for Testing and Materials* **58**, 955–970.

Ru, C.Q. and Aifantis, E.C. (1993). A simple approach to solve boundary value problems in gradient elasticity. *Acta Mechanica* **101**, 59–68.

Stolken, J.S. and Evans, A.G. (1998). A microbend test method for measuring the plasticity length scale. *Acta Materialia* **46**, 5109–5115.

Szekely, F. and Lendvai, J. Private communication.

Tsagrakis, I., Malmberg, T. and Aifantis, E.C. (1998). Gradient plasticity and size effects. *Proceedings of the 5th National Congress on Mechanics* (Edited by P.S. Theocaris et al.), Ioannina, Greece, 953–960.

Tsagrakis, I., Eleftheriadis, I., Konstantinidis, A. and Aifantis, E.C. (1999). Size effects in gradient elasticity and plasticity. *AUT Reports in Mechanics of Materials: AUT/MM4*, Aristotle University of Thessaloniki, Thessaloniki, Greece.

Vardoulakis, I. and Aifantis, E.C. (1991). A gradient flow theory of plasticity for granular materials. *Acta Mechanica* **87**, 197–217.

Vardoulakis, I. and Sulem, J. (1995). *Bifurcation Analysis in Geomechanics*, Blackie Academic and Professional, London.

Vardoulakis, I., Exadaktylos, G. and Kourkoulis, S.K. (1998). Bending of marble with intrinsic length scales: A gradient theory with surface energy and size effects. *Journal de Physique IV* **8**, 399–406.

Yang, J.F.C. and Lakes, R.S. (1981). Transient study of couple stress effects in compact bone: Torsion, *Journal of Biomedical Engineering* **103**, 275–279.

Zhu, H.T., Zbib, H.M. and Aifantis, E.C. (1997). Strain gradients and continuum modeling of size effect in metal matrix composites. *Acta Mechanica* **121**, 165–176.

*International Journal of Fracture* **95:** 315–324, 1999.


# Use of neural networks for fitting of FE probabilistic scaling model parameters

E.M.R. FAIRBAIRN[1], C.N.M. PAZ[1], N.F.F. EBECKEN[1] and F.-J. ULM[1,2]*
[1]*COPPE/UFRJ Programa de Eng. Civil, C.P. 68506, 21945-970, Rio de Janeiro, RJ, Brazil*
[2]*Laboratoire Central des Ponts et Chaussées, 58, Bd. Lefebvre, 75732 Paris Cedex 15, France*
*e-mail: ulm@mit.edu*

Received 2 November 1998; accepted in revised form 31 December 1998

**Abstract.** The probabilistic crack approach, based on the Monte Carlo method, was recently developed for finite element analysis of concrete cracking and related size effects. In this approach the heterogeneity of the material is taken into account by considering the material properties (tensile strength, Young modulus, etc.) to vary spatially following a normal distribution. N samples of the vector of random variables are generated from a specific probability density function, and the N samples corresponding to a simulation are functions of the mean value and of the standard deviation that define the Gauss density function. The problem is that these statistical moments are not known, *a priori*, for the characteristic volume of the finite elements used in the analysis. The paper proposes an inverse finite element analysis using neural networks for the determination of the statistical distribution parameters (e.g., for a normal distribution, the mean and the standard deviation) from a given response of the structure (for instance, an average load-displacement curve). From FE-analysis of 4-point bending beam tests, it is shown that the backanalysis technique developed in this paper is a powerful tool to determine the probabilistic distribution functions at the material level from structural tests for material volumes which are generally not accessible to direct testing.

**Key words:** Artificial neural networks, concrete cracking, inverse analysis, finite element back analysis, heterogeneity, probabilistic crack approach, tensile strength domain.

## 1. Introduction

Among several other relevant factors (such as water/cement ratio of the paste, casting and curing conditions, loading conditions, etc.) concrete cracking depends on the random distribution of constituents and initial defects. The heterogeneity governs the overall cracking behavior and related size effects on concrete fracture. The probabilistic crack approach, based on the direct Monte Carlo method, developed by Rossi and co-workers (Rossi and Richer, 1987; Rossi et al., 1994, 1996; Rossi and Ulm, 1997) takes this stochastic process into account in finite element analysis, by assigning randomly distributed material properties (tensile strength, Young's modulus) to both the solid elements and the contact elements interfacing the former (discrete crack approach). The stochastic process is introduced at the local scale of the material, by considering that cracks are created within the concrete with different energy dissipation depending on the spatial distribution of constituents and initial defects. The local material behavior in concrete is assumed to obey to a perfect brittle elastic behavior, so that the random distribution of local cracking energies can be replaced by a random distribution

* Correspondence Address: Massachusetts Institute of Technology, Department of Civil and Environmental Engineeering, Cambridge, MA 02139, U.S.A.

of local strengths. The crack-opening $[u]$ in the discrete crack approach is governed by the strength criteria

$$f = \max(f_\sigma, f_\tau); \qquad f_\sigma = \sigma_n - f_{\mathrm{ct}} \leqslant 0; \qquad f_\tau = |\tau|/k - f_{\mathrm{ct}} \leqslant 0, \tag{1}$$

where $\sigma_n = \boldsymbol{n} \cdot \boldsymbol{T}$ and $\tau = \boldsymbol{t} \cdot \boldsymbol{T}$ = the normal and the tangential components of the stress-vector $\boldsymbol{T}$ in the local base $(\boldsymbol{n}, \boldsymbol{t})$ of the contact element; $k \approx 5$ = a constant which relates the cohesion $c = kf_ct$ to the local – randomly distributed – tensile strength $f_ct$. Crack criteria (1) refer to the fact that concrete cracking corresponds to a mode-I mechanism in both tension and compression (Rossi et al., 1996). An irreversible fragile tensile behavior is assumed (i.e., $\sigma_n[u] = 0$).

The probabilistic model involves a number of mechanical properties of the material to be determined, which constitutes the modeling data. From a large number of direct tensile tests, it was found that a normal law describes rather well the experimental distribution (Rossi et al., 1994). The characteristics are the mean values $f_{\mathrm{ct},\mu}$ and $E_\mu$, and the standard deviation $f_{\mathrm{ct},\sigma}$ and $E_\sigma$, of the local tensile strength $f_{\mathrm{ct}}$ and the Young's modulus $E$, respectively; and the following analytical expressions were proposed

$$f_{\mathrm{ct},\mu} = 6.5C(V_t/V_g)^{-a}; \qquad f_{\mathrm{ct},\sigma}/f_{ct,\mu} = 0.35(V_t/V_g)^{-b} \tag{2}$$

$$E_\mu = E; \qquad E_\sigma/E = 0.15(V_t/V_g)^{-c}, \tag{3}$$

where $V_t$ = the volume of the two finite elements connected by a contact element; $V_g$ = the volume of the coarsest aggregate; $E$ = average Young's modulus which does not exhibit significant volume effects; and $a, b, c$ = constants related to the cylinder compressive strength $f_c$

$$\begin{aligned} a &= 0.25 - 3.6 \times 10^{-3}(f_c/C) + 1.3 \times 10^{-5}(f_c/C)^2 \\ b &= 4.5 \times 10^{-2} + 4.5 \times 10^{-3}(f_c/C) - 1.8 \times 10^{-5}(f_c/C)^2 \\ c &= 0.116 + 2.7 \times 10^{-3}(f_c/C) - 3.4 \times 10^{-6}(f_c/C)^2 \end{aligned} \tag{4}$$

with $C = 1$ MPa. In these expressions, the compressive strength $f_c$, represents the quality of the concrete matrix, while the volume of the coarsest aggregate $V_g$, refers to the *elementary material heterogeneity*.

Figure 1 shows, in a double logarithmic space, the tensile strength domain as a function of the volume ratio $V_t/V_g$ for a concrete of compressive strength $f_c = 60$ MPa, calculated from (2). The figure shows that the smaller the scale of observation (respectively the modeling scale), the larger the fluctuation of the local mechanical properties, and thus the (modeled) heterogeneity of the matter. In other words, the finer the mesh, the greater the modeled heterogeneity in terms of Young's modulus and tensile strength. In this way, the probabilistic approach aims at modeling size effects by bridging the gap between the strength theories and linear fracture mechanics: for small volume-ratios, the strength criteria (1) are enriched by the randomness of material strength (Weilbull, 1939); for larger sizes, both the mean value and the standard deviation decrease, i.e., the Weibull-type statistical size effects decrease. This is in agreement with Bažant's (1997) asymptotic analysis of quasi-brittle fracture. Provided that the elementary material volume (i.e., volume of the solid finite element $V_t$) remains small

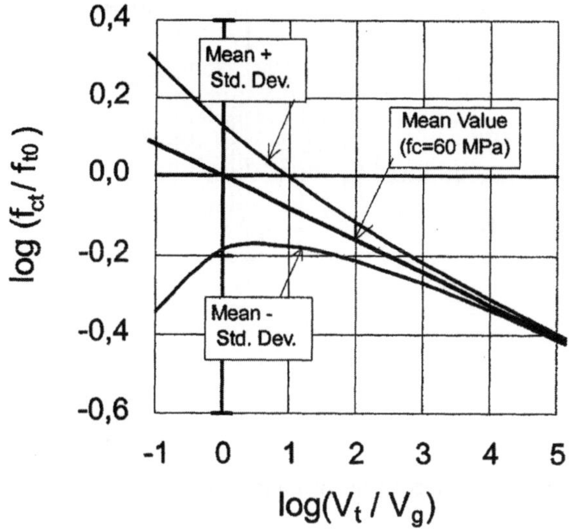


*Figure 1.* The tensile strength domain of concrete according to the probabilistic crack approach.

with respect to the structural volume, the numerical results are mesh independent (Rossi et al., 1996).

The empirical expressions (2)–(4) were calibrated to fit a number of material tests for different type of concretes and volumes. They give a first approximation to capture volume effects on concrete fracture; but they are not expected to be universal. Furthermore, little is known about the validity of these empirical formulae for small $V_t/V_g$ volume ratios which are typically applied in finite element structural analysis, and for which it is impossible to determine, by means of experimental tests, the probabilistic strength domain and its distribution within the bulk specimen. With respect to these shortcomings of the probabilistic crack approach, this paper presents an inverse analysis for the determination of the probabilistic distribution function for which the input is an experimental curve and the output is the spatial distribution. The feasibility of the inverse procedure has been shown recently (Fairbairn et al., 1998a). The present paper focuses on an optimized procedure based on the application of neural networks, which is suitable for finite element inverse analysis.

## 2. Problem statement

The inverse procedure will be developed here by using an illustrative example: a notched plain concrete 4-point bending beam test, carried out at TU Delft. The geometrical details of the test are shown in Figure 2a, together with the finite element discretization, composed of 2D elements interfaced by contact elements. The mesh has $m$ interface elements. Each interface element follows an elastic-brittle constitutive law (cf. Figure 2b) characterized by an individual tensile strength $f_{\mathrm{ct},i}$, and an individual Young's modulus, $E_i$.

Let these individual local tensile strengths and Young's moduli be normally distributed having the densities

$$g_f(f_{\mathrm{ct}}) = \frac{1}{f_{\mathrm{ct},\sigma}\sqrt{2\pi}} \exp\left[-\frac{1}{2}\left(\frac{f_{\mathrm{ct}} - f_{\mathrm{ct},\mu}}{f_{\mathrm{ct},\sigma}}\right)^2\right] \tag{5}$$

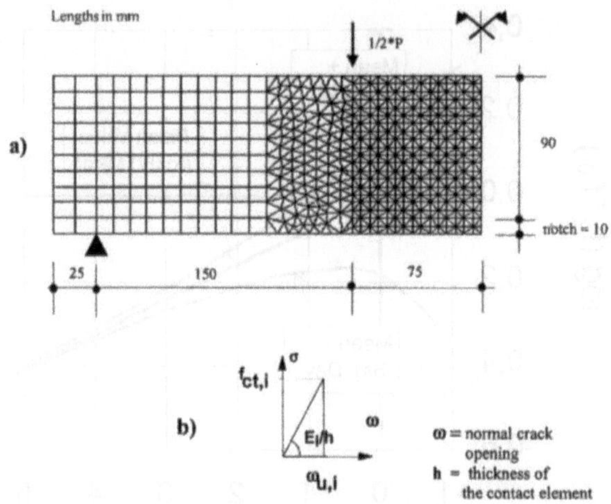


*Figure 2.* (a) TU-Delft beam discretized by finite elements, (b) Elastic-Brittle Contact law.

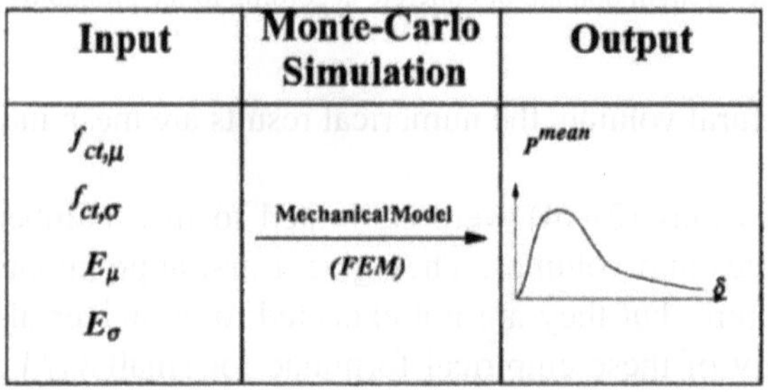


*Figure 3.* Sketch of the Direct Monte Carlo Simulation.

$$g_E(E) = \frac{1}{E_\sigma\sqrt{2\pi}} \exp\left[-\frac{1}{2}\left(\frac{E - E_\mu}{E_\sigma}\right)^2\right] \tag{6}$$

with $g_f(f_{\text{ct}})$ and $g_E(E)$ the density function for the tensile strength $f_{\text{ct}}$ and the Young's modulus $E$, respectively; while $x_\mu$ and $x_\sigma$ denote the mean and the standard deviation of the distribution of quantity $x$. For the problem at hand, it is possible to find a sample of $m$ pairs $(f_{\text{ct},i}, E_i)$, each pair corresponding to an interface element, by using a standard routine for generation of random numbers for a given normal distribution (e.g., Press et al., 1992). This allows for the computation of the load-displacement curve for the beam. The load-displacement curve represents one sample obtained by running the probabilistic model based FEM program (Fairbairn et al., 1997).

### 2.1. Direct Monte Carlo Simulation

At this point one can implement a Monte Carlo simulation: A number of $n$ samples is generated (for a given normal distribution) and the corresponding $P - \delta$ curves are computed. This direct Monte Carlo procedure is sketched in Figure 3. Let the $j$th sample correspond to the $j$th $P - \delta$ curve. This $j$th $P - \delta$ curve is composed of discrete values, $P_k^j$ and $\delta_k^j$, where the superscript $j$ indicates the sample and the subscript $k$ the discrete value of the $P - \delta$ curve. The

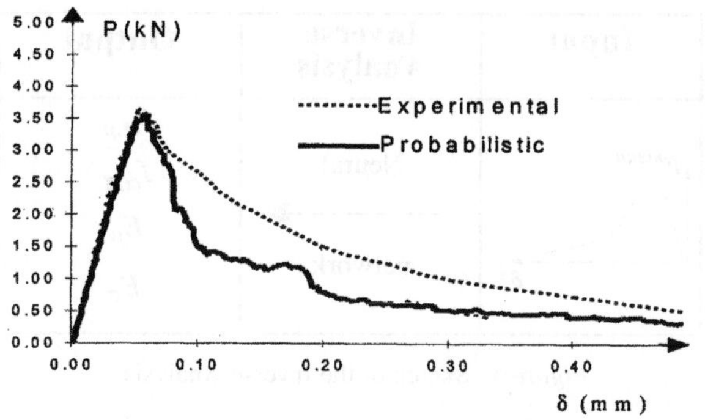


*Figure 4.* Result of the Direct Monte Carlo Simulation with the empirical distribution parameters $f_{ct,\mu} = 7.25$ MPa, $f_{ct,\sigma} = 2.76$ MPa, $E_\mu = 40,490$ MPa, and $E_\sigma = 7,080$ MPa [according Equations (2)–(4)].

same number of discrete $\delta$-values is assumed. The mean curve composed by pairs $(P_k^{\text{mean}}, \delta_k)$ then simply reads

$$P_k^{\text{mean},j} = \frac{1}{j}\sum_{l=1}^{j} P_k^l. \tag{7}$$

The Monte Carlo procedure is stopped when

$$|P_k^{\text{mean},j} - P_k^{\text{mean},j-1}| \leqslant \text{TOL1}, \tag{8}$$

where TOL1 is a prescribed tolerance to check the convergence of the procedure. Equation (7) and (8) constitute the Monte Carlo simulation, and with the convergence of the procedure, the total number of samples is set to $n = j$. This total number of samples $n$, corresponding to a Monte Carlo converged simulation, clearly depends on TOL1, which is a measure of the precision required by the analysis. It also depends on the heterogeneity of the material represented by the standard deviation. The more heterogeneous is the material the greater is the number of samples necessary to obtain a converged solution. Our experience in this field indicates that 25 to 50 samples are sufficient to obtain a converged $P - \delta$-curve.

For a given normal distribution characterized by the mean and the standard deviation of the tensile strength (i.e., the input of the system), the described direct Monte Carlo procedure delivers a $P-\delta$ mean curve which is characterized by the $P_k^j$ and $\delta_k^j$ - values. For the TU-Delft beam test at hand, Figure 4 presents the results of the converged Monte Carlo simulation. As a first approximation, Rossi's formulae (2)–(4) are used for the experimental TU-Delft beam maximum aggregate diameter of 8 mm, for which $f_{ct,\mu} = 7.25$ MPa, $f_{ct,\sigma} = 2.76$ MPa, $E_\mu = 40,490$ MPa, and $E_\sigma = 7,080$ MPa. Figure 4 shows also the experimental results. It is remarkable that this first approximation can be already considered as a very good result, as far as maximum load and the onset of structural softening is concerned. In return, we note some differences concerning the post-peak structural behavior. In fact, the FE probabilistic model with the given distribution values overestimates the structural brittleness, respectively underestimates the global fracture energy. This shortcoming merits further investigation, as shown in the sequel.

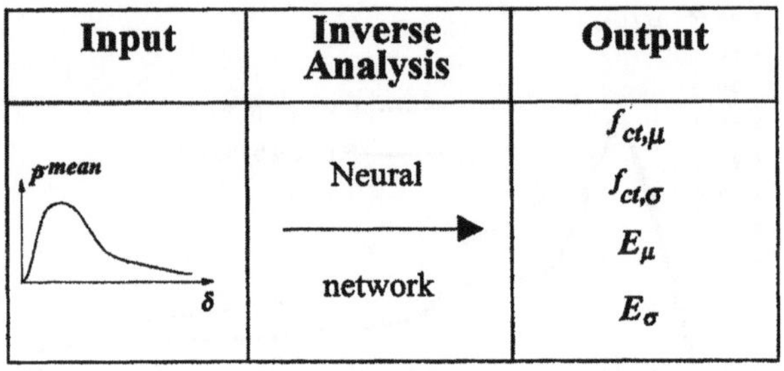


*Figure 5.* Sketch of the Inverse Analysis.

## 2.2. THE INVERSE PROBLEM: USING ARTIFICIAL NEURAL NETWORKS (ANN)

The inverse problem of the direct Monte Carlo simulation may be stated as follows (cf. Figure 5):

> For a given $P - \delta$-curve, find the normal distribution characterized by the mean, $f_{\mathrm{ct},\mu}$, and the standard deviation, $f_{\mathrm{ct},\sigma}$.

The feasibility of the procedure, together with the correspondence between the response curve and the pair of statistical moments (mean and standard deviation) has been shown in Fairbairn et al. (1998a), for a simplified mechanical model. The proposed algorithm was recently improved by the use of Artificial Neural Networks (ANN) (Fairbairn et al., 1998b). Here we show the use of Neural Networks to solve the stated inverse problem for finite element applications.

An ANN is a computational structure that is inspired by observed processes in natural networks of biological neurons in the brain. It consists of simple computational units, called neurons, that are highly interconnected. Each interconnection has a strength that is expressed by a number referred to as a weight (Klir and Yuan, 1995). The basic capability of neural networks is to learn patterns from examples. This is accomplished by adjusting the weights of given interconnections according to some learning algorithm (Benitez et al., 1997). The use of ANN to solve the inverse problem comprises thus two phases, a training phase and an exploitation phase:

- The training set is composed of input and output parameters. These are here a number of known $P - \delta$-curves (as input), together with the corresponding stochastic parameters ($f_{\mathrm{ct},\mu}$, $f_{\mathrm{ct},\sigma}$, $E_\mu$, $E_\sigma$) – as output. This training set is obtained by performing a number of direct Monte Carlo generation for given stochastic parameters on a given finite element mesh.
- Once trained, the ANN solves the inverse problem as depicted in Figure 5: for an input $P - \delta$-curve, the ANN determines the corresponding stochastic parameters for the finite element discretization.

In this way, the inverse anlysis gives access to the material parameters from structural experiments, and this irrespective of the structural volume, but only related to its finite-element discretization.

## 3. Example: the TU-Delft beam test

This inverse procedure will be applied to the TU-Delft beam shown in Figure 2a. The $m$ interface elements are placed in the region of the beam between the symmetry axis and the load axis, and the discretized material volume $V_t$ (i.e., the volume of two finite elements interfaced by the contact element) is constant. Therefore, the values of mean and standard deviation can be considered to be the same for all elements, and thus representative of the discretized material volume. The mesh pattern and refinement were chosen in such a way that the crack process independs on the orientation of finite elements. This mesh pattern has been used in the framework of FEM probabilistic analysis helping the capture of the physical mechanisms involved in the phenomenon (Rossi et al., 1996).

With respect to the good elastic results shown in Figure 4 in which the experimental and numerical curves almost coincide, it was decided to retain the values of $E_\mu = 40,490$ MPa, and $E_\sigma = 7,080$ MPa. In this way the only values to be determined by the inverse analysis are the mean and the standard deviation, $f_{\mathrm{ct},\mu}$ and $f_{\mathrm{ct},\sigma}$. A training set of pairs $(f_{\mathrm{ct},\mu}, f_{\mathrm{ct},\sigma})$ is then generated and the corresponding $P - \delta$ curves are determined by running the FEM probabilistic program. This training set is summarized in Table 1.

Due to the small number of examples, a special neural network (Time Delay Neural Network – TDNN) was adopted. TDNN is a more general form of Back Propagation employing this technique for setting weights between neurons. It also inherently considers time in its architecture. It can be viewed as a Back Propagation network where there are multiple connections from the input neurons to the output neurons. Each connection looks back over time and sets weights to minimize the Mean Squared Error of the overall network.

The TDNN used in this work has 31 inputs and one hidden layer with four neurons (2 *tanh* and 2 *linear*). The input layer corresponds to the patterns of the $P - \delta$-curves, i.e., 31 values for $P$ since $\delta$ values were the same for all curves. The output layer comprises 2 neurons using a linear transfer function with one connection each, i.e., the values of the mean and of the standard deviation for the tensile strength. A sketch of this neural network is given in Figure 6, where the neurons $x1$ to $x31$ are the input neurons, and $y1$, $y2$ are the output neurons.

Training of this network was considered complete, obtaining:

- Mean Squared Error on the training set = 0.0009
- Min. Mean Squared Error on the test set = 0.0014

This neural network was then used to determine the pair ($f_{\mathrm{ct},\mu} = 8.44$ MPa, $f_{\mathrm{ct},\sigma} = 6.41$ MPa) that corresponds to the experimental $P - \delta$-curve obtained in the TU-Delft experiment. Figure 7 shows the results of the Monte Carlo simulations (40 samples), and Figure 8 shows that the mean $P - \delta$-values obtained from the Monte Carlo simulation match the experimental results rather well, and greatly improve the results of the direct Monte Carlo simulation (cf. Figure 4) obtained with the empirical distribution functions (2)–(4).

## 4. Concluding remarks

The main points to be noted from this study are the following

(1) The proposed inverse analysis gives access to the statistical distribution of the material parameters which describe the heterogeneity of the matter constituting concrete in the

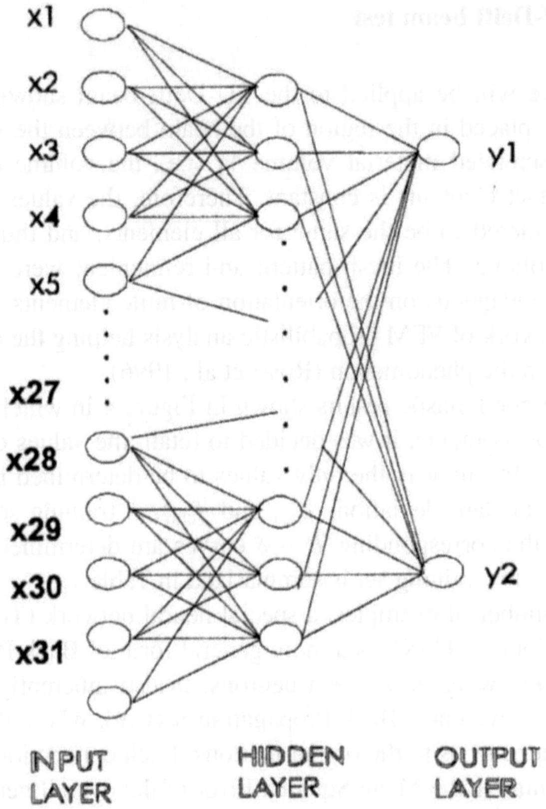


*Figure 6.* Sketch of the multilayered feedforward artificial neural network.

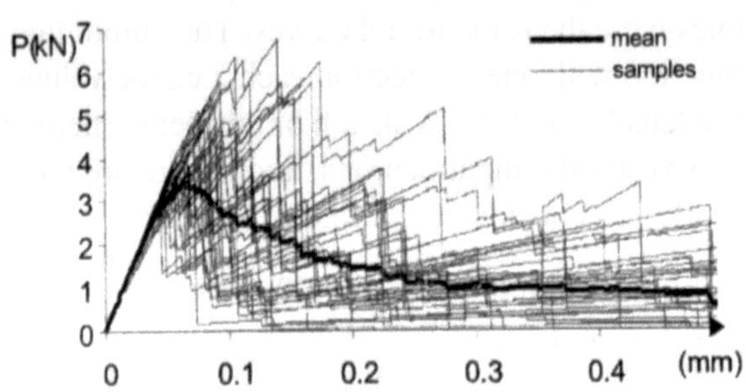


*Figure 7.* Monte Carlo simulation for the determined distribution paremeters $f_{ct,\mu} = 8.44$ MPa, $f_{ct,\sigma} = 6.41$ MPa, $E_\mu = 40,490$ MPa, and $E_\sigma = 7,080$ MPa [determined by Inverse Analysis].

*Table 1.* The training set for the ANN

| $f_{ct,\mu}$ [MPa] | $f_{ct,\sigma}$ [MPa] |
|---|---|
| 6.0 | 4.0 |
| 7.0 | 5.0 |
| 8.0 | 6.0 |
| 12.0 | 10.0 |
| 14.0 | 12.0 |

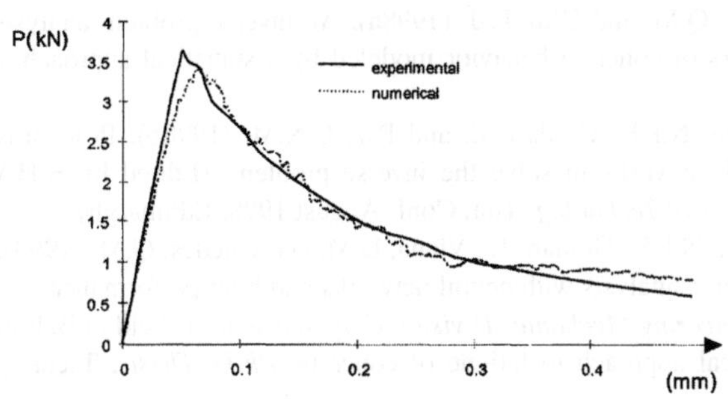


*Figure 8.* Comparison of experimental and numerical results for the determined distribution paremeters $f_{ct,\mu} = 8.44$ MPa, $f_{ct,\sigma} = 6.41$ MPa, $E_\mu = 40,490$ MPa, and $E_\sigma = 7,080$ MPa [determined by Inverse Analysis].

probabilistic crack approach. In this way, intrinsic material parameters related to the heterogeneity of concrete can be determined from structural test results.

(2) The statistical distribution is determined for a reference discretized material volume (i.e., the FE-volume of two triangle elements interfaced by a contact element). This allows for the determination of the probabilistic strength domain of concrete from structural test results, and this for material volumes (equal to the discretization volume), which are normally not accessible to experimental observation (i.e., direct tensile tests).

(3) The application of artificial neural networks can largely improve the solution of this inverse problem. The procedure is straightforward: for a given finite element discretization, a training set is generated using direct Monte Carlo simulation; the inverse problem then consists for the ANN to determine the stochastic parameters from an input load-displacement curve.

(4) The authors would like to emphasize that ANN can be a powerful tool for backanalysis of experimental data. In this way it should be mentioned that a research dealing with the identification of elastic and viscoelastic parameters for the concrete of an arch dam is presently being developed (Fairbairn et al., 1999), the first results indicating that the backanalysis method presented in this paper would not be restricted to the Rossi's model application.

## Acknowledgments

This work was developed within the framework of a scientific cooperation agreement CAPES COFECUB between COPPE/UFRJ and the Laboratoire Central des Ponts et Chaussées-LCPC/Paris.

## References

Benitez, J.M., Castro, J.L. and Requena, I. (1997). Are artificial neural networks black boxes? *IEEE Transactions on Neural Networks* **8**(5), 1156–1164.

Bažant, Z.P. (1997). Scaling of the quasi-brittle fracture: asymptotic analysis. *International Journal of Fracture* **83**(1), 19–44.

Fairbairn, E.M.R., Paz, C.N.M., Alvez, J.L.D. and Silva, R.C.C. (1997). Evaluation of constitutive relations for numerical modeling of concrete(in Portuguese). *Proceedings of XVIII CILAMCE*, Brasilia, Brazil, October 29–31.

Fairbairn, E.M.R., Guedes, Q.M. and Ulm, F.-J. (1998a). An inverse problem analysis for the determination of probabilistic parameters of concrete behavior modeled by a statistical approach. *Materials and Structures*, RILEM, in press.

Fairbairn, E.M.R., Ebecken, N.F.F., Goulart, E. and Paz, C.N.M. (1998b). Probabilistic modeling of concrete cracking: using neural networks to solve the inverse problem. (Edited by B.H.V. Topping), *Advances in Engineering Computational Technology* (Int. Conf. August 1998, Edinburgh).

Fairbairn, E.M.R., Ebecken, N.F.F., Goulart, E., Viana, D.M. and Guedes, Q.M. (1999). Durability assessment of an arch dam using inverse analysis with neural networks and high performance computing, abstract accepted to the *13th ASCE Engineering Mechanics Division Conference*, to be held in Baltimore, June 13–16.

Hordijk, D.A. (1991). Local approach to fatigue of concrete. *Ph.D. Thesis*, Technique Universitei Delft, The Netherlands.

Klir, J.J. and Yuan, B. (1995). *Fuzzy sets and fuzzy logic – theory and applications*. Binghamton.

Press, W.H., Teukolsky, S., Vetterling, W.T. and Flannery, B. (1992). *Numerical Recipes*, Cambridge University Press.

Rossi, P. and Richer, S. (1987). Numerical modeling of concrete cracking based on a stochastic approach. *Materials and Structures*, RILEM, **20**, 334–337.

Rossi, P., Wu, X., le Maou, F. and Belloc, A. (1994). Scale effect on concrete in tension. *Materials and Structures*, RILEM, **27**, 437–444.

Rossi, P., Ulm, F.-J. and Hachi, F. (1996). Compressive behavior of concrete: physical mechanisms and modeling. *Journal of Engineering Mechanics*, ASCE, **122**(11), 1038–1043.

Rossi, P. and Ulm, F.-J. (1997). Size effects in the biaxial behavior of concrete: physical mechanisms and modeling. *Materials and Structures*, RILEM, **30**, 210–216.

Weibull, W. (1939). A statistical theory of the strength of materials. *Proc. Royal Swedish Institute of Engineering Research*, **151**.

*International Journal of Fracture* **95:** 325–345, 1999.


# Scale effects on the *in-situ* tensile strength and fracture of ice. Part I: Large grained freshwater ice at Spray Lakes Reservoir, Alberta

J.P. DEMPSEY[1]*, S.J. DEFRANCO[2], R. M. ADAMSON[3] and S.V. MULMULE[4]
[1]*Department of Engineering, University of Cambridge, Trumpington Street, Cambridge CB2 1PZ, U.K. e-mail: jpd23@eng.cam.ac.uk*
[2]*Engineering Specialist, BP Amoco, 501 West Lake Park Blvd, Houston, TX 77253*
[3]*Clough, Harbour and Associates LLP, Engineers, Surveyors, Planners and Landscape Architects, 571 South Lake Boulevard, Suite B, Richmond VA 23236*
[4]*Hibbitt, Karlsson and Sorensen Inc., 1080 Main Street, Pawtucket, RI 02860-4847*

Received 9 October 1998; accepted in revised form 31 December 1998

**Abstract.** At lab-scale, issues such as inhomogeneity and polycrystallinity are especially important to the fracture of S1 freshwater ice. S1 freshwater ice is typically composed of large grains with predominantly vertical *c*-axes. Because of the very large grain sizes that one can encounter in S1 macrocrystalline ice sheets, it is essential that the effects of sample size on the fracture behavior be determined. In other words, are small scale (lab-scale) results applicable at larger scales (at the scale of ice-structure interactions, for instance)? To answer this question, a set of lab- to structural-scale ($0.34 < L < 28.64$ m) fracture tests were conducted on S1 freshwater lake ice at Spray Lakes, Alberta, using the base-edge-notched reverse-tapered plate geometry and covering a size range of 1:81. A Bažant-type size effect analysis of the measured fracture strengths (which do reveal a significant dependence on scale) is unexpectedly clouded by the fact that the data collected violates the associated scatter requirements, even though the size range tested is large. Moreover, via Hillerborg's fictitious crack model, large fracture energies were back-calculated (of order 20 J/m$^2$), but for miniscule process zone sizes; in addition, not all of the measured deformations for each test could be matched simultaneously. Apparently, these very warm S1 macrocrystalline lake ice experiments were dominated by nonlocal deformation and energy release rate mechanisms, in all likelihood brought about by grain boundary sliding. The reduced effectiveness of both the Bažant-type size effect analysis and Hillerborg's fictitious crack model is due mainly to the lack of crack growth stability achieved in the experiments. These unstable fractures truncated the fracture process. Given the irregular and large grain structure, the very warm ice temperatures, and the diffuse grain boundary surface energy, there is a marked dependence on specimen size to grain size ratio and distinctly non-unique pre-failure process zones occurred. Micromechanical simulations are required to resolve these coupled issues.

**Key words:** Lake ice, fracture, scale, polycrystallinity, anisotropy, S1 ice, grain brittleness.

## 1. Introduction

A major problem faced by engineers in the design of large structures, be they large rock or concrete containment vessels, earthquake-resistant bridges or buildings, damage resistant icebreakers, or ice-resistant offshore structures in polar regions, is the question of size versus strength. Do the mechanical properties at large scale differ significantly from those at small scale? Can laboratory-scale testing be used to predict properties at large scale? What particular

* *Permanent address*: Department of Civil and Environmental Engineering, Clarkson University, Potsdam, New York 13699-5710, U.S.A. e-mail: john.dempsey@clarkson.edu

problems are faced with large scale testing? How useful are the size-effect theories developed to date? The answers to these questions are key to cost effective structural design in the fields of ice engineering, concrete design and rock mechanics. For instance, the major deterrent to increased arctic oil and gas recovery is the lack of certainty regarding the appropriate ice design loads. In this context, the major task is to ascertain the material behavior of lake and sea ice on the scale of tens of meters up to kilometers (at the scale pertinent to the deformation). This topic itself has far reaching importance: ice forces encountered during offshore structure-ice interactions, loads on submarines while surfacing, the calving rate of icebergs, the design of icebreakers, and the ability of the northern countries to keep the Northern Sea Route open. The scale of interest is certainly huge; however, any conclusions obtained will clearly be of interest not only for arctic applications but also for the concrete and earthquake-resistant design communities.

A two-phase joint-industry-agency project (JIAP) was initiated in 1990 to calibrate a fracture theory for incorporation into probabilistic global ice load models. The first phase of the JIAP 'Large-Scale Ice Fracture Experiments' was completed in January, 1992 near Banff, Alberta (Kennedy et al., 1993). The second phase, conducted on first-year sea ice at Resolute, N.W.T. between April 17 to May 7, 1993, is reported in Part II of this paper (Dempsey et al., 1999). The primary goal of these freshwater experiments was to assess the feasibility of large-scale, *in-situ* full-thickness ice (freshwater and sea) fracture measurements. An abundance of data for large scale fracture tests on freshwater ice was recorded.

The information gathered in this two-phase study should help to link small scale laboratory experiments with large-scale events. In this context, it is interesting to note that nearly three decades ago, Weeks and Assur (1972) stated: '*We feel that an understanding of the scale effect in ice testing is essential before a thorough scientific basis can be developed for the utilization of small-scale testing in engineering design problems.*' An understanding of the scale effect is still being sought, primarily because large scale testing is fraught with difficulty and is also very expensive. Not many experiments can be done in any one program, and few field programs have been accompanied by the requisite subsidiary data, ... 'for a good reason: The collection of this type of information is time consuming' (Weeks and Assur, 1972).

The overall objective is to develop an applicable fracture mechanics for the fracture of ice; such a theory must in and of itself explain scale effects on the fracture strength (if observed). The emphasis on *in-situ* testing is very important: by testing the full thickness ice sheet, floating in its natural state, the resident thermal gradient is preserved. Because of this temperature gradient, the average ice temperature is in general really quite warm, when compared to most of the laboratory-scale tests completed to date. The fracture tests analyzed in this paper were completed under very warm conditions, with all of the ice being above $-5°C$. Such test temperatures would prove to be very difficult in a nonfloating environment; even though they are much needed, lab-scale fracture data at these warm ice temperatures is still rare. Note that the word 'warm' here is used in terms relative to the melting temperature. The amazing ability of ice to retain its structural integrity up to extremely high homologous temperatures has already been documented by Gandhi and Ashby (1979).

In ice engineering, many papers focusing on the role of fracture versus the ice forces encountered during ice-structure interaction have assumed, at the outset, the applicability of linear elastic fracture mechanics (LEFM). For this reason, it is important to also explore the requisite conditions for the applicability of the fracture toughness parameter, $K_{Ic}$.

Numerous factors affect the fracture behavior (deformations and failure loads) of prepared freshwater ice specimens such as: temperature, notch tip acuity, rate of loading and process zone size, specimen size and notch sensitivity, material anisotropy, and grain size effects and inhomogeneities such as grain boundaries. As one varies the specimen size and loading rate, there are three major interactive influences: homogeneity (specimen size to grain size), process zone size (zone of inelasticity ahead of the crack tip) and bulk creep. For severely undersized fracture tests, material homogeneity and small scale creep are both questionable assumptions. Small scale creep conditions hold when the nonlinear zone at the crack tip is sufficiently small (when compared with the crack length, the uncracked ligament, and any other geometric length quantities) such that the $K$-fields accurately model the stresses and deformations in an annular zone surrounding the crack tip. Linear elastic fracture mechanics (LEFM) for ice is based on the concept of small scale creep (this concept necessarily precludes loading rates that are too slow, the latter causing significant bulk creep). Because of the large grain sizes involved with ice, and especially S1 ice, the applicability of LEFM to ice presumes an adequate specimen size such that material homogeneity holds which in turn means that the ice can be treated as a continuum. As the specimen size increases, the material behavior is expected to become more homogeneous, and eventually the length of the fully developed process zone will be minor in comparison with the specimen size. With this in mind, it has become apparent that larger test sizes are required to truly model the behavior of the fracture of ice.

The process zone is a vague term used to refer to the inelastic zone ahead of the crack tip. The term crack tip refers to the 'visible' traction free crack tip cut in the ice. The effective crack tip in heterogeneous materials is not easily defined, particularly at the warm ice temperatures encountered in these experiments. The process zone is also not well defined, and as grain boundary sliding and facet cracking take place, stress re-distributions occur. If the fracture tests result in unstable crack growth initiation, the fracture process develops only partially (Van Mier, 1992; Van Vliet and Van Mier, 1998).

There is clearly a need for large scale testing to resolve some of the issues alluded to above. Although numerous factors affect the fracture behavior of ice, only a few of them can be investigated if restricted to the testing of laboratory sized specimens. Factors such as notch acuity, ice temperature and notch sensitivity can be studied independently. However, material anisotropy, specimen size effects, and the true nature of strain localization during fracture can only be addressed by testing large sized specimens. Other interesting aspects of large scale testing result from the possibility of obtaining $R$-curve information (DeFranco and Dempsey, 1992; 1994).

## 2. S1 macrocrystalline ice at Spray Lakes Reservoir

The initial site chosen for the freshwater ice experiments was Bearspaw Reservoir near Calgary, Alberta. The ice at this location was S2 freshwater ice with randomly oriented $c$-axes in the horizontal plane and grain sizes ranging from 1.5 cm to 5.0 cm. Also, the ice at Bearspaw was highly fractured because of water level changes for hydropower needs. Several experiments were completed at this site, but due to the unseasonably warm temperatures (rising up to 15°C during the day) sample preparation was extremely difficult and the overall success of the project was in jeopardy. To remedy the situation and save the project, the testing was moved to Spray Lakes Reservoir near Canmore, Alberta. There, S1 freshwater ice was encountered with very large grain sizes near the bottom of the sheet. In lake ice, grain sizes in the lower layer increase sharply when ice growth almost ceases as a result of warm air conditions (Butiagin,

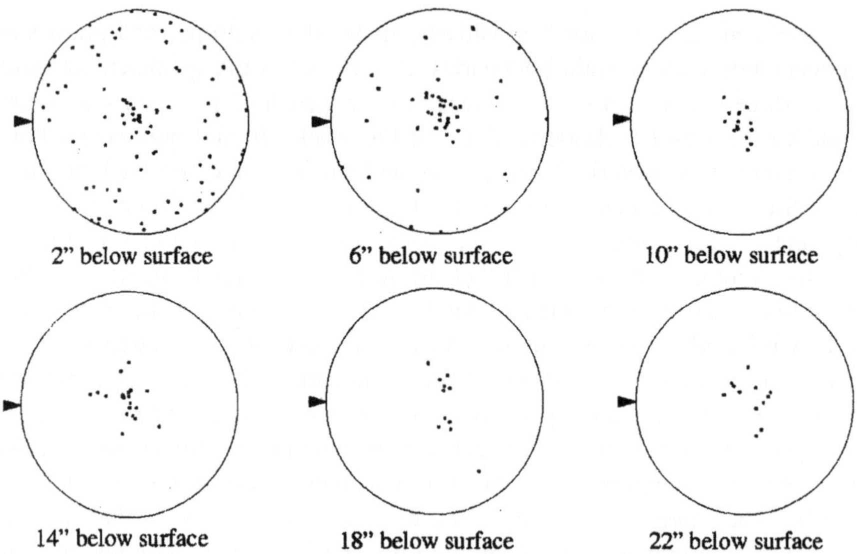


*Figure 1.* Schmidt net plots for the S1 freshwater ice at Spray Lakes.

1966). The ice at Spray Lakes Reservoir had limited fractures and large areas with no visible cracks. The ice sheet was 0.5 m thick with a near vertical $c$-axis orientation. While the grain shapes and sizes was rather irregular, the thickness of the ice was very uniform and very stable: the ambient weather was such that the ice was neither increasing nor decreasing in thickness during the test program.

Due to the warm temperatures at the Spray Lakes Reservoir test site, no ice characterization could be done on site. A large block of the ice from Spray Lakes Reservoir was shipped back to Clarkson University where detailed characterization of the ice was performed. To obtain a detailed characterization of the ice through the depth, 11 thin sections were made equispaced through the thickness. Grain size distributions and orientations were determined. Near the surface of the sheet the grains were about 0.5 cm to 1 cm in diameter without changing size for a depth of approximately 3 cm, with randomly oriented $c$-axes, mostly horizontal. Near the bottom of the middle third of the ice sheet, many of the grains had reached diameters of 20 cm with near vertical $c$-axes. For most of the thickness, except for the top few centimeters, the grains exhibited near vertical $c$-axes, classifying this ice as S1 macrocrystalline freshwater ice (Michel and Ramseier, 1971; Gow, 1986). Schmidt net plots (Figure 1) obtained from the thin sections reveal an increasing percentage of vertical $c$-axes with increasing depth into the ice sheet.

The task of establishing a realistic estimate of the grain size from one block of ice proved rather formidable; from a less than thorough grain size evaluation, the average through-the-thickness grain size is assumed from here on to be 15 cm. The plan area $A_p$ and through-the-thickness projected area $A_{tt}$ of each grain are approximated as $A_p = \pi d_{\mathrm{av}}^2/4$ and $A_{tt} = hd_{\mathrm{av}}$. There is a varied grain structure through the depth, with a large number of occluded grains (like small partial-depth inverted cones all originating from the top of the ice sheet) and others that grew the whole depth (like large full depth cones).

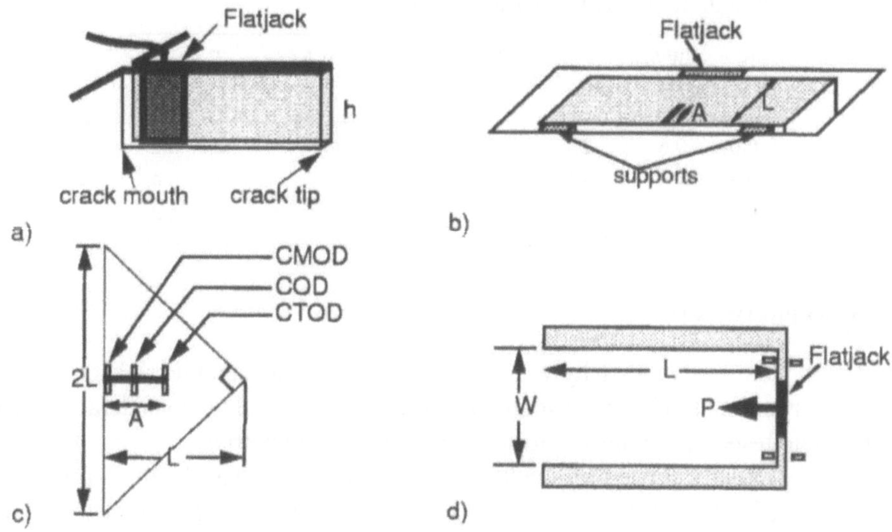


*Figure 2.* (a) Schematic of flatjack in the ice; (b) 3 point bend geometry; (c) Reversed-taper geometry – note the CMOD, COD and CTOD locations; (d) Cantilever column geometry ($L = 4W$).

## 3. Experimental procedure

All of the large scale experiments reported in this paper were *in-situ* tests enabling the testing of large sizes with resident thermal profiles. Experiments were prepared by first cutting the sample free from the parent ice sheet using chainsaws and/or a ditchwitch. The cut was made large enough to avoid refreezing between the specimen and the parent ice sheet. The openings were maintained with shovels and scrapers. While conducting the *in-situ* ice experiments, it was critical to ensure that the sample did not refreeze to the parent ice sheet.

Loading was applied to each test sample using a flatjack loading device (Shapiro et al., 1979) fabricated by Sandwell, Inc. The flatjack was a thin-walled steel bladder pressurized with either nitrogen gas or hydraulic fluid (Figure 2(a)). Tests performed with nitrogen gas were controlled by varying the flow of gas from a pressurized tank. These tests resulted in unstable fracture. A servo-hydraulic control system was used on several of the *in-situ* experiments controlling the load via a feedback signal from a displacement measuring gauge on the crack. These experiments generally involved multiple loadings and pseudo-stable crack propagation in the form of crack jumps and arrests. Displacement measuring devices, LVDT's and noncontacting sensors, were mounted at several points along the crack as shown in Figure 2(c). Note that the displacement gages were rigidly attached to the surface of the ice by screw clamps. These locations are labeled CMOD, COD and CTOD for the crack-mouth-, intermediate crack- and crack-tip-opening-displacements, respectively. On smaller experiments, the crack was not long enough to permit the placement of the COD gauge. The data from the displacement sensors and the pressure transducer were recorded on two computers equipped with A/D converters and an analogue tape backup system.

The flatjacks (and supports, if required) were scaled such that the contact pressure induced was at most of order 1 MPa. In this way, there was no microcracking or crushing in the vicinity of the applied load (and supports).

*Table 1.* In-situ elastic modulus of S1 ice.

| $W$ m | $W/d_{av}$ | $L/d_{av}$ | $W \times L/A_p$ | $W \times h/A_{tt}$ | $E$ GPa |
|---|---|---|---|---|---|
| 0.09 | 0.6 | 2.4 | 1.8 | 0.6 | 2.8, 3.2 |
| 0.27 | 1.8 | 7.2 | 16.5 | 1.8 | 6.1, 6.4, 7.2, 6.4 |

### 3.1. Beam experiments

The initial test plan was guided by experience gained in the laboratory back at Clarkson University. Conventional wisdom favored beam experiments; after considerable effort and expense, the field tests rapidly moved to self-equilibrated geometries.

The geometry initially chosen for the large scale experiments was the three-point bend test with dimensions of $0.17 \times 0.51$ m (Figure 2(b)). Each sample was prenotched with a crack length of 0.3 L, or 0.051 m. The crack tip was scribed with a fine tipped saw designed to create a very sharp crack. As discovered (Wei et al., 1991), the acuity of the crack tip has a significant influence on the failure load, especially in cold freshwater ice.

The very warm ice later tested at Spray Lakes Reservoir was much less affected by notch tip acuity. For example, scribing a through-the-thickness at the crack tip with a sharp blade and considerable hand pressure causes microcracking in cold freshwater ice. Because of this, it is thus relatively easy to generate a sharp crack front in cold freshwater ice. However, in the very warm ice tested at Spray Lakes Reservoir, the ice did not microcrack in the expected manner-it could not be made as sharp. This observation is noteworthy, given that stable cracking was not observed in the fracture tests conducted.

The first two beam experiments were carried out at Bearspaw reservoir. In experiment B2, the loading was controlled using the CTOD gauge for feedback and control resulting in crack propagation and arrest. The next two beams, B3 and B4, were tested at Spray Lakes. Unfortunately, an excessive amount of cutting was required to free each beam sample, given that the specimen size is measured by the length of the crack-parallel dimension (the beam depth), not the beam length. In addition, difficulties were encountered when applying the load to the sample. The flatjack was placed at the center of the beam between the sample and the parent ice sheet. On the opposite side two loading supports had to be placed at either end of the beam. The large space between the parent sheet and the beam coupled with the beam's ability to rotate slightly made it difficult to keep the supports and the flatjack in place. Moreover, the thermal gradient caused the initially vertically cut sides to melt with depth over time and cause slanted sides, thus causing problems with the flatjack loading (several of the flatjacks disappeared into the lake depths).

To obtain an estimate of the initial elastic modulus, cantilevered columns were tested by applying axial load as shown in Figure 2(d). A total of 6 tests were completed for two different sizes $W = 0.09$ m and 0.27 m, with $L = 4W$. Table 1 shows the initial elastic moduli obtained.

### 3.2. Reversed-taper geometry experiments

The abovementioned difficulties with the *in-situ* three-point bend tests provoked the adoption of the reversed-taper (RT) geometry (Figure 2(c)) (Mai et al., 1975; Dempsey et al., 1995).

*Table 2. In-situ* RT fracture of S1 ice: Short-time moduli.

| Test # | Air Temp °C | $L$ m | $A$ m | $A/d_{av}$ | $E_{COD}$ GPa | $E_{CMOD}$ GPa |
|---|---|---|---|---|---|---|
| RT4 | +0.4 | 0.34 | 0.095 | 0.6 | – | – |
| RT2 | −0.3 | 0.41 | 0.14 | 0.9 | – | 5.7 |
| RT5 | +0 | 1.04 | 0.39 | 2.6 | 2.8 | – |
| RT1 | 0 | 1.41 | 0.43 | 2.9 | 3.5 | 4.3 |
| RT7 | 0 | 3.18 | 0.99 | 6.6 | – | 2.9 |
| RT8 | −3 | 3.20 | 0.99 | 6.6 | 3.6 | 8.0 |
| RT3 | −0.6 | 4.42 | 1.23 | 8.2 | 5.7 | 7.6 |
| RT6 | +0 | 10.36 | 3.12 | 20.8 | 7.7 | 3.2 |
| RT9 | +0 | 28.64 | 8.98 | 59.9 | 10.0 | – |

This geometry had been previously used for lab-scale experiments at Clarkson University (DeFranco and Dempsey, 1994). Reducing the specimen width with increasing crack length $A$ is conducive to stable crack extension (at least in warm sea ice) given a rapid increase in compliance with crack extension. The use of the RT greatly reduced setup time as it required 40 percent less cutting and the loading setup improved considerably. Rather than loading the sample by placing the flatjack between the parent ice sheet and the sample, the flatjack was inserted directly into the pre-cut notch (see Figures 2(a) and 2(c)), eliminating the need for any supports. Nine RT tests, some with stable crack propagation and multiple loadings, yielded a specimen size range of 1 : 81.

### 3.3. *IN-SITU* RT FRACTURE DATA ANALYSIS

A total of 15 *in-situ* fracture and modulus experiments were completed on beam (BM) (Figure 2(b)), reversed-taper (RT) (Figure 2(c)), and cantilever column (CC) (Figure 2(d)) geometries. Most of the experiments were loaded to failure in less than 50 seconds. Tables 2 and 3 summarize the RT fracture data. The associated load *vs* COD plots indicate near linear behavior with no noticeable time dependent features. For each of the fracture experiments, the modulus was calculated from the displacement gauges at the CMOD and COD. The ambient temperature was too warm to rely on freezing the gage clamps to the ice sheet. The displacement gages were rigidly attached to the surface of the ice by screw clamps. The initial behavior of each experiment appeared to be nonlinear. This was due to flatjack effects: each flatjack took a while to establish full contact with both sides of the crack (even though each flatjack was fabricated to fit snugly in the chainsaw cut). The elastic moduli were determined from the more linear portions of the P-CMOD curves after these initial loading-related nonlinear portions. Due to the problems encountered with the field testing, the displacement records for some of the tests were not recorded. In these cases, the modulus could not be computed.

No discernible trend in the CMOD modulus was evident, and a considerable amount of scatter was present (Figure 3(a)). The scatter can be attributed to the effects of nonhomogeneity arising from the low crack size to grain size ratio. It can also be caused by the close proximity of the flatjack loading device. The COD modulus, which is farther from the loading, shows less scatter with a definite increasing trend with size. A size range of 1:81

*Table 3. In-situ* RT fracture of S1 ice: Fracture parameters.

| Test # | $L$ m | $L/h$ | $A$ m | $A/d_{av}$ | $\dot{K}$ kPa$\sqrt{m}$/s | CTOD$_i$ $\mu$m | $K_Q$ kPa$\sqrt{m}$ |
|---|---|---|---|---|---|---|---|
| RT4 | 0.34 | 0.68 | 0.095 | 0.6 | 5.13 | – | 134 |
| RT2 | 0.41 | 0.82 | 0.14 | 0.9 | 3.77 | – | 76 |
| RT5 | 1.04 | 2.08 | 0.39 | 2.6 | 3.71 | 5 | 190 |
| RT1 | 1.41 | 2.82 | 0.43 | 2.9 | 2.87 | 7 | 97 |
| RT7 | 3.18 | 6.36 | 0.99 | 6.6 | 37.8 | 8 | 193 |
| RT8 | 3.20 | 6.40 | 0.99 | 6.6 | 144 | 11 | 144 |
| RT3 | 4.42 | 8.84 | 1.23 | 8.2 | 4.79 | 8 | 144 |
| RT6 | 10.36 | 20.72 | 3.12 | 20.8 | 9.10 | 20 | 186 |
| RT9 | 28.64 | 57.28 | 8.98 | 59.9 | 57.8 | 25 | 297 |

was achieved by testing RT samples up to 28.64 m in length. Even at this size, the crack length to average grain size ratio, $A/d_{av}$, was only 59.9 because of the large grain sizes. The displacement measuring gauge at the crack tip was able to measure the deformation prior to crack propagation and indicate the exact time of initial macroscopic crack propagation. From this, the crack tip opening at failure was determined for each experiment. An increasing trend in the failure CTOD was noticed (see Figures 3(b) and 3(c)), indicating that specimen sizes were not reached at which the failure CTOD was independent of specimen size. Tables 2 and 3 list the short-time moduli and CTOD values at the initiation of crack growth, *viz.*, CTOD$_i$. All of the experiments were carried out at very warm temperatures (the ice was essentially just below freezing); laboratory experiments at these temperatures have not been reported. Ratios of crack length to average grain size ($A/d_{av}$) are also presented in these tables. For small test sizes, this ratio is very small. Table 3 includes a column that identifies the specimen size to ice thickness ratio, $L/h$. This aspect ratio is included, even though the nature of the RT test program emphasizes two-dimensional similarity. At the same time, for the low aspect ratios identified, particularly RT4 and RT2, the crack growth is likely to have distinct asymmetries along the crack front. While quasi-brittle fracture tests are not expected to depend on the thickness (length of the crack front), this topic needs to be investigated further (Van Vliet and Van Mier, 1998). The two-dimensional similarity is necessary; if the thickness of each specimen were to be scaled, one could be confronted by a dramatically different microstructure.

With regard to test RT9 (see Tables 2 and 3), a significant time period elapsed between sample preparation and final failure during which refreezing of the sides occurred. The sides had to be recut; moreover, after the test it was discovered that a large thermal crack was oriented roughly at right angles to the crack path, crossing the crack path a few meters ahead of the crack tip. The pre-cut crack grew parallel to its initial orientation until it reached the thermal crack, after which the test piece broke at the thermal crack. In some ways, this test was 'more valid' than the others because it was large enough to incorporate one of the many thermal cracks ever-present in lake ice. To serve as a reminder of these facts, the results of this particular experiment are plotted with an open circle around the associated data.

The influence of polycrystalline inhomogeneities and nonlocal inelastic deformations is very important when considering a material with a very large grain size as encountered in this

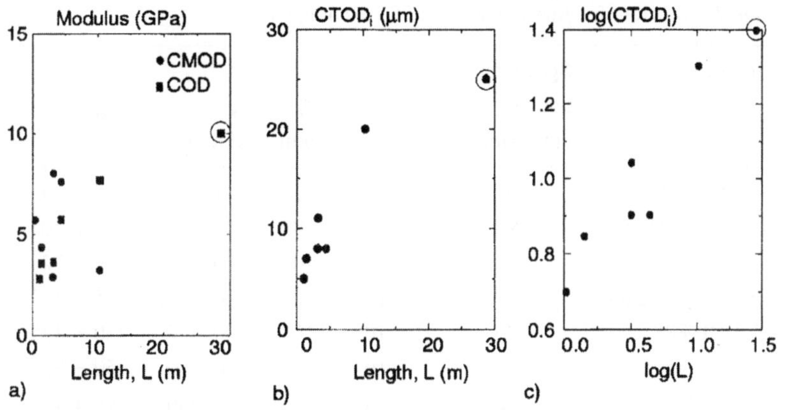


*Figure 3.* Effects of specimen size on (a) Short-time elastic modulus and (b) CTOD at the initiation of crack growth; (c) log-log plot of this CTOD.

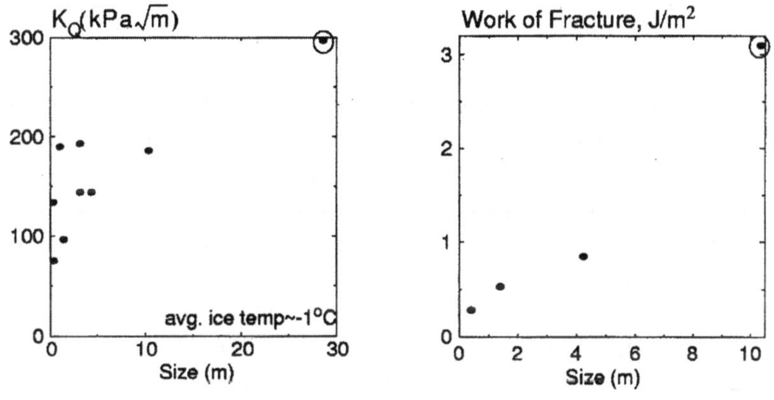


*Figure 4.* (a) Apparent fracture toughness vs size for in-situ experiments; (b) Fracture energy vs size.

field study. The inelasticity in the vicinity of the crack tip influences the behavior of the ice especially in the smaller samples. Other factors that affect the fracture of freshwater ice are thermal profile, notch tip acuity, loading rate and notch sensitivity. A large enough specimen must be tested to assume a homogeneous isotropic response. In this ice, the basal plane or weak plane is predominantly parallel to the ice surface. Because the ice is being fractured along stronger planes, any weakness almost has to be introduced by the grain boundaries (one major candidate at these very high homologous ice temperatures is liquidity on the grain boundaries), and therefore the grain size and number play an important role in the strength of the sheet.

### 3.4. Apparent Fracture Toughness versus Size

The fracture toughness is a measure of a material's ability to resist fracture and is represented by the critical stress intensity factor, $K_{Ic}$. Because no fracture toughness testing standard exists for ice and several assumptions are made to justify the applicability of LEFM to freshwater ice, it has been recommended (Dempsey, 1991) that the following definition be used for the critical stress intensity factor measured: $K_Q = K_{\text{apparent}}^{\text{initiation}}$ instead of the parameter $K_{Ic}$. Apparent fracture toughness values were computed from the load at the onset of cracking as determined from the CTOD gauge. The fracture toughness as a function of specimen size is

*Table 4.* Apparent fracture toughness of S1 ice at lab-scale.

| Source | Specimen size parallel to crack (mm) | Ice Temp. °C | $d_{av}$ mm | $\dot{K}$ kPa$\sqrt{\text{m}}$/s | $K_Q$ kPa$\sqrt{\text{m}}$ |
|---|---|---|---|---|---|
| Kollé (1981) | 50 | −17 | 10 | 1–10 | 240 ± 79 |
| Danilenko (1985) | 15–80 | −10 | 100 | 0.1–1 | 335 ± 50 |
| | 15–80 | −15 | 100 | 0.1–1 | 439 ± 72 |
| Dempsey, Nigam | 32 | −10 | 100 | 100–250 | 465 ± 182 |
| Cole (1988) | 48 | −10 | 100 | 100–250 | 280 ± 135 |
| | 64 | −10 | 100 | 100–250 | 250 ± 72 |
| | 80 | −10 | 100 | 100–250 | 282 ± 136 |
| Stehn, DeFranco | 100 | −16 | 10 | 5–50 | 126 ± 41 |
| Dempsey (1995a) | | | | | |

plotted in Fig. 4a. The fracture properties of S1 ice depend on the direction of the crack front relative to the direction of the *c*-axis. A higher apparent fracture toughness is observed when the crack front is parallel to the *c*-axis whereas fracture perpendicular to the *c*-axis (parallel to basal plane) results in a lower apparent fracture toughness. Table 4 lists observed values of the apparent fracture toughness in lab-scale tests with the crack front parallel to *c*-axis. The lab-scale data shown in Table 4 are extremely sensitive to specimen size; the local grain orientations are a primary cause of the scatter (Abdel-Tawab and Rodin, 1999; Ballarini et al., 1999), because the test temperatures are all relatively cold. For the large scale data plotted in Figure 4(a), the ice is very near its melting point, and therefore the associated lab-scale values are relatively low; as is clear, there is a distinct increase with specimen size.

For later use, note that in Figure 8 of Dempsey (1991), the nominal strength at failure was compared for the fracture and flexural four-point-bend tests conducted by Dempsey et al. (1988). These tests were found to be notch insensitive. For a test depth of 80 mm, the flexural strength varied between 0.5 and 2.0 MPa.

At the ice surface there is always a random c-axis alignment but within a certain short depth (of the order of 15 cm in this ice), a vertical alignment is established. In general only about 70 percent to 80 percent of the grains may have vertical alignment (these grains are typically larger than the other grains). If either the top layer with its random *c*-axis distribution becomes a significant part of the sample or there are too few grains in the sample, the predominant orientation of an individual sample may be quite different from that of the parent sheet. In a specimen containing too few grains the polycrystalline constraint is not fully mobilized. The above factors combined with the intrinsic rate effect in fracture, temperature effect, inelasticity in the vicinity of the crack tip and inadequate notch sensitivity produce a wide scatter in fracture results especially at the laboratory scale. In the present field experiments, a similar scatter is seen at smaller sizes but superimposed on this scatter there is a definite trend towards an increase in the apparent fracture toughness with increasing size, although no plateau is reached. Another independent indicator of this behavior is the crack-tip opening-displacement plotted in Figure 3(b) which is seen to increase monotonically with size. This is indicative of softening mechanisms (grain boundary sliding, dislocations, microcracking) dominating the failure of the smaller samples with a transition to brittle failure at the larger sizes. A plateau in the fracture toughness was not reached, because still larger sizes are required for this large grained S1 ice.

The work of fracture per unit crack area for a cracked body can be expressed as

$$G_{\mathrm{WF}} = \frac{1}{h(L-A)} \int_0^{\infty} P \, \mathrm{d}\delta_{LL} \tag{1}$$

where $P$ is the applied load normal to the crack and $\delta_{LL}$ is the load line displacement (LLD). Equation 1 represents the area under the load versus load-line displacement curve. It assumes that all of the work of the load is dissipated in creating the crack. For an ideally britttle material, the equation reduces to $G_{\mathrm{WF}} = P_i\delta_{LL}/2h(L-A)$, $P_i$ being the load at the initiation of crack growth. For the large scale experiments, very little nonlinearity was observed in the $P$ versus LLD curves, so the behavior was assumed to be linear. Numerous reasons account for excluding any area under the unloading portion of the curve. Once the ice sample fractures, the two halves are further pushed apart by the flatjack. Their motion is resisted by hydrodynamic forces as well as interaction with the parent ice sheet. This force is recorded and cannot be uncoupled from the unloading response of the ice sample. In addition, where load control was used for the fracture test, brittle cleavage fracture was observed. Figure 4(b) shows the work of fracture *vs* size as an increasing function. This tends to indicate that larger sizes are needed to reach a plateau fracture energy. These results are consistent with tendencies of the other material properties calculated (i.e. $\mathrm{CTOD}_i$, $K_Q$).

This work of fracture estimate neglects the post-peak behavior and is therefore, in all likelihood, much less than the true material fracture energy (Bažant, 1996). The post-peak behavior cannot be accurately described since stable crack growth was not obtained.

## 4. Scale effects on the nominal strength and brittleness

Surprisingly, very little attention has been paid to the study of the fracture behavior of S1 freshwater ice (Kollé, 1981; Danilenko, 1985; Dempsey et al., 1988; Dempsey and Wei, 1989; Stehn et al., 1994, 1995a, 1995b). The most recent studies (Dempsey et al., 1988; Dempsey and Wei, 1989; Stehn et al., 1994, 1995a, 1995b) appear to be the most comprehensive with investigations of the effects of crack *vs* *c*-axis orientation, grain size and specimen size on the fracture.

The fracture toughness expression for the RT geometry with partial crack face loading as applied by the flatjack is given as

$$K_Q = \frac{P_i}{wh} f(A/L)\sqrt{\pi A}, \tag{2}$$

where $w$ is the width of the flatjack; see Dempsey et al. (1995) for the expression for $f(A/L)$. The nominal strength provides a way of estimating the tensile strength of self-similar tests and evaluating the associated size effects. It is defined as the maximum nominal stress sustained at the crack tip at the initiation of crack growth. For the RT geometry, the nominal strength is computed by replacing the load from the flatjack with an equivalent load and moment at the center of the uncracked ligament. From this, the maximum nominal stress acting just ahead of the crack tip at the instant of unstable crack growth initiation may be computed. The actual tensile strengths ($\sigma_n^t$) measured can thus be found from the flatjack load via the following expression (see Figure 5(a))

$$\sigma_n^t = \frac{2P_i(A+2L)}{h(L-A)^2}. \tag{3}$$

Now, the nondimensional tensile strength $\sigma_n^t/\sigma_*$ can be written as

$$\sigma_n^t/\sigma_* = \frac{C_N}{\sqrt{L/l_{ch}}}, \tag{4}$$

in which $\sigma_*$ denotes a reference strength and $l_{ch}$ denotes the size-independent characteristic length

$$l_{ch} = \left(\frac{K_{Ic}}{\sigma_*}\right)^2 \tag{5}$$

while

$$C_N = \frac{2(a+2)}{(1-a)^2}\frac{(w/L)}{\sqrt{\pi a}f(a)}\frac{K_Q}{K_{1c}}. \tag{6}$$

The expressions in (4) to (6) reveal that, for a given value of $L/l_{ch}$, the most brittle or notch sensitive (Carpinteri, 1982; Dempsey, 1991) fracture response is to be had for crack lengths of $A/L = 0.6$, approximately. The tests were all conducted for $A/L = 0.3$, which is a marginally less notch sensitive geometry.

### 4.1. Size effect analysis

Figure 5(b) shows a plot of the nominal tensile strength vs sample size. The solid curve represents the size effect as predicted by LEFM. This prediction assumes a constant fracture toughness and self-similar geometries; that is the same geometry with the same nondimensional crack length, $a = A/L$. The curve is obtained by solving (2) and (3) for $\sigma_n^t$ assuming that LEFM is applicable for the largest specimen size tested; in other words only one data point provides the solid curve. As it must in this case, the nominal tensile strength $\sigma_n^t$ decreases as a function of $\sqrt{L}$, an LEFM size effect. The dashed line in Figure 5(b) represents the size effect law (SEL) proposed by Bažant (1984) for characterizing fracture of self-similar fracture geometries in a quasi-brittle material, *viz.*,

$$\sigma_n^t = \frac{B}{\sqrt{1+L/L_0}}. \tag{7}$$

The parameter $B$ in the size effect law (7) is closely linked to the tensile strength of the material while the parameter $L_0$ represents the transition size. The transition size marks the intersection of the asymptotes governing strength failures and fracture failures. For sizes smaller than this transitional size, strength governs while for larger sizes failure is more influenced by fracture. It has been proposed that LEFM is applicable at sizes 10 times the transition size. In carrying out the size effect law fit using nonlinear regression, it was seen that the $r^2$ factor of the fit was 0.6 which represented a poor fit. Even though the data violates the minimum scatter

*Table 5.* Parameters of the SEL.

| $r^2$ | $B$ MPa | Standard deviation for $B$ | $L_0$ m | Standard deviation for $L_0$ |
|---|---|---|---|---|
| 0.6 | 0.24 | 0.07 | 0.76 | 0.78 |

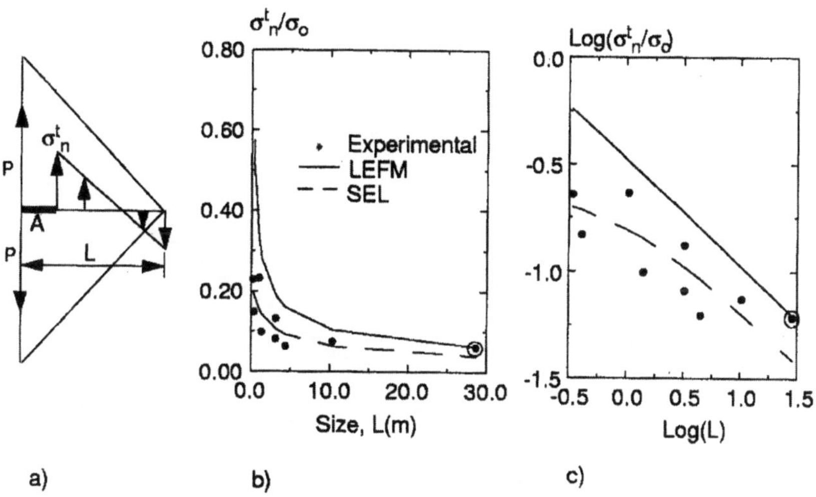


*Figure 5.* (a) Nominal stress, $\sigma_n^t$, at the crack tip; (b) Normalized nominal stress at failure for the in-situ fracture experiments; (c) Log-log plot of nominal failure stress versus size ($\sigma_o = 1$ MPa).

requirements recommended in Bažant et al. (1986), it is instructive to explore the statistics of the fit (given in Table 5).

The parameter $B$ has a reasonable standard deviation while parameter $L_0$ has a large standard deviation. It is known that such a violation of the scatter requirement is produced when other size effects are present in addition to the fracture effect due to fracture (Bažant et al., 1986). A thorough investigation of the lab-scale flexural strength of S1 freshwater ice was completed by Gow and Veda (1989). Intermediate-scale ($0.3 \times 0.3 \times 2.0$ m) flexural experiments on freshwater lake ice exhibited lower strengths than comparable smaller experiments on model S1 freshwater ice. These results are consistent with those presented here.

If an LEFM size effect exists, a linear scaling in the strength will be evident with a slope of $-1/2$ on a log-log plot of the nominal strength *vs.* size (Figure 5(c)). If other size effects are present, the scaling will be different. For an assumed tensile strength of $\sigma_* = B = 0.24$ MPa and a hypothetical value for $K_{Ic}$ of 240 kPa$\sqrt{\text{m}}$, the characteristic length $l_{ch}$ in (5) is of order 1 m, which is very large; this value suggests that the LEFM test size may be very large, tens of meters in fact.

## 4.2. Hillerborg's cohesive zone analysis

In the previous section, the size effect analysis has revealed that the data is too scattered: this result is rather unexpected and merits further analysis. In this context, it is next assumed that the rationale underlying the fictitious crack model (Hillerborg et al., 1976) is applicable to the very high temperature fracture of S1 macrocrystalline lake ice under investigation here.

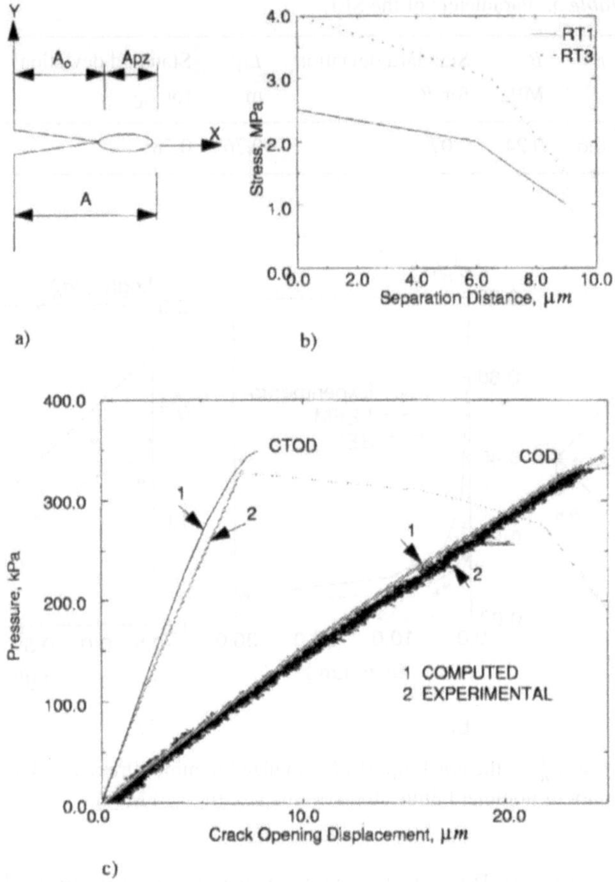


*Figure 6.* Fictitious crack model: (a) process zone model and coordinates; (b) stress-separation laws for RT1 ($L = 1.41$ m) and RT3 ($L = 4.42$ m); (c) experimental calibration on RT1 (the computed results follow the inverse procedure outlined in Dempsey and Mulmule (1997)).

Thus, if this ice behaves as a quasi-brittle material such as rock and concrete, strain softening would occur, caused by distributed cracking ahead of the crack. The fictitious crack model uses the concept of a stress-separation curve which is assumed to be a material property (see Mulmule and Dempsey, 1997 and 1998 for a thorough exposition of this method as applied in this section).

Efforts to apply this method to the problem at hand were hampered at first because the stress-separation curve for this warm macrocrystalline S1 lake ice was not known. Even if the stress-separation curve at lab-scale were known, this may not necessarily coincide with the stress-separation curve which would correspond to the behavior for *in-situ* full thickness lake ice. The procedure followed here was to back-calculate this information from the *in-situ* large-scale fracture experiments, using the load *vs* crack tip opening displacement and load *vs* crack mouth opening displacement data. Approximate stress-separation curves can thus be obtained by assuming various stress-separation curves and comparing the obtained load *vs* crack opening displacement plots with the experiments. If the experiment is under load control (as in most of the tests examined in this paper), then only a portion of the stress-separation curve can be obtained prior to the initiation of unstable crack growth.

The requirements imposed in order that the stress-separation curve approximate the true fracture behavior observed were:

- The peak load must be predicted correctly. In other words, under load control the unstable propagation of the crack should occur at the test peak load.
- The load-CMOD curve must be predicted correctly. This ensures that the correct bulk response as well as the nonlinearity due to the process zone is being modeled correctly.
- The load-CTOD curve must be predicted correctly. This puts stringent requirements on the validity of the cohesive stress distribution as well as the size of process zone.

The smallest test size for which the load-CMOD, load-COD and load-CTOD curves were available was RT1. The failure CTOD for RT1 was 7 $\mu$m while from RT9 it was known that the critical CTOD would be greater than or equal to 25 $\mu$m. As already noted, only a portion of the stress-separation can be obtained using RT1. In trying to match the load-CMOD and load-CTOD curves, a choice had to be made about the modulus of elasticity to be used for computing the crack opening profiles. After a few trials, it was realized that using $E_{\mathrm{COD}}$ resulted in very large values for the CMOD whereas using $E_{\mathrm{CMOD}}$ the COD values obtained were far below the experimental results. The COD gauge was closer to the crack tip hence it reflected the crack tip behavior better than the CMOD gage. For this reason, the intermediate modulus $E_{\mathrm{COD}}$ was used in carrying out trials aimed at determining the stress-separation curve. Load-CTOD and load-COD plots were used for calibration. Good results were obtained, finally, by assigning the stress separation curve shown in Figure 6(b). Figure 6(c) shows the comparison of the experimental load-COD curves with the constructed load-COD curves. The normalized process zone size obtained was 0.00057 $L$ which for RT1 was 0.8 mm. The load-CMOD curve, however, *could not be matched* and the disagreement was almost in the same ratio as $E_{\mathrm{COD}}$ to $E_{\mathrm{CMOD}}$. The tensile strength deduced from the stress separation curve was 2.5 MPa and the initial portion of the stress separation curve showed a pseudo-Dugdale type of behavior. The area enclosed under the stress-separation curve, the fracture energy, was 17 J/m$^2$ compared to 2.25 J/m$^2$ estimated using linear elastic fracture mechanics.

A similar procedure was adopted for RT3. The modulus value used for RT3 was 8.0 GPa. The resulting stress-separation curve is shown in Figure 6(b). The size of the process zone obtained was 0.0001 $L$ which for RT3 was 0.424 mm. The fracture energy obtained was approximately 31 J/m$^2$. There is again a large discrepancy in the fracture energy values obtained using the fictitious crack model and the fracture energy computed from linear elastic fracture mechanics even though the size of the process zone is exceedingly small relative to the specimen size. In addition, the tensile strength values obtained seem rather high.

In the large scale experiments, the mounts used for the CTOD gauges on the large scale tests were approximately 2 cm wide parallel to the crack. For this CTOD data to be representative, the size of the predicted process zone has to be much larger than the width of the CTOD mounts (Nanthikesan and Sunder, 1995). In these experiments, the process zone size was determined to be much less than the size of the mounts, rendering modeling based on the FCM unreliable. In addition, since the process zone size was found to be so small, the fictitious crack model is not necessary. The very small amount of nonlinearity found in the experiments can only be attributed to bulk (nonlocal) behavior. The fictitious crack model assumes a homogeneous response outside the process zone. Apparently, due to too few grains in each specimen tested, and the very high homologous test temperatures, the homogeneous response assumed was not observed. The process zone size $A_{\mathrm{pz}}$ is much, much less than the grain size, and yet there are significant nonlinear deformation mechanisms operative.

Both the size effect analysis and the fictitious crack model have proved less effective than expected.

## 5. Representative specimen sizes versus nonlocal deformation mechanisms

The other major freshwater ice type that forms on lakes is called S2 columnar ice, and has a much more regular structure. Much more data exists for the fracture of this ice (see Table 1 in Dempsey, 1991). If one compares the apparent fracture toughness of S2 columnar freshwater ice tested at a temperature of $-12°C$ on test dimensions of less than $35d_{av}$ (Dempsey, 1991; Dempsey et al., 1992) and in the form of thin floating double cantilever beams at a temperature of $-2°C$ and an *in-situ* test dimension of $372d_{av}$ (Bentley et al., 1989), one finds that the apparent fracture toughness increased by 20 to 65 percent for the larger test sizes (even though the fracture toughness increases with decreasing temperature). This result has intrigued Abdel-Tawab and Rodin (1993a, 1993b), and they attempted to evaluate the effect of polycrystalline inhomogeneity by calculating the jump in the $J$-integral value across the process boundary. According to a recent examination of this issue (Abdel-Tawab and Rodin, 1999; Ballarini et al., 1999), the size effect associated with polycrystalline inhomogeneity is weak, and 'polycrystalline inhomogeneities' induce statistical effects controlled by a small number of grains adjacent to the crack tip. The latter theoretical analyses neglected mechanisms such as grain boundary sliding and other coupled single crystal-cum-multiple crystal deformation mechanisms active at the very high homologous temperatures encountered in this study. The findings by Abdel-Tawab and Rodin (1999) and Ballarini et al. (1999) are expected to become more relevant at colder freshwater ice temperatures (below $-20°C$, or lower). The importance of temperature is emphasized by the tests completed by Parsons and Snellen (1985) on S2 freshwater ice, using the compact tension geometry. The test sizes (in meters) ranged between $0.45\times0.45\times0.90$ and $0.90\times0.90\times2.0$; the average fracture toughness obtained was $200\,\mathrm{kPa}\sqrt{\mathrm{m}}$, but the ice temperature ranged between $-15°C$ and $-20°C$.

Abdel-Tawab and Rodin (1999) introduced the dimensionless 'grain brittleness' ($GB$) number

$$GB = \frac{d_{av}}{r_y} \tag{8}$$

in which $r_y$ is the Irwin-type plastic zone radius. $GB$ is small for ductile materials, but large for brittle polycrystals. With regard to quasi-brittle materials, it seems more appropriate to replace $r_y$ with $A_{pz}$, the characteristic length defined in (5). Abdel-Tawab and Rodin (1999) and Ballarini et al. (1999) both found that 'the crack tip fields in materials whose $GB$ is large are controlled by a very small number of grains adjacent to the crack tip – the remaining grains merely serve as a loading device.' In terms of cracks in ice, the latter conclusion may well be true at cold ice temperatures, cold enough to restrict grain boundary sliding, for instance. Otherwise, if not only the $GB$ is large, but also grain boundary sliding is occurring, one suspects that a very much larger number of grains surrounding the crack tip will be required to establish stationary crack tip fields. This number is at present unknown.

Recently, Elvin (1996) showed that the number of grains required for the homogeneous elastic behavior of polycrystalline columnar S2 freshwater ice was at least 230, in the presence of either no grain boundary slip or free grain boundary sliding. A similar number had earlier been quoted by Laws and Lee (1989) in the context of thermal anisotropy and microcracking in polycrystalline ceramics. Elvin found that grain shape was more important for the case of free grain boundary sliding, and that the average homogenized moduli differed significantly depending on no slip or free.

It is evident from Table 1 that the sample size has a considerable influence on the observed values of the elastic modulus due to polycrystalline inhomogeneity and the very high ice temperature, which no doubt promoted grain boundary sliding and creep on grain boundaries (Hsia et al., 1991; Onck and Van der Giessen, 1997). In Section 6 of the paper by Hsia et al. (1991), a conical grain embedded into an annular surrounding under plane strain conditions is considered in order to understand grain boundary sliding in the three-dimensional configuration. The Figure 14 shown therein would be relevant here, but drawn upside down, while the plane strain conditions assumed are not applicable here. The macrocrystalline S1 grains of this lake ice could undergo incremental out-of-plane displacements; there are only a very few grains through-the-thickness. Note that, rather remarkably, the moduli increase by more than one hundred percent for a three-fold increase in the test size of the cantilever columns of ice (see Table 1). Onck and Van der Giessen (1997) demonstrated that an irregular grain structure both increases the creep rate enhancement due to grain boundary sliding and leads to a wider distribution of transverse facet stress levels (with a considerably higher average).

As argued by Ortiz and Suresh (1993), the surface energy of the grain boundaries introduces another length scale into the problem, and after microcrack-like discontinuities form on these boundaries, there will be a marked dependence on grain size.

Significantly, note that if one assumes that the average grain size was 15 cm in the tests reported in this paper, then the plan area of the largest specimen size tested ($L = 28.64$ m) harbored approximately 46,400 grains; even if the average grain size was more like 30 cm, the number is still as large as 11,600. In the context of the uniaxial compression-tension testing of polycrystalline ice, the above-quoted number of 230 grains refers to the number of grains in a cylindrical cross-section, for instance. Clearly, lab-scale uniaxial testing of S1 macrocrystalline ice is not feasible. Interestingly, the number 230 correlates almost exactly with the diameter of approximately $15d_{\mathrm{av}}$ recommended for the tension-compression testing of unnotched cylinders of ice based solely on experimental evidence (Jones and Chew, 1981; Earle et al., 1984). The associated cross-sectional area of concern for the RT *in-situ* fracture tests is the area of the uncracked ligament, which is found to span 131 grains. Given the reverse taper of the RT, the shortest ligament to the boundary from the crack tip is found to span 93 grains.

The question is: if not only the $GB$ is large, but also grain boundary sliding is occurring, and a macrocrack lies within the medium under consideration, then what are the number of grains required to fully establish the crack tip fields, especially if one is considering almost no polycrystal constraint in the thickness direction?

## 6. Requisite stability characteristics

If the crack growth were a self-similar process described by a law invariant with size, one could speculate on the existence of a universal fracture mechanism governed either by a constant fracture toughness or a constant material strength. S1 freshwater ice has two major length scales affecting its fracture behavior (assuming an adequate notch-tip acuity): ice thickness and grain size. The grain size relative to the other length scales appears to be the most influential scale. The average grain size for the ice was about 15 cm with crack lengths ranging from 0.14 to 8.98 m. Note, however, that it is somewhat misleading to speak of an average grain size for S1 freshwater ice. The grain size can increase dramatically with depth, with some grains measuring on the order of a meter near the bottom surface of the lake ice tested. To investigate the presence of nonuniversal scaling, the dependence of the CTOD at the initiation of crack growth ($CTOD_i$) versus the pre-notched crack length ($A$) is plotted in Figures 3(b) and 3(c). Figure 3(c) reveals no significant change in scaling. This is also consistent with the $K_Q$ results in Figure 4(a) which indicate that a scale invariant fracture toughness was not obtained. The effect of scaling on the nominal tensile strength (Figures 5(b) and 5(c)) is clouded by the wide scatter in the data.

This scatter is counter-intuitive for the large size range tested. There are two major factors:

- The material has a very irregular and large grain structure; the mechanics and physics of any intragranular crack growth is very different from any intergranular crack growth.
- Stable crack growth was not attained; crack growth initiation was always accompanied by unstable crack growth. No well-defined, unique process zone formed in any sense, rather, significant re-distributions in stress would have accompanied the noncontiguous grain boundary sliding and cracking. Such unstable crack growth initation is typically influenced by a high degree of scatter.

Underlying the choice of the RT fracture geometry for the experiments described in this paper was the desirability of achieving crack propagation in a controlled and stable manner; such control would have allowed a number of observations to be made just from one test. The motivation for the RT geometry was provided by Mai et al. (1975) who illustrated that reverse tapering greatly improved not only crack growth stability but also crack path stability. However, crack growth stability in freshwater ice is difficult to achieve, as is now well known (DeFranco and Dempsey, 1990, 1992 and 1994; Stehn et al., 1994). The concept of the $R$-curve has gained wide acceptance in characterizing the resistance of a material to fracture. In the context of ice, however, efforts to obtain $R$-curves have been hampered by the lack of stable crack growth. As is well known, to promote stable crack growth, displacement control rather than load control, can be used. Displacement control requires a certain amount of loop shaping for the material at hand; unfortunately, the large servohydraulic loading system used for Parts I and II never did evidence the degree of control desired. In the majority of the RT experiments where load control was used, the sample failed upon initial cracking. When displacement control was used to regulate the loading, semi-stable cracking was witnessed in the form of a series of crack jumps and arrests. The crack would jump up to 1m on some samples before arresting. This can be attributed to the significant compliance of the testing system and slow response time relative to the crack jumping speed; the latter feature resulted in ballooning flatjacks and associated unreliable estimates of the unloading behavior.

## 7. Conclusions

A field investigation of macrocrystalline S1 freshwater ice fracture was completed to determine the effects of specimen size on the fracture and tensile strength parameters. The *in-situ* fracture experiments involved S1 macrocrystalline lake ice just under the melting temperature, using the reversed-taper geometry. A size range of 1 : 81 was completed. Even though this is a very large size range, a typical size effect analysis proved less successful than expected, as did the application of a powerful cohesive zone model. Whilst miniscule fracture process zones were identified, the three different load versus crack-opening-displacement profiles could not all be matched simultaneously. The grain sizes were very large relative to the largest sample size tested, and the ice temperature was very, very close to the melting temperature; as a consequence, it appears that the effects of nonlocal deformation and energy release rate mechanisms dominated the scale dependencies on the measured tensile strength and fracture toughness. While a distinct size effect does exist, it is less pronounced than the size effect predicted by LEFM. Representative specimen sizes versus these nonlocal deformation mechanisms, in the presence of a macrocrack, appears to be an open issue. The underlying difficulty posed by the data gathered is the inherent scatter which resulted from the unstable cracking and associated truncation of the fracture process.

## Acknowledgements

Funding for this two part joint-industry-agency-project was provided by Amoco, Canmar (Canada), Mobil, National Energy Board (Canada), Texaco (Part I), Minerals Management Service and the Office of Naval Research under ONR's Sea Ice Mechanics Accelerated Research Initiative (SIMI), with Tom Curtin as Scientific Officer, and the Naval Research Solid Mechanics Program, with Y.D.S. Rajapakse as Scientific Officer [Grants N00014-90-J-1360, N00014-93-1-0714, N00014-96-1-1210]. The first author (J.P.D.) is grateful for being afforded the opportunity to complete Parts I and II of this paper while on sabbatical at the University of Cambridge (hosted by Professor A.C. Palmer), and supported by the European Marine Science and Technology-III Project LOLEIF. A thorough editorial review was provided by Jim Woodhouse. *Additional acknowledgements are included in Part II.*

## References

Abdel-Tawab, K. and Rodin, G.J. (1993a). On the relevance of linear elastic fracture mechanics to ice. *International Journal of Fracture* **62**, 171–181.

Abdel-Tawab, K. and Rodin, G.J. (1993b). Interpretation of results of the fracture toughness tests on ice. *Ice Mechanics – 1993* (Edited by J.P. Dempsey, Z.P. Bažant, Y.D.S. Rajapakse and S.S. Sunder), ASME AMD – Vol. 163, New York, 49–5.

Abdel-Tawab, K. and Rodin, G.J. (1999). Fracture size effects and polycrystalline inhomogeneity. *International Journal of Fracture* Vol ??, pp. 000–000.

Ballarini, R., Mullen, R.L. and Heuer, A.H. (1999). The effects of heterogeneity and anisotropy on the size effect in cracked polycrystalline films. *International Journal of Fracture* **95**, 19–39.

Bažant, Z. P. (1984). Size effect in blunt fracture: concrete, rock, metal. *ASCE Journal of Engineering Mechanics* **110**, 518–535.

Bažant, Z.P., Kim, J.K. and Pfeiffer, P.A. (1986). Nonlinear fracture properties from size effect tests. *ASCE Journal of Structural Engineering* **112**, 289–307.

Bažant, Z.P. (1996). Analysis of work-of-fracture method for measuring fracture energy of concrete. *ASCE Journal of Engineering Mechanics* **122**, 138–144.

Bentley, D.L., Dempsey, J.P., Sodhi, D.S. and Wei, Y. (1989). Fracture toughness of S2 columnar freshwater ice: large scale DCB tests. *Cold Regions Science and Technology* **17**, 7–20.

Butiagin, I.P. *Strength of Ice and Ice Cover*, Izdatel'stvo 'Nauka', Sibirskoe Otdelenie, Novosibirsk.

Carpinteri, A. (1982). Notch sensitivity in fracture testing of aggregative materials. *Engineering Fracture Mechanics* **16**, 467–481.

Danilenko, V.I. (1985). Determination of crack resistance ($K_{1c}$) of freshwater ice. *Mechanics of Solids* **20**, 131–136.

DeFranco, S.J. and Dempsey, J.P. (1990). Crack growth stability in S2 ice. *Proceedings of the 10th IAHR Symposium on Ice*, Helsinki University of Technology, Finland, Vol. 1, 168–181.

DeFranco, S.J. and Dempsey, J.P. (1992). Crack growth stability in saline ice. *Mechanics of Creep Brittle Materials-2* (Edited by A.C.F. Cocks and A.R.S. Ponter), Elsevier Applied Science. London, 25–36.

DeFranco, S.J. and Dempsey, J.P. (1994). Crack propagation and fracture resistance in saline ice. *Journal of Glaciology* **40**, 451–462.

DeFranco, S.J., Wei, Y. and Dempsey, J.P. (1991). Notch acuity effects on fracture toughness of saline ice. *Annals of Glaciology* **15**, 230–235.

Dempsey, J.P. (1991). The fracture toughness of ice. *Ice-Structure Interaction* (Edited by S.J. Jones, R.F. McKenna, J. Tillotson and I.J. Jordaan), Springer-Verlag, Berlin, 109–145.

Dempsey, J.P. and Wei, Y. (1989). Fracture toughness $K_Q$ and fractography of S1 type freshwater ice. *Advances in Fracture Research* (Edited by K. Salama, K. Ravi-Chandar, D.M.R. Taplin and P. Rama Rao), Pergamon Press, Vol. V, 3421–3428.

Dempsey, J.P., Adamson, R.M. and DeFranco, S.J. (1995). Fracture analysis of base-edge-cracked reverse-tapered plates. *International Journal of Fracture* **69**, 281–294.

Dempsey, J.P., Adamson, R.M. and Mulmule, S.V. (1999). Scale effects on the in-situ strength and fracture of ice. Part II: First-year sea ice at Resolute, N.W.T. *International Journal of Fracture* **95**, 347–366.

Dempsey, J.P., Nigam, D. and Cole, D.M. (1988). The flexure and fracture of macrocrystalline S1 type freshwater ice. *Proceedings of the 7th International OMAE Conference*, Vol. IV, 39–46.

Dempsey, J.P., Wei, Y. and DeFranco, S.J. (1992). Notch sensitivity and brittleness in fracture testing of S2 columnar freshwater ice. *International Journal of Fracture* **53**, 101–120.

Earle, E.N., Frederking, R.M.W., Gavilo, V.P., Goodman, D.J., Hausler, F.U., Mellor, M., Petrov, I.G. and Vaudrey, K. (1984). IAHR-recommendations on testing methods of ice. 4th Report of Working Group on Testing Methods in Ice. *Proceedings of the 7th IAHR Symposium on Ice*, Hamburg, Germany, Vol. IV, 1–41.

Elvin, A.A. (1996). Number of grains required to homogenize elastic properties of polycrystalline ice. *Mechanics of Materials* **22**, 51–64.

Gandhi, C. and Ashby, M.F. (1979). Fracture-mechanism maps for materials which cleave: F.C.C., B.C.C. and H.C.P. metals and ceramics. *Acta Metallurgica* **27**, 1565–1602.

Gow, A.J. (1986). Orientation textures in ice sheets of quietly frozen lakes. *Journal of Crystal Growth* **74**, 247–258.

Gow, A.J. and Veda, H.T. (1989). Structure and temperature-dependence of the flexural properties of laboratory fresh-water ice sheets. *Cold Regions Science and Technology* **16**, 249–269.

Hillerborg, A., Modeer, M. and Petersson, P.E. (1976). Analysis of crack formation and crack growth in concrete by means of fracture mechanics and finite elements. *Cement and Concrete Research* **6**, 773–782.

Hsia, K.J., Parks, D.M. and Argon, A.S. (1991). Effects of grain boundary sliding on creep-constrained boundary cavitation and creep deformation. *Mechanics of Materials* **11**, 43–62.

Jones, S.J. and Chew, H.A.M. (1981). On the grain-size dependence of secondary creep. *Journal of Glaciology* **27**, 517–518.

Kennedy, K., Blanchet, D., Prodanovic, A., Dempsey, J.P., DeFranco, S.J., Spencer, P.A. and Masterson, D. (1993). Large-scale ice fracture experiments. *Proceedings of the 12th International POAC Conference*, Hamburg, Germany, Vol. 2, 527–536.

Kollé, J.J. (1981). Fracture toughness of ice; crystallographic anisotropy. *Proceedings of the 6th International POAC Conference*, Quebec City, Vol. I, 366–374.

Laws, N. and Lee, J.C. (1989). Microcracking in polycrystalline ceramics: elastic isotropy and thermal anisotropy. *Journal of the Mechanics and Physics of Solids* **37**, 603–618.

Mai, Y.-W., Atkins, A.G. and Caddell, R.M. (1975). On the stability cracking in wedge tapered DCB specimens. *International Journal of Fracture* **11**, 939–953.

Michel, B. and Ramseier, R.O. (1971). Classification of river and lake ice. *Canadian Geotechnical Journal* **8**, 35–45.

Mulmule, S.V. and Dempsey, J.P. (1997). Stress-separation curves for saline ice using the fictitious crack model. *ASCE Journal of Engineering Mechanics* **123**, 870–877.

Mulmule, S.V. and Dempsey, J.P. (1998). A viscoelastic fictitious crack model for the fracture of sea ice. *Mechanics of Time-Dependent Materials* **1**, 331–356.

Nanthikesan, S. and Sunder, S.S. (1995). Tensile cracks in polycrystalline ice under transient creep. Part II – numerical simulations. *Mechanics of Materials* **21**, 281–301.

Onck, P. and Van der Giessen, E. (1997). Influence of microstructural variations on steady state creep and facet stresses in 2-D freely sliding polycrystals. *International Journal of Solids and Structures* **34**, 703–726.

Ortiz, M. and Suresh, S. (1993). Statistical properties of residual stresses and intergranular fracture in ceramic materials. *Journal of Applied Mechanics* **60**, 77–84.

Parsons, B.L. and Snellen, J.B. (1985). Fracture toughness of fresh water prototype ice and carbamide model ice. *Proceedings of the 8th International POAC Conference*, Narssarssuaq, Greenland, Vol. I, 128–137.
Shapiro, L.H., Hoskins, E.R., Nelson, R.D. and Metzner, R.C. (1979). Flatjack methods of in-situ measurements of the mechanical properties of sea ice. *ASME Journal of Energy Resources Technology* **101**, 196–202.
Stehn, L.M., DeFranco, S.J. and Dempsey, J.P. (1994). Fracture resistance determination of freshwater ice using a chevron notched tension specimen. *International Journal of Fracture* **65**, 313–328.
Stehn, L.M., DeFranco, S.J. and Dempsey, J.P. (1995a). Orientation effects on the fracture of pond (S1) ice. *Engineering Fracture Mechanics* **51**, 431–445.
Stehn, L.M., DeFranco, S.J. and Dempsey, J.P. (1995b). Specimen geometry effects on fracture of warm pond (S1) ice. *ASCE Journal of Engineering Mechanics* **121**, 16–25.
Van Mier, J.G.M. (1992). Scaling in tensile and compressive fracture of concrete. *Applications of Fracture Mechanics to Reinforced Concrete* (Edited by A. Carpinteri), Elsevier Applied Science, London, 95–135.
Van Vliet, M.R.A. and Van Mier, J.G.M. (1998). Experimental and numerical investigation of size/scale effects in concrete fracture. *Material Instabilities in Solids*. (Edited R. De Borst and E. Van der Giessen), John Wiley & Sons, Chichester, 185–206.
Weeks, W.F. and Assur, A. (1972). Fracture of lake and sea ice. *Fracture* (Edited by H. Liebowitz), Vol. VII, 879–978.
Wei, Y., DeFranco, S.J. and Dempsey, J.P. (1991). Crack fabrication techniques and their effects on the fracture toughness and CTOD for fresh-water columnar ice. *Journal of Glaciology* **37**, 270–280.

*International Journal of Fracture* **95:** 347–366, 1999.


# Scale effects on the *in-situ* tensile strength and fracture of ice. Part II: First-year sea ice at Resolute, N.W.T.

J.P. DEMPSEY[1,*], R.M. ADAMSON[2] and S.V. MULMULE[3]
[1]*Department of Engineering, University of Cambridge, Trumpington Street, Cambridge, CB2 1PZ, U.K. e-mail: jpd23@eng.cam.ac.uk*
[2]*Clough, Harbour and Associates LLP, Engineers, Surveyors, Planners and Landscape Architects, 571 South Lake Boulevard, Suite B, Richmond VA 23236*
[3]*Hibbitt, Karlsson and Sorensen Inc., 1080 Main Street, Pawtucket, RI 02860-4847*

Received 9 October 1998; accepted in revised form 31 December 1998

**Abstract.** A set of lab- to structural-scale ($0.5 < L < 80$ m) *in-situ* full thickness (1.8 m) fracture tests were conducted on first-year sea ice at Resolute, N.W.T. using self-similar (plan view) edge-cracked square plates. With a size range of 1:160, the data is used, via size effect analyses, to evaluate the influence of scale effects on the fracture behavior of sea ice over the range $10^{-1}$ m (laboratory) to 100 m and to predict the scale effect on tensile strength up to $\approx$1000 m. Details of this large-scale sea ice fracture test program are presented in this paper. The experimental results are presented as well as the fracture modeling of the data. The influence of scale on the ice strength and fracture toughness is dramatic. The applicability of various size effect laws are investigated and criteria for LEFM test sizes are presented. For the thick first-year sea ice tested, the size-independent fracture toughness is of order 250 kPa$\sqrt{\text{m}}$, not the 115 kPa$\sqrt{\text{m}}$ that is commonly used. The number of grains spanned by the associated test piece is 200, much larger than the number 15 typically quoted for regular tension-compression testing. The size-independent fracture energy is 15 J/m$^2$, while the requisite LEFM test size for the edge-cracked square plate geometry (for loading durations of less than 600 s and an average grain size of 1.5 cm), is 3 m square. Size effect analyses of sub-ranges of the data show that unless the specimen sizes tested are themselves sufficiently large, the true nature of the scale effect is not revealed, which was a concern raised by Leicester 25 years ago. In the case of the fracture tests reported in this paper, based on the lab-scale and field-scale strength data measured between 0.1 and 3 m and using Bažant's size effect law, it is possible to accurately predict the tensile strengths for all of the remaining tests, up to and including 80 m.

**Key words:** Sea ice, scale effects, size effect laws, LEFM, tensile strength, first-year sea ice, structural-scale, *in-situ*, homogeneity.

## 1. Introduction

An understanding of the fracture behavior of sea ice is sought at many scales. The lab-scale brittle failure of ice under compression is accompanied by a plethora of cracking scenarios, as are the structural-scale interactions of ice sheets with offshore structures, ships, icebreakers, coastal structures, submarines and tankers. At the mesoscale, the accurate incorporation of ridging, rafting and leading into granular models is seen as critical to the determination of ice strength on a geophysical scale. At the mesoscale and geophysical scale, the formation of leads is not understood, while the ice strength has been inferred to be of the order of tens of kPa (Dempsey, 1996). Given that lab-scale sea ice strengths are of the order of one MPa, there

* *Permanent address*: Department of Civil and Environmental Engineering, Clarkson University, Potsdam, NY 13699-5710. e-mail: john.dempsey@clarkson.edu

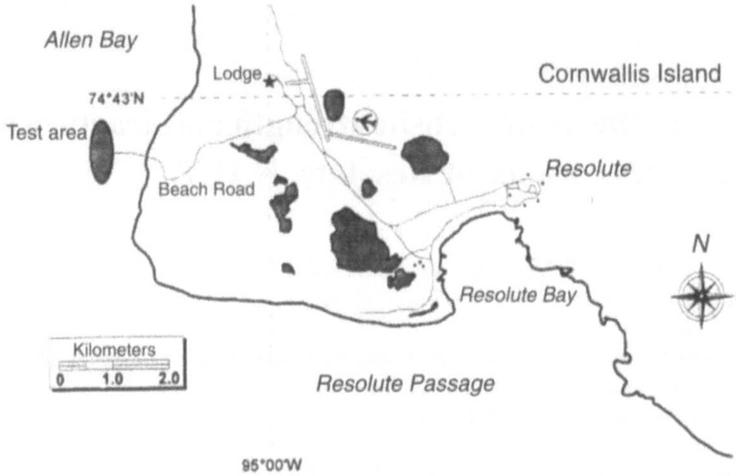


*Figure 1.* Map of Resolute Bay indicating the test site.

is a huge scale effect that needs to be explained. Of equal importance is the scale effect on the apparent fracture toughness (or fracture energy) that occurs from the scale of grain boundary facet cracks up to lab-scale. This paper will concentrate on the scale effects measured from lab-scale to structural-scale.

The link between the compressive strength of ice and its fracture toughness is rather tenuous, although Schulson (1997) has argued that the brittle compressive strength should be directly proportional to an expression of the form $K_{Ic} f(\mu, R)/\sqrt{d}$, in which $K_{Ic}$ is the fracture toughness, $\mu$ is the friction coefficient, $d$ is the length of the parent sliding crack, and $R$ is the combined stress ratio. Bažant (1997) has taken a look at models to examine the global mechanism of compression failure, and has outlined those associated with size effects versus those that are not. If size effects are predicted, the fracture energy is involved.

The range of scales over which the *in-situ* strength and fracture toughness of ice is sought is similar to that encountered for materials such as rock, coal and concrete. Without reviewing the associated literature, the papers by Bienawski (1968), Leicester (1973), and Heuze (1980) are especially noteworthy. In fact, a portion of Leicester's abstract written 25 years ago echoes a crucial finding of this paper: *"A particularly dangerous aspect noted of the size effect is that it may not occur unless member sizes are sufficiently large and consequently may not appear in scaled-down laboratory testing."* In other words, while it has long been emphasized that the size range must cover an adequate size range, as in 1:10, for instance, it has been more or less been overlooked, apparently, that the sizes themselves must surpass some critical size.

A two-phase joint-industry-agency project (JIAP) was initiated in 1990 to calibrate a fracture theory for incorporation into probabilistic global ice load models. The first phase of the JIAP 'Large-Scale Ice Fracture Experiments,' completed in January, 1992 near Banff, Alberta, is reported in Part I of this paper (Dempsey et al., 1999). This paper discusses the second phase, a series of *in-situ* large-scale full-thickness fracture tests conducted on first-year sea ice at Resolute, N.W.T. (see Figure 1) between April 17 to May 7, 1993 (Kennedy et al., 1994; Adamson et al., 1995; Mulmule et al., 1995; Wei et al., 1995; Dempsey, 1996). In this paper, the phrase 'large-scale' denotes full thickness (1.8 m) specimen sizes which ranged in crack-parallel size between 0.5 and 80 m; the results and conclusions may well be applicable up to

*Table 1.* Compact Tension Fracture Tests at Resolute, N.W.T. (Parsons et al., 1986)

| Test ID | Size $L$, m | Thickness $h_*$, m | $L/h_*$ | $A/L$ | $L/d_{av}$ | $\dot{K}$ kPa$\sqrt{m}$/s | $K_Q$ kPa$\sqrt{m}$ |
|---|---|---|---|---|---|---|---|
| 77 | 0.70 | 1.24 | 0.6 | 0.23 | 47 | 30 | 175 |
| 78 | 0.70 | 1.25 | 0.6 | 0.37 | 47 | 74 | 116 |
| 84 | 0.75 | 1.25 | 0.6 | 0.55 | 50 | 85 | 172 |
| 104 | 1.06 | 1.12 | 1.0 | 0.34 | 71 | 45 | 187 |
| 96 | 1.07 | 1.20 | 0.9 | 0.41 | 71 | 51 | 163 |
| 65 | 1.37 | 1.25 | 1.1 | 0.18 | 91 | 26 | 138 |

ice pieces of the order of several hundred meters or more, until other length scale(s) dominate. The test program was designed to explore the fracture characteristics of sea ice.

The two papers by Adamson et al. (1995) and Mulmule et al. (1995) document the experiments and analyses completed by that date for six field trips investigating the fracture of ice; Parts I and II of this paper are fully detailed accounts of the first two field programs. The crack opening displacements measured here for sea ice were accompanied by a significant degree of creep; for this reason, Mulmule and Dempsey (1997, 1998, 1999a) developed a viscoelastic version of Hillerborg's fictitious crack model (see Section 4.2 of Part I, Dempsey et al., 1999) to apply to the data obtained. This alternative method of analysis to the size effect type of analysis, along with the information presented in this paper, allows one to postulate LEFM size requirements for the fracture testing of sea ice (Mulmule and Dempsey, 1999b).

The fracture testing of sea ice began in earnest more than twenty years ago, and has included many different laboratory tests, as well as *in-situ* tests of various sizes, orientations and rates (Vaudrey, 1977; Urabe et al., 1980, 1983; Shapiro et al., 1981; Urabe and Yoshitake, 1981a, 1981b; Timco and Frederking, 1983; Goldstein and Osipenko, 1983a, 1983b; Parsons et al., 1986, 1987, 1988, 1992, 1993; Shen and Lin, 1986; Urabe and Inoue, 1986, 1987; Tuhkuri, 1987; Shen et al., 1991; Danilenko and Rogachko, 1991; DeFranco and Dempsey, 1991, 1992, 1994; DeFranco et al., 1991; Bentley, 1992; Williams et al., 1993; Lazo, 1994; Stehn, 1994; Adamson et al., 1997; LeClair et al., 1997; Vincent and Dempsey, 1998). Adamson et al. (1995) present a table (Table 8, p. 123) in which the test geometry, the specimen size, grain size, crack length, orientation, ice temperature and fracture toughness values are reported for 15 of the above papers. It is outside the scope of this paper to review all of the data provided in these papers.

Note, however, that many of the above investigations assumed that if a *linear* load-CMOD plot has been measured, then the associated Mode I fracture toughness measured could be reported as a material property, and identified as $K_{Ic}$. While it is understood that the fracture toughness $K_{Ic}$ may change with factors such as ice temperature, rate of loading, salinity, and ice thickness, it should not vary with specimen size, or crack length. As a result of this one misconception, the applicability of LEFM has been assumed in many instances, whereas the suitability of a one parameter fracture mechanics for ice (that is, $K_{Ic}$) has not yet been demonstrated.

In a closely related study to that of this paper, Parsons et al. (1986) examined the *in-situ* fracture toughness of sea ice in Allen Bay, near Resolute, N.W.T. between April 9 and May 3,

1985. This group used the compact tension fracture geometry, and examined vertical cracks propagating downwards, crack front parallel to the ice surface (VV), through-the-thickness cracks propagating horizontally (VH), and horizontal cracks propagating horizontally (HH); the pre-cut crack planes (VV and VH) and crack front (HH) were oriented parallel, at 45°, and perpendicular to the predominant *c*-axis alignment observed in the ice sheet. The results of most relevance for this study are the 'V.H. Perpendicular to *c*-axis' results presented on page 363, Parsons et al. (1986). In the notation of the present paper, these results are provided for later comparison in Table 1 ($h_*$ is the thickness in the growth direction, measured from the top surface). These VH tests by Parsons et al. (1986) were chainsawed out and tested, being exposed on the specimen sides to the ambient temperature. Each test took of the order of 3–4 hours; the crack front would have been dry in each case, unlike the *in-situ* tests. It should be noted that in 1985, the upper half of the ice sheet tested by Parsons et al. (1986) evolved through an upward temperature change (the top surface changed from −27°C to −12°C), whilst for the tests reported in this paper, only the upper quarter warmed up, with the top surface changing from −20°C to −12°C (see Figure 2 of Wei et al., 1995).

Most of the experiments carried out to obtain the fracture toughness of ice have been performed on lab-sized specimens. Because of the material anisotropy, nonhomogeneity of the ice due to large grain sizes, brine drainage channels and the very high homologous temperature at which fracture tests are usually carried out, the fracture toughness obtained at lab sizes can not be considered as a true material property. Dempsey (1991) defined an apparent fracture toughness $K_Q$ as obtained from lab tests and advocated the use of larger sized specimens. With this in mind, large scale *in-situ* experiments coupled with small scale field and laboratory tests were completed on first-year sea ice.

The fracture of ice is influenced by the salinity and temperature, notch tip acuity, rate of loading, material anisotropy, grain size effects and other inhomogeneities such as grain boundary sliding and brine drainage channels, specimen size and notch sensitivity, and inelasticity ahead of crack tip. Each of these factors above are examined below as related to the current investigations. A full thickness ice sheet has both a temperature and salinity profile through the depth: the isolated effect of either parameter is not obtainable. The issue of crack tip acuity for saline ice was explored by DeFranco et al. (1991) who found a decrease in $K_Q$ with increase in crack tip acuity. Notch tip acuity was maintained in these experiments by sharpening and scribing the cracks with a special apparatus. The quantification of rate effects demands a large number of tests to be carried out over a range of rates. Large scale testing with its time consuming test setups and limited time precluded any such possibility. Notch sensitivity for first-year sea ice was experimentally investigated by Parsons et al. (1993). The length of the pre-cut crack was always such that the specimens were notch sensitive and failure occurred by fracture.

Due to the large grain size of ice and anisotropy of grains, the crack size to grain size and the ligament size to grain size ratio of the fracture specimen dictate the size of specimen if micromechanical simulations are to be avoided. For sea ice, Abdel-Tawab and Rodin (1993a, 1993b) estimated that the minimum notched specimen size necessary to preserve homogeneity would be of the order of 400 grain diameters. In this context, note that the ratio of specimen size to the average grain size in the experiments reported in this paper have ranged from 33 to 5333.

Although many issues affect the fracture toughness of ice, only a few can be addressed if one is restricted to the testing of laboratory sized specimens. Factors such as notch acuity, ice temperature, salinity and notch sensitivity can be studied independently. However, material anisotropy, specimen size effects, and the true nature of strain localization during fracture can only be addressed by testing large sized specimens. *In-situ* testing also incorporates the actual temperature, grain size and salinity profiles through the thickness. Furthermore, the testing of ice at ice temperatures above −5°C is ideal in the field, but fraught with difficulties in the lab.

## 2. Experiments

The site chosen for the large scale sea ice experiments was Resolute, Northwest Territories in Canada. It was chosen primarily for the availability of logistical support and the growth of uniform, thick sea ice. The history of the temperatures, snow cover and average ice thickness in Resolute Bay served to indicate the most favorable time period and location for testing. April 17 – May 8, 1993 was the chosen time frame for the experimental program and the testing area was established in Allen Bay, approximately 6 km from the lodge (Figure 1). The ice was first-year sea ice with a thickness of approximately 1.8 meters. The depth of the snow cover on the surface of the ice sheet ranged from 5 cm to 40 cm. Temperatures during the program typically remained between −15°C and −20°C. More details of the site and conditions can be found in Kennedy et al. (1994).

Ice characterization and micrography were conducted at intervals through the thickness of the sheet. Approximately 14 horizontal thin sections were done specifically for grain size and orientation studies. The average grain size was about 1.5 cm. The salinity was maximum (7.8°/∘∘) near the top surface and a minimum (3.3°/∘∘) near the bottom, with most change occurring in the top 20 cm. The density of the ice sheet was fairly uniform at 900 kg/m$^3$ for the top 1.4 m, and then decreased linearly to 850 kg/m$^3$ in the bottom portion. The grain size varied from approximately 5 mm at the top to 20 mm at a depth of 80 cm, and then did not change markedly. Thermistors were placed at depths of 20, 70, 120 and 170 cm, and were accurate (supposedly) to 0.1°C. The ice temperature varied almost linearly with depth, with the surface temperature increasing from approximately −17 to −12°C over the duration of the test program. The complete details of the characterization and microstructure are given in the paper by Wei et al. (1995), which includes the salinity, density, temperature and grain size profiles versus depth in the ice sheet.

### 2.1. *IN-SITU* FRACTURE AND FLEXURE EXPERIMENTS

All large scale experiments were *in-situ* tests enabling the testing of large sizes with resident thermal profiles. Experiments were prepared by first cutting the sample free from the parent ice sheet using a R-100 DitchWitch equipped with special cutting teeth. It achieved cutting speeds of approximately 90 meters per hour, creating a slot about 15 cm wide. A second smaller Ditchwitch was used to cut the crack in the specimens. This machine cut a notch 1.6 cm wide, enough to insert the flatjack loading device. Because it was a narrow cut, it had to be constantly cleaned to prevent refreezing.

Load was applied to the test sample using the flatjack loading device (Shapiro et al., 1979) inserted in the pre-cut crack. The flatjack was a thin-walled steel bladder pressurized with either nitrogen gas or hydraulic fluid. Tests performed with nitrogen gas were controlled by varying the flow of gas from a pressurized tank and generally resulted in unstable fracture. A servo-hydraulic control system was used on several of the *in-situ* experiments controlling the load via a feedback signal from a displacement measuring gauge on the crack. These experiments generally involved multiple loadings and stable crack propagation in the form of

*Table 2.* *In-situ* fracture tests at Resolute, N.W.T. (I)

| Test ID | Test size m | Crack length m | Control | Mean ice temperature °C | $E_{COD}$ GPa | $E_{CMOD}$ GPa |
|---|---|---|---|---|---|---|
| SQ10 | 0.5 | 0.26 | Load | −15 | — | 2.0 |
| SQ12 | 0.5 | 0.25 | CMOD | −14 | — | 4.0 |
| SQ2 | 0.9 | 0.28 | Load | −13 | — | 7.0 |
| SQ1 | 1.0 | 0.3 | Load | −9 | — | 8.2 |
| SQ8 | 3.0 | 0.9 | Load | −15 | 7.6 | 8.0 |
| SQ9 | 3.0 | 0.9 | CMOD | −15 | 5.2 | 5.8 |
| SQ15 | 3.0 | 0.9 | CMOD | −17 | — | — |
| SQ3 | 10.0 | 3.0 | Load | −13 | — | — |
| SQ4 | 10.0 | 5.0 | Load | −13 | 6.1 | 6.8 |
| SQ5 | 30.0 | 9.0 | Load | −13 | 5.0 | 7.0 |
| SQ6 | 30.0 | 9.0 | Load | −14 | 5.2 | 6.4 |
| SQ7 | 30.0 | 9.0 | Load | −15 | 11.5 | 8.7 |
| SQ11 | 30.0 | 9.0 | NCTOD | −14 | — | 2.0 |
| SQ14 | 30.0 | 9.0 | NCTOD | −12 | — | — |
| SQ13 | 80.0 | 24.0 | Load | −12 | 3.9 | 4.7 |

*Table 3.* *In-situ* fracture tests at Resolute, N.W.T. (II)

| Test ID | Size $L$ m | $L/h$ | $A/L$ | $L/d_{av}$ | $\dot{K}$ kPa$\sqrt{m}$/s | $K_Q$ kPa$\sqrt{m}$ | CTOD$_i$ $\mu$m |
|---|---|---|---|---|---|---|---|
| SQ10 | 0.5 | 0.3 | 0.5 | 33 | 220 | 143 | 9 |
| SQ12 | 0.5 | 0.3 | 0.5 | 33 | 0.26 | 120 | 12 |
| SQ2 | 0.9 | 0.5 | 0.3 | 60 | 0.42 | 216 | — |
| SQ1 | 1.0 | 0.6 | 0.3 | 67 | 0.48 | 223 | 6.5 |
| SQ8 | 3.0 | 1.7 | 0.3 | 200 | 0.4 | 213 | 36 |
| SQ9 | 3.0 | 1.7 | 0.3 | 200 | 5.14 | 192 | 45 |
| SQ15 | 3.0 | 1.7 | 0.3 | 200 | 15.0 | 226 | — |
| SQ3 | 10.0 | 5.6 | 0.3 | 667 | 0.45 | 224 | — |
| SQ4 | 10.0 | 5.6 | 0.5 | 667 | 0.32 | 250 | — |
| SQ5 | 30.0 | 16.7 | 0.3 | 2000 | 0.21 | 220 | 40 |
| SQ6 | 30.0 | 16.7 | 0.3 | 2000 | 0.22 | 233 | 35 |
| SQ7 | 30.0 | 16.7 | 0.3 | 2000 | 0.46 | 238 | 35 |
| SQ11 | 30.0 | 16.7 | 0.3 | 2000 | 1944 | — | 14 |
| SQ14 | 30.0 | 16.7 | 0.3 | 2000 | 8.33 | 250 | — |
| SQ13 | 80.0 | 44.4 | 0.3 | 5333 | 0.17 | 250 | 39 |

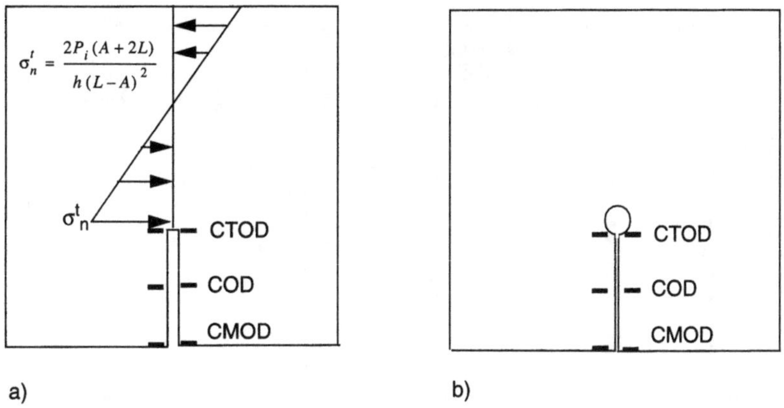


*Figure 2.* The fracture and strength test geometries, respectively: (a) edge-notched square plate; (b) key-holed square plate.

*Table 4.* Flexural strength of first-year sea ice obtained using the key-holed square plate geometry (Figure 2(b))

| Test ID | Test size m | Crack length m | Control | Mean ice temperature °C | Flexural strength MPa |
|---|---|---|---|---|---|
| KH1 | 3.0 | 1.5 | Load | −6 | 0.40 |
| KH1 | 3.0 | 1.5 | Load | −6 | 0.47 |

crack jumps and arrests. Displacement measuring devices, LVDT's and noncontacting sensors, were mounted at several points along the crack as shown in Figure 2(a). These locations are labeled CMOD, COD and CTOD for the crack-mouth-, intermediate crack- and crack-tip-opening-displacements, respectively. At each point, two displacement gauges were used in parallel, an LVDT and a noncontacting displacement gauge (in Figure 7, p. 125 of Adamson et al. 1995, a representative plot of the load versus CMOD and crack extension, CMOD and CTOD versus time for test SQ7 – see Tables 2 and 3 – is provided). The noncontacting displacement gauge had a finer resolution, but went out of range much earlier; the LVDT would continue measuring, providing a continuous record of the crack opening deformation. On smaller experiments, the crack was not long enough for the COD gauge to be setup. On larger samples, it was possible to place a fourth set of displacement transducers between the COD and the CTOD, labeled as the NCTOD. The data from the displacement sensors and the pressure transducer were recorded digitally on two computers equipped with A/D converters. One computer was set up to provide real time viewing of the pressure and displacements as well as to record the data. The other computer was dedicated to data acquisition and sample rates up to 5 kHz were possible for 8 channels. Recording the data directly to the computer made it possible to do preliminary analysis and determine if any changes to the test plan were required.

The square plate geometry (Figure 2(a)) was used for the fracture experiments. The sample size was varied from 0.5 m on a side to 80 m (Table 2). The testing of *in-situ* flexure beams

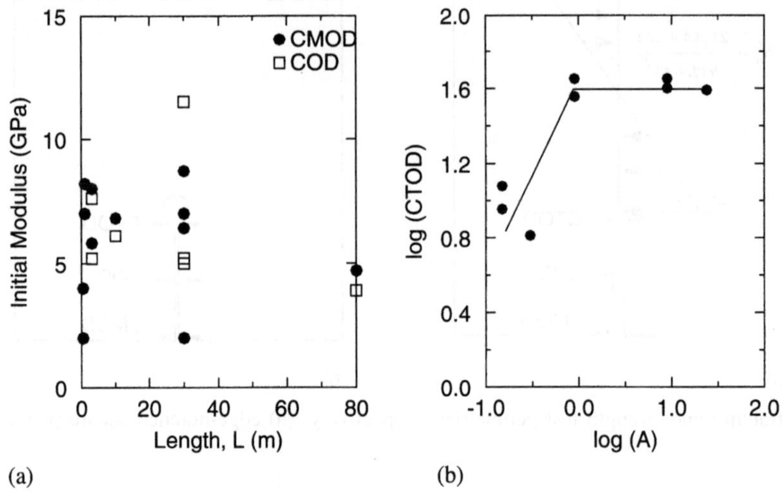


*Figure 3.* Effects of specimen size on (a) Short-time moduli and (b) CTOD at the initiation of crack growth.

(FL) proved to be a difficult task. One beam was tested in flexure at Resolute. The test was unsuccessful and required an excessive amount of preparation. It was found that test specimens using self-equilibrated loading such as the edge-notched square plate were inherently easier to setup, minimizing the preparation time. This encouraged the use of the square plate keyhole geometry (KH) shown in Figure 2(b). This was a flexure test similar to the square plate fracture tests with a 20 cm hole bored at the crack tip. The displacement gauges were placed at points on the crack, similar to the fracture tests. A finite element analysis was conducted using the applied loading to find out the failure stresses. The flexural strength values obtained from keyhole tests are given in Table 4.

### 2.2. Small-scale isothermal fracture and flexure experiments

A complementary small-scale program was carried out during the large-scale testing (Williams et al., 1993). The three-point bend geometry was used for both fracture and flexure experiments. Sixty fracture experiments as well as 42 flexure experiments were completed varying the size, depth in the sheet and orientation. In this paper, only the fracture tests with the same orientation as the large scale experiments are reported for comparative purposes. Most of the beams were machined to dimensions of 1.2 m $\times$ 0.1 m $\times$ 0.1 m and tested over a span of 0.80 m with a crack length of 0.05 m (giving $L/d_{av} < 7$). The fracture tests were loaded at a rate of $\dot{K} = 200\,\text{kPa}\sqrt{\text{m}}/\text{s}$. The load and CMOD were recorded providing fracture toughness and modulus values for each of the fracture tests (Table 5).

## 3. Large-scale experimental results

The experimental results are summarized in Tables 2 and 3. Since the data covered such a large size range, the effect of size on the results provides valuable insight. Initial (short-time) moduli were computed from the CMOD and COD records and are plotted vs size in Figure 3. A fair amount of scatter is evident in the results, an expected outcome from sea ice testing. No discernible trend in the moduli is evident. The CTOD values at the instant of growth of the pre-cut crack, $\text{CTOD}_i$ on the other hand, show a distinct change in behavior above crack

*Table 5.* Small Scale Isothermal Sea Ice Experiments at Resolute, N.W.T. (Williams et al., 1993)

| Depth in ice sheet m | Ice temperature °C | Short-time modulus, $E$ GPa | $K_Q$ kPa$\sqrt{m}$ |
|---|---|---|---|
| 0.05 | −14.8 | 8.06 | 75 |
| 0.05 | −17.4 | 5.25 | 77 |
| 0.05 | −18.0 | 3.87 | 86 |
| 0.15 | −20.5 | 4.21 | 77 |
| 0.15 | −20.4 | 4.93 | 91 |
| 0.20 | −15.7 | 5.51 | 83 |
| 0.55 | −18.2 | 5.85 | 74 |
| 0.71 | −16.1 | 4.36 | 64 |
| 0.71 | −16.8 | 5.19 | 53 |
| 1.05 | −17.1 | 6.45 | 66 |
| 1.05 | −16.8 | 4.43 | 59 |
| 1.22 | −16.5 | 5.12 | 69 |
| 1.22 | −17.7 | 7.54 | 67 |
| 1.40 | −15.1 | 4.58 | 72 |
| 1.40 | −16.7 | 4.17 | 77 |
| 1.40 | −16.5 | 4.90 | 90 |
| 1.59 | −15.9 | 5.04 | 88 |
| 1.59 | −16.7 | 4.32 | 64 |
| 1.75 | −14.7 | 4.30 | 58 |

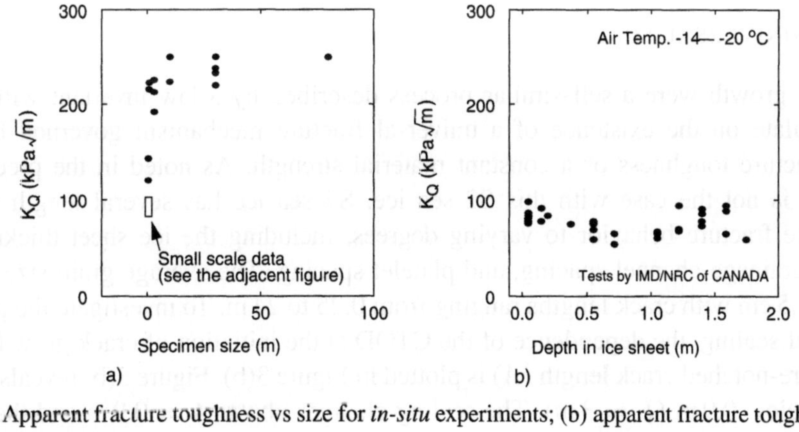


*Figure 4.* (a) Apparent fracture toughness vs size for *in-situ* experiments; (b) apparent fracture toughness vs depth in sheet for isothermal small scale experiments.

lengths of 0.9 m ($A/d_{\mathrm{av}} = 60$); for a crack length of 0.9 m, the test specimen dimension was 3 m (for which $L/d_{\mathrm{av}} = 200$).

The stress intensity factor for an edge-notched square plate subjected to loading on a portion of the crack was computed following the weight function procedure described in Dempsey et al. (1995); the necessary weight function details are provided in Appendix I of Mulmule and Dempsey (1999a). Because of the uncertainties in applying LEFM to fracture testing of sea ice as alluded to earlier, $K_Q = K_{\text{apparent}}^{\text{initiation}}$ will be used throughout this paper to represent the apparent fracture toughness. The apparent fracture toughness values were computed from the load at the onset of cracking as determined from the CTOD gauge; the CTOD gauge was set to measure $\mu$m, and thus at the initiation of growth of the pre-cut crack, this gage would almost instantaneously go out of range. The fracture toughness as a function of size is plotted in Figure 4(a) and shows a rise in toughness with size and seems to approach a plateau $K_Q$ value above a size of about 3 m. This is indicative of softening mechanisms (grain boundary sliding, dislocations, microcracking) around the crack tip dominating the failure of the smaller samples with a transition to brittle failure at the larger sizes.

### 3.1. Slow crack velocities in sea ice

During these lab- to structural-scale fracture tests (3 m square and 30 m square, 1.8 m thick) conducted at Resolute, N.W.T. in 1993 (see Adamson et al., 1995), the crack growth characteristics were measured acoustically by Xie and Farmer (1993). It was observed that the initial crack growth included a few infrequent meter-scale jumps, followed ultimately by a series of centimeter-scale, almost continuous, stick-slip failures until a visual fracture formed on the ice surface. Although the small-scale (grain size) crack jumps registered speeds of order 400 m/s, the overall averaged crack velocity was of order 10 m/s. Similar slow crack velocities in sea ice were first measured by Parsons et al. (1987). The slow velocities observed may well be because of energy being dissipated in the vicinity of the crack tip due to movement of the brine through the extended brine drainage networks. This idea is supported by the fact that faster crack velocities are observed at colder temperatures. An interesting area of research is to now explain these slow crack velocities in sea ice through an appropriate fracture mechanics treatment (a suitable viscoelastic/viscoplastic constitutive sea ice model will no doubt be required as well). These slower than expected crack velocities may be connected with the transitions observed with ice-structure indentation velocity.

### 3.2. Non-universal scaling

If the crack growth were a self-similar process described by a law invariant with size, one could speculate on the existence of a universal fracture mechanism governed by either a constant fracture toughness or a constant material strength. As noted in the preceding discussions, it is not the case with this S2 sea ice. S2 sea ice has several length scales that influence the fracture behavior to varying degrees, including the ice sheet thickness, grain size, brine drainage channel spacing, and platelet spacing. The average grain size for the ice was about 1.5 cm with crack lengths ranging from 0.25 to 24 m. To investigate the presence of nonuniversal scaling, the dependence of the CTOD at the initiation of crack growth ($\text{CTOD}_i$) versus the pre-notched crack length ($A$) is plotted in Figure 3(b). Figure 3(b) reveals no change in scaling below 0.9 m ($L = 3$ m). The scaling changes above $A = 0.9$ m and the CTOD at the initiation of crack growth becomes invariant with size (over the size range studied). These results are consistent with findings by Hatton et al. (1994) in a study of fracture in the lava flows of the Krafla fissure swarm. This is also consistent with the $K_Q$ results in Figure 4(a) which seem to indicate a scale invariant fracture toughness above $L = 3$m.

## 4. Size effect laws

The results of the fracture tests can be characterized using the nominal tensile strength $\sigma_n^t$ derived from the failure load in the square plate fracture tests, Figure 2(a). The nominal tensile strength obtained from fracture tests shows a decrease with an increase in test specimen size. Linear elastic fracture mechanics predicts this decrease to have a $1/\sqrt{L}$ dependence, $L$ being the characteristic dimension of the specimen. Fracture experiments conducted on ice show a dependence of nominal strength which is milder than $1/\sqrt{L}$ for relatively small sizes but does tend to this limit when the specimen sizes become large. The concept of small and large sizes will be explored later. The reason for this deviation from the $1/\sqrt{L}$ dependence is the nonlinearity ahead of the crack (induced because of microcracking, grain boundary separation, and other small flaws ahead of the crack) which limits the applicability of linear elastic fracture mechanics. There are similarities with the fracture of concrete, where a number of approaches have been proposed to model the change in nominal tensile strength with size. Prominent among these are the size effect laws proposed by Bažant (1984), Carpinteri et al. (1993), Bažant et al. (1991) and Kim et al. (1993). Jenq and Shah (1985) developed a two parameter fracture model which yields size independent fracture properties. Nallathambi and Karihaloo (1986) proposed an effective crack model from similar considerations.

The size effect law proposed by Bažant (1984), hereafter called the BZ law, provides a smooth transition from failures which are closer to strength failures at small sizes to failures controlled by linear elastic fracture mechanics at large sizes. The BZ expression for the failure strength is asymptotic to the tensile/flexural strength at small sizes and to LEFM at large sizes. The BZ law is presented below. For detailed analysis as well as comparisons with other numerical models the reader is referred to Planas and Elices (1991) and to Smith (1995).

Consider now a crack in an elastic strain softening material having an initial length of $A_0$ and an equivalent length at crack growth instability to be $A$. Let $\sigma_n^t$ be a nominal tensile strength which can be used to both characterize the strength prior to unstable fracture and to compare the peak tensile strengths of test specimens of various sizes. The initial length of the crack $A_0$ and the length of crack at crack growth instability $A$ are normalized by a characteristic dimension of the specimen $L$ and written as $a_0$ and $a$. The critical-stress-intensity-factor can be expressed in the general form $K_{\mathrm{Ic}} = \sigma_n^t \sqrt{L} \hat{f}(a)$. However, for very large test sizes, it may be assumed that the fracture energy $G_{\mathrm{Ic}}$ and the short-time modulus of elasticity $E$ are related to the critical-stress-intensity-factor by $G_{\mathrm{Ic}} E = K_{\mathrm{Ic}}^2$. That is, $G_{\mathrm{Ic}} E = \sigma_n^{t2} L \hat{f}^2(a) = \sigma_n^{t2} L g(a)$. Now let $A = A_0 + c_f$, in which $c_f$ is an equivalent elastic length for the process zone (the zone of nonlinearity at the crack tip). As in the treatment by Bažant (1992), this $c_f$ is an equivalent length for the process zone only for a very large specimen size (one that may be analyzed using LEFM) and as such is considered a material property. By expanding $g(a)$ in a Taylor series in terms of $c_f/L$ and recognizing that the fracture energy equals $G_{\mathrm{Ic}}$ for this very large specimen size, it follows that

$$G_{\mathrm{Ic}} E = \sigma_n^{t2} L \left( g(a_0) + \frac{c_f}{L} g'(a_0) + \cdots \right). \tag{1}$$

Rearranging equation (1) gives

$$\sigma_n^t = \frac{\sqrt{G_{\mathrm{Ic}} E}}{\sqrt{c_f g'(a_0)} \sqrt{1 + \frac{g(a_0)}{c_f g'(a_0)} L}} = \frac{\Sigma}{(1 + L/B)^{1/2}}. \quad \text{BZ} \tag{2}$$

The right-hand side of (2) is the form of the BZ law. Many approximations were made in obtaining this form as explained below. The expansion of in (1) is strictly valid only for small $c_f/L$. The length of the each process zone that forms in each test is assumed to evolve stably and to scale self-similarly with specimen size (strictly speaking, the above Taylor series expansion is accurate only if the constant $c_f/L$ is small relative to the length $a = a_0 + c_f/L$). Further, the equation $G_{\mathrm{Ic}}E = K_{\mathrm{Ic}}^2$ is valid only for LEFM (which necessitates that the specimen size be very large) but is an approximation for smaller sized specimens. Planas and Elices (1991) showed that results from this BZ law closely follow the Dugdale formulation. The constant $\Sigma$ has the dimensions of stress and can be related to the tensile strength of the material being studied by making the fracture energy $G_c$ dependent on the tensile strength. Bažant and Pfeiffer (1987) introduced two generalizations of the BZ law

$$\sigma_n^t = \frac{\Sigma}{(1 + L/B + C/L)^{1/2}}, \quad \text{BZ1} \tag{3}$$

$$\sigma_n^t = \frac{\Sigma}{(1 + (L/B)^F)^{1/2F}}. \quad \text{BZ2} \tag{4}$$

The generalization in BZ1 is supposed to help model the data better for smaller sizes while $F$ in BZ2 influences the transition from near strength failures at small sizes to LEFM failures at larger sizes. The higher the value of $F$, the faster the transition from strength to LEFM.

The BZ law uses data from the fracture of self-similar specimens. As the specimen size increase, the BZ, BZ1 and BZ2 law follow LEFM. Unnotched specimens made of quasi-brittle material (material which undergoes microcracking) show a constant residual strength for large specimen sizes. This occurs because the zone of microcracking attains a size independent constant length. Another reason for specimen behavior to deviate from LEFM is a change in the failure mechanism which requires extra energy input. To account for this behavior Bažant et al. (1991) and Kim et al. (1993) modified the BZ law by including a constant term as a residual strength. For purpose of identification, here it is referred to as the KSL law

$$\sigma_n^t = \frac{\Sigma}{(1 + L/B)^{1/2}} + D. \quad \text{KSL} \tag{5}$$

Carpinteri et al. (1993) formulated a size effect law from fractal considerations (called the multi-fractal scaling law, MFSL). It is useful for unnotched geometries which reach a size independent residual nominal tensile strength. For very small sizes the MFSL predicts a large tensile strength which is consistent with the notion that as a smaller and smaller specimen size is chosen, the number of inherent flaws reduce. Though this law may not be ideally suited for studying the fracture of self-similar specimens (which the BZ law is), it has been used to study notched specimens (Carpinteri et al., 1995a, 1995b). The form of Carpinteri's MFSL is given here by

$$\sigma_n^t = D(1 + 10^B/L)^{1/2}. \quad \text{MFSL} \tag{6}$$

In all the above size effect laws, $\Sigma$, $B$, $C$, $D$ and $F$ are the undetermined constants which are obtained by fitting the size effect expression to the experimental data.

The effective crack models of Jenq and Shah (1985) and Nallathambi and Karihaloo (1986) make use of the difference between loading and unloading compliances or deviation of the crack opening displacement from the results of LEFM to obtain size independent fracture parameters. During the current experiments, a large amount of creep was observed in the load

*Table 6.* Size effect law parameters for various size ranges

| FIT | Range m | Law | Σ MPa | *B* m | *C* m | *D* MPa | *F* | $r^2$ |
|---|---|---|---|---|---|---|---|---|
| FIT1 | 0.1–80 | BZ | 0.59 | 0.37 | — | — | — | 0.98 |
| | | BZ1 | 0.74 | 0.21 | 0.06 | — | — | 0.98 |
| | | BZ2 | 0.54 | 0.38 | — | — | 1.5 | 0.98 |
| | | KSL | 0.59 | 0.39 | — | −0.01 | — | 0.98 |
| | | MFSL | — | 0.68 | — | 0.1 | — | 0.79 |
| FIT2 | 0.5–80 | BZ | 0.68 | 0.26 | — | — | — | 0.97 |
| | | BZ1 | 2.4 | 0.02 | 5.85 | — | — | 0.97 |
| | | BZ2 | 0.38 | 0.71 | — | — | 466 | 0.98 |
| | | KSL | 0.68 | 0.26 | — | −0.0 | — | 0.97 |
| | | MFSL | — | 1.8 | — | 0.04 | — | 0.96 |
| FIT3 | 0.1–3 | BZ | 0.59 | 0.36 | — | — | — | 0.96 |
| | | BZ1 | 0.93 | 0.13 | 0.14 | — | — | 0.96 |
| | | BZ2 | 0.53 | 0.37 | — | — | 1.9 | 0.97 |
| | | KSL | 0.76 | 0.95 | — | −0.21 | — | 0.98 |
| | | MFSL | — | −0.24 | — | 0.21 | — | 0.73 |

versus crack opening displacement plots which added an additional compliance to the crack opening displacement. A quick unload-reload was attempted during the fracture tests prior to failure in an effort to obtain the unloading compliance. However, the loading device could not unload quickly enough which prevented an elastic response to unloading. Hence, the latter two models have not been used in the analysis of the data under discussion in this paper.

Size effect laws were fitted to the entire size range on which the tests were performed as well as subranges of the data. An average result from the small scale testing program was included in the analysis which extended the size range to 0.1–80 m. The undetermined constants were determined by using Tablecurve (an automated nonlinear curve fitting program that uses the 64-bit Levenberg–Marquardt algorithm). The data range on which the curve fits were carried out were 0.1–80 m (FIT1), 0.5–80 m (FIT2) and 0.1–3 m (FIT3) respectively. FIT1 was carried out to obtain the size effect law parameters for the experiments conducted. FIT2 was carried out only on the large scale data to assess the importance of small scale data on the undetermined parameters. FIT3 was carried out on the size range from the smallest to a size at which the crack-tip-opening-displacement at the initiation of crack growth became size independent (3 m) indicating self-similar fracture above that size.

Table 6 shows the obtained parameters. Σ represents the nominal strength value for small specimens and $D$ represents the residual nominal strength value. $B$ and $C$ are the lengths which set the point of intersection of the two asymptotes and the size can be decided by comparing the specimen size to $B$. In general $B$ is termed as the transition size and the size at which LEFM may be applied can be taken to be $10\times B$. $F$ represents an unknown parameter to determine the sharpness of the transition to LEFM. The $r^2$ value presented in Table 6 is an indication of the goodness of the fit. The $r^2$ values for the fits which were carried out

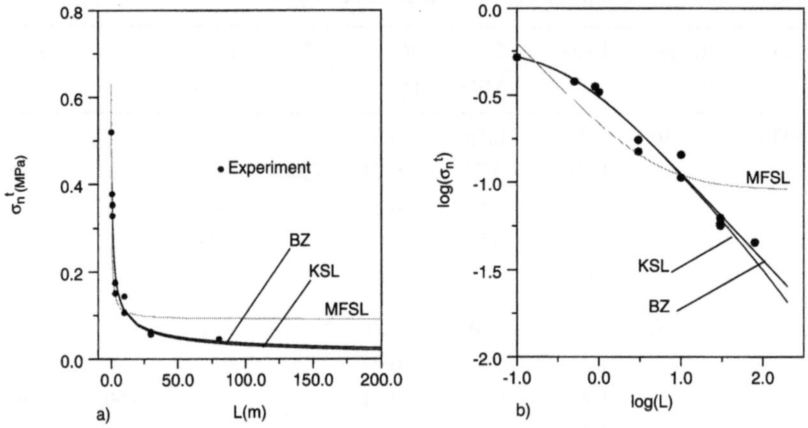


*Figure 5.* Size effect laws for fits carried out over a range of 0.1–80 m. (a) The nominal tensile strength vs the specimen size (b) Same plot plotted on a log-log scale.

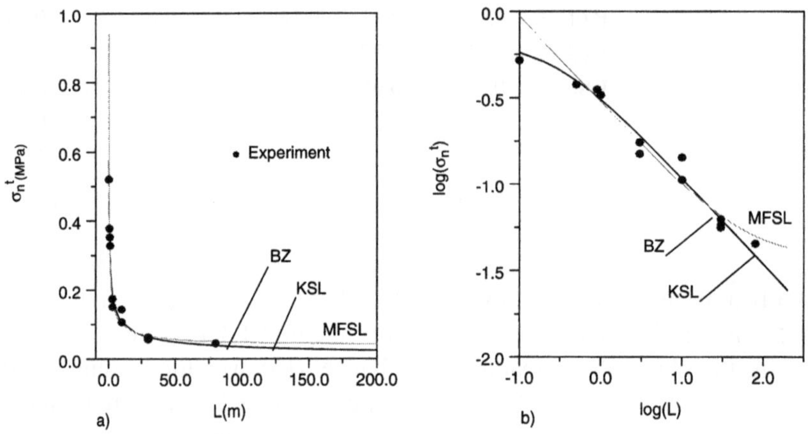


*Figure 6.* Size effect laws for fits carried out over a range of 0.5–80 m. (a) The nominal tensile strength vs the specimen size (b) Same plot plotted on a log-log scale.

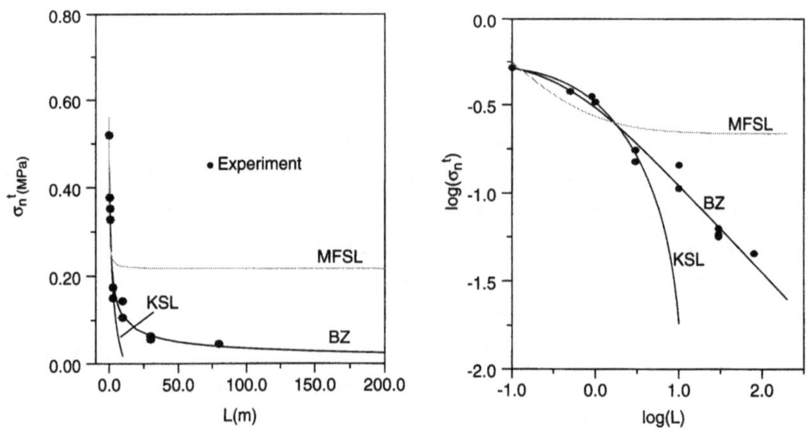


*Figure 7.* Size effect laws for fits carried out over a range of 0.1–3 m. Note the plot covers entire data range from 0.1–80 m. Experimentsl results over 0.1–80 m. (a) The nominal tensile strength vs the specimen size (b) Same plot plotted on a log-log scale.

on the partial data sets are given, though they only determine the goodness of the fit to the partial range and do not indicate how well the fit compares outside the calibrated data range. It should be noted that the results from the small scale tests are from a 4-point bend geometry and the smallest size tested in the large scale program (0.5 m) has a crack length of 0.25 m giving a nondimensional crack length $a = 0.5$ rather than $a = 0.3$ which has been used for the rest of the sizes. The obtained fits are plotted in Figures 5, 6 and 7. The BZ1 and BZ2 laws have not been plotted to reduce the clutter since for FIT1 they produce results which are indistinguishable from BZ, while for other fits the behavior can be easily discerned by comparing the parameters with the BZ law.

FIT1 for the BZ, BZ1, BZ2 and KSL laws produce very similar results as indicated by almost constant $r^2$ values. The $C$ value from the BZ1 law is small hence the effect of including an extra constant on the overall goodness of the fit is not much. However, this may be due to the fact that much of the data is large scale. Bažant and Pfeiffer (1987) found the value of $F$ to be approximately 1 for a number of geometries. The $F$ value for the BZ2 law is larger than 1 indicating that the square plate geometry is probably more conducive in promoting a sharper transition to LEFM behavior ($F$ is purely a function of geometry). The KSL law follows the BZ law parameters closely because the BZ law is the KSL law with $D = 0$. It should be noted that the KSL law consistently produced a negative residual stress. The MFSL clearly is not applicable for self-similar fracture specimens as indicated by a low $r^2$ value. If the small scale values are removed as in FIT2, the BZ parameters are affected but BZ1 and BZ2 are drastically changed. The BZ1 law, which is supposed to improve the modeling of the small scale data, now suffers from the lack of proper small scale data while BZ2 suffers because the small scale asymptote is not well defined. The MFSL is strongly influenced by the small scale data and after this small scale data removal performs very well, as can be seen from Figure 6. Looking at FIT3, the BZ, BZ1 and BZ2 laws are almost restored to the FIT1 status. *Evidently, the size range must range up to and include a particular size to obtain a realistic understanding of the fracture behavior and tensile strength at failure*. The KSL law is now strongly influenced and shows a large negative residual stress. The MFSL residual value is sensitive to the large scale data included in the fit and shows a large residual stress.

Substituting the parameter values obtained from the size effect law fit for the BZ law in equation (2), an approximation for the fracture energy $G_{\mathrm{Ic}}$ can be determined. The relationship between the parameters and the fracture energy involves the short-time modulus of elasticity. The modulus of elasticity computed from the experimental data is presented in Table 2. Using an average value of 4–6 GPa, the fracture energy is found to be 8–12 J/m$^2$ and the equivalent length of the nonlinear zone, $c_f$ to be approximately equal to 0.1–0.15 m. The LEFM metals standard ASTM E399 recommends that all length parameters in fracture specimens for metals should be greater that $2.5K_{\mathrm{Ic}}^2/\sigma_y^2$. This is done to ensure small scale yielding conditions and the satisfaction of LEFM. Typical plastic zone sizes at crack tips in metals is $0.11K_{\mathrm{Ic}}^2/\sigma_y^2$ (Broek, 1982). In other words, all length parameters of the fracture test specimen should be about 25 times the process zone size. This approach predicts that the crack length as well as the ligament length both should be around 2.5 m. However, these requirements are excessive as can be seen from the experimental observations. Figure 3(b) shows a distinct change in scaling at around the 3 m specimen size. Following the arguments presented in Dempsey (1996), the size requirements on first-year sea ice fracture specimen in terms of grain size can be written as

$$A > 60d_{\mathrm{av}}; \qquad L > 200d_{\mathrm{av}}. \tag{7}$$

In the context of the fictitious crack model, it is conventional to define a characteristic length $l_{ch}$ introduced by Hillerborg et al. (1976). The characteristic length is given by

$$l_{ch} = \frac{EG_{\mathrm{Ic}}}{(\sigma_t^{\mathrm{ref}})^2} = \left(\frac{K_Q^{\infty}}{\sigma_t^{\mathrm{ref}}}\right)^2. \tag{8}$$

The characteristic length is a material dependent quantity. As is evident from Table 3 and Figure 4, for the 1.8 m thick first-year sea ice at Resolute, $K_Q^{\infty} \approx 250\,\mathrm{kPa}\sqrt{\mathrm{m}}$, while Table 4 provides estimates for $\sigma_t^{\mathrm{ref}}$. In addition, analyses of the data in this paper using the fictitious crack model (Mulmule and Dempsey, 1997; Mulmule and Dempsey, 1998, 1999a, 1999b) have revealed that a suitable estimate for the reference tensile strength of sea ice (for plates of ice ranging in size between 3 and 80 m) is 0.5 MPa. With this information, given that the fracture toughness levels out in the vicinity of the 3 m specimen size, it is found that $l_{ch} = 0.25\,\mathrm{m}$, $L/l_{ch} = 12$, and the necessary LEFM size is provided by the criterion that

$$L > 12 l_{ch} = 12\frac{EG_{\mathrm{Ic}}}{(\sigma_t^{\mathrm{ref}})^2} = 12\left(\frac{K_Q^{\infty}}{\sigma_t^{\mathrm{ref}}}\right)^2. \tag{9}$$

The LEFM size requirements for the fracture testing of sea ice is studied in more depth in Mulmule and Dempsey (1999b).

The fracture information presented here is specific to these tests which had approximately constant failure times of about 600 s. Though the specimen geometry is not expected to have any effect on these parameters since they have been derived for a large specimen size, the loading rate may have some influence. The results presented here are for the initiation of crack growth only and do not reflect the final failure loads. In many cases, a stick-slip crack advance was seen. The failure load was greater than the crack initiation load. Evidently there was an increase in fracture energy with the growth of the crack.

### 4.1. ICE THICKNESS TO SPECIMEN SIZE RATIO, $L/h$

This specimen sizes fractured at Resolute were self-similar in plan only: only 2D similarity was maintained. For the small specimen sizes, the length of the crack front $h$ exceeded the specimen size $L$. This does raise the importance of the aspect ratio $L/h$. For this reason, this ratio is included in Tables 1 and 3. Lazo (1994) varied the ice thickness from 2.5 cm to 30 cm and found no significant variation in the fracture toughness of saline ice. However, the characteristic length $l_{ch}$ may vary with ice thickness, and the LEFM size may be dependent on this ratio as well. Further tests are required to resolve this issue.

## 5. Conclusions

The test data and size effect analyses for an *in-situ* field program of lab- to structural-scale full ice sheet thickness fracture tests for sea ice are presented. The apparent fracture toughness measured shows a plateau with respect to the specimen size indicating that a size independent fracture toughness was reached. A nonself-similar failure process is also indicated by the attainment of an invariable crack-tip-opening-displacement. Various parameters such as the size-independent fracture energy as well as equivalent size of the nonlinear zone ahead of the crack were deduced. Criteria for the LEFM test size are proposed. The ability of Bažant's size effect law to predict accurate strengths for test sizes 27 times larger than the largest test size used in the fit is noteworthy, and reflects that no additional length scale had an influence on the fracture behavior at these larger sizes.

## Acknowledgements

The experiments in this two-part study were supported by a joint-industry-agency project organized by Denis Blanchet (Amoco) and the first author (JPD) on behalf of Canadian Marine Drilling Limited (Canmar), a business unit of the Amoco Canada Petroleum Company specializing in arctic offshore drilling and marine activities. The experiments were planned and implemented by JPD, SJD and RMA (Part II). Canmar was represented in the field by Kurt Kennedy; Kelly Mamer (Canmar) assisted with the data acquisition. Logistics were managed by Peter Jess and the loading system by Sandwell, Inc (Bill Graham, Dan Masterson, John Robertson and Paul Spencer). The Institute of Marine Dynamics, National Research Council of Canada provided additional logistical support (Part I), data acquisition backup (Part II), high speed video measurements (Part II), ice characterization (Part I), and completed a small scale test program (Part II). In this context, the authors acknowledge the support provided by Bruce Parsons (Part I), Trent Slade and Michelle Johnston (Parts I and II), Mary Williams, Craig Kirby and Bob Gagnon (Part II). Paul Spencer (with Sandwell, Inc.) and Max Coon and Skip Echert (Northwest Research Associates) provided a continuous record of the ice sheet temperature profile with depth (Part II). Yingchang Wei (Clarkson University) and Michelle Johnston (NRC, Canada) determined the salinity, density, orientations, and grain size profiles for the first-year sea ice at Resolute (Part II). Yunbo Xie and David Farmer (Institute of Ocean Sciences, Sydney, BC) measured the crack velocities (Part II). *Additional acknowledgements are included in Part I.*

## References

Abdel-Tawab, K. and Rodin, G.J. (1993a). On the relevance of linear elastic fracture mechanics to ice. *International Journal of Fracture* **62**, 171–181.

Abdel-Tawab, K. and Rodin, G.J. (1993b). Interpretation of results of the fracture toughness tests on ice. *Ice Mechanics – 1993* (Edited by J.P. Dempsey, Z.P. Bažant, Y.D.S. Rajapakse and S. Shyam Sunder), ASME AMD, Vol. 163, New York, 49–59.

ASTM-E399 (1981). *Standard Test Methods for Plane Strain Fracture Toughness of Metallic Materials*. American Society for Testing and Materials, Philadelphia.

Adamson, R.M., Shapiro, L.H. and Dempsey, J.P. (1997). Core-based SCB fracture of aligned first-year sea ice (Phases III and IV). *ASCE Journal of Cold Regions Engineering* **11**, 30–44.

Adamson, R.M., Dempsey, J.P., Mulmule, S.V., DeFranco S.J. and Xie, Y. (1995). Large-scale in-situ ice experiments. Part I: Experimental aspects. *Ice Mechanics – 1995* (Edited by J.P. Dempsey and Y.D.S. Rajapakse), ASME AMD, Vol. 207, New York, 107–128.

Bažant, Z.P. (1984). Size effect in blunt fracture: Concrete, rock, metal. *ASCE Journal of Engineering Mechanics* **110**, 518–535.

Bažant, Z.P. (1992). (ed). *Fracture Mechanics of Concrete Structures*, Elsevier Applied Science.

Bažant, Z.P. and Xiang, Y. (1997). Size effect in compression fracture: Splitting crack band propagation. *ASCE Journal of Engineering Mechanics* **123**, 162–172.

Bažant, Z.P., Kim, J.K. and Pfeiffer, P.A. (1986). Nonlinear fracture properties from size effect tests. *ASCE Journal of Structural Engineering* **112**, 289–307.

Bažant, Z.P. and Pfeiffer, P.A. (1987). Determination of fracture energy properties from size effect and brittleness number. *ACI Materials Journal* **84**, 463–480.

Bažant, Z.P., Bai, S.P. and Gettu, R. (1993). Fracture of rock, effect of loading rate. *Engineering Fracture Mechanics* **45**, 393–398.

Bažant, Z.P., Kazemi, M.T., Hasegawa, T. and Mazars, J. (1991). Size effect in Brazilian split-cylinder tests: Measurments and fracture analysis. *ACI Materials Journal* **88**, 325–332.

Bentley, D.L. (1992). Fracture of first-year sea ice: Preliminary results. *Proceedings of the 11th International OMAE Conference*, Tokyo, Japan, Vol. IV, 343–348.

Bentley, D.L., Dempsey, J.P., Sodhi, D.S. and Wei, Y. (1989). Fracture of columnar freshwater ice from large scale DCB tests. *Cold Regions Science and Engineering* **17**, 7–20.

Bienawski, Z.T. (1968). The effect of specimen size on compressive strength of coal. *International Journal of Rock Mechanics and Mining Science* **5**, 325–335.

Broek, D. (1982). *Elementary Engineering Fracture Mechanics*, Martinus-Nijhoff.

Carpinteri, A., Chiaia, B. and Ferro, G. (1994). Multifractal scaling law for the nominal strength variation of concrete. *Size Effect in Concrete Structures* (Edited by H. Mihashi, H. Okamura and Z.P. Bažant), E&FN Spon. London, 193–205.

Carpinteri, A., Chiaia, B. and Ferro, G. (1995a). *Multifractal Scaling Law: An Extensive Application to Nominal Strength Size Effect of Concrete Structures*. Departmental Report, Politechnico Di Torino, Dipartimento Di Ingegneria Strutturale, Italy.

Carpinteri, A., Chiaia, B. and Ferro, G. (1995b). Size effects on nominal tensile strength of concrete structures: Multifractality of material ligaments and dimensional transition from order to disorder. *Materials and Structures* **28**, 311–317.

Danilenko, V.I. and Rogachko, S.I. (1991). The experimental determination of Okhotsk sea ice fracture toughness. *Proceedings of the 11th International POAC Conference*, St John's, Newfoundland, Vol. I, 293–304.

DeFranco, S.J. and Dempsey, J.P. (1991). Crack growth stability in saline ice. *Mechanics of Creep Brittle Materials – 2* (Edited by A.C.F. Cocks and A.R.S. Ponter), Elsevier Applied Science, 25–36.

DeFranco, S.J. and Dempsey, J.P. (1992). Nonlinear fracture analysis of saline ice: Size, rate, and temperature effects. *Proceedings of the Eleventh IAHR Ice Symposium*, Banff, Alberta, Vol. 3, 1420–1435.

DeFranco, S.J. and Dempsey, J.P. (1994). Crack propagation and fracture resistance in saline ice. *Journal of Glaciology* **40**, 451–462.

DeFranco, S.J., Wei, Y. and Dempsey, J.P. (1991). Notch acuity effects on fracture toughness of saline ice. *Annals of Glaciology* **15**, 230–235.

Dempsey, J.P. (1991). The fracture toughness of ice. *Ice-Structure Interaction* (Edited by S.J. Jones, R.F. McKenna, J. Tillotson and I.J. Jordaan), Springer-Verlag, Berlin, 109–145.

Dempsey, J.P. (1996). Scale effects on the fracture of ice. *The Johannes Weertman Symposium* (Edited by R.J. Arsenault, D.M. Cole, T. Gross, G. Kostorz, P.K. Liaw, S. Parameswaran and H. Sizek), The Minerals, Metals and Materials Society, Warrendale, Pennsylvania, 351–361.

Dempsey, J.P., Adamson, R.M. and DeFranco, S.J. (1995). Fracture analysis of base-edge-cracked reverse-tapered plates. *International Journal of Fracture* **69**, 281–294.

Dempsey, J.P., Wei, Y. and DeFranco, S.J. (1992). Notch sensitivity and brittleness in fracture testing of S2 columnar freshwater ice. *International Journal of Fracture* **53**, 101–120.

Dempsey, J.P., DeFranco, S.J., Adamson, R.M. and Mulmule, S.V. (1999). Scale effects on the in-situ tensile strength and fracture of ice. Part I: Large grained freshwater ice at Spray Lakes Reservoir, Alberta. *International Journal of Fracture* (this issue), 325–345.

Goldstein, R.V. and Osipenko, N.M. (1983a). Fracture mechanics and some questions of ice fracture. *Mechanics and Physics of Ice* (Edited by R.V. Goldstein), Nauka, Moscow, 31–62 (in Russian).

Goldstein, R.V. and Osipenko, N.M. (1983b). Some aspects of fracture mechanics of ice cover. *Proceedings of the 7th International POAC Conference*, Helsinki, Finland, Vol. 3, 132–143.

Hatton, C.G., Main, I.G. and Meredith, P.G. (1994). Non-universal scaling of fracture length and opening displacement. *Nature* **367**, 160–162.

Heuze, F.E. (1980). Scale effects in the determination of rock mass strength and deformability. *Rock Mechanics* **12**, 167–192.

Jenq, Y.S. and Shah, S.P. (1985). Two parameter fracture model for concrete. *ASCE Journal of Engineering Mechanics* **111**, 1227–1241.

Karihaloo, B.L. and Nallathambi, P. (1987). Notched beam test: Mode I facture toughness. Draft report to RILEM Committee 89-FMT, Fracture Mechanics of Concrete: Test Method.

Kennedy, K.P., Mamer, K.J., Dempsey, J.P., Adamson, R.M., Spencer, P.A. and Masterson, D.M. (1994). Large-scale ice fracture experiments: Phase 2. *Proceedings of the Twelfth IAHR Ice Symposium*, Trondheim, Norway, Vol. 1, 315–324.

Kim, J.K., Park, Y.D. and Eo, S.H. (1993). Size effect in concrete specimen with dissimilar initial cracks. *Size Effect in Concrete Structures* (Edited by H. Mihashi, H. Okamura and Z.P. Bažant), E&FN Spon. London, 181–192.

Lazo, J. (1994). *Fracture of Saline Ice: Thickness, Orientation and Reversed-Direct Tension Testing*. M.S. Thesis, Clarkson University.

LeClair, E.S., Adamson, R.M. and Dempsey, J.P. (1977). Core-based SCB fracture of aligned first year sea ice (Phase VI). *ASCE Journal of Cold Regions Engineering* **11**, 45–58.

Leicester, R.H. (1973). *Effect of Size on the Strength of Structures*. CSIRO Australian Forest Products Laboratory, Division of Building Research Technological Paper No. 71, 1–13.

Mulmule, S.V. and Dempsey, J.P. (1997). Stress-separation curves for saline ice using the fictitious crack model. *ASCE Journal of Engineering Mechanics* **123**, 870–877.

Mulmule, S.V. and Dempsey, J.P. (1998). A viscoelastic fictitious crack model for the fracture of sea ice. *Mechanics of Time-Dependent Materials* **1**, 331–356.

Mulmule, S.V. and Dempsey, J.P. (1999a). Scale effects on sea ice fracture. *Mechanics of Cohesive-Frictional Materials* **4** (In press).

Mulmule, S.V. and Dempsey, J.P. (1999b). LEFM size requirements for the fracture testing of sea ice. *International Journal of Fracture* (In press).

Mulmule, S.V., Adamson, R.M. and Dempsey, J.P. (1995). Large-scale in-situ fracture experiments. Part II: Modeling aspects. *Ice Mechanics – 1995* (Edited by J.P. Dempsey and Y.D.S. Rajapakse), ASME AMD-Vol. 207, New York, 107–128.

Nallathambi, P. and Karihaloo, B.L. (1986). Determination of specimen-size independent fracture toughness of plain concrete. *Magazine of Concrete Research* **38**, 67–76.

Parsons, B.L., Snellen, J.B. and Hill, B. (1986). Physical modeling and the fracture toughness of sea ice. *Proceedings of the 5th International OMAE Conference*, Tokyo, Japan, Vol. IV, 358–364.

Parsons, B.L., Snellen, J.B. and Hill, B. (1987). Preliminary measurements of terminal crack velocity in ice. *Cold Regions Science and Technology* **13**, 233–238.

Parsons, B.L., Snellen, J.B. and Muggeridge, D.B. (1988). The initiation and arrest stress intensity factors of first-year columnar sea ice. *Proceedings of the 9th IAHR Ice Symposium*, Sapporo, Japan, Vol. I, 502–512.

Parsons, B.L., Williams, F.M., Everard, J. and Slade, T. (1992). The influence of aging on the strength of prepared cracks in first year sea ice. *Proceedings of the 13th International OMAE Conference*, Tokyo, Japan, Vol. IV, 357–363.

Parsons, B.L., Williams, F.M., Everard, J. and Slade, T. (1993). Notch sensitivity of first year sea ice. *ASCE Journal of Engineering Mechanics* **119**, 1303–1313.

Planas, J. and Elices, M. (1991). Nonlinear fracture of cohesive materials. *International Journal of Fracture* **51**, 139–157.

Schulson, E.M. (1997). The brittle failure of ice under compression. *Journal of Physical Chemistry B* **101**, 6254–6258.

Shapiro, L.H., Hoskins, E.R., Nelson, R.D. and Metzner, R.C. (1979). Flatjack methods of in-situ measurements of the mechanical properties of sea ice. *ASME Journal of Energy Resources Technology* **101**, 196–202.

Shapiro, L.H., Metzner, R.C. and Johnson, J.B. (1981). *Fracture Toughness of Sea Ice*. Report submitted to the Shell Development Company.

Shen, W. and Lin, S.Z. (1986). Fracture toughness of Bohai Bay sea ice. *Proceedings of the 5th International OMAE Symposium*, Tokyo, Japan, Vol. IV, 354–357.

Shen, W., Lin, S.Z., Gu, P. and Zhou, X.A. (1991). The study of fracture toughness of Bohai sea ice. *Ice-Structure Interaction* (Edited by S.J. Jones, R.F. McKenna, J. Tillotson and I.J. Jordaan), Springer-Verlag, Berlin, 147–164.

Smith, E. (1995). The size effect expression for an elastic-softening material. *Mechanics of Materials* **19**, 261–270.

Stehn, L. (1994). Fracture toughness and crack-growth of brackish ice using chevron-notched specimens. *Journal of Glaciology* **40**, 415–426.

Timco, G.W. and Frederking, R.M.W. (1983). Flexural strength and fracture toughness of sea ice. *Cold Regions Science and Technology* **8**, 35–41.

Tuhkuri, J. (1987). The applicability of LEFM and the fracture toughness of sea ice. *Proceedings of the 9th International POAC Conference*, Fairbanks, Alaska, Vol. I, 21–32.

Urabe, N. and Inoue, M. (1986). Mechanical properties of Antarctic sea ice. *Proceedings of the 5th International OMAE Symposium*, Tokyo, Japan, Vol. IV, 303–309.

Urabe, N. and Inoue, M. (1987). Mechanical properties of Antarctic sea ice (II). *Proceedings of the 6th International OMAE Symposium*, Houston, Texas, Vol. IV, 239–244.

Urabe, N. and Yoshitake, A. (1981a). Strain rate dependent fracture toughness ($K_{IC}$) of pure ice and sea ice. *Proceedings of the 6th IAHR Ice Symposium*, Quebec City, Vol. II, 551–563.

Urabe, N. and Yoshitake, A. (1981b). Fracture toughness of sea ice – In-situ measurement and its application. *Proceedings of the 6th International POAC Conference*, Vol. I, 356–365.

Urabe, N., Iwasaki, T. and Yoshitake, A. (1980). Fracture toughness of sea ice. *Cold Regions Science and Technology* **3**, 29–37.

Urabe, N., Yoshitake, A., Iwasaki, T. and Kawahara, M. (1983). Parameters on fracture strength of sea ice. *ASME Journal of Energy Resources Technology* **105**, 12–16.

Vaudrey, K.D. (1977). *Ice Engineering-Study of Related Properties of Floating Sea Ice Sheets and Summary of Elastic and Viscoelastic Analyses.* Technical Report 860, Civil Engineering Laboratory, Naval Construction Battalion Center, Port Hueneme, California.

Vincent, M.R. and Dempsey, J.P. (1998). Fracture energy of saline ice. *Ice in Surface Waters* (Edited by H.T. Shen), Balkema, Rotterdam, Vol. I, 567–573.

Wei, Y., Johnston, M. and Dempsey, J.P. (1995). A grain multipication mechanism for the formation of transition zones in first-year sea ice. *Cold Regions Science and Technology* **23**, 367–375.

Williams, F.M., Kirby, C. and Slade, T. (1993). *Strength and Fracture Toughness of First-Year Arctic Sea Ice.* Report Tr-1993-12, Institute for Marine Dynamics, NRC Canada.

Xie, Y. and Farmer, D.M. (1993). *The Sound of Ice Fracturing Due to Propagating Tensile Failure*. ONR Report.

*International Journal of Fracture* **95:** 367–378, 1999.


# Size effect and inverse analysis in concrete fracture

J. PLANAS*, G.V. GUINEA and M. ELICES
*Dep. de Ciencia de Materiales, E.T.S. de Ingenieros de Caminos, Universidad Politécnica de Madrid, Spain*

Received 8 December 1998; accepted in revised form 31 December 1998

**Abstract.** This paper summarizes the basic experimental and numerical results supporting an easy procedure to determine up to two fracture parameters based on numerically computed size effect curves. Furthermore, it supplies closed-form expressions to determine the initial linear segment approximation of the (stress vs. crack opening) softening curve of cohesive crack models for concrete, based only on the peak loads determined in splitting-tension (Brazilian) tests and in three-point-bending tests on notched specimens. Knowledge of the initial segment, although not enough to describe all the fracture process of concrete structures, is enough to predict the fracture behavior of unnotched concrete structures prior and around the peak load. The same is true for notched structures provided their size is less than a limiting size, approximately defined in the paper.

**Key words:** Size effect, fracture parameters, inverse analysis, size effect, three-point-bend specimen, cohesive crack, softening curve.

**Abbreviations:** ASTM – American Society for Testing and Materials; CMOD – crack mouth opening displacement; GBF – general bilinear fitting; RILEM – International Union of Testing and Research Laboratories for Materials and Structures.

## 1. Introduction

One of the main purposes of laboratory tests is to provide data from which material properties can be determined. The process of extracting information about the material from raw test data is called, in general, inverse analysis or data reduction. Processes of inverse analysis may range from simple (e.g., in quasi-static tensile tests of strain-hardening metallic materials, for which the stress and strain are essentially uniform) to extremely complex (e.g., in impact tests, which include very inhomogeneous stress and strain fields, together with high-rate and inertia effects).

It is obvious that the inverse analysis is strongly related to the underlying model assumed for material behavior. What may not be so obvious is that for a given model (dependent on a number of parameters or functions to be determined), various techniques of inverse analysis can exist that may or may not be equivalent. If these techniques are robust, then they will yield similar, although not identical, results. The results are not necessarily identical because the underlying model is only an approximation of the actual material behavior, and each technique assigns different weights to the various test data. Moreover, if a particular technique is not robust (i.e., ill-posed), it will tend to increase the errors in the data and produce results with a wide uncertainty interval. If a technique is strongly ill-posed, then the results may be unreliable.

* Correspondence Address: Dep. Ciencia de Materiales, ETS Ing. Caminos' Ciudad Universitaria s/n, 28040-Madrid, Spain. e-mail: jplanas@mater.upm.es

Recently, a strong interest has developed for formulating tests and inverse methods to determine the quasi-static fracture behavior of concrete, particularly, the stress vs. crack opening softening curve to be used within the frame of the fictitious or cohesive crack model (in the following, the expression *softening curve* is used, for short, to mean stress vs. crack opening curve). In a recent international workshop (Kitsutaka and Mihashi, 1998), the possibility was discussed of developing standard tests and methods of inverse analysis that would yield accurate information on the softening curve of concrete.

From the analysis of the methods available in the literature, it turns out that there are two groups of inverse analysis techniques: (1) those that use the complete load-displacement (or load-CMOD) curve of one single specimen size and shape as data; and (2) those that use the peak loads of specimens of various sizes or shapes as data.

Within the first group there are two distinct kinds of approaches: (1) those that define *a priori* the softening curve in terms of certain parameters, typically by assuming a bilinear shape (depending on 4 parameters), and then use some optimization technique to fit the parameters to the data (Roelfstra and Wittmann, 1986; Ulfkjær and Brincker, 1993); and (2) those that use points on the load-displacement (or load-CMOD) curve to determine a corresponding set of points on the softening curve, as in the so-called polylinear approach (Kitsutaka, Kamimura and Nakamura, 1994; Uchida et al., 1995; Kitsutaka, 1995, 1997). A thorough analysis of the reliability of these methods has not yet been performed but it is known that in the first kind of approach the results are sensitive to the objective function (the function to be minimized; see Ulfkjær and Brincker, 1993) and that in the second kind of approach, the initial part of the curve (particularly the tensile strength), is highly unreliable if no further data are available (Uchida and Barr, 1998). A recent proposal suggests taking the tensile strength equal to the splitting tensile strength instead of trying to infer it from the load-displacement (or load-CMOD) curve alone (Kitsutaka, Kurihara and Nakamura, 1998).

In this paper the techniques that use the peak loads of specimens of different sizes or shapes, and so incorporate, explicit or implicitly, *size and geometry effects*, are addressed.

The size effect, i.e., the variation of the strength of geometrically similar structures with their size, is one of the most important consequences of fracture phenomena (ACI Committee 446, 1992; Bažant and Planas, 1998).

Although considering the size effect is essential in structural analysis an design, it is also helpful in experimental characterization of fracture behavior. This was first used by Bažant to determine fracture parameters $K_f$ (fracture toughness for infinite size) and $c_f$ (effective crack extension for infinite size) by fitting the peak-load values for various sizes of specimens to a size effect law (Bažant, 1984). In Bažant's method, developed in a series of increasingly refined papers (Bažant, Kim and Pfeiffer, 1986; Bažant and Kazemi, 1990a, 1990b) and adopted as a RILEM Recommendation (RILEM, 1990b), it is specified that a minimum of three sizes of specimens be tested to carry out the determination. However, strictly speaking, only two sizes are required because only two parameters are to be determined from experiments, and so two equations are sufficient to solve for the parameters.

Recently it became apparent that, for any model in which the peak load depends on only two fracture parameters, it is *in principle* possible to determine the parameters if the peak load is measured for two different geometries, which differ in shape, size or both. Although the two-geometry approach has been developed independently by various authors based on specific models (Planas, Guinea and Elices, 1994; Guinea, Planas and Elices, 1994b; Bažant, 1996; Bažant and Li, 1996; Tang et al., 1995; Tang, Ouyang and Shah, 1996), it is possible to formulate it in a general form. This is done in Section 2, which introduces the general aspects

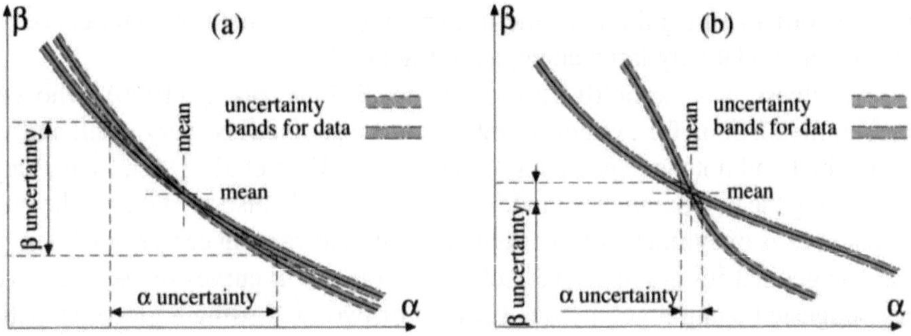


*Figure 1.* System of equations based on two-geometry peak loads: (a) ill-posed system; and (b) well-posed system.

of the subject and shows that the two geometries have to be selected carefully to get reliable results.

Based on this, a specific two-geometry set is analyzed in Section 3, which seems particularly useful for concrete: the first geometry is the well-known cylinder subjected to diametrical compression, and the second is the three-point-bend notched beam. It is shown that with this set, the tensile strength $f_t$ and the horizontal intercept $w_1$ of the initial linear approximation of the softening curve can be reliably determined. Furthermore, closed-form expressions are presented, valid over a wide range of notch-to-depth ratios, which are handy for practical use.

The initial linear portion of the softening curve given by $f_t$ and $w_1$ is enough to make a number of interesting engineering predictions, mainly of peak loads and the behavior close to the peak load of unnotched structures and of notched structures of sizes that are not too large. However, for predicting the post-peak behavior or for large notched structures, the tail of the softening curve is also required. A simple method to determine the additional two parameters (assuming a bilinear approximation) to get a good approximation of the full curve is given elsewhere (Guinea, Planas and Elices, 1994b; Bažant and Planas, 1998, Section 7.3.4).

## 2. Bases for the two-geometry approaches

Consider a fracture model depending on two parameters only, say $\alpha$ and $\beta$, and two different geometries, say 1 and 2. The peak load corresponding to each geometry ($P_{u1}$ and $P_{u2}$) can be put in the form of the equations

$$P_{u1} = \phi_1(\alpha, \beta) \tag{1}$$

$$P_{u2} = \phi_2(\alpha, \beta). \tag{2}$$

If such equations are known, and $P_{u1}$ and $P_{u2}$ are measured with *sufficient accuracy*, then one can solve from the above equations for $\alpha$ and $\beta$.

However, although apparently simple, the method is not free from problems, both fundamental and practical. On the fundamental side, the main problem is that the above equations tend to be ill-posed if no special care is taken. To see this, it is essential to note that, given $P_{u1}$ and $P_{u2}$, Equations (1) and (2) represent two curves in the $\alpha - \beta$ plane, whose intersection is the solution for the values of the parameters (Figure 1). If the geometries are not *very* different, the curves tend to intersect at a very acute angle, as in Figure 1a, and the system is ill-posed, because small uncertainties in the data of $P_{u1}$ and $P_{u2}$ can generate large uncertainties in the solution, as illustrated in Figure 1a. The system is well-posed if the uncertainties in the results are of the same order of magnitude as the uncertainties in the data, which occurs only if the curves intersect at a relatively large angle, as in Figure 1b.

This ill-posedness was explicitly recognized by Bažant and Li (1996), who gave some practical recommendations for extreme combination of geometries (very small and very large size, or unnotched and notched specimens). Tang et al. (Tang et al., 1995; Tang, Ouyang and Shah, 1996), using the two parameter model of Jenq and Shah, also recognized the problem for splitting-tension tests of cylinders of a certain diameter and varying center-notch length. Since, in this case, the geometries are really very close and the $\alpha - \beta$ curves intersect at a very small angle, they suggested using 3 geometries (instead of two) and using a global statistical treatment of the data. In the approach of Guinea, Planas and Elices (1994b) and Planas, Guinea and Elices (1994) based on the cohesive crack model, the two geometries were selected so that the system of equations woul be well-behaved: the (unnotched) splitting-tension test of cylinders and the three-point-bend test of notched beams (a method further discussed in Section 3). A similar concept was used for PMMA by Gomez (1998), with cracked and notched specimens with large notch-tip radius.

An additional fundamental problem arises for cohesive crack models. A general cohesive crack model is not defined by two parameters but by a function, i.e., the number of parameters is, in principle, infinite. In practice, the number of parameters is drastically reduced by assuming *a priori* a certain shape for the function. For example, bilinear softening curves, with 4 independent parameters, are widely used (for a review, see Bažant and Planas, 1998, pp. 167–170). However, as discussed in Section 3 in more detail, for unnotched specimens and notched specimens of sizes that are not very large, the peak load is controlled by only two parameters, namely, the tensile strength and the slope of the first (initial) segment of the softening curve, and so these parameters can be easily obtained using the two-geometry approach.

On the practical side, one must recognize that, although Equations (1) and (2) exist, their explicit exact expressions are available only for few simple models and geometries. For almost any fracture model, these equations must be inferred from numerical calculations. Since the computation is usually *direct*, i.e. given $\alpha$ and $\beta$ one numerically calculates the peak loads, some kind of inverse analysis must be carried out. One way to do so is the iterative approach, in which the values of $\alpha$ and $\beta$ are modified until the predictions are close enough to the measured values.

This method has the advantage that it can, in principle, be applied to any pair of geometries. However, it has two main drawbacks: first, the convergence of the method relies on the initial guess of the values of $\alpha$ and $\beta$ and on a (usually arbitrary) convergence criterion, which makes it difficult to assess the uncertainty of the final fit; second, the iterative loops include two numerical calculations of the peak loads, which highly increases the computational cost.

In another approach proposed by the authors the most complex part of the numerical calculations in the inverse analysis is carried out *a priori* to get approximate closed-form expressions for Equations (1) and (2), or even better, for their solution for $\alpha$ and $\beta$, in terms of the peak loads. The main advantages of this method are that the computations of the peak loads are performed only once and for all and the solution is straightforward. Most importantly, the accuracy of the approximating closed-form expressions can be assessed quite reliably. The drawback is that the range of shapes and sizes for which closed-form expressions are available at a given time is obviously restricted. However this is not a strong drawback for standardization purposes, where such restrictions are not only permissible but also advisable.

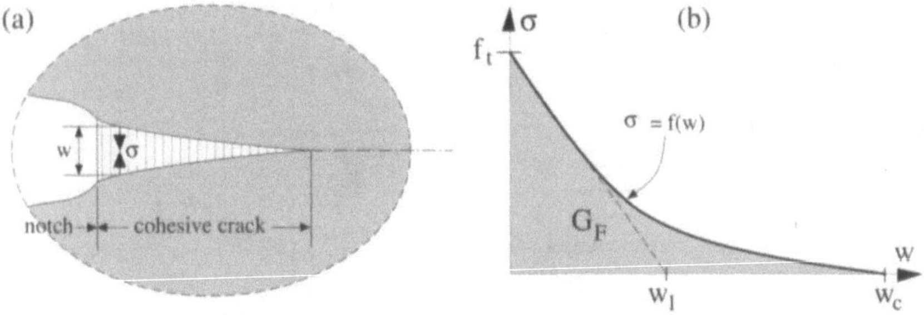


*Figure 2.* Cohesive crack in mode I: (a) sketch of stress and crack opening distributions over cohesive zone; and (b) softening curve and its initial linear approximation.

## 3. Determination of the initial lineal portion of the softening curve of a cohesive crack model

In this section a summary of some background material is presented to justify the structure of the equations governing the size effect and of their inverse functions. Then we give the basic results supporting the use of the splitting tensile strength as a good approximation for the tensile strength, followed by an inverse equation to determine the other fracture parameter from the peak load of laboratory-size notched beams. We close the section with a summary of the steps needed to determine the initial linear portion of the softening curve.

### 3.1. Background

For concrete and other quasibrittle materials, the simplest model still capturing most of the essential aspects of fracture is the fictitious or cohesive crack model. We recall here that in this model it is generally assumed that the material displays a linear elastic isotropic behavior until the maximum principal stress reaches the tensile strength $f_t$. At this moment a cohesive crack forms normal to the maximum principal stress direction. The cohesive crack opens but keeps transferring stress between its faces. For mode I (relative movement of the crack faces normal to the crack plane) and monotonous crack opening (Figure 2a), the cohesive stress $\sigma$ is assumed to be a unique function of the crack opening $w$:

$$\sigma = f(w), \tag{3}$$

where $f(w)$ is called the softening function which, for concrete, is continuously decreasing with increasing $w$ down to zero for a certain crack opening $w_c$ (Figure 2b).

The area bounded by the softening curve and the axes equals the fracture energy $G_F$, a fundamental fracture parameter for very large cracks in very large bodies. However, for relatively small notched structures, or unnotched structures of any size, the peak load happens to occur before much softening takes place anywhere in the body, and thus a linear approximation of the initial part of the curve, as shown by the dashed line in Figure 2b, is enough to describe the fracture behavior up to and including the peak (Planas and Elices, 1992; Guinea, Planas and Elices, 1994a,b; Planas, Guinea and Elices, 1995, 1997).

This property is quantitatively shown in Figure 3 for beams in three-point-bending (Figure 3a). A sketch of the stress distribution along the center cross section is shown in Figure 3b at peak load; as indicated, $\Delta\sigma$ is the maximum amount of softening at peak load. Figure 3c shows the maximum range of softening (dashed zone) that has occurred in the material in

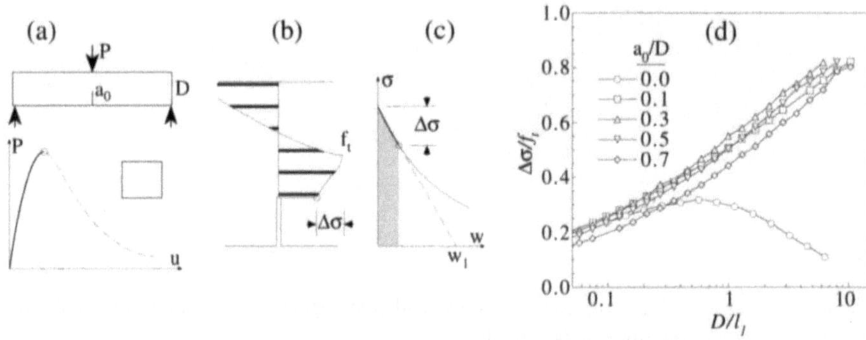


*Figure 3.* Initial linear approximation and peak load: (a) Geometry and situation; (b) stress profile and definition of maximum stress softening $\Delta\sigma$; (c) segment of softening curve 'activated' at peak load; and (d) maximum amount of softening at peak load (relative to the tensile strength) as a function of the beam depth $D$ for various initial relative notch depths. (Adapted from Planas, Guinea and Elices, 1995, 1997.)

loading the specimen up to the peak; the open circle corresponds to the point with maximum softening (the notch tip). Figure 3d shows the maximum amount of softening at peak load (relative to the tensile strength) as a function of the beam depth $D$ for various initial relative notch depths. The curves were computed using the discrete smeared-tip method, implemented in the *Splitting-Lab* program (see, e.g., Bažant and Planas, 1998, Sections 7.4.4–7.4.9; also Planas, Guinea and Elices, 1999). Note that in the horizontal axis, the beam depth is divided by the characteristic length $\ell_1$ defined as

$$\ell_1 = \frac{E' w_1}{2 f_t}, \tag{4}$$

where $w_1$ is the horizontal intercept of the initial linear approximation (Figures 2b or 3c), and $E'$ is the effective modulus, equal to the elastic modulus $E$ for plane stress and to $E/(1-\nu^2)$ for plane strain, with $\nu =$ Poisson's ratio.

From the curves in Figure 3d we see that the maximum amount of softening never exceeds $0.33 f_t$ for unnotched beams (whatever the size) and $0.73 f_t$ for $D \leqslant 3\ell_1$ and notched beams. Since, for concrete, bilinear approximations of the softening curve have been shown to give satisfactory results for practical purposes, and, furthermore, the first segment of such bilinear curves extends typically down to $\sigma = 0.25 f_t$ (i.e. $\Delta\sigma = 0.75 f_t$),[1] we may conclude that the peak load of beams in bending (with or without notches) is controlled by the initial linear approximation of the softening curves at least up to beam depths as large as $3\ell_1$. For ordinary concrete $\ell_1 \approx 150$ mm, and the corresponding maximum beam depth is 450 mm. This means that for most laboratory tests the peak load is completely controlled by the initial portion of the softening curve, and thus by the parameters $f_t$ and $w_1$.

To find closed-form expressions for the inverse analysis, it is essential to know the structure of the equations governing the size effect. It is possible to show that, whenever the peak load is controlled by $f_t$ and $w_1$, the nominal strength $\sigma_{Nu}$ of geometrically similar structures is given by

$$\frac{\sigma_{Nu}}{f_t} = \varphi\left(\frac{D}{\ell_1}\right), \tag{5}$$

[1] The only exception is Petersson's (1981) bilinear softening, whose first segment extends down to $0.33 f_t$. This was the first proposal of a bilinear approximation for concrete and has been superseded by others following more recent and extensive experimental results; see Bažant and Planas (1998), pp. 168–169.

where $\varphi$ is a dimensionless function, $D$ is a characteristic dimension of the structure (cylinder diameter, beam depth), and $\ell_1$ is the characteristic material length defined in (4).

The nominal strength $\sigma_{Nu}$ is conventionally defined for each type of geometry as the peak load $P_u$ divided by an area related to the geometry of the specimens (Bažant and Planas, 1998, pp. 11–13). The conventions used here for the splitting-tension test and the three-point-bend beam tests are given later.

### 3.2. The split-cylinder or prism test

The splitting-tension test is based on the the fact that when an *elastic* disc is loaded on two diametrically opposite points by concentrated forces, a uniform distribution of tensile transverse stress appears along the loading axis. The procedure has been standardized for concrete testing in most countries and its result is generally accepted as an estimate of the actual tensile strength of concrete. In some countries, where cubic specimens are used for standard compression tests, the splitting-tension test has also been standardized for cubes, or, more generally, prisms of square cross section (square prisms).

Here we take the nominal strength $\sigma_{Nu}$ as identical to the splitting tensile strength $f_{st}$

$$\sigma_{Nu} \equiv f_{st} = \frac{2P_u}{\pi DL}, \tag{6}$$

where $L$ is the length of the specimen and $D$ the diameter (for cylinders) or the side of the cross section (for square prisms).

Although ideally $f_{st} = f_t$ (for elastic perfectly brittle materials), $f_{st}$ actually depends on the specimen size $D$ and on the details of the boundary conditions, particularly on the relative width $h/D$ of the load bearing strips. The results of a recent research on the splitting-tension tests, are summarized in Figure 4 (Rocco, 1996; Rocco et al., 1999).

Figure 4a shows the results of displacement-controlled experiments (symbols) and numerical simulations (lines) for split cylinders of various diameters and various values of the relative width $h/D$ of the load bearing strips. The agreement between experiments and theory is good, and the influence of the diameter $D$ and of $h/D$ very important, especially for small sizes. However, for $D \geqslant 0.5\ell_1$ and $h/D \leqslant 0.08$ the splitting tensile strength differs from the true tensile strength by only a few percent. Qualitatively similar results are found for square prisms (Figure 4b), although the value of $f_{st}$ is larger than for cylinders of the same dimensions.

One can therefore conclude that the splitting-tension test, particularly cylinder splitting, gives results close to the actual tensile strength if properly used. The conditions include utilizing not only relatively narrow load bearing strips and large sizes but also a careful control of the loading rate. The effect of the loading rate was recently detected (Rocco, 1996; Rocco et al., 1999) by comparing the results of stable tests (in which stability was achieved by controlling the opening rate of the crack) and unstable tests (load- or displacement-controlled tests). The results showed that (1) unstable tests produced values of the splitting strength larger than the stable tests; (2) the difference in strength from both types of tests depends on the loading rate of the unstable tests and on the relative width of the load-bearing strips; (3) the larger the loading rate and the relative width of the load-bearing strips, the larger the splitting strength of the specimen; and (4) the strength in unstable tests approaches the strength in stable tests if the loading rate tends to zero. For practical purposes, and based on the limited amount of data currently available, for $h/D = 0.08$, the loading rate should be limited to about $f_t/16$

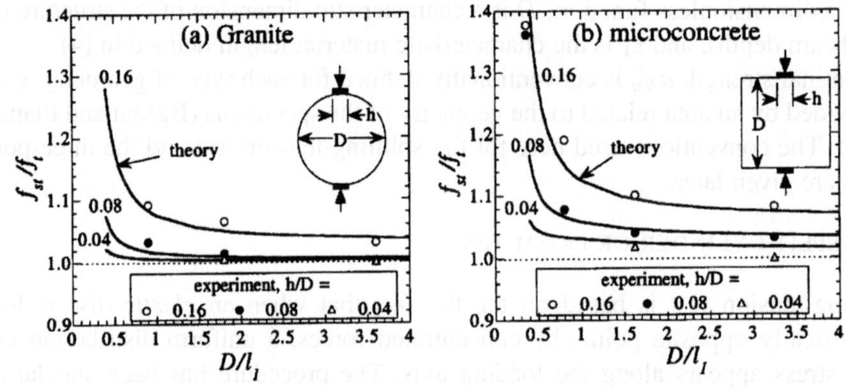


*Figure 4.* Ratio of the splitting tensile strength and the actual tensile strength as a function of size for various relative widths of the load bearing strips, (a) for granite cylinders, and (b) for microconcrete square prisms. The symbols correspond to displacement-controlled experimental data and the lines are numerical predictions using the cohesive crack model. (Adapted from Rocco, 1996.)

per minute (0.25 MPa/min for a concrete of 4 MPa in tensile strength) to get a difference in strength between stable and unstable tests less than 1 percent. The above conditions contradict the ASTM Standard E 496 where wider strips ($h/D \approx 0.16$) and larger load-rates (0.69 to 1.38 MPa/min) are specified.

### 3.3. THREE-POINT-BEND TEST ON NOTCHED BEAMS

For the three-point-bend test on notched beams, we define the nominal strength as identical to the flexural strength $f_f$ defined as

$$\sigma_{Nu} \equiv f_f = \frac{3P_u S}{2BD^2}, \tag{7}$$

where $D$ and $B$ are the beam depth and thickness, respectively, and $S$ the loading span.

According to the discussion in Section 3.1, for notched specimens with $D \leqslant 3\ell_1$, the flexural strength is related to the beam depth by an equation of the form of (5). Let us assume that such an equation is known and that $f_t$ can be approximated by the splitting strength $f_{st}$ (according to the discussion in the previous subsection). Then, if $f_{st}$ is measured in splitting-tension tests and $f_f$ in bending tests, then we can solve for $\ell_1$ from (5). Here we present an explicit (closed-form) approximation for such a process, for beams with a relative loading span $S/D = 4$ and a relatively wide range of notch-to-depth ratios.

The approximation is based on the previous direct analysis of Planas, Guinea and Elices (1997). In that work, founded on interpolation between the asymptotic limits for small and large sizes, a closed-form expression was obtained for the flexural strength, approximate within ±1% for any notch-to-depth ratio between 0.3 and 0.7. The expression reads

$$f_f = \eta_0 f_t \left[\frac{1 + 4.23\sqrt{\kappa D/\ell_1}}{1 + 0.622\sqrt{\kappa D/\ell_1}} + 2.7\frac{\kappa D}{\ell_1}\right]^{-1/2}, \tag{8}$$

where $\kappa$ and $\eta_0$ are functions of the notch-to depth ratio $\alpha_0 = a_0/D$ given by

$$\kappa = 1 - \alpha_0^{1.45}, \qquad \eta_0 = 3(1 - \alpha_0)^2. \tag{9}$$

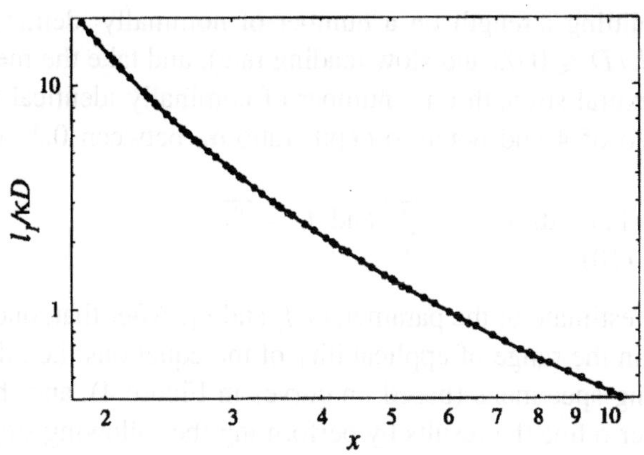


*Figure 5.* Bilogarithmic plot of $\ell_1/\kappa D$ vs. $x$; the symbols are computed ('exact') values and the curve is the approximate Equation (10).

Unfortunately, this expression cannot be explicitly solved for $\ell_1$. Therefore, an analysis similar to that carried out in the previous work for the direct problem (determine $f_f/f_t$ as a function of $D/\ell_1$ and $\alpha_0$), has been carried out here for the inverse problem (determine $\ell_1/D$ as a function of $f_t/f_f$ and $\alpha_0$). The following simple expression has been found to be accurate to within $\pm 1.5\%$ for $0.1 \leqslant \ell_1/D \leqslant 10$ and $0.3 \leqslant \alpha_0 \leqslant 0.7$

$$\ell_1 = \kappa D \left[ \frac{13.11}{(x-1)^2} + \frac{2.68}{x} \right], \tag{10}$$

where

$$x = \left( \eta_0 \frac{f_t}{f_f} \right)^2 \tag{11}$$

and $\kappa$ and $\eta_0$ are the functions of $\alpha_0$ displayed in (9). The bilogarithmic plot in Figure 5 shows the computed values (symbols) of $\ell_1/\kappa D$ vs. $x$ together with the approximation (10).

Once $\ell_1$ is determined, $w_1$ follows from (4) as

$$w_1 = \frac{2 f_t \ell_1}{E'}, \tag{12}$$

where $E'$ is the effective elastic modulus. $E'$ can be determined from standard compression tests or, better, from the initial compliance of the load-CMOD curve. This is readily done based on existing formulas for the compliance (see, e.g., RILEM, 1990a, or Bažant and Planas, 1998, Section 5.5.2).

### 3.4. SUMMARY OF THE PROCEDURE FOR INVERSE ANALYSIS

The foregoing results are useful to set a procedure for inverse analysis based on splitting strength and three-point bend tests on notched specimens. The procedure is similar to that given in Planas, Guinea and Elices (1994) and summarized in Bažant and Planas (1998), Section 7.3.2, except that the range of applicability of the present equations is much wider. The overall procedure is as follows:

1. Determine the splitting strength on a number of nominally identical cylinder or prism specimens (with $h/D \leqslant 0.08$ and slow loading rate), and take the mean, say $\overline{f_{st}}$.
2. Determine the flexural strength on a number of nominally identical notched beams with span-to-depth ratio of 4 and notch-to-depth ratio $\alpha_0$ between 0.3 and 0.7, and take the mean, say $\overline{f_f}$.
3. Compute $x$ from (11) setting $f_f = \overline{f_f}$ and $f_t = \overline{f_{st}}$.
4. Compute $\ell_1$ from (10).

This gives the first estimate of the parameters $f_t$ and $\ell_1$. After that, one must *at least* check that the tests are within the range of applicability of the equations, i.e., that (a) $D/\ell_1$ is large enough for the splitting specimens (based on curves in Figure 4), and (b) $D/\ell_1 < 3$ for the beams. One can further refine the results by performing the following steps

5. Compute $D/\ell_1$ for the splitting specimens.
6. Use the curves in Figure 4 to get the theoretical ratio $\rho_1 = f_{st}/f_t$ for the value of $D/\ell_1$ computed in the previous step.
7. Repeat steps 3 and 4 setting $f_t = \overline{f_{st}}/\rho_1$.

The steps 5–7 can be repeated again to increase the accuracy of the solution. However, owing to the limited accuracy of the equations, further iterations become useless as soon as the variations in the results become less than 1 or 2 percent.

## 4. Concluding remarks

In this paper, a two-geometry method based on measuring only peak loads has been presented. Compared to others, this method is very simple to apply, since it rests on closed-form expressions.

These closed-form expressions have a limited range of applicability: notched beams with a span-to-depth ratio equal to 4 and notch-to-depth ratio between 0.3 and 0.7. However, this range seems sufficient for most laboratory standards. Similar formulas could, certainly, be developed for other geometries such as wedge-splitting of notched cubes or cylinders.

The accuracy of the present closed-form equation is $\pm 1.5\%$, but it could be easily improved down to $\pm 0.5\%$ by letting the constants in the equation depend linearly on the notch-to-depth ratio. This would complicate the equation, however, without a clear benefit (the scatter in the data will be always well over 1.5 percent), and so the simplest equation is preferred.

This method yields only two parameters of the softening curve, namely $f_t$ and $w_1$, which determine uniquely the initial linear approximation of such curve. This is enough for many practical applications, particularly those that concern structural peak loads, but for other applications, particularly those that seek the far post-peak structural behavior, the complete softening curve is required.

The method proposed in this paper can be easily extended to obtain the complementary information required to determine the tail of the softening curve. For this purpose, it is enough to make the three-point-bend tests *stable* (by using, for example, a stiff machine or CMOD control). Then the peak load is used together with the splitting strength to get $f_t$ and $w_1$ as described here, and the extra data in the full load-CMOD curve is used to get the tail of the softening curve.

One simple method of obtaining the tail of the curve is the so-called general bilinear fitting (GBF) approach in which two further parameters are obtained in closed form from the area and the far post-peak trend of the load-CMOD curve (Guinea, Planas and Elices, 1994b; also described in Bažant and Planas, 1998, Section 7.3.4).

Further possibilities, however, exist. For example, the two-geometry method could be advantageously combined with the polylinear method of inverse analysis. Since, as pointed out in the introduction, the latter method does not yield reliable results for the initial part of the softening curve, the initial linear approximation could be first determined using the method described here and next the polylinear approach would be used to get the remaining part of the curve.

## Acknowledgements

The authors gratefully acknowledge financial support for this research provided by Comisión Interministerial de Ciencia y Tecnología, Spain, under grants MAT94-0120-C03 and MAT97-1022.

## References

ACI Committee 446 (1992). Fracture mechanics of concrete: concepts, models and determination of material properties. *Fracture Mechanics of Concrete Structures* (Edited by Z. P. Bažant), Elsevier Applied Science, London, 1–140.

Bažant, Z.P. (1984). Size effect in blunt fracture: concrete, rock, metal. *Journal of Engineering Mechanics-ASCE* **110**, 518–523.

Bažant, Z.P. (1996). Size effect aspects of measurement of fracture characteristic of quasibrittle material. *Advanced Cement Based Materials* **4**, 128–137.

Bažant, Z.P. and Kazemi, M.T. (1990a). Determination of fracture energy, process zone length and brittleness number from size effect, with application to rock and concrete. *International Journal of Fracture* **44**, 111–131.

Bažant, Z.P. and Kazemi, M.T. (1990b). Size effect in fracture of ceramics and its use to determine fracture energy and effective process zone length. *Journal of the American Ceramics Society* **73**(7), 1841–1853.

Bažant, Z.P. and Li, Z. (1996). Zero-brittleness size-effect method for one-size fracture test of concrete. *Journal of Engineering Mechanics-ASCE* **122**(5), 458–468.

Bažant, Z.P. and Planas, J. (1998). *Fracture and Size Effect in Concrete and Other Quasibrittle Materials.* CRC Press, Boca Raton, Florida.

Bažant, Z.P., Kim, J-K. and Pfeiffer, P.A. (1986). Nonlinear fracture properties from size effect tests. *Journal of Structural Engineering-ASCE* **112**, 289–307.

Gomez, F.J. (1998). *Un Criterio de Rotura en Sólidos Entallados.* Dep. Ciencia de Materiales, Universidad Politecnica de Madrid, ETS de Ingenieros de Caminos, Ciudad Universitaria, 28040 Madrid, Spain. (*A Fracture Criterion for Notched Solids,* in Spanish.)

Guinea, G.V., Planas, J. and Elices, M. (1994a). Correlation between the softening and the size effect curves. *Size Effect in Concrete Structures* (Edited by H. Mihashi, H. Okamura and Z. P. Bažant), E and FN Spon, London, 233–244.

Guinea, G.V., Planas, J. and Elices, M. (1994b). A general bilinear fit for the softening curve of concrete. *Materials and Structures* **27**, 99–105.

Kitsutaka, Y. (1995). Fracture parameters of concrete based on poly-linear approximation analysis of tension softening diagram. *Fracture Mechanics of Concrete Structures*, Vol. 1 (Edited by F. H. Wittmann), Aedificatio Publishers, Freiburg, Germany, 199–208.

Kitsutaka, Y. (1997). Fracture parameters by poly-linear tension softening analysis. *Journal of Engineering Mechanics-ASCE* **123**(5), 444–450.

Kitsutaka, Y and Mihashi, H. (1998). *Proc. FRAMCOS 3 Preconference Workshop on Quantitative Evaluation Methods for Toughness and Softening Properties of Concrete,* October 11-12, Gifu, Japan.

Kitsutaka, Y., Kamimura, K. and Nakamura, S. (1994). Evaluation of aggregate properties on tension softening behavior of high strength concrete. *High Performance Concrete* (Edited by V. M. Malhotra), ACI SP149, 711–727.

Kitsutaka, Y., Kurihara, N. and Nakamura, S. (1998). Evaluation method of tension softening properties. *Proc. FRAMCOS 3 Preconference Workshop on Quantitative Evaluation Methods for Toughness and Softening Properties of Concrete* (Edited by Y. Kitsutaka and H. Mihashi), Gifu, Japan, 58–63.

Petersson, P.-E. (1981). *Crack Growth and Development of Fracture Zone in Plain Concrete and Similar Materials.* Report No. TVBM-1006, Division of Building Materials, Lund Institute of Technology, Lund, Sweden.

Planas, J. and Elices, M. (1992). Shrinkage eigenstresses and structural size-effect. *Fracture Mechanics of Concrete Structures* (Edited by Z.P. Bažant), Elsevier Applied Science, London, 939–950.

Planas, J., Guinea, G.V. and Elices, M. (1994b). *SF-2. Draft Test Method for Linear Initial Portion of the Softening Curve of Concrete.* Report No. 94-jp03, Departamento de Ciencia de Materiales, ETS de Ingenieros de Caminos, Universidad Politécnica de Madrid, Ciudad Universitaria sn. 28040 Madrid, Spain. (Contribution to the Japan Concrete Institute International Collaboration Project on Size Effect in Concrete Structures. Reprinted in *Proc. FRAMCOS 3 Preconference Workshop on Quantitative Evaluation Methods for Toughness and Softening Properties of Concrete* (Edited by Y. Kitsutaka and H, Mihashi), October 11-12, 1998, Gifu, Japan, pp. 82–85.)

Planas, J., Guinea, G.V. and Elices, M. (1995). Rupture modulus and fracture properties of concrete. *Fracture Mechanics of Concrete Structures* (Edited by F. H. Wittmann), Aedificatio Publishers, Freiburg, Germany, 95–110.

Planas, J., Guinea, G.V. and Elices, M. (1997). Generalized size effect equation for quasibrittle materials. *Fatigue and Fracture of Engineering Materials and Structures* **20**(5), 671–687.

Planas, J., Guinea, G.V. and Elices, M. (1999). Integral equation method for modeling cracking in concrete. *Computational Fracture in Concrete Technology* (Edited by M. H. Aliabadi and A. Carpinteri), Computational Mechanics Publications, Southampton, pp. 103–131.

RILEM (1990a). Determination of fracture parameters ($K_{Ic}^s$ and CTOD$_c$) of plain concrete using three-point bend tests. *Materials and Structures* **23**, 457–460. (RILEM Draft Recommendation, TC 89-FMT Fracture Mechanics of Concrete–Test methods.)

RILEM (1990b). Size-effect method for determining fracture energy and process zone size of concrete. *Materials and Structures* **23**, 461–465. (RILEM Draft Recommendation, TC 89-FMT Fracture Mechanics of Concrete–Test methods.)

Rocco, C.G. (1996). *Influencia del Tamaño y Mecanismos de Rotura del Ensayo de Compresión Diametral.* Doctoral Thesis. Dep. Ciencia de Materiales, Universidad Politecnica de Madrid, ETS de Ingenieros de Caminos, Ciudad Universitaria, 28040 Madrid, Spain. (*Size-Dependence and Fracture Mechanisms in the Diagonal Compression Splitting Test*, in Spanish.)

Rocco, C., Guinea, G.V., Planas, J. and Elices, M. (1999). Size Effect and boundary conditions in the Brazilian test: Experimental verification. *Materials and Structures,* in press.

Roelfstra, P.E. and Wittmann, F.H. (1986). Numerical method to link strain softening with failure of concrete. *Fracture Toughness and Fracture Energy of Concrete* (Edited by F. H. Wittmann), Elsevier Science, Amsterdam, 163–175.

Tang, T., Ouyang, C. and Shah, S.P. (1996). A simple method for determining material fracture parameters from peak loads. *ACI Materials Journal* **93**(2), 147–157.

Tang, T., Ouyang, C., Libardi, W. and Shah, S.P. (1995). Determination of $K_{Ic}^s$ and $CTOD_c$ from peak loads and relationship between two-parameter fracture model and size effect model. *Fracture Mechanics of Concrete Structures*, Vol. 3 (Edited by F.H. Wittmann), Aedificatio Publishers, Freiburg, Germany, 135–144.

Uchida, Y. and Barr, B.I.G. (1998). Tension softening curves of concrete determined from different test specimen geometries. *Fracture Mechanics of Concrete Structures*, Vol. 1 (Edited by H. Mihashi and K. Rokugo), Aedificatio Publishers, Freiburg, Germany, 387–398.

Uchida, Y., Kurihara, N., Rokugo, K. and Koyanagi, W. (1995). Determination of tension softening diagrams of various kinds of concrete by means of numerical analysis. *Fracture Mechanics of Concrete Structures*, Vol. 1 (Edited by F.H. Wittmann) Aedificatio Publishers, Freiburg, Germany, 17–30.

Ulfkjær, J.P. and Brincker, R. (1993). Indirect determination of the $\sigma$–$w$ relation of HSC through three-point bending. *Fracture and Damage of Concrete and Rock* (Edited by H.P. Rossmanith), E and FN Spon, London, 135–144.

*International Journal of Fracture* **95:** 379–390, 1999.


# Size effect in shallow and deep notched quasi-brittle structures

B. L. KARIHALOO
*Division of Civil Engineering, Cardiff School of Engineering, Cardiff University, Queen's Buildings, PO Box 686, Newport Road, Cardiff CF2 3TB, U.K. e-mail:karihaloob@cardiff.ac.uk*

Received 6 July 1998; accepted in revised form 28 December 1998

**Abstract.** The nominal strength of a quasi-brittle structure is known to vary with its size. If the structure undergoes large stable crack growth prior to failure or if it contains a large pre-existing crack, then the failure load is known to approach the asymptotic limit of linear elastic fracture mechanics (LEFM) for large structures from below. In this paper, the size effect is studied on a particular structural geometry containing a crack which can be relatively shallow or deep. The study is conducted within the framework of the fictitious crack model for the fracture of quasi-brittle materials. By allowing for the redistribution of the stresses in the fracture process zone (FPZ), the essential result of the size effect is confirmed. However, it is shown that this result can only be obtained from tests on specimens whose size exceeds a certain minimum value depending on the material, so that at failure the fully developed FPZ is contained wholly within the test specimen. Moreover, the minimum size of the test specimen is shown to increase as the depth of the pre-crack is reduced, thus requiring specimens of very large sizes to obtain valid results from tests on specimens with very shallow pre-cracks.

**Key words:** Size effect, concrete beams, shallow notch, deep notch, minimum specimen size.

## 1. Introduction

The scaling problem related to the effect of structural size on its nominal strength has a long and often controversial history, as may be judged from a recent review paper by Bazant and Chen (1997). These authors identified three basic theories of scaling in solid mechanics:

(i) the Weibull statistical theory of random strength (Weibull, 1939);
(ii) the theory of stress redistribution and fracture energy release caused by large cracks (Bazant, 1984); and
(iii) the theory of crack fractality (Carpinteri, 1994a, 1994b; Carpinteri and Ferro, 1994).

Leicester (1969, 1973) seems to have been the first to investigate the effect of size on the strength of notched structures made from a quasi-brittle material, such as concrete, and to propose a size effect less severe than that predicted by the Griffith theory (i.e., with $s < 1/2$)

$$(\sigma_N)_u = A_1/D^s. \tag{1}$$

Here, $A_1$ is a constant, $(\sigma_N)_u$ is the nominal stress at failure of a structure of specified shape and loading condition, and $D$ is a charateristic size of the structure. Leicester also identified two fundamental causes of size effect in real structures as

(i) the material heterogeneity, and
(ii) the occurrence of notches and other discontinuities in the flow of stress.

The size effect due to the heterogeneity in the material strength can be explained by Weibull statistics. The size effect due to the occurrence of notches was explained by Leicester using the stress intensity factor concept, but he ignored the redistribution of stresses ahead of the notch tip. Bazant (1984), using the energy release rate concept, proposed the formula

$$(\sigma_N)_u = A\left(1+\frac{D}{B}\right)^{-(1/2)}, \tag{2}$$

where $A$ and $B$ are positive constants. Bazant's formula reduces to the linear elastic fracture mechanics (LEFM) limit as $D \to \infty$, whereas that of Leicester reaches this limit for slit-like cracks ($s = 1/2$). In fact, Bazant's formula been established by Taylor's expansion from this asymptotic limit (Karihaloo, 1995, 1997).

Many quasi-brittle structures are known to fail at crack initiation, although the process zone is well developed. For such structures the nominal failure stress was found to approach the LEFM limit for $D \to \infty$ from above

$$(\sigma_N)_u = \left(A+\frac{B}{D}\right)^{1/2}. \tag{3}$$

Here, $A$ and $B$ are positive constants different from those in (2). This formula was obtained by Carpinteri and his co-workers using multifractal concepts (Carpinteri, 1994a; 1994b; Carpinteri and Ferro, 1994). Bazant (1997a, 1997b) also obtained a size effect formula for initially-unnotched structures using the same concepts as for the notched structures. That formula however differs from (3) in that the exponent is slightly different from 1/2.

It is not the intention of this paper to dwell on the merits or otherwise of the size effect formulae (2) and (3). Readers interested in learning more about them will find plenty of works in the literature (e.g. (Bazant, 1997a, 1997b, 1998; Carpinteri and Chiaia, 1998)). We will instead aim at establishing the conditions for the applicability of the size effect formula (2), based on the original idea of Leicester (1969; 1973) and its subsequent generalisation by Karihaloo (1995; 1997) to quasi-brittle materials described by the fictitious crack model. We will show on the three-point notched bend (TPB) geometry that the size effect formula (2) generally holds, provided that the size of the TPB specimen exceeds a certain minimum value which depends on the test material. This minimum size ensures that at failure the fully developed FPZ is contained wholly within the specimen. The approach to be taken in this paper allows us to express explicitly the size of the FPZ in a TPB specimen of finite size in terms of its asymptotic value for an infinitely large TPB structure.

It will be shown that the minimum depth of the TPB test specimen increases as the depth of the pre-crack is reduced, so that very deep specimens are required for obtaining valid results from tests on specimens with very shallow pre-cracks. This may offer a clue as to why the size effect formula (3) differs from the formula (2). The results of the present study will show that in order to resolve the current controversy in the literature on the different size effects in notched and unnotched quasi-brittle structures, very large unnotched TPB specimens would need to be tested.

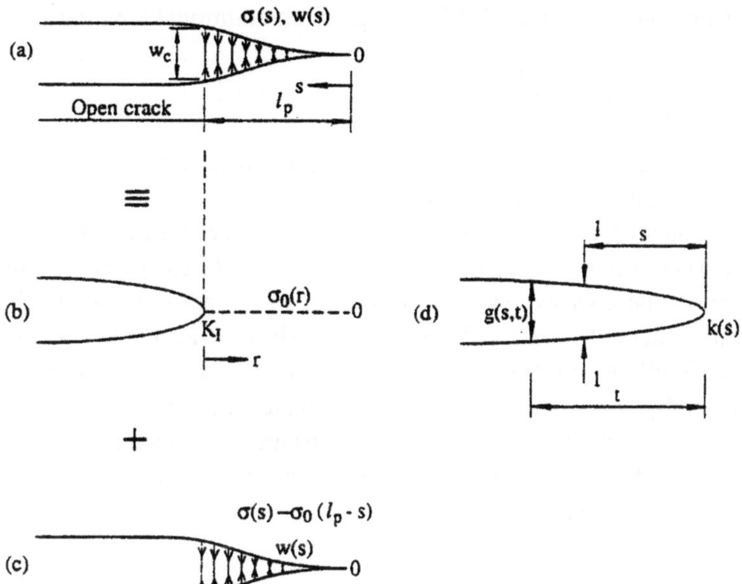


*Figure 1.* A traction-free crack with a FPZ represented by a fictitious crack of length $l_p$, showing the co-ordinate system.

## 2. Mathematical formulation

Under tensile and/or flexural loading, quasi-brittle materials develop a diffuse process zone before the formation of a traction-free crack. This zone can be commensurate with the size of a small test specimen. Within this zone the elastic asymptotic expansion is no longer sufficient. It is now necessary to consider also the nonsingular terms in the expansion of the normal stress $\sigma_{yy}$ ahead of a pre-existing crack tip (which we assume is unaffected by its other tip)

$$\sigma_{yy}(r) \equiv \sigma_0(r) = \frac{K_{\mathrm{I}}}{\sqrt{2\pi r}}\left[1 + A_1\left(\frac{r}{\beta}\right) + A_2\left(\frac{r}{\beta}\right)^2 + A_3\left(\frac{r}{\beta}\right)^3 + \cdots\right], \tag{4}$$

where $r$ is the distance ahead of, and in line with, the crack tip, $\beta$ is a parameter with the dimension of length, and $A_1, A_2, \ldots$ are constants which depend on the applied loading and the geometry of the structure.

Following Karihaloo (1995; 1997), we decompose a traction-free crack with a fracture process zone (FPZ) of length $l_p$ into a traction-free crack with stress intensity factor $K_{\mathrm{I}}$ at the tip and the stress distribution (4) ahead of it and the FPZ with the stress $[\sigma(s) - \sigma_0(l_p - s)]$ (Figure 1). The displacement across the FPZ is $w(s)$, where $s$ is the distance measured from the tip of FPZ. The faces of FPZ are assumed to close smoothly to account for the finite tensile strength of the material. In other words, the stress intensity factor must vanish at the tip of FPZ. The displacement of the cohesive crack faces $w(s)$ can be calculated from the singular integral equation

$$\frac{4}{\pi E}\int_0^{l_p} \log\left|\frac{\sqrt{s}-\sqrt{t}}{\sqrt{s}+\sqrt{t}}\right| \sigma(s)\,\mathrm{d}s + \frac{8K_{\mathrm{I}}\sqrt{t}}{E\sqrt{2\pi}}\left(1 + \frac{l_p}{2\beta} - \frac{t}{3\beta}\right) = w(t). \tag{5}$$

The smooth closure condition of the FPZ faces gives an additional singular integral equation

$$K_{\mathrm{I}}\left(1+\frac{l_p}{2\beta}\right)-\sqrt{\frac{2}{\pi}}\int_0^{l_p}\frac{\sigma(s)}{\sqrt{s}}\,\mathrm{d}s=0. \tag{6}$$

Only the first two terms of the expansion (4) are retained in (5) to get an indication of the influence of the first nonsingular term. Even then this will be only an approximation because of the use of the truncated weight function $k(s)$. The singular integral equations (5)–(6) have to be solved for $\sigma(s)$ and $K_{\mathrm{I}}$ for given $l_p$ or for $\sigma(s)$ and $l_p$ for given $K_{\mathrm{I}}$. In both cases it is necessary to know the cohesive stress (tension softening) relation $\sigma(w)$. The solution can only be obtained numerically, as has been done by Horii et al. (1989) for a linear tension softening relation. Karihaloo (1995; 1997) used an inverse procedure, first proposed by Smith (1974; 1994), whereby $w(s)$ is judiciously approximated in the process zone $0 \leqslant s \leqslant l_p$ instead of $\sigma(w)$ and the integral equations (5) and (6) are solved analytically for $\sigma(s)$ and $l_p$ for prescribed $K_{\mathrm{I}}$. The distance $s$ is then eliminated from the assumed $w(s)$ and the calculated $\sigma(s)$ to establish the $\sigma(w)$ relation.

The choice of $w(s)$ in the form

$$w(s)=\sum_{n=0}^{N}a_n s^{n+(3/2)}, \tag{7}$$

permits the solution of (5) and (6) to be obtained analytically. In (7), $n \geqslant 0$, and $a_n$ are constants to be determined. Note that each term in (7) meets the requirements $w(0)=0$ and $\mathrm{d}w/\mathrm{d}s=0$ at the FPZ tip. Now, following the procedure outlined in (Karihaloo, 1995), and denoting $s/l_p=y$ and $t/l_p=x$, it can be shown that for a generic term of (7)

$$w(y) = a_n l_p^{n+(3/2)} y^{n+(3/2)}, \tag{8}$$

$$\sigma(y) = \frac{Ea_n(n+(3/2))}{4\pi} l_p^{n+(1/2)}\sqrt{1-y}\;J_n(y)+\frac{K_{\mathrm{I}}\sqrt{(1-y)l_p}}{\beta\sqrt{2\pi}}, \tag{9}$$

$$K_{\mathrm{I}} = \frac{1}{4}\sqrt{\frac{2}{\pi}}Ea_n(n+\tfrac{3}{2})l_p^{n+1}\int_0^1\frac{x^{n+(1/2)}}{\sqrt{1-x}}\,\mathrm{d}x, \tag{10}$$

where

$$J_n(y)=\int_0^1\frac{x^n}{(x-y)}\sqrt{\frac{x}{1-x}}\,\mathrm{d}x. \tag{11}$$

We now consider explicitly three terms in (7) and write

$$w(s)=a_0 s^{3/2}+a_1 s^{5/2}+a_2 s^{7/2}. \tag{12}$$

From (8) and (9) it follows that

$$w(y) = y^{3/2}l_p^{3/2}(a_0+a_1 l_p y+a_2 l_p^2 y^2), \tag{13}$$

$$\sigma(y) = \frac{E}{8\pi}l_p^{1/2}\sqrt{1-y}\left[3a_0J_0(y)+5a_1l_pJ_1(y)+7a_2l_p^2J_2(y)\right] + \frac{K_{\mathrm{I}}\sqrt{(1-y)l_p}}{\beta\sqrt{2\pi}}. \tag{14}$$

The above three-term approximation was also considered in (Karihaloo, 1997) but with $\beta = \infty$. Here we take a finite $\beta$ into account. Following the procedure outlined in (Karihaloo, 1997), it can be shown that

$$\frac{\sigma}{f_t} = (1-y)^{3/2}(1+\tfrac{3}{2}y), \tag{15}$$

$$\frac{w}{w_c} = y^{3/2}\frac{[35+y(7-6y)(4+A)]}{(39+A)}, \tag{16}$$

where

$$A = \frac{32K_{\mathrm{I}}}{3a_0 E\beta\sqrt{2\pi}}, \tag{17}$$

and $w_c = w(1)$, $f_t = \sigma(0)$. $w_c$ is the critical crack opening and $f_t$ the tensile strength of the material. It only remains to determine the unknown constant $a_0$, for which we again follow the procedure outlined in (Karihaloo, 1997). We finally arrive at the following

$$\frac{l_p}{l_{p\infty}} = (1-\tfrac{1}{2}\lambda)^{-1}, \tag{18}$$

$$\frac{a_0}{a_{0\infty}} = (1-\tfrac{1}{2}\lambda)^{3/2}, \tag{19}$$

$$\frac{K_{\mathrm{I}}}{K_{\mathrm{I}\infty}} = (1-\tfrac{1}{2}\lambda)^{1/2} \tag{20}$$

where we have denoted

$$\lambda = \frac{l_{p\infty}}{\beta}. \tag{21}$$

All quantities identified with the subscript $\infty$ refer to $\beta = \infty$ or $\lambda = 0$, i.e. to the singular term only in the asymptotic expansion (4). They can be expressed in terms of the known material properties, such as the tensile strength $f_t$, Young's modulus $E$, and the critical crack opening displacement $w_c$ (Karihaloo, 1995)

$$l_{p\infty} = \frac{14}{39}\frac{Ew_c}{f_t}, \tag{22}$$

$$a_{0\infty} = \frac{5}{2}\left(\frac{f_t}{E}\right)^{3/2}\sqrt{\frac{39}{14w_c}}, \tag{23}$$

$$K_{\mathrm{I}\infty} = \tfrac{3}{8}E\sqrt{\frac{\pi}{2}}a_{0\infty}l_{p\infty}. \tag{24}$$

It may well be argued that the tension softening diagram (15)–(16) obtained by the inverse procedure is not representative of concrete. However, for the purposes of the present analysis the real shape of the tension softening diagram is rather unimportant. As will become clear later, the important factor is the form of the relations (18)–(24). Their form remains unchanged when a tension softening relation more representative of concrete is chosen. This can be easily confirmed by comparing these relations with those obtained by Horii et al. (1989) who integrated equations (5) and (6) numerically for a commonly-used $\sigma(w)$ approximation for concrete.

## 3. Three-point notched bend beam

In order to be able to use the above results for a beam of finite size, we use the approximation for finite bodies suggested by Horii et al. (1989). Accordingly, we approximate the stress intensity factor corresponding to the stress distribution (4) by

$$K_{\mathrm{I}}(r) = -\int_0^r k(s)\sigma_{yy}(r-s)\,\mathrm{d}s, \tag{25}$$

where the weight function $k(s) = -\sqrt{2/(\pi s)}$. It is clear that in general the weight function $k(s)$, as indeed $\sigma_0(r)$ in (4), depends in a complex manner on the geometry, crack size and loading. However, Horii et al. (1989) have shown that the approximate $k(s)$ used in (25) gives a good estimate of the parameter $\beta$ in (4). Substituting (4) into (25) and performing the indicated integrations gives (with $A_1$ normalised to unity)

$$\begin{aligned}\frac{K_{\mathrm{I}}(r)}{K_{\mathrm{I}}} &\approx 1 + \frac{1}{2}\left(\frac{r}{\beta}\right)\\ &= 1 + \tfrac{1}{2}\lambda\left(\frac{r}{l_{p\infty}}\right).\end{aligned} \tag{26}$$

Next, the ratio

$$\frac{K_{\mathrm{I}}(r)}{K_{\mathrm{I}}} = \frac{Y\left(\frac{a+r}{D}\right)}{Y\left(\frac{a}{D}\right)}\sqrt{\frac{a+r}{a}} \tag{27}$$

is set up using the geometry function $Y(a/D)$ appropriate to the three-point notched bend beam. For a TPB with span to depth ratio of 4

$$Y(\alpha) = \frac{[1.99 - \alpha(1-\alpha)(2.15 - 3.93\alpha + 2.7\alpha^2)]}{(1+2\alpha)(1-\alpha)^{3/2}}, \tag{28}$$

where $\alpha = a/D$. The geometry function (28) has the advantage that it can handle relatively shallow as well as deep cracks ($0 < \alpha \leqslant 0.6$) with high accuracy (error <0.2 percent). The ratio (27) is expanded in Taylor's series about $r = 0$ and only the first two terms are retained

$$\frac{K_{\mathrm{I}}(r)}{K_{\mathrm{I}}} \approx 1 + \left(\frac{1}{2\alpha} + \left.\frac{Y'(\alpha)}{Y(\alpha)}\right|_{r=0}\right)\left(\frac{r}{D}\right), \tag{29}$$

where $Y'(\alpha) = \partial Y(\alpha)/\partial(r/D)$. For a general dimensionless function of the form (28), it is easily verified that

$$\left.\frac{\partial Y\left(\frac{a+r}{D}\right)}{\partial\left(\frac{r}{D}\right)}\right|_{r=0} \equiv \left.\frac{\partial Y\left(\frac{a+r}{D}\right)}{\partial\left(\frac{a}{D}\right)}\right|_{r=0}, \tag{30}$$

and so on for the higher derivatives, so that the derivatives in (29) can be replaced with derivatives with respect to $a$ taken at $r = 0$.
It is now possible to identify the parameter $\lambda$ for the TPB beam by comparing the like terms in (26) and (29)

$$\lambda = \frac{l_{p\infty}}{\beta} = \left(\frac{1}{\alpha} + 2\frac{Y'(\alpha)}{Y(\alpha)}\right)\left(\frac{l_{p\infty}}{D}\right), \tag{31}$$

where $Y'(\alpha) = \partial Y(\alpha)/\partial\alpha$. Substitution of (31) into (20) gives

$$\frac{K_{\mathrm{I}}}{K_{\mathrm{I}\infty}} = \left[1 - \left(\frac{1}{2\alpha} + \frac{Y'(\alpha)}{Y(\alpha)}\right)\left(\frac{l_{p\infty}}{D}\right)\right]^{1/2}. \tag{32}$$

The ratio (32) is for one and the same beam geometry. The only difference in the calculation of $K_{\mathrm{I}}$ from that of $K_{\mathrm{I}\infty}$ being the inclusion of the first nonsingular term in the stress distribution (4) ahead of the pre-existing crack of depth $a$. It therefore also gives the ratio of the corresponding nominal failure stresses

$$\frac{(\sigma_N)_u}{(\sigma_N)_{u\infty}} = \left[1 - \left(\frac{1}{2\alpha} + \frac{Y'(\alpha)}{Y(\alpha)}\right)\left(\frac{l_{p\infty}}{D}\right)\right]^{1/2}. \tag{33}$$

For future use, we quote here the corresponding ratios according to the size effect formula (2)

$$\frac{(\sigma_N)_u}{(\sigma_N)_{u\infty}} = \left[1 + \frac{g'(\alpha)}{g(\alpha)}\frac{c_f}{D}\right]^{-(1/2)}. \tag{34}$$

Here, the ratio $g'(\alpha)/g(\alpha)$ is related to the ratio $Y'(\alpha)/Y(\alpha)$ of the TPB as follows

$$\frac{g'(\alpha)}{g(\alpha)} = \frac{1}{\alpha} + 2\frac{Y'(\alpha)}{Y(\alpha)}. \tag{35}$$

The same relation also holds for the compact tension geometry. $c_f$ has the same meaning as $l_{p\infty}$ in (33), i.e., it is the size of the FPZ in an infinitely large body for which the singular term alone in (4) is sufficient to describe the stress state ahead of a pre-existing crack. $(\sigma_N)_{u\infty}$ refers to the LEFM asymptotic value that works out to be $\sqrt{(EG_f)/(Dg(\alpha))}$.

The size effect formula (33) can be rewritten in terms of (35) to read

$$\frac{(\sigma_N)_u}{(\sigma_N)_{u\infty}} = \left[1 - \frac{1}{2}\frac{g'(\alpha)}{g(\alpha)}\frac{l_{p\infty}}{D}\right]^{1/2}. \tag{36}$$

The size effect formulae (34) and (36) look somewhat similar and indeed predict a similar effect when $D/l_{p\infty}$ is large. In the limit as $D/l_{p\infty} \to \infty$, both formulae give the LEFM value, as they should. However, the manner in which (36) was developed has some distinct advantages. First, we are able to express explicitly the development of the FPZ in a finite size

TPB specimen in terms of the size of the FPZ in an infinitely large specimen. For this, we need only to rewrite (18), using (31)

$$l_p = l_{p\infty}\left[1 - \frac{1}{2}\frac{g'(\alpha)}{g(\alpha)}\frac{l_{p\infty}}{D}\right]^{-1}. \tag{37}$$

The second advantage of the present approach is that it is possible to determine the minimum depth of the TPB specimen that will ensure that at failure the fully developed FPZ is contained wholly within the specimen. For this we set $l_p = D_{\min}$ in (37) to get

$$\frac{D_{\min}}{l_{p\infty}} = 1 + \frac{1}{2}\frac{g'(\alpha)}{g(\alpha)}. \tag{38}$$

Equation (38) shows that the minimum TPB specimen depth depends on the material being tested through $l_{p\infty}$ and the depth of the pre-crack $\alpha$.

## 4. Results and discussion

The three-point notched bend beam was chosen because for this geometry an accurate compliance function (28) is available which is valid from the very shallow to the very deep notches ($0 < \alpha \leqslant 0.6$). For other geometries, such as the compact tension, no compliance functions of equal accuracy exist, particularly in the shallow end of the notch depth range.

The ratios $Y'(\alpha)/Y(\alpha)$ and $g'(\alpha)/g(\alpha)$ for the compliance function (28) of a TPB with span to depth ratio of 4 are listed in Table 1 for the range $0.05 \leqslant \alpha \leqslant 0.5$, together with the minimum specimen depth $D_{\min}$ according to (38).

*Table 1.* Ratios $Y'(\alpha)/Y(\alpha)$, $g'(\alpha)/g(\alpha)$, and $D_{\min}/l_{p\infty}$ for TPB notched beam with span to depth ratio 4.

| $\alpha$ | $Y(\alpha)$ | $Y'(\alpha)$ | $Y'(\alpha)/Y(\alpha)$ | $g'(\alpha)/g(\alpha)$ | $D_{\min}/l_{p\infty}$ |
|---|---|---|---|---|---|
| 0.05 | 1.8624 | −2.0070 | −1.0776 | 17.8448 | 9.9224 |
| 0.10 | 1.7856 | −1.0952 | −0.6133 | 8.7734 | 5.3867 |
| 0.15 | 1.7505 | −0.3275 | −0.1871 | 6.2925 | 4.1463 |
| 0.20 | 1.7514 | 0.3554 | 0.2029 | 5.4058 | 3.7027 |
| 0.25 | 1.7854 | 1.0010 | 0.5618 | 5.2236 | 3.6118 |
| 0.30 | 1.8516 | 1.6527 | 0.8926 | 5.1185 | 3.5593 |
| 0.35 | 1.9515 | 2.3563 | 1.2074 | 5.2719 | 3.6360 |
| 0.40 | 2.0890 | 3.1677 | 1.5163 | 5.5366 | 3.7683 |
| 0.45 | 2.2713 | 4.1629 | 1.8328 | 5.8878 | 3.9439 |
| 0.50 | 2.5102 | 5.4553 | 2.1732 | 6.3467 | 4.1734 |

The variation of the minimum size $D_{\min}$ of the TPB specimen with the pre-crack size $\alpha$ is shown in Figure 2. Note that $D_{\min}/l_{p\infty}$ is nearly constant in the range $0.2 \leqslant \alpha \leqslant 0.4$, with the least value of about 3.5 at $\alpha \approx 1/3$.

The development of the size of the FPZ, $l_p$, with the TPB specimen size $D$ is depicted in Figure 3 for several values of the pre-crack size $\alpha$. In view of the fact the $g'(\alpha)/g(\alpha)$ is nearly

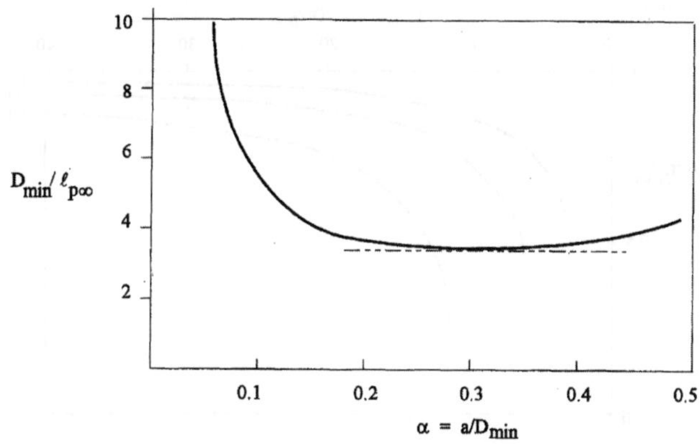


*Figure 2.* Variation of the minimum depth $D_{\min}$ with pre-crack size $\alpha$. The broken line is tangent to the curve.

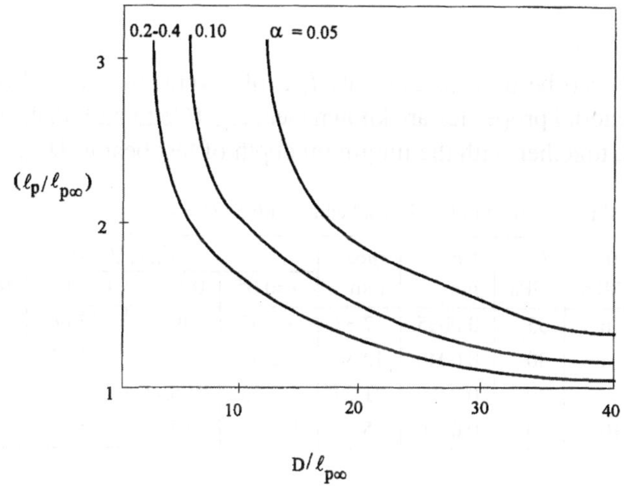


*Figure 3.* Variation of the size of FPZ with depth for several values of pre-crack size $\alpha$. A single representative curve has been drawn in the range $0.2 \leqslant \alpha \leqslant 0.4$.

constant in the range $0.2 \leqslant \alpha \leqslant 0.4$ (see Table 1), only a single representative curve has been drawn for this range.

An important result to flow from the size effect formula (36) is that the nominal strength of beams can be determined only when the ratio $D/l_{p\infty}$ exceeds $0.5\, g'(\alpha)/g(\alpha)$. In other words, depending upon the quasi-brittle material being tested (i.e. the value of $l_{p\infty}$) the depth of the test beam must exceed $0.5\, [g'(\alpha)/g(\alpha)]l_{p\infty}$. However, as shown above (formula (38)), the minimum depth of the TPB specimen should be even larger than this value to accommodate fully the FPZ at failure. For this reason, the size effect curves shown in Figure 4 corresponding to (36) extend down to $D_{\min}/l_{p\infty}$ only. Again, because of the near constancy of $g'(\alpha)/g(\alpha)$ in the range $0.2 \leqslant \alpha \leqslant 0.4$, only one representative curve has been drawn for this range.

It will be noted from Figure 2 and Table 1 that $D_{\min}/l_{p\infty}$ increases with decreasing notch depth, so that very deep beams are required, if very shallow pre-cracks are to be introduced into them. It is worth noting that the size effect formula (34) for notched beams does not impose any limit on the depth of the test TPB specimen, although it has consistently been argued, without proof, that the depth should exceed three times the maximum grain size of the quasi-brittle test material.

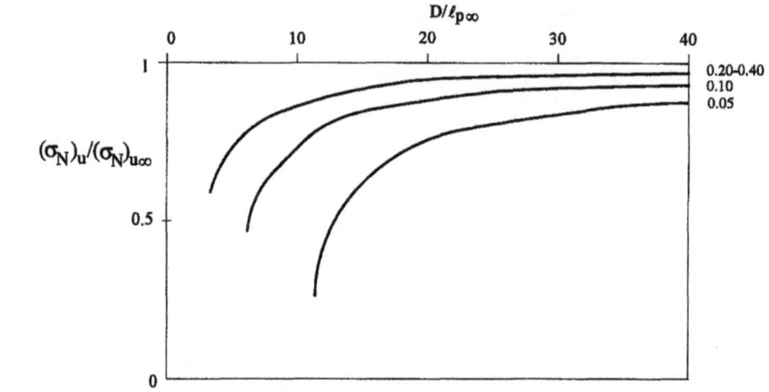


*Figure 4.* Variation of the nominal failure strength with depth for several values of the pre-crack size. In the range $0.2 \leqslant \alpha \leqslant 0.4$, the variation of the nominal failure strength is nearly the same. The curves extend down to $D_{\min}$ only.

The relation (22) can be used to calculate $l_{p\infty}$ of several mortar and concrete mixes for which the typical material properties are known (see, e.g. (Gettu and Shah, 1992)). The results are listed in Table 2, together with the minimum depth of test beams $D_{\min}$.

*Table 2.* Minimum depth of TPB test beams with $S/D = 4$.

| Mix | $f_t$ | $E$ | $w_c$ | $l_{p\infty}$ | $D_{\min}$ (mm) | | | |
|---|---|---|---|---|---|---|---|---|
| | MPa | GPa | mm | mm | $\alpha$=0.05 | 0.1 | 0.2–0.4 | 0.5 |
| 1 | 6 | 35 | 0.0035 | 7.33 | 67.61 | 39.48 | 25.66 | 30.59 |
| 2 | 4 | 30 | 0.0046 | 12.38 | 122.84 | 66.69 | 43.33 | 51.67 |
| 3 | 3 | 30 | 0.0200 | 71.79 | 712.33 | 386.71 | 251.27 | 299.61 |
| 4 | 10 | 40 | 0.0110 | 15.79 | 156.67 | 85.06 | 55.27 | 65.90 |

Mixes 1 to 4 in Table 2 refer respectively to hardened cement paste, coarse mortar, normal concrete, and high strength mortar. The minimum sizes established by formula (38) serve only as a guide. They should be considered together with other limiting factors, such as the maximum grain size. Moreover, as mentioned above, the numerical factor in (22) can be expected to change somewhat, depending upon the shape of the tension softening diagram.

## 5. Conclusions

A scaling law for the nominal strength of TPB notched beams has been established on the basis of the singular and first nonsingular terms in the asymptotic stress distribution ahead of a pre-existing crack in a quasi-brittle material. The TPB geometry was chosen for this derivation because of the availability of the geometry function that is accurate over a wide range of notch depths – from the very shallow to the very deep.

The scaling law is similar to the well-known formula (34) in that it predicts that the nominal failure strength of notched beams approaches the LEFM asymptotic limit for large beams from below. However, as it was derived based on the fictitious crack model, it reveals several features of practical importance which the formula (34) either implied without proof or imposed on the TPB test geometry without a formal physical basis.

Thus, the scaling law (36) establishes on physical grounds a lower limit on the size of the TPB test specimen. This limit depends upon the material being tested and the pre-crack size. It will ensure that at failure the FPZ will be fully developed and wholly contained within the specimen.

The study shows that the minimum size of the TPB specimen required for performing a valid test increases as the depth of the pre-crack is reduced. Thus, for example, to perform a valid test on normal concrete, the minimum depth of the TPB specimen with a very shallow pre-crack ($\alpha = 0.05$) must be around 712 mm (see Table 2). The loaded span required would be around 2848 mm. It would be a formidable task to test such a massive plain concrete beam. On the other hand, a valid test could be performed on a TPB specimen of reasonable dimensions by choosing a deeper pre-crack.

It would appear from Table 2 that the minimum depth of an unnotched three-point bend specimen made of normal concrete would exceed that of a TPB specimen with $\alpha = 0.05$. It is tempting to suggest that the violation of this minimum depth requirement may be the reason why the measured nominal strength of unnotched three-point bend beams would appear to follow the formula (2), rather than (34) or (36). The present study may have shown a way forward to resolving this controversy. As a start, tests may be conducted on unnotched bend specimens made from hardened cement paste for which the minimum dimensions are likely to be within manageable limits (see Table 2).

It should be mentioned that the qualitative results of the present study would remain unaffected by changes in the tension softening diagram. Only the numerical factors in the relations (18)–(24) would be slightly altered. Likewise, the numerical factor in (26) would change slightly when an exact weight function is used in place of the approximate one in expression (25)

Finally, it should be mentioned that it is quite conceivable that scaling laws similar to (36) can be established from geometries other than the TPB, when accurate geometry functions over a wide range of notch depths (especially near its shallow end) become available in the future for those geometries.

## References

Bazant, Z.P. (1984). Size effect in blunt fracture: concrete, rock, metal. *ASCE Journal of Engineering Mechanics* **110**, 518–535.

Bazant, Z.P. (1997a). Scaling of quasi-brittle fracture: asymptotic analysis. *International Journal of Fracture* **83**, 19–40.

Bazant, Z.P. (1997b). Scaling of quasibrittle fracture: hypotheses of invasive and lacunar fractality, their critique and Weibull connection. *International Journal of Fracture* **83**, 41–65.

Bazant, Z.P. and Chen, E.P. (1997). Scaling of structural failure. *Applied Mechanics Reviews* **50**, 593–627.

Bazant, Z.P. (1998). Modelling of compressive strain softening, fracture and size effect in concrete. *Proceedings of European Conference on Computational Methods for Concrete Structures, EUROC-98* (Edited by R. de Borst, N. Bicanic, H. Mang and G. Meschke), Balkema, Rotterdam **1**, 249–264

Carpinteri, A. (1994a). Fractal nature of material microstructure and size effects on apparent mechanical properties. *Mechanics of Materials* **18**, 89–101.

Carpinteri, A. (1994b). Scaling laws and renormalization groups for strength and toughness of disordered materials. *International Journal of Solids and Structures* **31**, 291–302.

Carpinteri, A. and Ferro, G. (1994). Size effect on tensile properties: a unified explanation based on disorder and fractality of concrete microstructure. *Materials and Structures* **27**, 563–571.

Carpinteri, A. and Chiaia, B. (1998). Embrittlement and decrease of apparent strength in large-sized structures. *EUROC-98* (Edited by R. de Borst, N. Bicanic, H. Mang and G. Meschke), Balkema, Rotterdam, **2**, 715–730.

Gettu, R. and Shah, S.P. (1992). Fracture mechanics and high strength concrete. *High Performance Concrete and Applications* (Edited by S.P. Shah and S.H. Ahmed), Edward Arnold, UK, 1–75.

Horii, H., Hasegawa, A. and Nishino, F. (1989). Process zone model and influencing factors in fracture of concrete. *Fracture of Concrete and Rock* (Edited by S.P. Shah and S.E. Swartz), Springer-Verlag, New York, 205–219.

Karihaloo, B.L. (1995). *Fracture Mechanics and Structural Concrete*, Addison Wesley Longman, UK, 151–165.

Karihaloo, B.L. (1997). Scaling effects in the fracture of quasi-brittle materials and structures. *Proceedings of IUTAM Symposium on Nonlinear Analysis of Fracture* (Edited by J.R. Willis), Kluwer Academic, Dordrecht, 13–22.

Leicester, R.H. (1969). The size effect of notches. *Proceedings of the Second Australasian Conference on Mechanics of Materials and Structures*, Melbourne, 4.1–4.20.

Leicester, R.H. (1973). Effect of size on the strength of structures. *CSIRO Division of Building Research Paper 71*, Melbourne, 1–13.

Smith, E. (1974). The structure in the vicinity of a crack tip: a general theory based on the cohesive zone model. *Engineering Fracture Mechanics* **6**, 213–222.

Smith, E. (1994). The elastically equivalent softening zone size for an elastic-softening material I: power law softening behaviour; II: a simple piece-wise softening law. *Mechanics of Materials* **17**, 363–378.

Weibull, W. (1939). A statistical theory of the strength of materials. *Proceedings of the Royal Swedish Institute for Engineering Research* **153**, 1–55.

*International Journal of Fracture* **95:** 391–404, 1999.


# Size effect on the concrete cone pull-out load

J. OŽBOLT, R. ELIGEHAUSEN and H.W. REINHARDT
*Institut für Werkstoffe im Bauwesen, University of Stuttgart, 70550 Stuttgart, Germany*
*e-mail: ozbolt@iwb-uni.stuttgart.de*

Received 3 November 1998; accepted in revised form 18 December 1998

**Abstract.** In the present paper the failure mechanism and size effect of the concrete cone resistance is reviewed and studied. The influence of material and geometrical parameters on the failure mode and size effect is investigated. In the numerical studies the smeared crack finite element analysis, based on the microplane material model for concrete, was used. Both, experimental and numerical results show that there is a strong size effect on the nominal concrete cone pull-out strength. It is demonstrated that besides the embedment depth the scaling of the head of the stud as well as the scaling of the concrete member influence the nominal strength.

**Key words:** Size effect, fastening technique, concrete, headed stud, pull-out, microplane model, finite elements.

## 1. Introduction

In engineering practice, headed anchors are often used to transfer loads into reinforced concrete members. Experience, a large number of experiments as well as numerical studies of anchors of different sizes confirm that fastenings are capable to transfer a tension force into a concrete member without using reinforcement. Provided the steel strength of the stud as well as the load bearing area of the head are large enough, a headed stud subjected to a tensile load normally fails by cone shaped concrete breakout. A typical concrete cone observed in experiments is show in Figure 1. The failure is due to the failure of concrete in tension by forming a circumferential crack (Eligehausen et al., 1997). In the test shown in Figure 1, the concrete block has been reinforced with a slight surface reinforcement. However, tests without surface reinforcement showed the same cone formation and the same strength as well.

To better understand the crack growth and to predict the concrete cone pull-out failure load of headed studs for different embedment depths a number of experimental and theoretical studies have been carried out (Ottosen, 1981; Bode and Hanenkamp, 1985; Ballarini et al., 1986; Eligehausen and Sawade, 1985; Eligehausen and Ožbolt, 1990; Ožbolt, 1995). Summarising these activities it can be said that the experimental results for headed anchors show a significant size effect on the concrete cone pull-out failure strength. Furthermore, it has been demonstrated that numerical finite element studies based on macroscopical constitutive models according to the conventional plasticity or elasticity (strength) theory are not capable to predict the behavior of anchors as observed in the experiments (Eligehausen and Ožbolt, 1990; Ožbolt, 1995). Therefore, more sophisticated numerical fracture analysis needs to be carried out in which the employed computational model should account for the concrete strength as well as for the equilibrium between the structural energy release rate and concrete energy consumption capacity. The pull-out from a concrete block does not represent a pure mode-I failure case *a priori*. Consequently, to model the entire load history correctly the concrete

*Figure 1.* Typical concrete breakout cone obtained in the tests (Eligehausen et al., 1997).

behavior needs to be represented realistically not only for dominant tensile loads but also for general three-dimensional combinations of stresses and strains.

In the present study the size effect on the concrete cone capacity is reviewed and the results of additional studies are shown and discussed. The test data are compared with numerical results which are obtained using a finite element code that is based on a smeared type of the fracture analysis and mixed constraint microplane model for concrete (computer code MASA; Ožbolt et al., 1997). The influence of the material and geometrical parameters on the failure load and size effect is investigated. Besides the pull-out geometries which are typically used in the engineering practice, where the head of the stud is not scaled proportional, studs of different sizes with proportional scaling of the head are investigated as well. Finally, the size and scaling of the concrete block is studied.

## 2. Experimental evidence of the size effect

The concrete resistance of the headed stud relies only on the concrete cone tensile resistance (no reinforcement). Therefore, to design safe and economical structures it is important to fully understand the failure mechanism and to know how the variation of the material and geometrical properties influence the cone pull-out failure capacity. The first experiments in which the size effect on the concrete cone breakout strength has systematically been investigated were performed by Bode and Hanenkamp (1995). The embedment depth was varied between 40 and 150 mm. Subsequently, a number of experiments were carried out (see Eligehausen et al., 1997). The embedment depth was varied between $h_{ef} = 130$ and 520 mm. However, the size of the concrete specimen was not systematically scaled. Therefore, the additional experiments were performed by Eligehausen et al. (1992), with the variation of the embedment depth between $h_{ef} = 50$ mm and 450 mm (see Figure 2). In these experiments, the concrete properties have been kept constant while the diameters of the circular support were scaled proportionally to the embedment depth. The distance between the support reaction and the outer edges of the member was kept constant and equal to 100 mm (see Figure 2). Furthermore, in the tests the size of the heads was such that the pressure under the head was

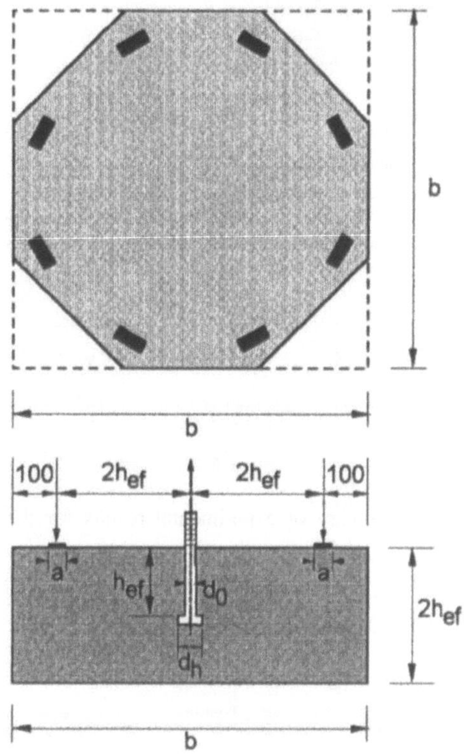


*Figure 2.* Typical pull-out test geometry for the headed stud anchors (Eligehausen et al., 1992).

approximately constant for different embedment depths. In one test serie, the diameter of the head was scaled proportionally to the embedment depth which leads to a smaller concrete pressure under the head at failure with increasing embedment depth.

The nominal strength for a number of experimental results are summarised in Figure 3. The measured nominal concrete cone strengths normalised with respect to the concrete cube compressive strength $f_{cc} = 33$ MPa (normalising factor $= (33/f_{cc})^{1/2}$ because the maximum load is proportional to $(f_{cc})^{1/2}$ (Eligehausen et al., 1997) and because the reference concrete has a strength of 33 MPa), for the size range $h_{\text{ef}} = 40$ to 520 mm, are plotted versus the embedment depth in a log-log scale. The nominal strength $\sigma_N$ is calculated as the ultimate load $P_U$ divided by the area of a circle with a radius equal to the embedment depth i.e. $\sigma_N = P_U/(h_{\text{ef}}^2 \pi)$. Comparing the test data with the ACI design formula (ACI standard, 1985) which does not consider the size effect it is obvious that the experimental results for larger embedment depths are lower than the predicted values. In the same figure the design formula which is based on linear elastic fracture mechanics (LEFM) is also plotted (CEB, 1990). This curve exhibits a good agreement with the experimental results for the whole size range. The fact that the experimental results agree with the LEFM based design formula means that the size effect on the nominal pull-out strength is strong since this formula predicts the maximal possible size effect (Reinhardt, 1981; Bažant, 1984).

To find out the reason for the size effect the crack development as well as the distribution of the stresses along the crack surface were measured in test with $h_{\text{ef}} = 130$ mm, 350 mm and 520 mm. Figure 4 shows the position of the strain gages for the anchor with an embedment depth of $h_{\text{ef}} = 520$ mm. They were placed along the assumed concrete cone crack

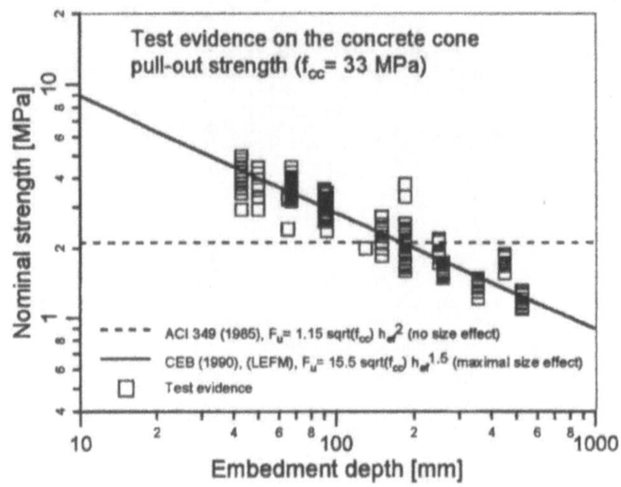


*Figure 3.* Concrete cone breakout – summary of experimental results for the nominal pull-out strength and comparison with design formulas.

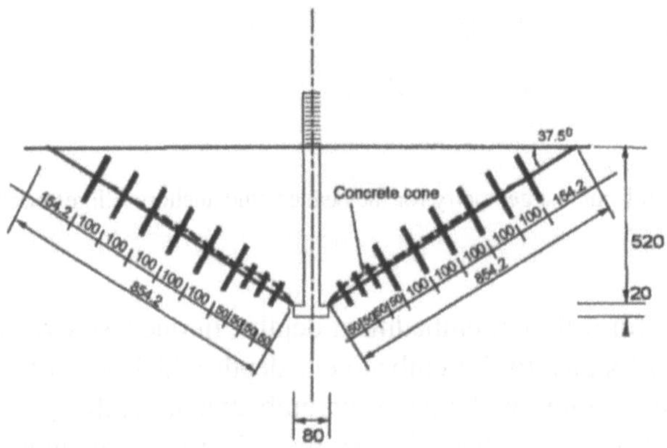


*Figure 4.* Distribution of strain gages along the concrete cone surface in order to measure the strain and stress distribution for different load stages, $h_{ef} = 520\,\text{mm}$ (Eligehausen and Sawade, 1989).

surface which was verified in a previous test (for detailes of the experimental technique see Eligehausen and Sawade (1989)). In Figure 5(a) the strains normal to the crack surface at 30 percent and 90 percent of the ultimate load are plotted. Cracking started at about 25 percent of the peak load at the head. At 90 percent of the ultimate load the crack length reached approximately 35 percent of the total crack length at failure. The corresponding stresses perpendicular to the crack surface are plotted in Figure 5(b). The test data clearly show a stable crack growth i.e. with increase of the crack length the resistance increases as well and reaches the maximum value at a critical crack length of approximately $l_{cr} = 0.4 l_{tot}$. As it is well known from LEFM, if the structure exhibits a stable crack growth the size effect on the ultimate resistance is strong and the nominal strength is more sensitive to the variation of the fracture energy than to the variation of the concrete tensile strength (Ožbolt, 1995).

The embedment depths investigated in the tests cover most of the practical applications. As shown in Figure 3, for this range the size effect on the nominal pull-out strength is strong and close to the LEFM prediction formula. The question is whether the same trend may be expected for larger embedment depths. Furthermore, in the experiments the influence of the

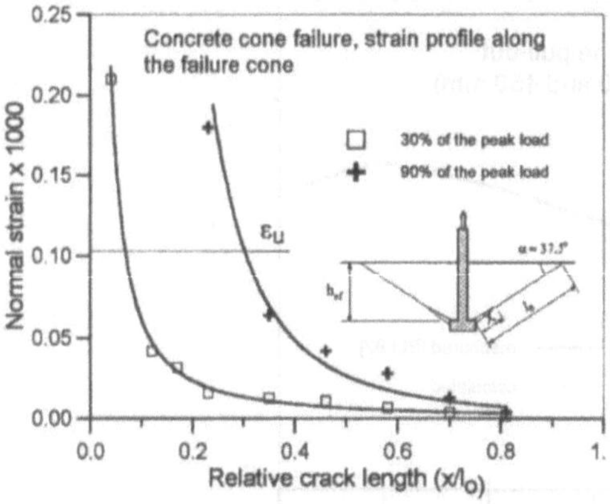


*Figure 5(a).* Distribution of strains along the concrete cone surface at 30 percent and 90 percent of the ultimate load, $h_{ef} = 520\,\text{mm}$ (Eligehausen and Sawade, 1989); $\varepsilon_u$ stays for the equivalent strain at crack opening.

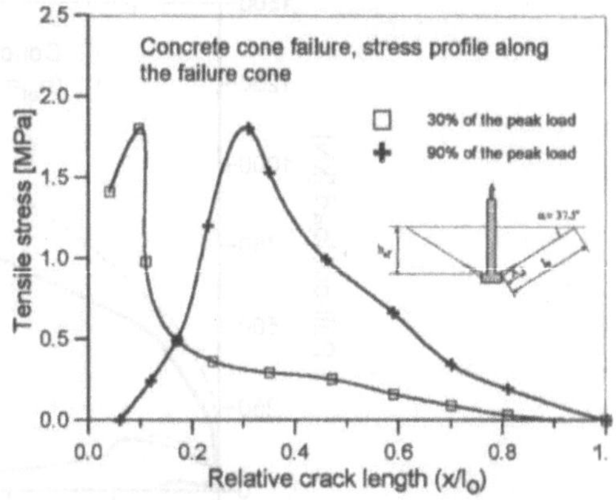


*Figure 5(b).* Distribution of stresses along the concrete cone surface at 30 percent and 90 percent of the ultimate load, $h_{ef} = 520\,\text{mm}$ (Eligehausen and Sawade, 1989).

concrete fracture properties (tensile strength and concrete fracture energy) and geometry on the pull-out capacity was not systematically investigated. Therefore, to study the influence of the above parameters and to better understand the experimental results further numerical studies are carried out.

## 3. Numerical studies – failure mechanism and size effect

In the numerical analysis three different series of geometries are investigated. Two of them were tested by Eligehausen and Sawade (1989) and Eligehausen et al. (1992) (see Figure 2). The third geometry is taken from the RILEM TC 90 Round robin proposal (1990). The numerical results are first compared with the test data. Subsequently, the influence of the geometry i.e. the scaling of the test specimen as well as the influence of material properties on the pull-out resistance is investigated.

The spatial discretization is performed by axisymmetrical finite elements. Four node quadrilateral elements with four integration points (linear strain field) are used. The cracking and damage phenomena are modelled by employing the smeared crack approach. As a constitutive law for concrete the mixed formulation of the microplane model was adopted (Ožbolt et al., 1997). To assure the objectivity of the analysis and to prevent so-called spurious mesh sensitivity two approaches were alternatively used: (1) the crack band approach (Bažant and Oh, 1983) and (2) the nonlocal integral approach based on the interaction of microcracks (Ožbolt and Bažant, 1996).

The typical specimen geometry was the same as in the experiments (see Figure 2). In all calculations pulling of the anchor was performed by prescribing displacements at the bottom of the stud. Contact between steel stud and concrete existed only under the head of the stud. To account for the restraining effect of the embedded anchor, the displacements of the concrete surface along the steel stud in the vicinity of the head were fixed in direction perpendicular to the load direction. The supports were fixed in both vertical (loading) and horizontal direction.

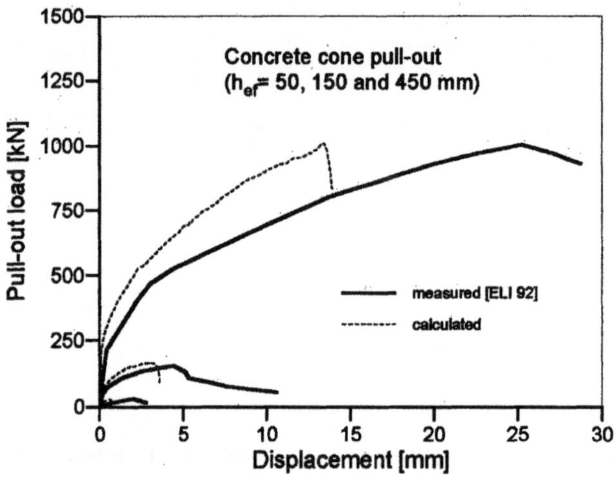


*Figure 6.* Measured and calculated load-displacement curves for three embedment depths: $h_{ef} = 50, 150$ and 450 mm.

The distance between the support and the anchor was assumed to be $2h_{ef}$ (see Figure 2) so that an unrestricted formation of the failure cone was possible.

Figure 6 shows the comparison between calculated and measured load-displacement curves for three embedment depths. It can be seen that the calculated peak loads agree well with the experimental data. However, the calculated displacement are smaller than the measured ones. This may be due to the fact that the interfacial transition zone between steel head and concrete is weak and less stiff than the bulk concrete. This phenomenon has not been taken into account. Then may also be a contribution by the reinforcement which has been neglected. The reinforcement may be activated at large displacement which does not influence the maximum load but can have a favourable effect on the maximum displacement. The general behaviour and the failure mechanism of the cone forming has been predicted very well.

### 3.1. Influence of the concrete fracture properties on failure load

The concrete cone breakout resistance relies on the concrete fracture properties. Therefore, it is important to know how macroscopical fracture properties of concrete influence the concrete cone pull-out resistance. A numerical analysis was carried out for the geometry proposed by RILEM TC 90 (1990). A parameter study of headed stud specimens with $h_{ef} = 450$ mm was performed as follows: (1) for constant $G_F = 0.08$ N/mm, the tensile strength was varied from 2.4 to 3.6 MPa and (2) for constant $f_t = 2.8$ MPa, the concrete fracture energy was changed from 0.08 to 0.14 N/mm. The calculated nominal pull-out strengths are plotted in Figure 7 as a function of the tensile strength and fracture energy, respectively. As can be seen, for the embedment depth of $h_{ef} = 450$ mm the nominal strength is practically independent of the tensile strength (Figure 7(a)). However, Figure 7(b) shows approximately a square root dependency between the nominal pull-out strength and the concrete fracture energy. The same result has been found by Eligehausen and Sawade (1989), by analytical studies based on the LEFM and by tests on headed studs pulled out from a glass specimen (Sawade, 1994).

The above results clearly show that for pull-out of headed stud concrete cracking is an important aspect of the resistance mechanism. Namely, in contrast to a number of structures

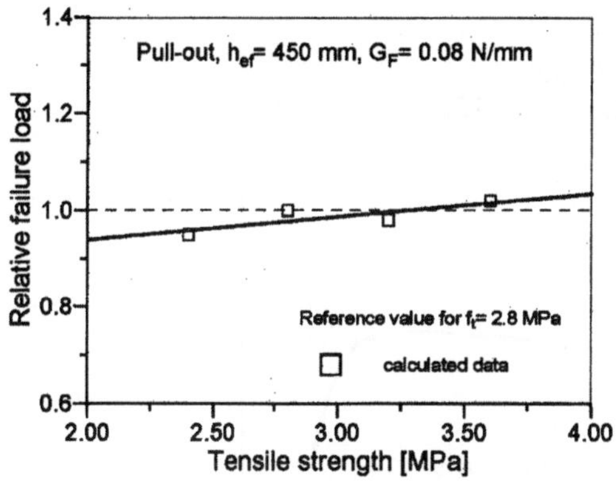


*Figure 7(a).* Nominal pull-out strength as a function of concrete tensile strength (Ožbolt, 1995).

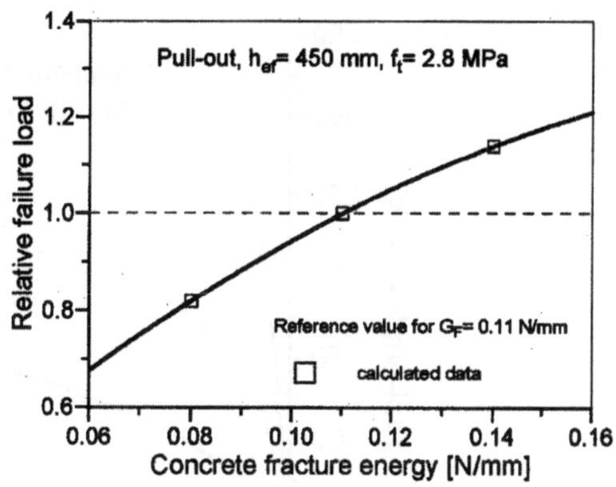


*Figure 7(b).* Nominal pull-out strength as a function of concrete fracture energy (Ožbolt, 1995).

which rely only on the material strength this type of structures relies also on the energy consumption capacity of concrete.

### 3.2. Failure mechanism for different head sizes

The influence of the head size on the concrete cone pull-out response and failure mode was studied for the geometry shown in Figure 2 (Eligehausen et al., 1992). The material properties and geometry were the same as in the experiment. For a constant embedment depth ($h_{ef} =$ 150 mm) the size of the head was varied. Three different head sizes were investigated: small ($d_h - 0.25h_{ef}$), medium ($d_h = 0.30h_{ef}$) and large ($d_h = 0.56h_{et}$).

The calculated load-displacement curves are plotted in Figure 8. A large head exhibits a stiffer response and a higher pull-out capacity than a small head. There are two reasons for this. The first is due to the fact that a larger head generates a larger concrete cone surface for the same embedment depth and therefore a higher pull-out capacity and stiffer response. The second reason, which is important for relatively small heads, is due to the shearing effect which influences the failure mode and reduces the embedment depth. To demonstrate this, in Figure 9 the crack patterns for three different head sizes are shown. The figure shows that in the case of a small head there is a strong shearing effect. The crack starts to propagate almost vertically and before the mode-I failure is activated the embedment depth is reduced. By increasing the head the shearing is less pronounced and for a relatively large head there is almost no shearing with crack growth which is for the whole load history approximately of mode-I type. A similar influence of the head size on the load-displacement behavior and the failure load was found in tests by Furche (1994).

### 3.3. Size effect

#### 3.3.1. *General*

As shown above, the series of experimental data as well as the numerical results for headed anchors indicate that there is a significant size effect on the pull-out capacity. To investigate the phenomena in more detail a number of theoretical studies have been performed in the past (Eligehausen et al., 1997). Besides already available results additional studies are carried out.

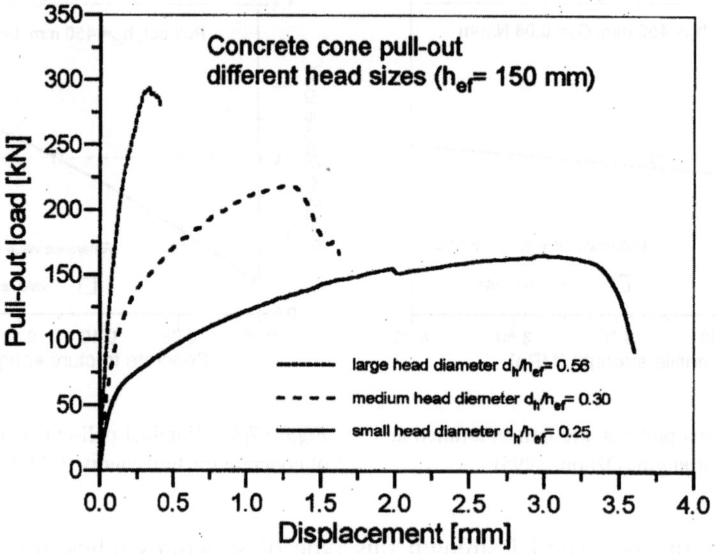


*Figure 8.* Calculated load-displacement curves for three different head sizes: small, medium and large.

In Figure 10 the nominal pull-out strength is plotted as a function of the embedment depth. The failure loads are normalised to a cube compressive concrete strength $f_{cc}$ = 33 MPa. For most of the plotted test data the size of the head was relatively small ($d_h/h_{ef}$ = 0.20 – 0.25) and it was scaled such that the pressure under the head of the stud at peak load was approximately constant for all embedment depths. As can be seen, the test data agree well with the design formula based on LEFM, what means that for the test size range there is a strong size effect on the pull-out capacity. To check whether the same tendency is to be expected for larger embedment depths a numerical analysis (Ožbolt, 1995) was carried out for the size range $h_{ef}$ = 50 to 2700 mm. In the analysis the geometry proposed by RILEM TC 90 (1990) was used. The ratio head diameter to embedment depth was $d_h/h_{ef}$ = 0.3 for all $h_{ef}$. The numerical results shown in Figure 10 are fitted with Bažant's size effect formula (Bažant, 1984). As can be seen, the value of the so-called characteristic embedment depth $h_{ef,0}$ is relatively small ($h_{ef,0} \approx 13$ mm) what is a clear indication for a relatively strong size effect. Namely, $h_{ef,0} \approx 13$ mm means that already for an embedment depth larger than 13 mm the size effect should be closer to the LEFM prediction than to the prediction according to the theory of plasticity. It can be seen that the CEB design formula based on LEFM agrees well with Bažant's size effect formula in the whole size range. Obviously, according to the numerical results for embedment depths larger than available from the current test data a strong size effect may be expected as well.

The numerical results show that for $h_{ef} \to \infty$ the nominal strength coincides with the LEFM prediction and Bažant's size effect formula (see Figure 10; Ožbolt, 1995). This means that there is a strong localisation of damage and stable crack growth. To confirm this in Figure 11 the calculated relative critical crack length at ultimate load is plotted against the embedment depth (Ožbolt, 1995). The relative critical crack length is measured as the ratio between the crack length at peak load and the maximum crack length of the concrete cone, under the assumption that the final crack is inclined with an average angle of $\alpha = 37.5$ degree. As can be seen from Figure 11, for smaller embedment depths ($h_{ef}$ = 50 to 150 mm) the relative critical length is approximately equal to 0.4. However, for $h_{ef} \to \infty$ it approaches a

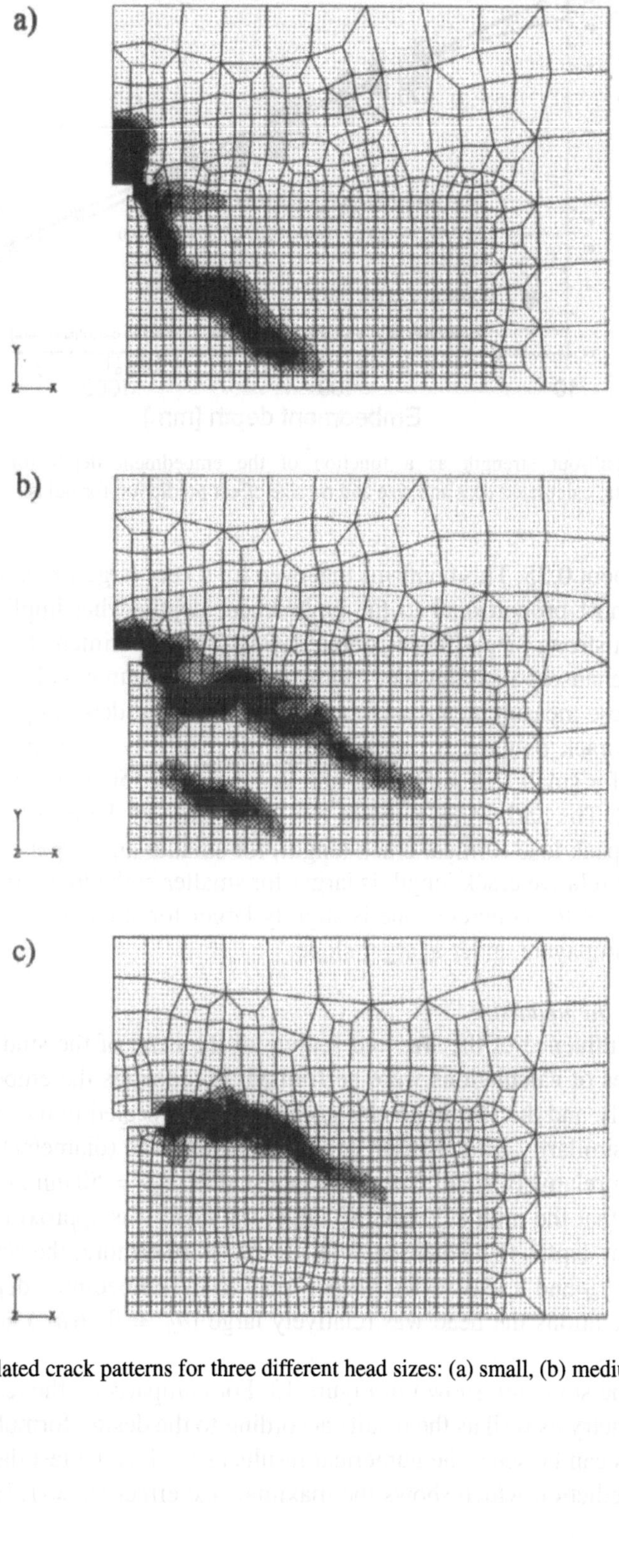


*Figure 9.* Calculated crack patterns for three different head sizes: (a) small, (b) medium and (c) large.

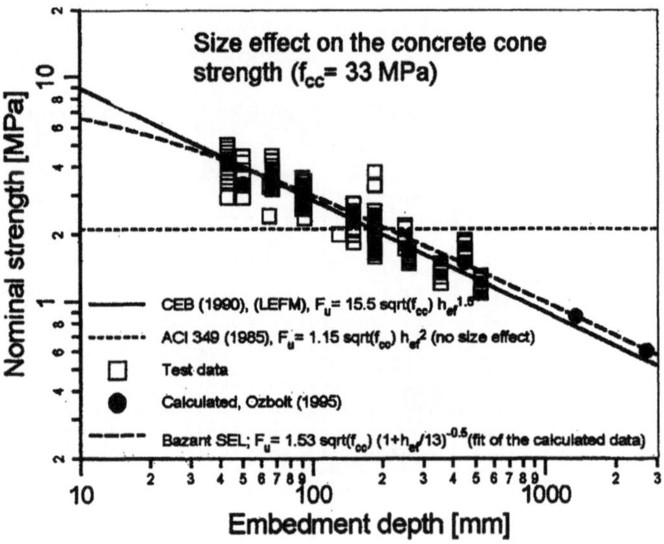


*Figure 10.* Nominal pull-out strength as a function of the embedment depth for a broad size range (10-3000 mm) – test data, calculated data and size and no size effect prediction formulas.

constant value of about 0.25. This confirms that even for a very large structure the crack length at peak load is almost proportional to the embedment depth, what implies stable cracking and consequently a strong size effect (Ožbolt, 1995). The experimental data for the critical crack length (Eligehausen and Sawade, 1989) confirm the numerical results, although as aforementioned, these measurements exist only for embedment depths up to 520 mm.

The calculated crack patterns for specimens with relatively small proportionally scaled heads at peak load for small ($h_{ef} = 50$ mm) and larger ($h_{ef} = 150$ mm) embedment depth are shown in Figure 12. The dark zones indicate the localisation of damage (cracks). Comparing the crack length at peak load (critical crack length) for smaller and larger embedment depth it can be seen that the relative crack length is larger for smaller embedment depth. Furthermore, the average angle $\alpha$ of the concrete cone is slightly larger for smaller embedment depths i.e. the crack tends to propagate under a larger slope.

3.3.2. *Influence of the head size*

To investigate the influence of the size and scaling of the head of the stud on the size effect three different series of calculations were performed. In all series the embedment depth was scaled proportionally and the specimen size (geometry) was scaled in the same way as in the experiment i.e. no similarity (Eligehausen et al., 1992). The size (diameter) of the head of the stud was first taken relatively small (for the smallest size, $h_{ef} = 50$ mm, $d_h = 0.25\,h_{ef}$) and it was scaled such that the ultimate pressure under the head was approximately constant for all three embedment depths (Eligehausen et al., 1992). Furthermore, the size of the head was taken as $d_h = 0.30\,h_{ef}$ and it was scaled proportional to the embedment depth. Finally, in the third series of calculations the head was relatively large ($d_h = 0.56\,h_{ef}$) and scaled with the embedment depth.

The results of the study are shown in Figure 13. For comparison, the test data for the first series of head geometry as well as the results according to the design formula based on LEFM are also plotted. As can be seen, the numerical results as well as the test data slightly deviate from the LEFM prediction which shows the maximal size effect ($\beta = 1.5$). The power fit of

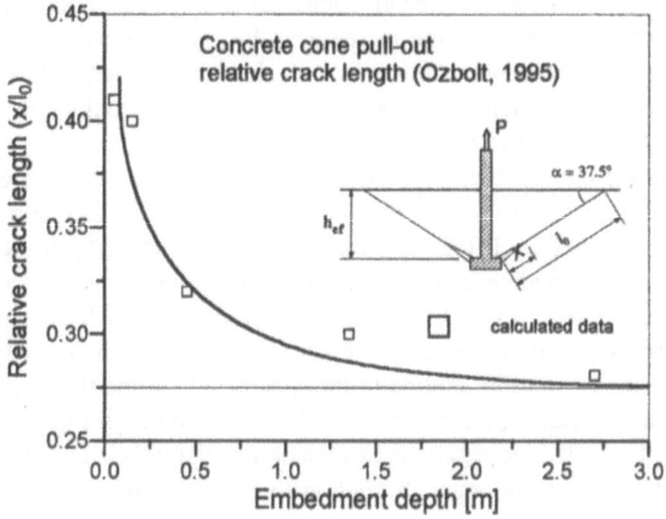


*Figure 11.* Calculated relative crack length at peak load (Ožbolt, 1995).

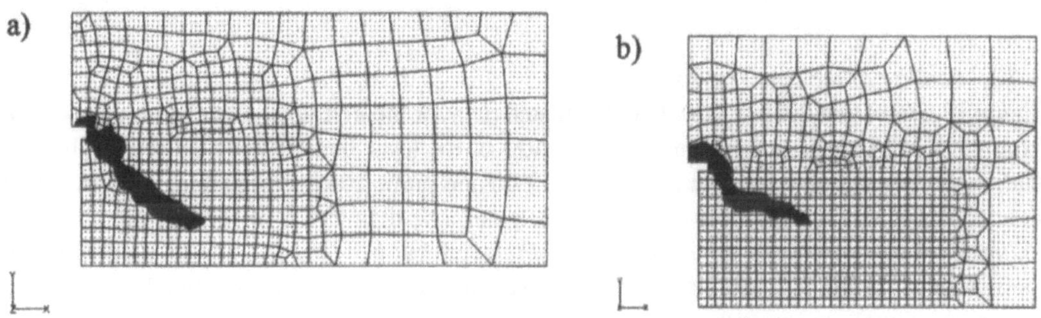


*Figure 12.* Calculated crack patterns short after the peak load (approximately $0.95P_U$) for: (a) $h_{\mathrm{ef}} = 50\,\mathrm{mm}$ and (b) $h_{\mathrm{ef}} = 150\,\mathrm{mm}$.

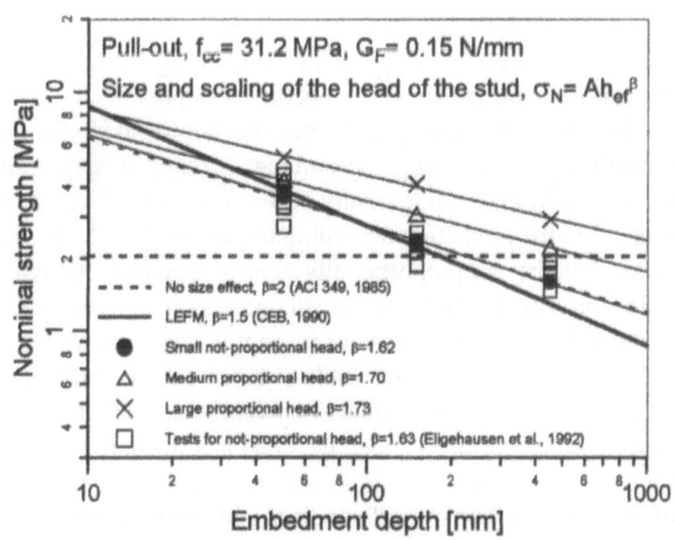


*Figure 13.* Size effect on the nominal pull-out strength as a function of the head size diameter.

the calculated and test data for small heads yields an exponent of approximately $\beta = 1.6$. Furthermore, it is obvious that with the increase of the head diameter the nominal strength increases for all embedment depths. However, it is clear that for larger heads the size effect tends to be less pronounced. Namely, for smaller heads $\beta$ is approximately equal to 1.6 and for relatively large heads equal to 1.7.

The explanation for the fact that with increase of the head size the size effect tends to be lower can be found in a slightly different failure mechanism for geometries with different head sizes, as discussed before. If the head size is small and especially if it does not increase proportionally with the embedment depth, there is a strong inhomogeneity of the strain field (localisation of damage due to the cutting effect of the head) which does not decrease significantly with increase of the embedment depth. Consequently, the damage is localised in the volume whose size is small in relation to the specimen size and the response close to the LEFM prediction i.e. the damage zone compared to the embedment depth is small. On the contrary, when the head size increases the localisation of damage is less pronounced and the damage volume at peak load is larger in relation to the embedment depth. Due to this, the nonlinear effect on the crack growth is stronger and the size effect deviates from the LEFM prediction.

### 3.3.3. *Scaling of the concrete specimen size*

Figure 13 shows that the test as well as calculated data for the first series of calculations deviate from the LEFM prediction as well as from most of the tests. This may be caused by the fact that, in the experiments (Eligehausen et al., 1992) and numerical investigations, the specimen size was assumed as shown in Figure 2, that means the diameter $b$ of the test member (see Figure 2) was not proportional to the embedment depth but $b = 4h_{ef} + 200$ mm. To investigate the influence of the nonproportional member scaling, the analysis for the member geometry with proportional scaling of the specimen was carried out. The basic geometry and the material properties were the same as in the experiment except that the total specimen height was taken as $h = 3h_{ef}$ (in the test $h = 2h_{ef}$).

In Figure 14 the nominal strength is plotted as a function of the embedment depth for the proportional and nonproportional scaling of the member geometry. Furthermore, the test data (nonproportional scaling) as well as the prediction according to the LEFM design formula are also shown. As can be seen, for smaller embedment depths ($h_{ef} = 50$ and 150 mm) the member geometry does not significantly influence the failure load. On the contrary, for a large embedment depth ($h_{ef} = 450$ mm) the nominal strength is approximately 20 percent higher for the specimen with nonproportionally scaled geometry. The size effect is stronger and practically identical to the prediction according to LEFM ($\beta = 1.51$, see Figure 14).

It is worth noting that the nonproportionally scaled specimen exhibits higher nominal strength. One would not expect this from the strength point of view since the nonproportionally scaled specimen has a smaller diameter $b$ and it is higher. However, from the fracture mechanics point of view, a larger diameter leads to better lateral confinement and therefore stiffer response. In such a body the accumulated energy is concentrated in a smaller volume since the elastic deformations out of the concrete cone are smaller. This is important for the structures in which the concrete fracture energy governs the behaviour, and it generally leads to the reduction of the strength. The overall conditions of proportionally scaled specimen are close to the main assumption of the LEFM and therefore the size effect should also be closer to the prediction according to the LEFM.

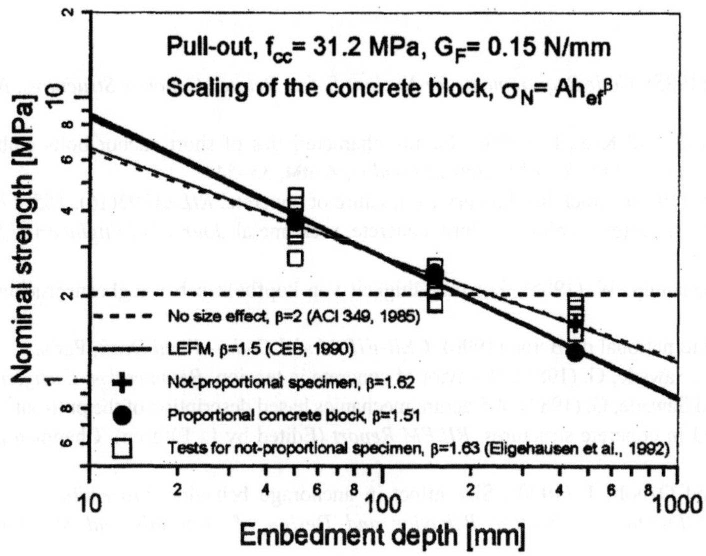


*Figure 14.* Size effect on the nominal concrete cone pull-out strength as a function of the size (scaling) of the concrete member.

## 4. Conclusions

- The experimental results as well as the numerical simulations confirm that a headed stud embedded into a plain concrete block is able to transfer a tensile force into the concrete utilising only the tensile resistance of the concrete with no need for reinforcement. The main reason for this is a stable crack growth which assures insensitivity of the structural response on the variation of the concrete tensile strength.
- Calculated and measured load-displacement curves for headed anchors with relatively small head sizes exhibit rather ductile response. When the head size increases the ductility of the response decreases, however, the concrete breakout resistance increases.
- The failure mode depends on the size of the head. For anchors with relatively small heads a strong shearing under the head can be observed before the typical mode-I crack opens. This reduces the effective embedment depth and therefore the resistance as well. For anchors with relatively large heads the crack growth is close to the mode-I failure type for the whole load history.
- As a consequence of the stable crack growth in the pre-peak regime the experimental and numerical results show a strong size effect on the nominal concrete breakout strength for a broad range of embedment depths. The size effect depends on the scaling of the head of the stud as well as on the scaling of the concrete member. For small heads and proportional scaling of the whole geometry (embedment depth, head size and member geometry) the size effect is strong and close to the prediction according to the LEFM. On the contrary, when the ratio head size to embedment depth is large or if the concrete member is not scaled proportional to the embedment depth, the size effect is weaker and it deviates from the prediction based on the LEFM.

## References


ACI Standard 349 (1985). *Code Requirements for Nuclear Safety Related Concrete Structures*, Appendix B – Steel Embedments.

Ballarini, R., Shah, S. and Keer, L. (1986). Failure characteristics of short anchor bolts embedded in a brittle material. *Proceedings of the Royal Society of London, A* **404**, 35–54.

Bažant, Z.P. and Oh (1983). Crack band theory for fracture of concrete. *RILEM* **93**(16), 155–177.

Bažant, Z.P.(1984). Size effect in blunt fracture: concrete, rock, metal. *Journal of Engineering Mechanics, ASCE*, **110**(4), 518–535.

Bode, H. and Hanenkamp, W. (1985). Zur tragfähigkeit von kopfbolzen bei zugbeanspruchung. *Bauingenieur*, 361–367.

Committee Euro-International du Beton (1990). *CEB-FIP Model Code – Final Draft*, Paris.

Eligehausen, R. and Sawade, G. (1985). Behavior of concrete in tension. *Betonwerk + Fertigteil Technik* **5** and **6**.

Eligehausen, R. and Sawade, G. (1989). A fracture mechanics based description of the pull-out behavior of headed studs embedded in concrete structures. *RILEM Report* (Edited by L. Elfgren), Chapman and Hall, London, 281–299.

Eligehausen, R. and Ožbolt, J. (1990). Size effect in anchorage behavior. *Proceedings of the 8th European Conference on Fracture – Fracture Behavior and Design of Materials and Structures*, Torino, Italy, 2671–2677.

Eligehausen, R., Bouška, P., Červenka, V. and Pukl, R. (1992). Size effect of the concrete cone failure load of anchor bolts. *Fracture Mechanics of Concrete Structures (FRAMCOS 1)* (Edited by Bažant, Z.P.), Breckenridge, Elsevier Applied Science, London, New York, 517–525.

Eligehausen, R. Mallee, R. and Rehm, G. (1997). *Befestigungstechnik*, Ernst and Sohn, Berlin, Germany.

Furche, J. (1994). *Zum Trag- und Verschiebungsverhalten von Kopfbolzen bei zentrischem Zug*, Dissertation, University of Stuttgart.

Ottosen, N.S. (1988). Nonlinear finite element analysis of pull-out tests. *Journal of Structural Division* **107**(4), 591–603.

Ožbolt, J. (1995). *Massstabseffekt in Beton- und Stahlbeton- konstruktionen*, Postdoctoral Thesis, University of Stuttgart, Germany.

Ožbolt, J. and Bažant, Z.P. (1996). Numerical smeared fracture analysis: nonlocal microcrack interaction approach. *IJNME* **39**(4), 635–661.

Ožbolt, J., Li, Y.-J. and Kozar, I. (1997). Microplane model for concrete – mixed approach. Submitted for publication in *IJSS*.

Reinhardt, H.W. (1981). Similitude of brittle fracture of structural concrete. *IABSE* Colloquium on Advances in Reinforced Concrete, Delft, 175–184.

RILEM TC 90 – FMA (1990). Round robin analysis of anchor bolts – invitation. *Materials and Structures* **23**, 78.

Sawade, G. (1994). *Ein energetisches Materialmodell zur Berechnung des Tragverhaltens von zugbeanspruchtem Beton*, Dissertation, University of Stuttgart.

*International Journal of Fracture* **95:** 405–433, 1999.


# Fracture and size effect characters of cemented sand

S. STURE, A. ALQASABI and M. AYARI
*Department of Civil, Environmental, and Architechtural Engineering, University of Colorado, Campus Box 428, Boulder, Colorado 80309-0428, U.S.A. e-mail: stein.sture@colorado.edu*

Received 5 August 1998; accepted in revised form 31 December 1998

**Abstract.** In the past three decades, fracture mechanics (FM) of brittle and quasibrittle materials have developed mainly in concrete applications and has yet to gain its rightful place in the geotechnical field. There are many examples in the geotechnical literature, especially those related to brittle and stiff soils, where traditional approaches of analysis have proven to be inadequate. While geotechnical problems are inherently complex in nature, the use of computational fracture mechanics have been shown to provide a powerful tool that could be used to investigate and solve many application problems in geomechanics and geotechnical engineering. This paper addresses the application of FM concepts and theories in analysis of cemented soils. In addition to theoretical aspects, experiments were conducted to evaluate the application of FM to cemented soils. Three-point bending beam tests with crack mouth opening displacements (CMOD) conducted on cemented sand samples showed that fracture parameters, such as CMOD, indeed could play an important role in investigation of such soils. Using this unambiguous material parameter, field engineers might have a reliable measure that could prove to be useful in stability assessment of earth structures and soil structure system. By studying size effect on cemented sand, strong relationship was established between critical CMOD and failure, which might be a very useful index and analysis tool in geotechnical engineering practice.

**Key words:** Fracture mechanics, cemented soil, geotechnical engineering, brittle behavior.

## 1. Introduction

The load-displacement relationships of stiff soils, e.g. overconsolidated clays (OC) and cemented sands (CS), have been known to differ dramatically from those of the more 'ductile' soils, e.g. normally consolidated soils and loose sands. Stiff soils exhibit more brittle behavior, which is characterized by a distinct peak shear strength followed by a decrease until residual shear strength is attained. Many researchers have described such behavior as 'strain softening', see, for example, Rizkallah (1977). There is no single cause for this behavior. However, for stiff OC, the presence of flaws, i.e. fissures, cracks, and/or joints, is frequently cited among of the most prominent factors leading to strain softening behavior, see Lo (1970, 1972). That is due to the fact that these flaws serve as stress concentration points where local failure may occur. This failure would propagate in the intact material which lead to a reduction in the overall strength of the material. Numerous studies of natural slopes, which are mainly stiff fissured OC, have shown that their failure mechanics do not correspond to that described in the classical slope stability literature, see Terzaghi (1936), and Skempton (1964). The presence of fissures, cracks, and/or joints has usually been cited as the source of trouble in these slopes. Terzaghi (1936) described a sequence of probable events, initiating from tension cracks that might lead to complete failure. Gould (1960) documented several cases of slope failures in stiff OC clays and Shales. He argued that the main reason for those failures was the degradation of shear strength due to excessive shear deformation and movement. Skempton (1964) and

Bjerrum (1967) cited planes of 'weakness' in slopes of stiff soils as the initial stage that lead to total collapse. Morgenstern and Tchalenko (1967a, b) showed that, outside the failure planes developed in failing slopes, the soil mass was almost intact, e.g. no failure or visible deformation had taken place in them. Rizkallah (1977) presented similar results in his study of failure planes in stiff clays using an electron microscope. He showed that the failure planes were concentrated in very thin bands in the samples. The concept of 'progressive failure' in soils was introduced by Bjerrum (1967), when he studied slope failures in natural slopes. This concept is based on the notion that the presence of some types of planes of weakness that would initiate a sequence of events, which eventually lead to the complete failure of such slopes.

In solids, it has been long observed experimentally, as well as shown theoretically, that the presence of any type of flaw in the material fabric leads to reduction in measured overall strength. Griffith (1921) formulated his energy balance theorem of crack propagation based on this observation. It has been, nonetheless, common practice in geotechnical engineering to ignore this fact and treat soils as continous material and analyze their behavior solely from the strength of material point of view. While the approximation of soils as continuous materials might be acceptable with normally consolidated soils, it may lead to unacceptable errors, when dealing with stiff, cemented, or overconsolidated soils. That is due to the fact that, it is highly probable that these soils may contain some type of flaws that act as stress concentration zones, where failure is likely to be initiated.

To the best of our knowledge, it was Skempton (1964) who first suggested the applicability of FM to analysis and interpretations of fissured clays. Bishop (1967) and Bjerrum (1967) gave more evidence supporting Skempton's observation. In recent years, other researchers in the geotechnical community have conducted several studies on application of FM theories and concepts in the analysis of soils. However, work is still needed in this field, and that is the motivation behind this contribution.

## 2. Why fracture mechanics in stiff soils?

### 2.1. Stiff clays

Cracking is a fracture problem that is not adequately treated by the strength of material (SOM) approach. That is due to the fact that in fracture, the stress singularity at the tip of the crack is the overriding criterion for initiation of crack propagation. While, in the SOM approach, the criterion for tensile failure is governed solely by the tensile strength of the material that is, theoretically speaking, greatly exceeded at the tip of the crack. The stress concentration effect on the behavior of stiff clays soil masses was highlighted by Bjerrum (1967), when he introduced the progressive failure concept. However, he concentrated on the influence of pore pressure effects on initiating such failure. There was, nonetheless, compelling evidence of fracture-like behavior for stiff soils that was left out for no apparent reason. Skempton (1964), in his Rankine lecture about the long-term slope stability problem, stated: *'Now it is well known that the strength of material is greatly diminished by the presence of microscopic cracks, holes and other imperfections. In the simplest terms these act as stress concentrators, and they are responsible for fracture taking place at an average stress that is far less than the ideal strength of the material. And there would seem to be no reason to support that the macroscopic fissures, joints, and slickenslides, present in so many clays, do not also act as discontinuous planes of weakness – for it is unlikely that the strength on fissures or joints can be appreciably higher than the residual value.'* Such powerful remark passed without much follow-up from the geotechnical research community. However, knowing that FM as a well-defined and applicable science was not fully developed till the late 60's and early 70's, it was expected that the geotechnical community at large would not adopt such an undeveloped, or not understood technique.

Griffith developed the fundamental theory of fracture mechanics (Griffith, 1921). When measuring tensile strength of glass, he observed a broad discrepancy between the results of small and large samples. He formulated his energy balance theorem describing the effect of flaws on the stability of the material response. His theory then became the foundation for the whole FM discipline. Similar trend, of size effect on strength, could also be found in stiff natural soils. In soils, Bishop and Little (1967) showed that the measured shear strength values of London Clay, investigated in a 2 ft. by 2 ft. *in situ* shear box, were 55 percent of those measured from samples of core samples of 1.5 in. in diameter. Similar strength reduction phenomenon was also reported for Boston Blue Clay by Simons (1967). The relation between the of size of samples and the measured shear strength of OC clays was thoroughly documented by Lo (1970, 1972) and many others.

Skempton (1966) cited the formations of tensions fractures in formations of stiff soils and rocks. Further evidence of fracture-like behavior of stiff soil was documented by Bjerrum et al. (1972). He suggested that, without properly knowing of the fracture behavior of clays, all field permeability measurements of clay deposits should be taken with caution. He cited that due to hydrofracturing, recorded field permeability values could be higher than the actual values, when a critical pressure is reached. He concluded that when that pressure is reached, existing cracks in the clay would open which lead to creation of channels for the water to escape through thus leading to incorrect measurement of permeability values.

### 2.2. CEMENTED SANDS

Evidence of fracture-like behavior of slopes in cemented sands is scattered in the literature. No attempt was made to study the fracture behavior of such soil masses, nor any attempt was made to use FM concepts in the analysis of such slopes. The first comprehensive study of the characteristics of cemented sand (from strength of materials (SOM) point of view) was conducted by Sitar et al. (1980), in their study of cemented sand slopes located on the California coast and in Guatemala. Eyewitness reports of splitting of large pieces of cemented sands, due to earthquakes and gravity loads were reported. Evidence of failure by fracture were cited as probable reasons for such failures, as shown by this quote from their report: *'Field Observations of the types of failure slopes in the loess soils (Lutton (1969) and in the volcanic ash indicate that tensile splitting is an important mechanism involved in the slope failures. Therefore, application of conventional slip circle type of stability analyses is inadequate and some other approach to slope stability analyses in the weakly cemented soils is needed.'* In the report, it was shown that the slightest cementation in the sands changed the stress strain response to a more brittle behavior with distinct maximum strength values followed by strength reduction. Such behavior coupled with the often reported presence of tension cracks at the top of the slopes' crests is a clear indication that the failure of the slopes is more than likely due to a fracture-like failure. This indication is reinforced with the many citings of large chunks of cemented sands falling during earthquakes and/or due to erosion in slopes' toe areas.

Suggestions of a fracture-like behavior in cemented sands were cited by Butterfield (1986) and Wroth (1986). They observed that the recorded resistance decreased with time while holding flat Dilatometer (DMT) instruments stationary in dry sand deposits. Since the dissipation of pore water pressure was ruled out as a possible reason, it is more likely that the blade-like instrument actually cut into the deposit creating some form of cracks which propagated in the deposit leading to redistribution of stress and nonuniform displacement and the gradual decrease of resistance. Habersfield and Johnston (1989) reported that theoretical results indicated that pressuremeter probes would eventually lead to tensile failures around the probe in weak rock. Since the mechanical behavior of such rocks should be similar to that of stiff cemented sand, the observation should also hold for cemented sands. They reported that field testing conducted by Briaud (1986) reported formation of radial cracks in clay.

## 3. Research motivation

The observation of an apparent paradox between the results of maximum shear strength values measured in the laboratory and the values back-calculated from failed slopes might be attributed to fracturing. Researchers have always noted that the laboratory measured values of shear strength were much higher than those obtained by back-calculation of safety factor of failed slopes in naturally fissured soil. It has been suggested that due to stress concentration and shear displacement, soil would reach its peak strength then weaken to a residual value, see Skempton (1964). This behavior is similar in concept to strength reduction observed in solids due to the presence of flaws.

Palmer and Rice (1973), to the best of our knowledge, were the first to propose consistent theoretical treatment of stiff clays using concepts of FM. They used the concept of the $J$-integral to study the propagation characteristics of long shear bands in infinite slopes. However, their study lacked experimental validation. Many researchers followed their work in studying applications of FM theories and concepts in many analysis of stiff soils.

The field, however, still lacks some of the basic information needed to conduct a comprehensive study. Namely, the fracture characteristics of soils need to be evaluated in more accurate experimental procedures. For example, to the best of our knowledge, Saada et al. (1985), Lee et al. (1988), Fang et al. (1989), Fang H.-Y. (1994), and Harison et al. (1994) are the only researchers who have published measurements of fracture toughness of stiff cohesive soils. This parameter is essential to conducting comprehensive FM study of stiff soils. Furthermore, all of the above mentioned studies have limitations either with regard to the way the tests were conducted or the reliability of the results in general.

In this research, an experimental program is designed to measure the fracture toughness of an artificially cemented sand. A new model for post-peak prediction of cemented sand behavior is proposed and imitations of application of LEFM in analyzing cemented sands are then discussed.

## 4. FM applications in geotechnical engineering

As mentiond earlier, the first attempt to apply concepts of FM in geotechnical engineering analysis was by Palmer and Rice (1973). They derived the $J$-integral for an infinite slope in over-consolidated soils. The formulation was used to predict the initiation and the extent of a straight shear band in a sloping soil mass. Although the approach used was very limited with regard to simplifying assumptions, the researchers showed that concepts of FM could potentially be used in analyzing cracking problems in brittle soils. No attempt was made in this research to measure any fracture parameter for such soils. Saada et al. (1985) were the first to publish measurements of fracture parameters for soils. They presented an experimental procedure to measure fracture toughness parameters for OC clays. Concepts of thermodynamics and FM were then used to conduct slope stability analysis in infinite slopes. However, the suggested method was cumbersome and the results are biased to the way the person conducting the test. O'Rourk and Crespo (1988) published some results of fracture toughness parameters for cemented volcanic ash material. However, no details were given regarding the testing procedure. Vallejo (1985, 1986, 1987, 1989) used concepts of LEFM in a variety of predictions involving geotechnical engineering problems. Crack initiation direction was predicted using LEFM concepts for OC clays samples with a variety of boundary loading conditions. Predictions of hydraulic fracturing in clay barriers (inside earth dams) was also conducted. In all of Vallejo reports, no suggestions were made on how to measure at pertinent fracture parameters.

Lee et al. (1988) published first measurements of Mode II critical energy release $G_c$ for cohesive soils. Compact tension (CT) specimens were used in the experimental program. One of the shortcomings in the reported experimental procedure was that the crack length during testing was obtained visually via a magnifying glass. Any possible effects of the presence of an invisible crack advancement, which is referred to as the fracture process zone in the concrete fracture mechanics literature, ahead of the crack was unaccounted for in the results. Results of fine-grained cohesive soils using the same procedure used by Lee et al. were also reported by Fang et al. (1989) and by Fang (1994).

Cai et al. (1990) used principles of LEFM to analyze the stability of vertical slopes of stiff OC clays with tension cracks at the top of the crest. The basic mechanism of 'slope' and 'base' types of failures were successfully depicted. Design charts, which consisted of relations between the critical strain energy density $(SED)_c$ criterion vs slope height were presented. The contribution of water filled tension cracks in stability of open cuts was also investigated. However, no experimental procedure was presented to measure the suggested $(SED)_c$ soil criterion.

Haberfield and Johnston (1989) presented an extensive experimental program using a three-point bending (TPB) beam setup to measure fracture toughness of synthetic soft rock. The effects of various testing variables (e.g. water content, loading rate, specimen size, and crack depth) were investigated. Even though the crack mouth opening was measured using clip gauge, no post peak results were reported. Test results showed that, even with different water content, the obtained fracture toughness was constant for $a/W$ (crack depth to total sample depth) values between 0.1 to 0.6. The results confirmed that the fracture toughness decreases, for the range used in the testing program, with water contents. The researchers indicated that effects of loading rate were minimal as long as enough time was allowed for pore pressure to dissipate. Harison et al. (1994) presented results of fracture tests conducted on cohesive soils using ring tests. The testing configuration consisted of a soil core specimen with an inner hole in the middle subjected to loading similar to the Brazilian test. The soil specimen would be prenotched from the inside diameter. The fracture toughness would be calculated using the diagram of applied load vs. the displacement of the loading rings. The specimen would be first loaded till the prenotch would propagate, then a sequence of unloading–reloading would be initiated. Based on the area under the load–unload–reload graph, the researchers devised a procedure to calculate the fracture toughness of the sample. Relationships between the fracture toughness values, water content and loading rate were presented. The results here confirmed those reported by Haberfield and Johnston (1990), which indicated that fracture toughness decreases with increasing water contents and is almost constant with loading rates less than 0.02 in/min.

## 5. Experimental program

Due to lack of standards in fracture testing of cemented sands, and due to the similarity in the fabric between the cemented sands and concrete, available fracture tests in the area of concrete fracture mechanics were reviewed to select the best experimental procedure to be used in this research. The most widely used test in concrete fracture testing is the splitting compact tension (CT) specimen. The test provides many advantages, most of which is the extended length of the uncracked ligament of the specimen. However, soils are far weaker in tension than concrete, and applying loads to the CT specimen might lead to unfavorable indentations in the area near points of loading of the specimen during testing. Lee et al. (1988) reported similar problems when testing cohesive compacted soils. Such inelastic zones would lead to dissipation of energy which otherwise would lead to unwanted cracking.

When searching the literature, one can see that there is not much previous research in the area of fracture testing of soils. This statement is even more accurate when addressing properties of cemented sands. To the best of our knowledge, no previous studies have been conducted on cemented sands' fracture behavior. There were, though, other studies conducted on cohesive soils as described previously.

As was reported above, Lee et al. (1988) published values of mode I critical energy release rate ($G_c$) for compacted cohesive soils using CT specimens and using the compliance method. In this method, values of specimen compliance during testing were obtained for each specimen. Several readings of cracking force and crack length would be then plotted to obtain relationships of cracking force, and compliance against crack length. Values of crack length were obtained visually using a traveling microscope that was mounted on top of the specimen. Change of compliance variation with change of crack length was then obtained numerically. The following expression was then used to calculate $G_c$

$$G_c = \frac{1}{2t} F_c^2 \frac{d\lambda}{dc}, \tag{1}$$

where $t$ is the specimen thickness, $\lambda$ is the compliance, and $F_c$ is the cracking force measured at a crack length $c$. The researchers reported the presence of indentations in the contact area of the load pins, which indicated that plastic energy was dissipated in that area. Also, the researchers relied on visual means (magnifying glass) to measure the crack length. It is well known fact that for such cohesive materials, some type of invisible crack advancement (which is called fracture process zone) would take place ahead of the crack, which deems any visual measure of the crack length inaccurate. Due to the fact that crack length were not measured correctly, any subsequent derivations of compliance values have to be considered with caution. This problem was compounded even more, when the derivative of compliance with respect to change in crack length was obtained, and used to obtain $K_{\mathrm{I}c}$ values.

Saada et al. (1985) reported a different procedure to obtain $K_{\mathrm{I}c}$ values for overconsolidated clays. Hallow cylindrical samples, with complex geometry were used to conduct the tests. Evaluation of fracture parameters, such as length of critical crack (which was used to calculate the fracturing force) was cumbersome and did not address or account for any post-peak response. Also, in order to obtain the $K_{Ic}$ values, the researchers relied on visual inspection of the specimen itself to obtain what was termed as (stable crack length). Evaluation of such values would always be a subjective matter that eventually made this procedure highly biased to the person who conducts the tests. It is, therefore, concluded that the proposed procedure is not reliable. Harison et al. (1994) used hollow disk samples to evaluate $K_{Ic}$ values of compacted cohesive samples using ring tests. The procedure seems to be promising and could be performed at most geotechnical laboratories. However, the proposed procedure relies mainly on calculating areas under the curve of load vs. displacement of the loading rings. This meant the error due to bias in personal preference in choosing the areas would greatly affect the calculated results of such tests. Also, the procedure did not allow for any post peak response measurement. With more refinements the proposed procedure, nonetheless, has many advantages that could make it a plausible alternative.

After reviewing the available test procedures for $K_{Ic}$ evaluation, the three-point bending beam (TPB) tests was found to be the most suitable test for cemented sands for the following reasons:

- loading configuration in the test minimizes any indentations on the sample, which in turn concentrates the energy dissipation in the crack tip area;
- the ability to monitor and record the post-peak response of the specimen;
- the availability of closed form solutions for such a configuration; and
- the simplicity of the required testing equipment, where one can perform such tests virtually in any civil engineering laboratory.

### 5.1. Design of the experimental program

Samples of cemented sands could possibly be obtained by extraction from natural undisturbed blocks, or they could be prepared in the laboratory. Obtaining natural undisturbed samples is not easy. Properties of such samples are very sensitive to any type of disturbance, which leads to scatter in the measured properties of the soil, which eventually would complicate the interpretation of any testing. Fracture properties are particularly sensitive to any defects in the samples. Haberfield and Johnston (1989) reported that fracture testing of soft rocks, which were conducted on natural samples, produced such a wide scatter that made any numerical verifications unattainable. Moreover, the cementing agents in the field varied quite significantly from location to location that would increase the number of variables in the specimens to be tested. In addition, the objective of this research is towards understanding the fundamental fracture behavior of cemented sands. Due to the above, it was decided that the laboratory prepared samples would be used.

To create cemented sand in the laboratory, several cementing agents were considered. Regular Portland cement is a viable choice in this matter. However, this would require monitoring the mixing percentages and the effects of temperature and humidity of the mix. A more readily available cementing agent was used instead. Ordinary household sugar was used as a cementing agent. Mould (1983) successfully used sugar solution to prepare cubical samples of sand to study stress induced anisotropy. Sugar provides an ideal cementing agent for the following reasons:

- very easy to handle;
- gives very competent samples; and
- can be cured at room temperature.

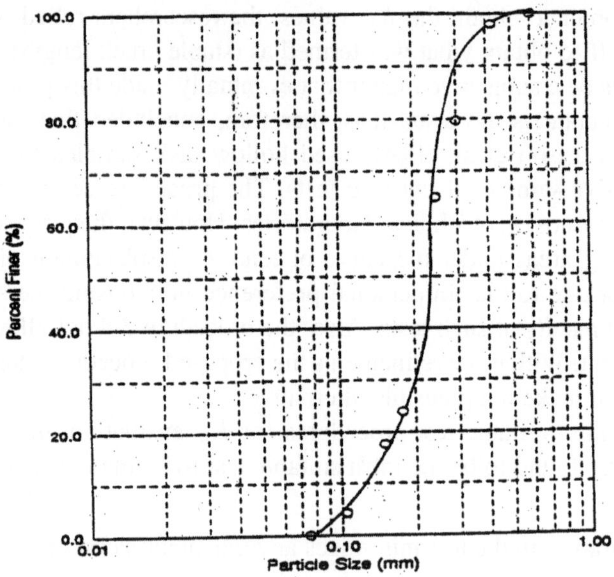


*Figure 1.* Sieve analysis for F-75 Ottawa Sand, after Sture et al. (1994).

### 5.2. MATERIAL

Throughout this research, fine grained uniformly graded Ottawa sand was used. Sieve analysis of the sand is shown in Figure 1. The sand has specific gravity ($G_s = 2.65$). The Unified Soil Classification of the sand is SP. It has maximum and minimum void ratios of 0.805 and 0.486 respectively. The sand is commercially available by the name F-75, and can be procured from 'U.S. Silica Industrial Company'.

### 5.3. MOULD DESIGN

To ensure plane strain conditions for metals, ASTM standard (1994) states that width ($B$) : depth ($W$) : span length ($S$) ratio has to be 1 : 2 : 8. Due to a lack of standardization for cemented sand testing, the dimensions of the single edge cracked beam (SECB) sand-beam specimens, to be used in the TPB tests, had to be experimented with. In order to create suitable beam specimens, the following constraints had to be considered:

- They should provide enough uncracked length to enable a study of the post peak response of the material.
- They should be large enough to provide sufficient homogeneity to minimize the effect of the expected irregularities in the fabric during the raining of the sand.
- They should be relatively easy to handle and transport with minimum effects due to dead loads on the imperfections (which were expected no matter how carefully the sand was rained).
- The shape of the sample should be relatively easy to reproduce to minimize the irregularities due to variation in preparation procedure.

Haberfield and Johnston (1990) showed that for TPB tests of SECB of soft rocks the fracture toughness remained constant for values of $a/W$ between 0.2 to 0.6. With the above concerns in mind, the dimensions of the beam samples were chosen to be at least 200 mm long, 45 mm wide, and 50 mm deep. Additional length of the sample was deemed necessary to account for the expected imperfections, due to extraction of the sand.

A Plexiglas mould, which consisted of a rectangular-shaped upper part and a base, was designed and erected. The upper part of the mould was designed to have four sections separated by three-cross bars to hold four SECB samples in each mix. Each of the four section of the mould had the following dimensions: 254 mm long, 45 mm wide, and 50 mm deep.

## 5.4. SAMPLE PREPARATION

### 5.4.1. *Raining technique*

The stress-strain relationship of the loose sand was known to be greatly affected by few traces of cementation. Sitar et al. (1980) reported that for weakly cemented sands, traces of cementation produced a marked brittleness in the behavior of such sands. Therefore, it was decided that the only objective of sample preparation was to consistently reproduce similar samples for further testing. An arbitrary height of 10 in. was chosen for sand raining. Dry sand was rained in the Plexiglas mould through a 0.5 in. cone. A no. 40 sieve was placed at the upper neck of the cone to provide uniform raining. After raining, the excess sand was removed and the sand was leveled at the top of the mould. The mould (with the dry sand) was then weighed. This technique produced fairly consistent values of dry density of the sand ($\gamma_d = 1.59\text{–}1.61\ \text{gm/cm}^3$).

### 5.4.2. *Cemented sand mixing*

Since no prior fracture testing of sugar cemented sands was previously reported, some experimentation was carried out to choose a suitable percentage of water–sugar mixture. The objective here was to be able to create as competent samples as possible with as minimum drying time as possible. Pilot specimens of sands with water–sugar percentages of 5, 10, 15, 20, 25, and 30 percent were prepared and monitored for several days. Drying time was recorded until complete drying was reached. After several trials, 10, 15, and 20 percent initial mixing sugar contents were found to produce representative samples with reasonable drying times. The sugar–water mixture, with the required mixing percentage, was first prepared in a separate large container. Regular tap hot water was mixed with the required sugar percentage. The mixture was stirred thoroughly to ensure that all the sugar was completely dissolved in the water. After sand raining, the mould (with the dry sand) was then carefully immersed in the sugar–water mixture. Time was allowed for all entrapped air in the sand to escape. The mould was then removed from the mixing container and carefully weighed. Samples of wet sands were then taken immediately to obtain the percentage of the water–sugar solution in the wet sand immediately after mixing. Values of this percentages ranged between 18.2–19.5 percent.

After 12 hours, the cemented sand beams were extracted from the Plexiglas mould. They were thereafter left to dry completely under room temperature. During this period, the sugar–water content of the sample was monitored. Complete drying was assumed, when the sugar–water mixture value showed no change in 24 hours. Measured percentages of water–sugar solution after drying were between 0.10–0.15 percent. The drying time for the samples varied depending on the initial sugar percentage in the water–sugar solution. Samples of 10 percent initial sugar percent dried completely after about 5 days. While samples of 15 percent initial sugar dried after about 8 days. Whereas samples of 20 percent dried after about 10 days. After drying, the cemented sand beams were prenotched (to a specified length), using a thin hacksaw blade, at the middle of the sample, to form a crack where the clip gauge would be mounted during testing.

At the initial stages of drying, the cemented sand beam samples were extremely sensitive to the slightest form of loading. The samples remained very sensitive to loading until about 75 percent drying was achieved. At that stage, the beam sample could be turned around to ensure complete drying, without any significant indentation on the sample. After complete drying, the same samples showed marked increases in strength. This allowed the final trimming of each beam sample to be done using the metal hacksaw blade.

### 5.4.3. *Problems with sample extraction*

At the start of the experimental program, the Plexiglas mould was erected as one unit. The base and the upper part assembly were glued together. Therefore, in order to extract the cemented sand beams, the entire mould (with the sand beams) would need to be turned upside down to let the sand beams fall out under their own weight. No pressing or extrusion device could be used due to the extreme sensitivity of the wet sand to any kind of loading. Using this procedure, it was found that extracting the intact sand beams from the mould was not the easiest thing to do. Several lubrication agents were used to retrieve the sand beams intact. With all the lubrication agents used, samples were almost always damaged completely or cracked during extraction. Any unwanted cracking in the sample meant another area of imperfection was introduced in the sample. Since the main fracture test's objective is to measure the strength of the cracked material properties, any unwanted cracks would definitely alter the test results. Therefore, any unwanted cracks (in areas other than the designated area) deemed the samples useless. Therefore, a new mould where all the sides were not glued together but rather inserted into each other was designed. The long sides of the new mould were screwed to the base. The shorter sides were then inserted in the edges of the long sides by special grooves made in the long sides' bars. The crossbars of the mould (the inside bars separating the four beam samples) were then inserted in special grooves made in the short sides of the mould. The design worked perfectly even without any lubrication. It produced almost completely intact samples. The samples were extracted by sequentially removing the sides of the mould after some drying time had elapsed (about 12 hours).

A total number of 23 sample beams of cemented sand were prepared. Table 1 shows their dimensions. Six samples of each sugar (initial sugar mixing) concentrations of 10 and 20 percent were prepared. The rest of the samples were prepared with initial mixing concentration of 15 percent concentration.

## 5.5. THREE-POINT BENDING (TPB) TESTS

The cemented sand beams where placed on a three point bending beam setup to be tested. Figure 2 shows a schematic illustration of the setup with magnified illustration of the clip gauge mounting assembly. To hold the clip gauge near the mouth of the notch in the cemented sand beam, Plexiglas knife edges were prepared and then glued to the specimen. They were glued near the mouth of the notch (with a distance equal to 7.5 mm separating them).

Prior to starting a test, the cemented sand beam was centered on two roller supports. When evaluating the fracture toughness of the beam, the distance between the two supports was taken to be the span length. Loading was applied directly above the notch in the beam to ensure Mode I fracture. An Instron testing machine was used to apply the load and a PC computer using the Labview computer software with data acquisition capabilities was used to control

*Table 1.* Dimensions of SECB samples.

| Sample no. (ISP) | Span length S (m) | Depth W (m) | Crack length a (m) | Width B (m) |
|---|---|---|---|---|
| 6-1 (10%) | 0.18 | 0.048 | 0.023 | 0.05 |
| 6-2 (10%) | 0.17 | 0.048 | 0.021 | 0.043 |
| 7-1 (10%) | 0.16 | 0.045 | 0.015 | 0.039 |
| 7-2 (10%) | 0.15 | 0.044 | 0.016 | 0.04 |
| 9-1 (10%) | 0.18 | 0.05 | 0.02 | 0.042 |
| 9-2 (10%) | 0.18 | 0.049 | 0.018 | 0.041 |
| 11-1(20%) | 0.13 | 0.05 | 0.029 | 0.046 |
| 11-2(20%) | 0.09 | 0.05 | 0.028 | 0.0448 |
| 16-1(20%) | 0.196 | 0.048 | 0.026 | 0.043 |
| 16-2(20%) | 0.196 | 0.048 | 0.027 | 0.04 |
| 16-3(20%) | 0.196 | 0.048 | 0.024 | 0.04 |
| 16-4(20%) | 0.196 | 0.048 | 0.0285 | 0.043 |
| 14-1(15%) | 0.18 | 0.048 | 0.027 | 0.043 |
| 14-3(15%) | 0.192 | 0.048 | 0.024 | 0.04 |
| 15-1(15%) | 0.192 | 0.049 | 0.0285 | 0.04 |
| 15-2(15%) | 0.192 | 0.049 | 0.027 | 0.04 |
| 15-3(15%) | 0.196 | 0.049 | 0.027 | 0.04 |
| 15-4(15%) | 0.196 | 0.048 | 0.025 | 0.043 |
| 17-1(15%) | 0.196 | 0.048 | 0.0265 | 0.043 |
| 17-2(15%) | 0.196 | 0.048 | 0.0275 | 0.04 |
| 17-3(15%) | 0.196 | 0.048 | 0.023 | 0.04 |
| 17-4(15%) | 0.196 | 0.048 | 0.025 | 0.043 |

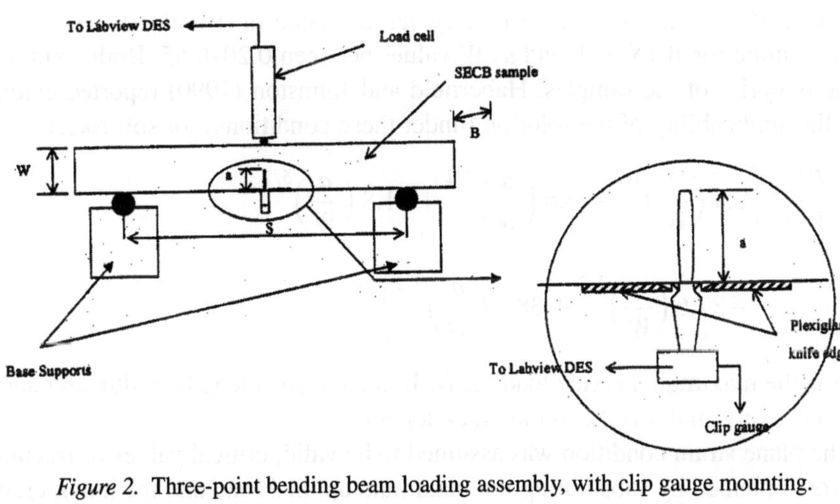


*Figure 2.* Three-point bending beam loading assembly, with clip gauge mounting.

loading. The applied load was measured using a load cell (Instron load cell, $P_{max} = 50$ lb.) Both the load cell and the clip gauge were calibrated prior to testing.

### 5.6. TPB TESTING PROCEDURE

For soft synthetic rocks, Haberfield and Johnston (1990) showed that for a loading rate $< 0.10$ mm/min., fracture toughness measurements were not affected by change of rate. Therefore, an arbitrary loading rate of 0.002 in/min (0.0051 mm/min) was used throughout the TPB testing program described here. Loading was applied with closed loop CMOD monitoring. CMOD values at the initial stages of loading would increase almost linearly which indicated that no crack advancement took place. In order to be able to sustain stable crack growth, the instance of crack growth had to be predicted while testing. At the stage where the crack began to propagate, values of CMOD would increase at a much faster rate. At that stage, loading would be reversed. Cycles of unloading–reloading were generated using this technique.

The technique proved to be successful and the post-peak response of at least three samples in each concentration group were monitored and recorded.

### 5.7. POST-PEAK RESPONSE

Cycles of loading–unloading–reloading were conducted to study the post-peak response of the cemented sands. The loading would be increased till excessive displacement readings of the clip gauge were observed. Then, loading would be reversed until zero load cell reading was reached. The cycle was repeated till either the sample would completely fail (by splitting) or till the load-carrying capability of the sample diminished to very small values. At that stage the crack length in the sample was assumed to reach the total depth of the sample. Figures 3–6 show the load versus CMOD readings of four SECB samples (6-1), (15-4), (15-3), and (16-4), respectively. These diagrams represent typical TPB results of the three different concentration groups (10, 15, and 20 percent, respectively) used in this research.

### 5.8. EVALUATION OF $K_{Ic}$ FOR CEMENTED SAND

The results of all the TPB tests were used to calculate the fracture toughness values ($K_{Ic}$) of the cemented sand for the three different concentrations. The closed form solution, shown in (2), to evaluate $K_I$ for the specimen geometry can be found in (Broek, 1991). The proposed solution is accurate for $W/S = 4$, and $a/W$ values between 0.20–0.55. Both conditions were met for the majority of the samples. Haberfield and Johnston (1990) reported evidence that supported the applicability of the solution (under these conditions) for soft rocks.

$$K = \frac{PS}{BW^{3/2}}\left[2.9\left(\frac{a}{W}\right)^{1/2} - 4.6\left(\frac{a}{W}\right)^{3/2} + 21.8\left(\frac{a}{W}\right)^{5/2} - 37.6\left(\frac{a}{W}\right)^{7/2} + 38.7\left(\frac{a}{W}\right)^{9/2}\right], \quad (2)$$

where $P$ is the maximum applied load, $S$, $B$, $W$ are the span length, width, and depth of the sample, respectively, and $a$ is the initial crack length.

Since the plane strain condition was assumed to be valid, critical values of fracture toughness $K_{Ic}$ were obtained by substituting the maximum applied load and the initial crack length of the sample in (2). Table 2 shows the values used in (2) for all the SECB samples and the

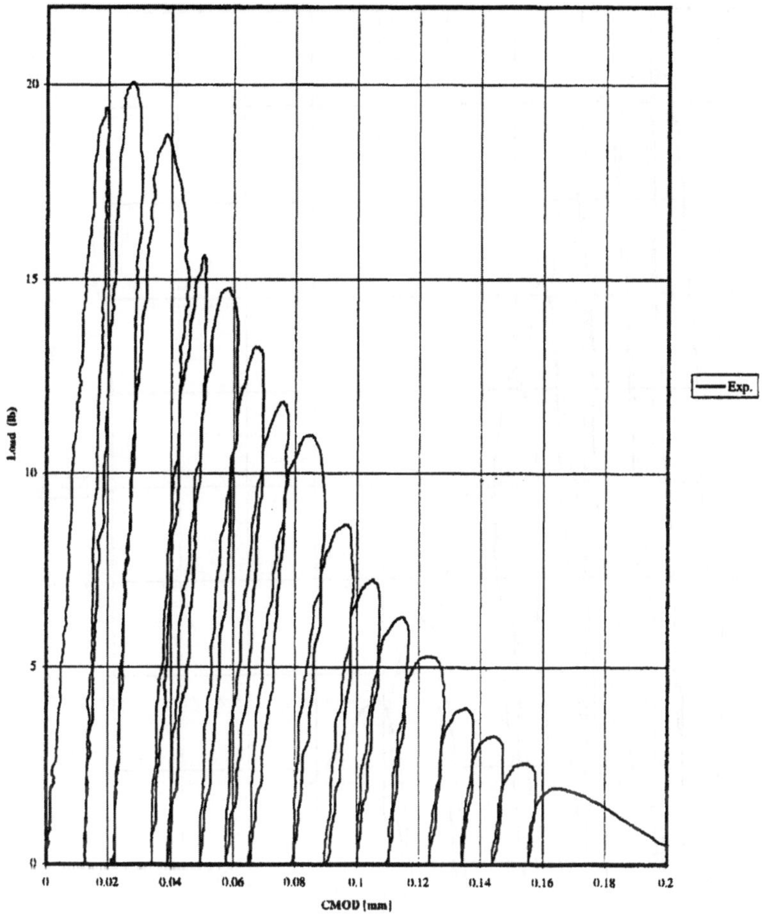


*Figure 3.* Load vs CMOD for sample (6-1) with 10 percent concentration.

calculated values of $K_{Ic}$. Figure 7 shows the calculated values of $K_{Ic}$ vs the initial sugar mixing percentages. From the it is clear that the fracture toughness of the cemented sand increases with increasing percentage of sugar content. An empirical relationship between initial sugar percent and $K_{Ic}$ was found to be as follows

$$K_{Ic} = 0.188\ (\text{ISP})^2 - 0.19237\ (\text{ISP}) + 13.491, \tag{3}$$

where (ISP) is the initial mixing sugar percent in the sand.

## 5.9. DISCUSSION

A magnifying glass was used to check for visible traces of cracking in the samples prior to failurc. All samples showed absolutely no visual indication of cracking before failure. Samples fracture was straight upward to the direction of the point load. The crack was visible only after failure and the samples showed no signs of degradation in the values of the unloading slopes (in the load vs CMOD diagrams) until the very late stages of the tests; see Figures 3–6.

The above mentioned observations, namely the invisibility of the crack advancement and the unchanged response of the slope at the load–unload cycles, indicated that the elastic mod-

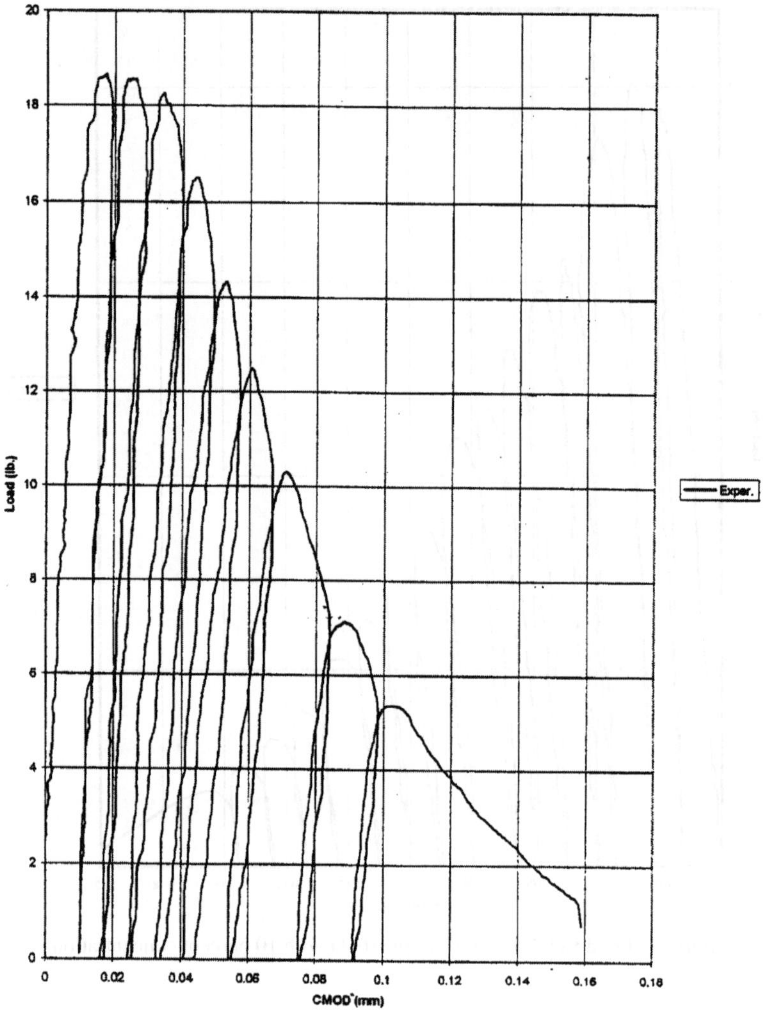


*Figure 4.* Load vs CMOD for sample (15-4) with 15 precent concentration.

uli of the samples were not affected till the late stages of the tests. Evidence of specimen degradation was clearly observed with the decrease in the load carrying capacity of the samples. The load vs CMOD graphs show that the cemented sands has linear loading/unloading branches until the advanced stages of specimen degradation. The load–unload slopes also remained almost the same until the late stages of the test. For example, the slope of the load–unload–reload response in sample (6-1), did not show any significant change until about the CMOD value reached about 0.10 mm, which was about 50 percent of the total CMOD recorded for that test. The load carrying capacity of the samples reached about 7 lbs, which constituted about 33 percent of the total load carrying capacity of the sample.

Even with the apparent unaffected elastic modulus, diagrams of load versus CMOD of all samples indicated some type of permanent offset in the values of CMOD in each load–unload–reload cycle. The permanent offset could be attributed to the progressive loss of cohesive forces among the cemented sands particles. The loss of cohesion would deter complete closure of the crack, when the sample is unloaded, leading to a permanent offset.

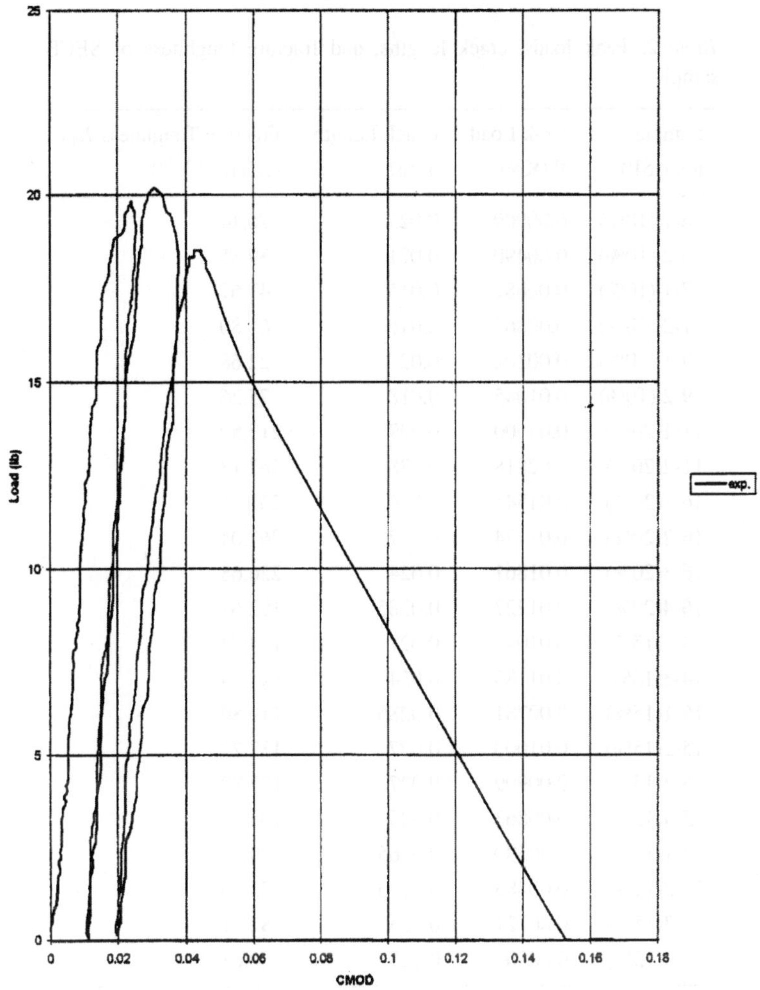


*Figure 5.* Load vs CMOD for sample (15-3) with 15 percent concentration.

## 6. FEM simulation of TPB tests

Using concepts of linear elastic fracture mechanics (LEFM) and the finite element method (FEM), TPB tests were simulated. The pre- and post-peak predictions by LEFM were compared with the corresponding actual TPB tests results.

Results generated in the above FEM simulation steps of TPB tests for samples (6-1), (15-4), (15-3), and (16-4), consistently showed that at the initial stages of the tests, LEFM predictions were in good agreement with the actual test results. Sufficiently reasonable agreement between the two results extended just before the peak loads were reached. Taking the initial notch length as the actual crack length, values of $K_{Ic}$ are determined for the above samples and are shown in Table 3.

Although, the predicted initial load vs CMOD was properly captured, the predictions were far less satisfying after peak loads were reached. This was, nonetheless, expected for the following observations:

(1) The presence of the process zone produced conservative predicted values of $K_{Ic}$.

*Table 2.* Peak loads, crack lengths, and fracture toughness of SECB samples.

| Sample no. (ISP) | Peak Load P (KN) | Crack Length a (m) | Fracture Toughness $K_{Ic}$ ($KN\,m^{-(3/2)}$) |
|---|---|---|---|
| 6-1 (10%) | 0.00909 | 0.023 | 76.14 |
| 6-2 (10%) | 0.00490 | 0.021 | 39.85 |
| 7-1 (10%) | 0.00681 | 0.015 | 47.67 |
| 7-2 (10%) | 0.00863 | 0.016 | 61.80 |
| 9-1 (10%) | 0.00304 | 0.02 | 22.68 |
| 9-2 (10%) | 0.01045 | 0.018 | 75.26 |
| 11-1(20%) | 0.02500 | 0.029 | 217.50 |
| 11-2(20%) | 0.02818 | 0.028 | 162.13 |
| 16-1(20%) | 0.01545 | 0.026 | 201.13 |
| 16-2(20%) | 0.01774 | 0.027 | 267.04 |
| 16-3(20%) | 0.01861 | 0.024 | 226.68 |
| 16-4(20%) | 0.01227 | 0.0285 | 192.67 |
| 14-1(15%) | 0.01045 | 0.027 | 134.44 |
| 14-3(15%) | 0.01185 | 0.024 | 141.34 |
| 15-1(15%) | 0.00781 | 0.0285 | 119.80 |
| 15-2(15%) | 0.01000 | 0.027 | 137.27 |
| 15-3(15%) | 0.00909 | 0.027 | 127.39 |
| 15-4(15%) | 0.00863 | 0.025 | 104.72 |
| 17-1(15%) | 0.00590 | 0.0265 | 79.74 |
| 17-2(15%) | 0.00483 | 0.0275 | 75.46 |
| 17-3(15%) | 0.00727 | 0.023 | 82.91 |
| 17-4(15%) | 0.00500 | 0.025 | 60.63 |

*Table 3.* Experimental and FEM $K_{Ic}$ for four SECB samples.

| Sample no. | Exp. $KN\,m^{-(3/2)}$ | FEM $KN\,m^{-(3/2)}$ |
|---|---|---|
| 6-1 | 79.19 | 76.15 |
| 15-4 | 104.73 | 100.25 |
| 15-3 | 127.39 | 122.17 |
| 16-4 | 192.67 | 181.1 |

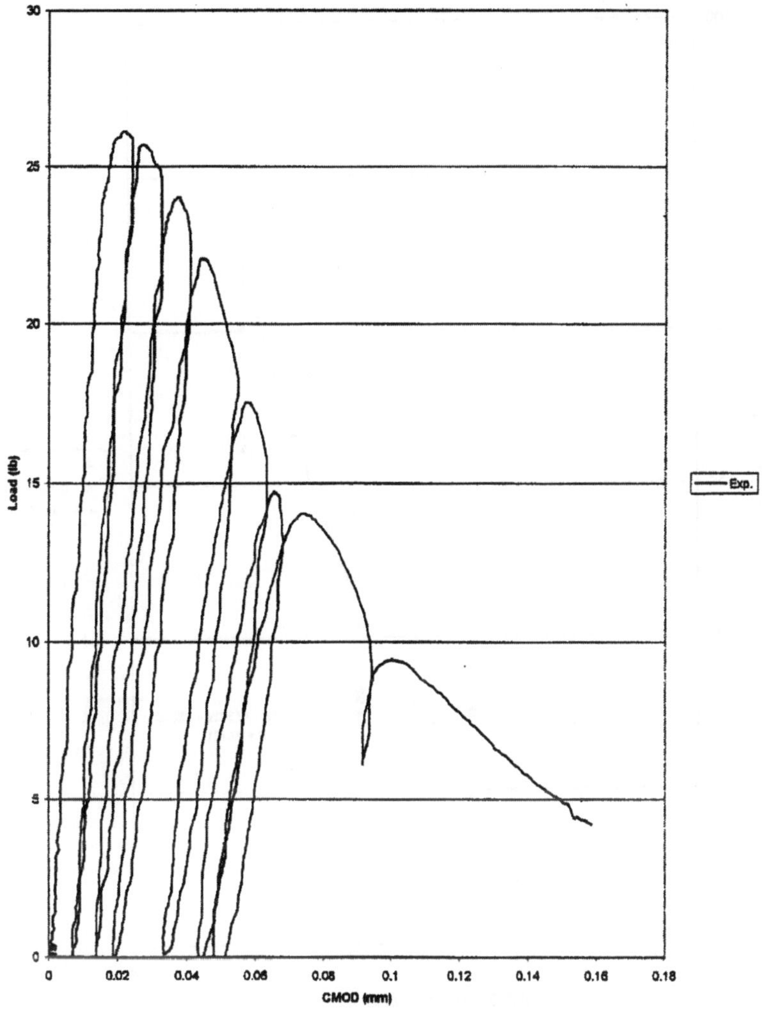


*Figure 6.* Load vs CMOD for sample (16-4) with 20 percent concentration.

(2) As was noted earlier, permanent offsets in CMOD were observed with each additional loading cycle. This permanent offset accumulated with each unload–reload cycle, which meant extra deviation from the linear elastic solution obtained from FEM.

In order to be able to accurately model the post peak response, two approaches were investigated. The first approach was by assuming continuum damage accumulation ahead of the crack. The second approach involved modeling the observed permanent accumulation of the CMOD values after peak resistance. We will discuss both approaches in the following sections.

### 6.1. Continuum damage approach

After peak load was reached, one can argue that the elastic modulus of the material ahead of the crack is somewhat reduced due to damage accumulation in the material. Therefore, in this approach it was assumed that the permanent CMOD offset (or permanent set upon unloading of the specimen) was caused by the permanent damage generated by the microcracking taking

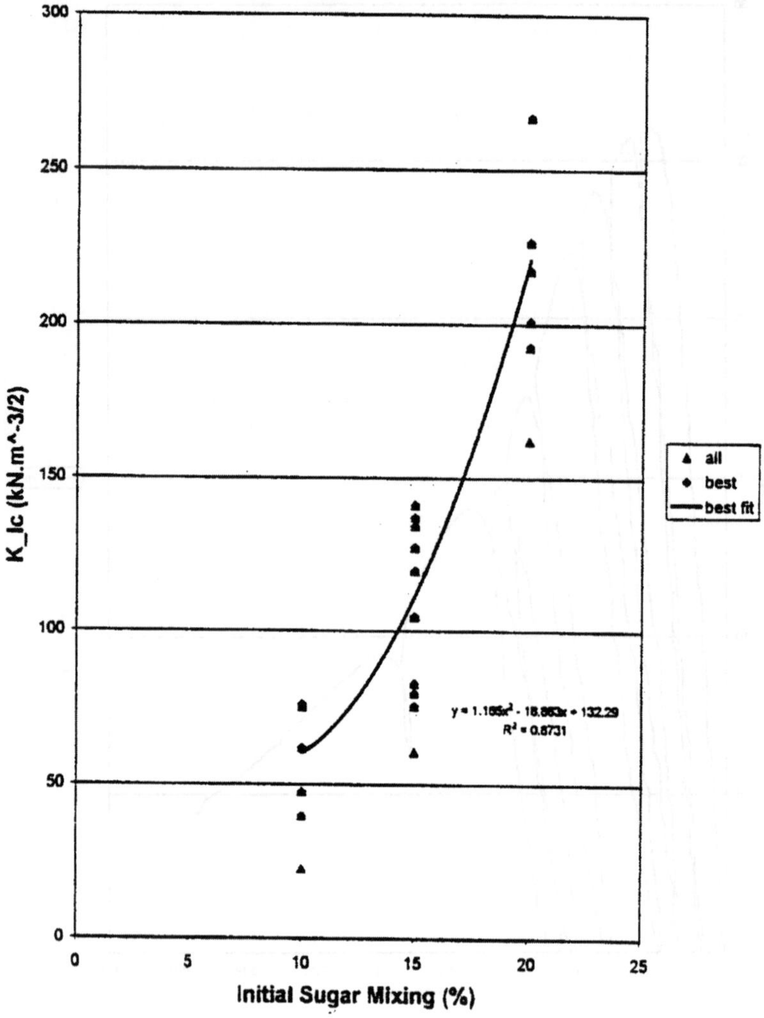


*Figure 7.* $K_{Ic}$ vs initial mixing sugar percent.

place in the material. This was modeled by progressively degrading the elastic modulus of the elements immediately in the vicinity of the crack tip in the FE mesh. This approach seemed logical at first, since the advancement of the crack was supposed to destroy the cohesive forces among the sand particles. FEM simulation was then conducted using the same procedure mentioned earlier. Since the continuum damage was assumed to take place only after peak loads were reached, the necessary simulation steps were conducted for crack lengths greater than the initial ones.

This approach was used in analyzing the results of sample (6-1). The elastic modulus values were assumed to decrease linearly with CMOD or crack extension in two stages. Both values were less than the calculated value of the test, which was ($E_{\mathrm{eff}} = 3.17 \times 10^6$ KPa). The first stage of reduced ($E$) value was $E_1 = 2 \times 10^6$ KPa, while the second stage further reduced $E$ to $E_2 = 2 \times 10^3$ KPa.

For the remaining of the material, $E_{\mathrm{eff}}$ was kept at the original value. In all trials, the post peak response values were calculated using the same procedure outlined earlier. The degradation of $E$ consistently underestimated the measured CMOD values. Figure 9 shows

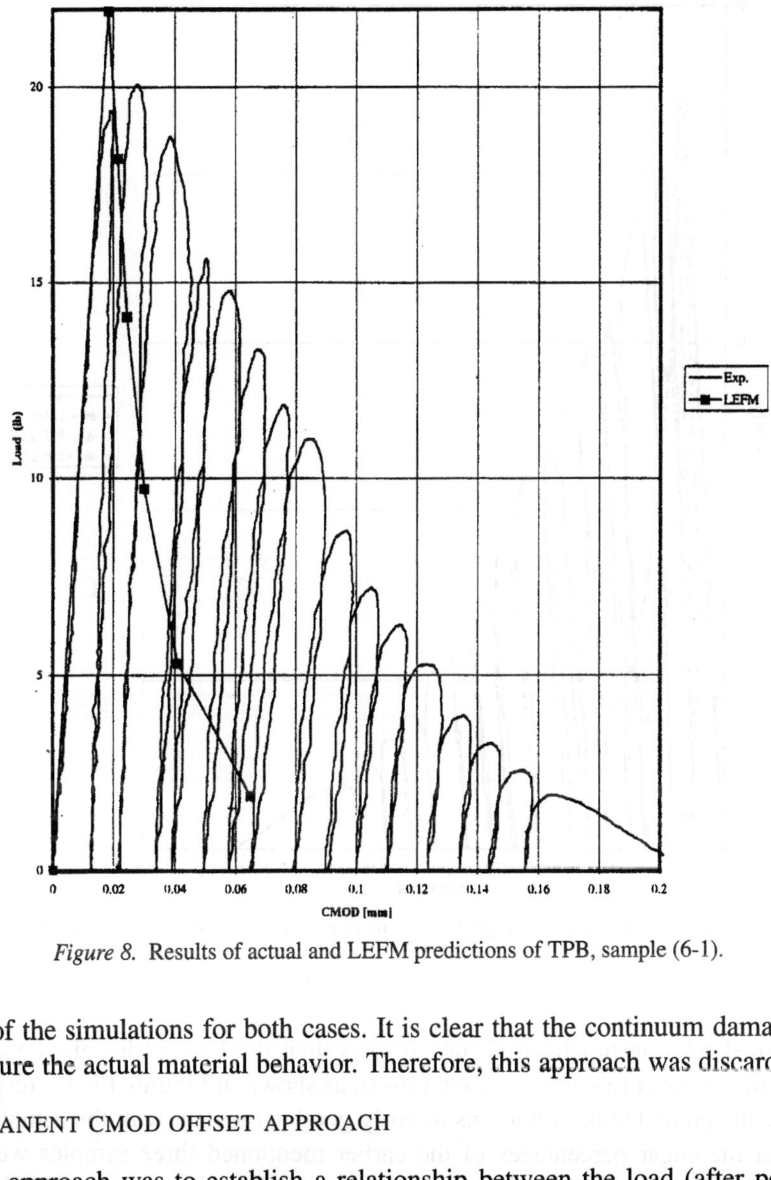


*Figure 8.* Results of actual and LEFM predictions of TPB, sample (6-1).

the results of the simulations for both cases. It is clear that the continuum damage approach did not capture the actual material behavior. Therefore, this approach was discarded.

## 6.2. PERMANENT CMOD OFFSET APPROACH

The second approach was to establish a relationship between the load (after peak) and the permanent CMOD offset. Using the results of sample (6-1) a relationship was obtained by measuring the difference between the calculated CMOD (by FEM) and the measured value corresponding to the percentage decrease in the load-carrying capacity of the sample in the actual test. The difference between the actual results and LEFM predictions of CMOD values of sample (6-1), were taken from Figure 11 and plotted for other tests.

It was found that the relationship to be perfectly linear, as it is clearly shown in Figure 10. The equation describing this relationship could be written as follows:

$$(\text{CMOD})_{\text{of}} = 0.1094\,\text{LF} - 0.0007, \tag{4}$$

where $(\text{CMOD})_{\text{of}}$ is the permanent CMOD offset; LF is the load factor, i.e., $P_{cr}^{pk} - P_{cr}/P_{cr}^{pk}$; $P_{cr}^{pk}$ is the computed critical load at peak, i.e., with the initial crack length; and $P_{cr}$ is the computed critical load at that crack length.

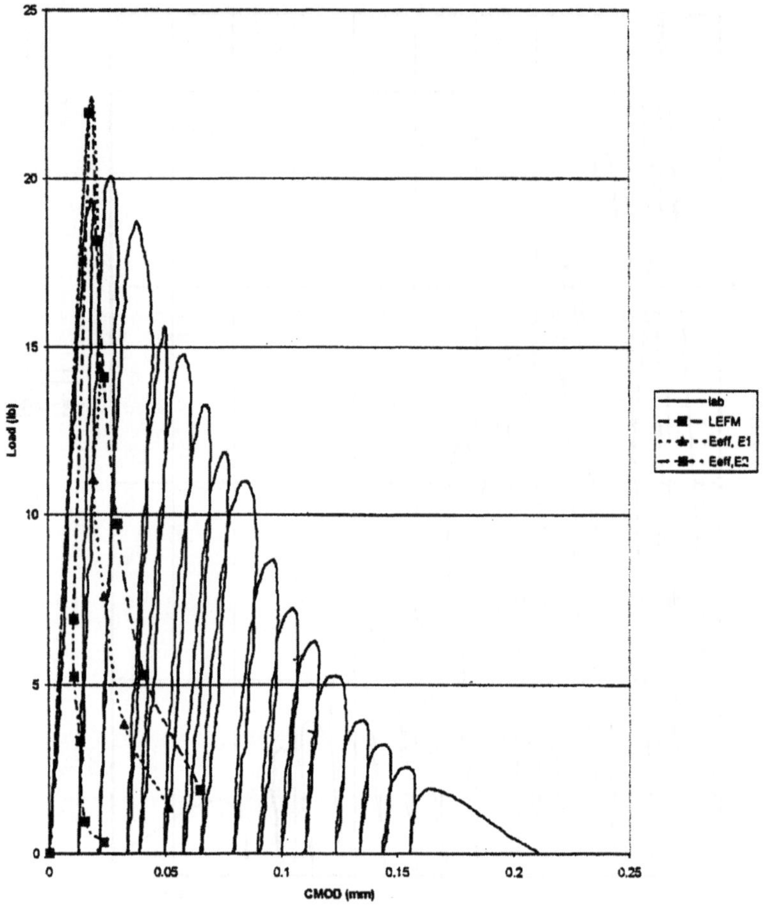


*Figure 9.* Continuum damage predictions of post-peak response, sample (6-1).

The obtained relationship from Figure 10 was then used to predict the behavior of the cemented sand samples (15-4), (15-3), and (16-4), as shown in Figures 12–14, respectively. In these figures, the predicted behavior was in fairly good agreement with the actual test results. Knowing that the sugar percentages of the earlier mentioned three samples were different from that of sample (6-1), this simple procedure suggested that the permanent offset may be a genuine material property. Moreover, it suggested that regardless of its ori ginal resistance to fracture, the cemented sand looses its fracture resistance in the same way when it is cracked.

Jenq and Shah (1985) suggested similar fracture parameters for concrete. They suggested that fracture properties of concrete could be characterized by two parameters, namely a critical stress intensity factor $K_{Ic}$ and a critical crack mouth opening ($CMOD_c$). By examining the experimental results of TPB tests in Figure 8, one can observe that after reaching the peak load values, CMOD values increase without any loss in the load carrying capacity of the sample creating a plateau. The plateau extends until what seemed to be a critical CMOD value is reached. Then, the sample starts to lose its strength and a reduction in the load carrying capacity is observed. Based on the obtained results, it seemed that it is too early to speculate about the actual critical CMOD value for cemented sand. However, the initial results indicate that such a value could be established for the cemented sand material.

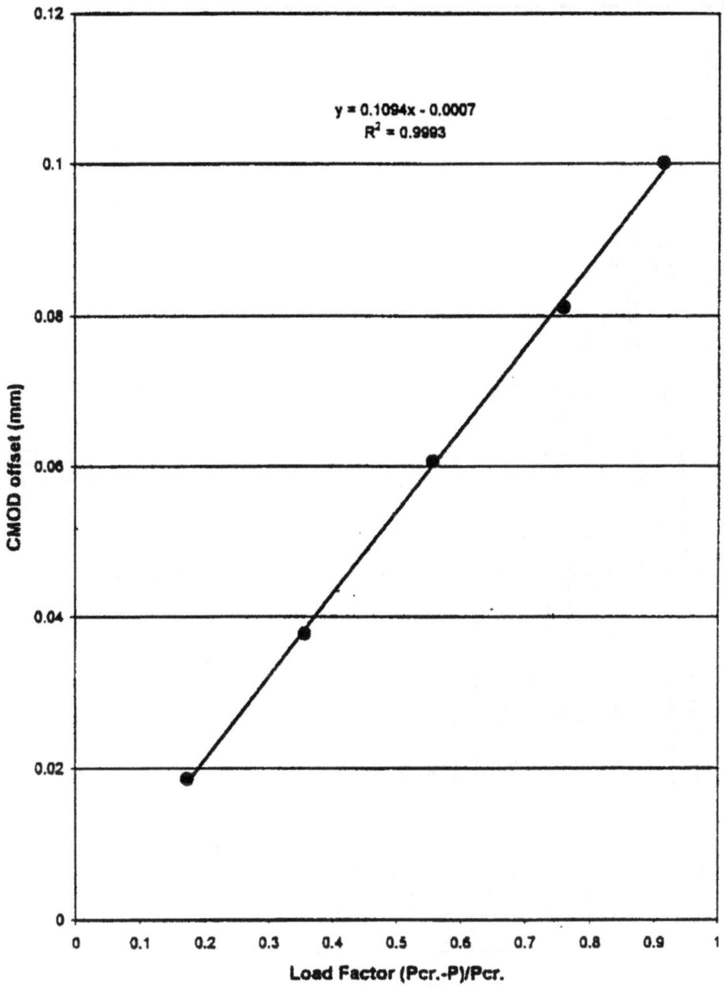


*Figure 10.* Permanent offset CMOD values for sample (6-1).

## 7. Size effect formulation based on critical CMOD model

The property of the permanent set of CMOD which was identified in the previous section seems to provide a promising capability for assessing slope stability. It is however required to perform additional experimental tests to investigate its validity for various specimen sizes and types.

This section highlights the size-effect character of a CMOD-based failure criterion and follows the early work of Bazant. For simplicity, yet without loss of generality, we use a simplified loading condition of a finite width cracked plate. It will be shown that the permanent set in CMOD observed in the TPB tests could be incorporated in a generalized formulation to extend the LEFM concept to be applied to materials similar to cemented sands. The theoretical formulation will include comparison between the predicted failure loads in each of the following three theoretical models: the strength of material (SOM), LEFM, and a critical CMOD model.

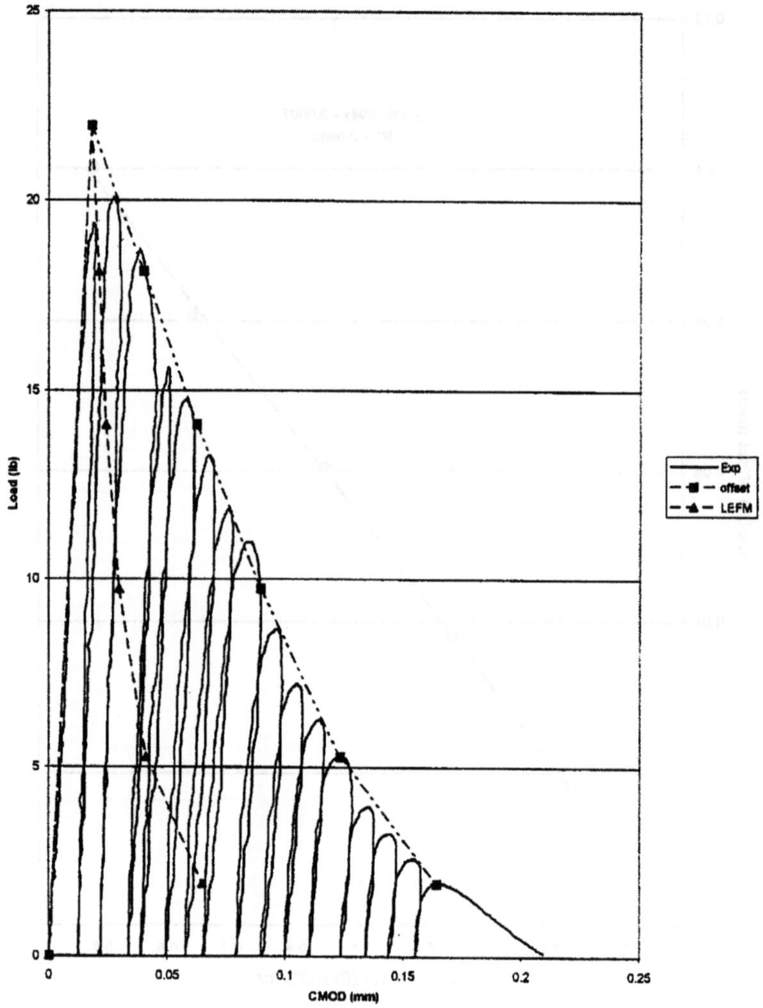


*Figure 11.* Difference between aActual and predicted CMOD values for sample (6-1).

## 7.1. STRENGTH OF MATERIALS (SOM) APPROACH

Consider a plate of finite thickness $t$ and dimensions $2b$, for the width and $2a$ for the crack length. The plate is remotely loaded by $P$. In the strength of material approach it is postulated that failure in the plate occurs, when the stress in the ligament exceeds the material uniaxial (size-independent) tensile strength ($f_t$) and the critical load is computed from

$$\log\left(\frac{\sigma}{f_t}\right) = \log\xi = \text{const.}, \tag{5}$$

where $\xi = (1 - (a/b)) = \text{const.}$ The critical remote stress that ultimately produces failure will be the same for all geometrically similar plates. Thus, the critical load evaluated using the SOM approach is size independent when cracks are present.

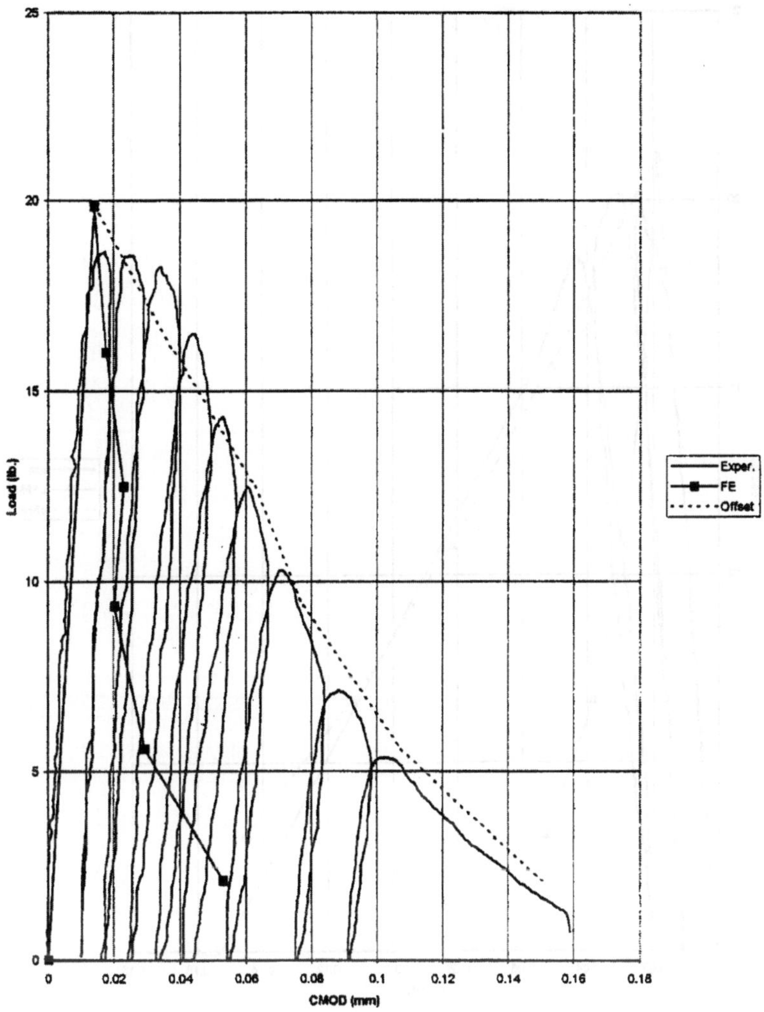


*Figure 12.* Permanent CMOD offset predictions of TPB test, sample (15-4).

## 7.2. LINEAR ELASTIC FRACTURE MECHANICS (LEFM) APPROACH

For the same example plate, using LEFM, failure occurs when $K_{\mathrm{I}}$ (at the crack tip) reaches $K_{\mathrm{I}c}$ (Mode I plane strain fracture toughness) which is also a material property and considered to be size-independent (see Shah et al. (1995)). For the same cracked plate problem, we have

$$\log\left(\frac{\sigma_{cr}^{k}}{f_t}\right) = -\tfrac{1}{2}\log(k) + \log\lambda, \tag{6}$$

where $\lambda$ is a constant, obtained by taking an arbitrary reference specimen for which $\sigma_{cr} = \lambda f_t$, and $k$ is a geometrical similarity variable (i.e., dimensionally similar plates with $a/b =$ const., crack width equal to $2(ka)$ and plate width equal to $2(kb)$). The intersection point, where the SOM and LEFM criteria predict equal value of strength corresponds to the specimen with a $k$ value equal to $(\lambda/\xi)^2$.

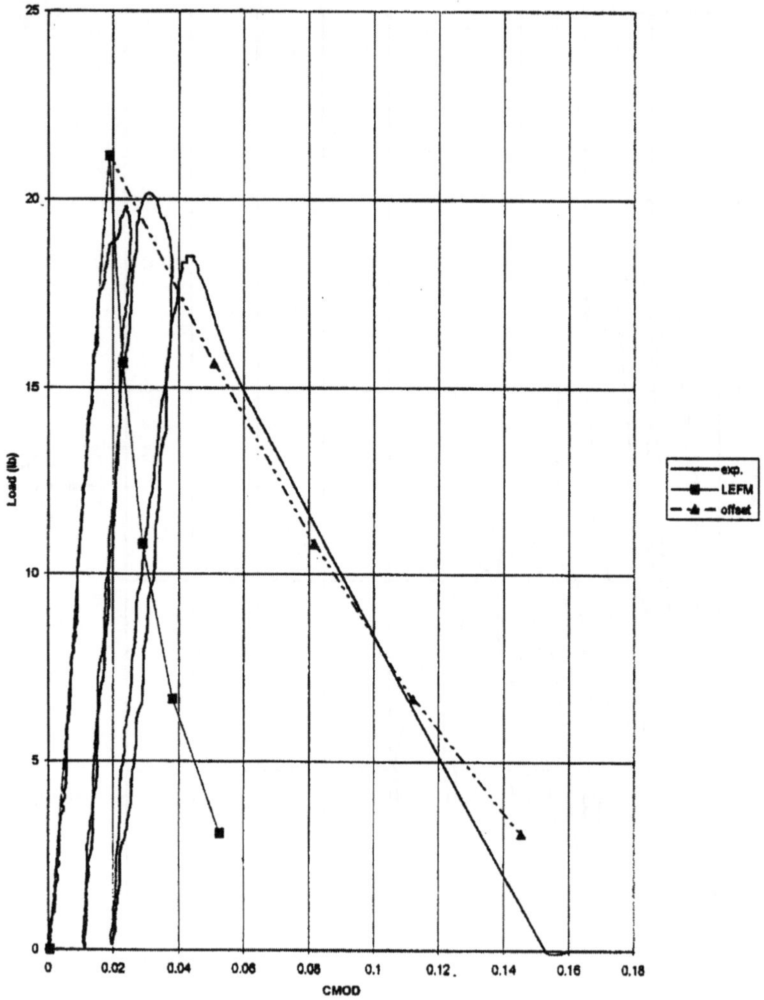


*Figure 13.* Permanent CMOD offset predictions of TPB test, sample (15-3).

### 7.3. FRACTURE MECHANICS APPROACH WITH CMOD FAILURE CRITERION

In the following, it is shown that the permanent CMOD also has a size effect similar to that of the fracture mechanics approach. We first assume a decomposition of the form

$$(\mathrm{CMOD})_T = (\mathrm{CMOD})_e + (\mathrm{CMOD})_p, \tag{7}$$

where $(\mathrm{CMOD})_T$ is the total crack mouth opening displacement; $(\mathrm{CMOD})_e$ is the elastic crack mouth opening displacement, predicted by LEFM; and $(\mathrm{CMOD})_p$ is the permanent crack mouth opening displacement given by

$$(\mathrm{CMOD})_p = \Lambda_1 \sigma + \Lambda_2, \tag{8}$$

where $\Lambda_1$ and $\Lambda_2$ are material constants obtained from simple CMOD-controlled test. Thus, we may write

$$(\mathrm{CMOD})_T = \frac{4\sigma a}{E} g_2 \left(\frac{a}{b}\right) + \Lambda_1 \sigma + \Lambda_2. \tag{9}$$

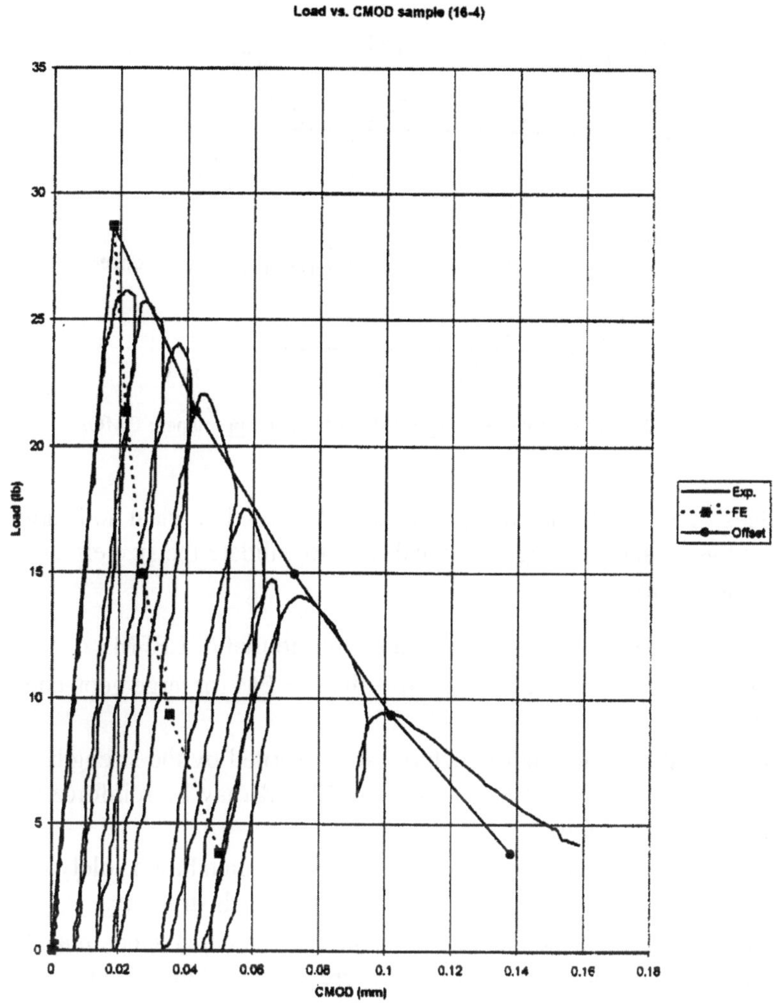


*Figure 14.* Permanent CMOD offset predictions of TPB test, sample (16-4).

Stating that failure occurs when $(\text{CMOD})_T$ reaches $(\text{CMOD})_{cr}$, we obtain

$$(\text{CMOD})_{cr} = \sigma_{cr}\left[\left(\frac{4a}{E}g_2\left(\frac{a}{b}\right)\right) + \Lambda_1\right] + \Lambda_2. \tag{10}$$

Again, for geometrically similar plates, Equation (10) becomes

$$(\text{CMOD})^k_{cr} = \left[\sigma^k_{cr}\left(k\frac{4a}{E}g_2\left(\frac{a}{b}\right) + \Lambda_1\right) + \Lambda_2\right] \tag{11}$$

and substituting for the reference case $\sigma_{cr} = \lambda f_t$, we finally have

$$\log\left(\frac{\sigma}{f_t}\right) = \log\lambda + \log(1+\Gamma) - \log(k+\Gamma). \tag{12}$$

Figure 15 shows diagrams of the failure criteria using the CMOD approach, strength of material, and LEFM criteria. It is clear from the figure that the CMOD criterion produces

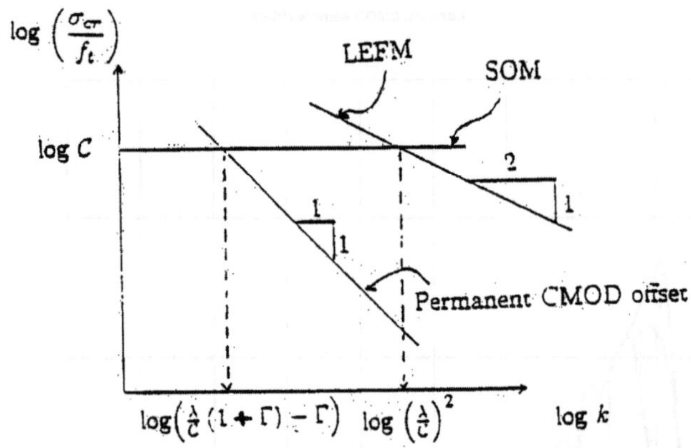


*Figure 15.* Failure criteria using SOM, LEFM, and permanent CMOD.

decreasing values of $\sigma_{cr}$ with increasing values of $k$. This is a clear indication of the size effect. From the figure, and pending additional tests to validate the size effect, we may reach the following conclusions:

(1) The CMOD criterion shows a size effect similar to that of LEFM.
(2) The CMOD criterion is more conservative than LEFM for large geotechnical (or other) structures.
(3) The concept stating that small structures are governed by the strength of material criterion and large structures are governed by LEFM may be reinforced by statements limiting the crack opening.
(4) The CMOD criterion also shows that laboratory test data based solely on strength may grossly over-estimate the load carrying capacity of such structures.
(5) In practical application, the stability of slopes in cemented sands may be assessed by monitoring crack openings at the top of the crest of the slope. The CMOD could be the closest variable that is easily observed and could give some indication on slope stability.

## 8. Conclusions

The fracture behavior of an artificially cemented sand was investigated. Three-point bending (TPB) tests were conducted on a single-edge-cracked-beam (SECB) cemented sand samples. While LEFM predictions are adequate in the initial stage of the response, deviation from LEFM becomes important in the post-peak regime. Consistent measurements of the plane strain fracture toughness $K_{Ic}$ of cemented sand were obtained and a relationship between degree of cementation and $K_{Ic}$ was established.

LEFM with continuum damage modelling (ahead of the crack) failed to capture the post-peak regime. Before complete failures, all SECB samples showed no visible signs of crack propagation. However, crack propagation was indicated by the decline in the load-carrying capacity of the sample. A fracture process zone ahead of the crack was assumed to form and actual separation of crack surfaces followed it prior to failure.

Elastic continuum damage coupled with LEFM failed to predict the inelastic post-peak behavior of SECB samples.

A permanent CMOD offset is identified in the response after each load–unload–reload cycle. A model, based on permanent CMOD offset, was shown to provide excellent predictions of the post-peak regime.

For a CMOD-based criterion it is shown that a size-effect character exists for geometrically similar structures. such model is actually more conservative than LEFM, when used in design of large geotechnical structures. The CMOD-based size effect law showed that design values of strength, based solely on laboratory test specimens, would grossly over-estimate the load carrying capacity of larger structures.

Further studies of fracture behavior of cemented sands are needed. The effects of variables such as water content (for moist sand), other cementation agents, and confining pressure should be of interest for engineers who use FM parameters in design of structures of cemented sands. The effect of confinement should be of particular interest, since it is well establish fact that soils characteristics are affected by confining stress. In order to study such effect, special testing chamber where confinement could be regulated has to be designed and tested.

## References


Anderson, T.L. (1995). *Fracture Mechanics, Fundamentals and Applications*, CRC Press, Ann Arbor, (2nd edn).

Ayari, M.L. (1994). *Simex, User's Manual*, University of Manitoba, Canada.

Babuska, I. and Miller, A. (1984). The post-processing approach to finite element method – Part II: The calculation of stress intensity factors. *International Journal for Numerical Methods in Engineering* **20**, 111–129.

Barsoum, R.S. (1974). Application of quadratic isoparametric finite element in linear fracture mechanics. *International Journal of Fracture* **10**, 603–605.

Bishop, A.W. (1967). Shear strength of OC clays. *Proceedings of the Geotechnical Conference* **2**, 153–154.

Bishop, A.W. and Little, A.L. (1967). The influence of the size and orientation of the sample on the apparent strength of London clay at Malden, Essex. *Proceedings of the Geotechnical Conference* **1**, 89–96.

Bjerrum, L. (1967). Progressive failure in slopes of OC plastic clay and clay shales. *Journal of Soil Mechanics and Foundations, ASCE* **93** (SM5), 3–49.

Bjerrum, L., Nash, J.K.T.L., Kennard, R.M. and Gibson, R.E. (1972). Hydraulic fracturing and field permeability testing. *Géotechnique* **22**(2), 319–332.

Briaud, J.L. (1986). Pressuremeter and foundation design, *ASCE Specialty Conference – The Use of In Situ Tests in Geotechnical Engineering*, Blacksburg, VA, June, 74–85.

Broek, D. (1991). *Elementary Engineering Fracture Mechanics*, Kluwer Academic Publishers (4th edn).

Butterfield, (1986), Personal communication with Stein Sture.

Cai, W.M., Murti, V. and Valliappan, S. (1990). Slope stability analysis using fracture mechanics approach. *Theoretical and Applied Fracture Mechanics* **12**, 261–281.

Erdogan, F. and Sih, G.C. (1963). On the crack extension in plates under plane loading and transverse shear. *Journal of Basic Engineering* **85**, 519–527.

Fang, H.-Y. (1994). Cracking and fracture behavior of soils. *ASCE, Geotechnical Special Publication No. 43 – Fracture Mechanics, Applied to Geotechnical Engineering, Oct. 9–13* (Edited by L.E. Vallejo and R.Y. Liang), Atlanta, Georgia, 102–117.

Fang, H.-Y., Mikrudis, G.K. and Pamukcu, (1989). Fracture behavior of compacted fine-grained soils. *ASTM STP 1020 – Fracture Mechanics, Perspectives and Directions* (Edited by R.P. Wei and R.P. Gangloff) Philadelphia, 659–667.

Gould, J.P. (1960). A study of shear failure in certain tertiary marine sediments. *Proceedings of ASCE Conference on the Shear Strength of Soils*, Boulder, Colorado, 615–641.

Griffith, A.A. (1921). The phenomena of rupture and flow in solids. *Philosophical Transaction of the Royal Society of London A* **221**, 163–197.

Haberfield, C.M. and Johnson, I.W. (1989). Relationship between fracture toughness and tensile strength for geomaterials. *Proceedings of 12th International Conference of Soil Mechanics and Foundations*, Rio de Janeiro, Brazil, Vol. I, 47–52.

Haberfield, C.M. and Johnson, I.W. (1990). Determination of the fracture toughness of a saturated soft rock. *Canadian Geotechnical Journal* **27**(3), 276–284.

Harison, J.A. and Hardin, B.O. (1994). Cracking in clays: Solutions to problems in earth structures. *International Journal for Numerical and Analytical Methods in Geomechanics* **18**, 467–484.

Harison, J.A., Hardin, B.O. and Mahboub, K. (1994). Fracture toughness of compacted cohesive soils using ring test. *Journal of Geotechnical Engineering, ASCE* **120**(5), 872–889.

Henshell, R.D. and Shaw, K.G. (1975). Crack tip elements are unnecessary. *International Journal of Numerical Methods in Engineering* **9**(3), 495–509.

Hussain, M.A., Pu, S.L. and Underwood, J.H. (1974). Strain energy release rate for a crack under combined mode I and mode II. *Fracture Analysis, ASTM, STP* **560**, 2–28.

Ingraffea, A.R. and Manu, C. (1980). Stress-intensity factor computations in three dimensions with quarter-point elements. *International Journal for Numerical Methods Engineering* **15**(10), 1427–1445.

Irwin, G.R. (1957). Analysis of stresses and strains near the end of a crack traversing a plate. *Journal of Applied Mechanics, Transaction ASME* **24**, 361–364.

Jenq, Y.S. and Shah, S.P. (1985). A two parameter fracture model for concrete. *Journal of Engineering Mechanics Division, ASCE* **111**(4), 1227–1241.

Lajtai, E.Z., Carter, B.J. and Ayari, M.L. (1990). Brittle fracture criteria for compression. *Engineering Fracture Mechanics* **37**(1), 59–74.

Lee, F.-H., Lo, K.-W. and Lee, S.-L. (1988). Tension crack development in soils. *Journal of Geotechnical Engineering, ASCE* **114**(8), 915–930.

Lo, K.Y. (1970). The operational strength of fissured clays. *Géotechnique* **20**(1), 57–74.

Lo, K.Y. (1972). An approach to the problem of progressive failure. *Canadian Geotechnical Journal* **9**, 407–429.

Lutton, R.J. (1969). Fractures and failure mechanics in loess and applications to rock mechanics. *Research Report S-69-1, US Army Corps of Eng. Waterways Experiment Station, Corps of Eng.*, Vicksburg, Miss.

Mitchell, J.K. (1976). *Fundamentals of Soil Behavior*, John Wiley and Sons, New York.

Morgenstern, N.R. and Tchalenko, J.S. (1967a). Microscopic structures in Kaolin subjected to direct shear. *Géotechnique* **17**, 309–328.

Morgenstern, N.R. and Tchalenko, J.S. (1967). Microstructural observations on shear zones from slip in natural clays. *Proceedings of Geotechnical Conference* **2**, 146–152.

Mould, J.C., Jr. (1983). *Stress Induced Anisotropy and the Evaluation of Multi-surface Elasto-plastic Material Model*, PhD thesis, University of Colorado, Boulder.

O'Rourk, T.D. and Crespo, E. (1988). Geotechnical properties of cemented volcanic soil. *Journal of Geotechnical Engineering, ASCE* **114**(10), 1126–1147.

Owen, D.R.J. and Hinton, E. (1983). *Engineering Fracture Mechanics: Numerical Methods and Applications*, Pineridge Press, Swansea, UK.

Palmer, A.C. and Rice, J.R. (1973). The growth of slip surface in the progressive failure of OC clay. *Proceedings of the Royal Society of London* **A332**, 527–548.

Rice, J. (1968). A path independent integral and the approximation of strain concentration by notches and cracks. *Journal of Applied Mechanics, Transaction of ASME* **35**, 379–386.

Rizkallah, J. (1977). Stress-strain behavior of fissured stiff clays. *9th International Conference on Soil Mechanics and Foundations* **1**, 267–270.

Saada, A.S., Chudnovsky, A. and Kennedy, M.R. (1985). A fracture mechanics study of stiff clays. *Proceedings of the 11th International Conference in Soil Mechanics and Foundations* **2**, 637–640.

Saouma, V., Broz, J.J. and Brühwiler, E. (1991). Effect of aggregate and specimen size on fracture properties of dam concrete. *Journal of Materials in Civil Engineering, ASCE* **3**(3), 204–218.

Shah, S.P., Swartz, S.E. and Ouyang, C.-S. (1995). *Fracture Mechanics of Concrete*, John Wiley and Sons, New York.

Shih, C.F., de Lorenzi, H.G. and German, M.D. (1976). Crack extension modeling with singular quadratic isoparametric elements. *International Journal of Fracture* **12**, 647–651.

Sih, G.C. (1973). *Hand Book of Stress-Intensity Factors for Researchers and Engineers*, Institute of Fracture and Solid Mechanics, Lehigh University, Bethlehem, Pa.

Sih, G.C. (1974). Strain energy factors applied to mixed mode crack problems. *International Journal of Fracture* **10**, 305–000.

Sih, G.C., Paris, P.C. and Erdogan, F. (1962). Crack-tip intensity factors for plane extension and plate bending problems. *Journal of Applied Mechanics, Transaction of ASME* **29**, 306–312.

Sih, G.C., Paris P.C. and Irwin G.R. (1965). On cracks in rectilinearly anisotropic bodies. *International Journal of Fracture Mechanics* **1**(3), 189–203.

Sih, G.C. and Liebowitz, H. (1968). Mathematical theories of brittle fracture. *Fracture an Advanced Treatise* (Edited by H. Liebowitz) **2**, 68–188.

Simons, N.E. (1967). Discussion on shear strength of stiff clay. *Proceedings of the Geotechnical Conference* **2**, 159–160.

Sitar, N., Clough, G.W. and Bachus, R. (1980). *Behavior of Weakly Cemented Soil Slopes under Static and Seismic Loading Conditions*, final report prepared for the United States Geological Survey, Department of Interior. (June).

Skempton, A.W. (1964). Long-term stability of clay slopes. *Géotechnique* **14**, 77–101.

Skempton, A.W. (1966). Some observations on tectonic shear zones. *Proceedings of 1st International Congress on Rock Mechanics*, Lisbon, Portugal, 329–335.

*Standard Test Method for Plane Strain Fracture Toughness of Metallic Materials*, ASTM Designation E399-74, Part 10, ASTM Annual Standards.

Sture, S., Swanson, R., Al-Shibli, K., Jeremic, B. and Zhang, R. (1994). *Mechanics of Granular Materials: Experiments*, University of Colorado Research Report, NASA No. NAS8-38779, August, 305.

Taylor, D.W. (1948). *Fundamentals of Soil Mechanics*, John Wiley and Sons.

Terzaghi, K. (1936). Stability of slopes of natural clay. *Proceedings of 1st International Conference of Soil Mechanics and Foundations* **1**(G-7), 161–165.

Tracey, D.M. (1977). Discussion – On the use of isoparametric finite elements in linear fracture mechanics – R.S. Barsoum. *International Journal for Numerical Methods in Engineering* **11**, 401–402.

Vallejo, L.E. (1985). Fissure interaction and progressive failure of slopes. *Proceedings of 11th International Conference of Soil Mechanics and Foundations*, San Francisco **4**, 2353–2356.

Vallejo, L.E. (1986). Mechanics of crack propagation in stiff clays. *Proceedings of ASCE Specialty Conference (Geotechnical Aspects of Stiff and Hard Clays)*, Seattle, 14–27.

Vallejo, L.E. (1987). The influence of fissures in stiff clay subjected to direct shear. *Géotechnique* **37**(1), 69–82.

Vallejo, L.E. (1989). Fissure parameters in stiff clays under compression. *Journal of Geotechnical Engineering, ASCE* **115**(9), 1303–1317.

Westergaard, H.M. (1939). Bearing pressure and cracks. *Journal of Applied Mechanics, Transaction of ASME* **61**, A49–A53.

Williams, M.L. (1957). On the stress distribution at the base of a stationary crack. *Journal of Applied Mechanics, Transaction of ASME* **24**(109), 109–114.

Wroth, C.P. (1986). Personal communication with Stein Sture.

Zienkiewicz, O.C. and Taylor, R.L. (1989). The finite element method, Vol. 1: *Basic Formulations and Linear Problems*, McGraw Hill, London (4th edn).

*International Journal of Fracture*

# AUTHOR INDEX, VOL. 95 (1999)

MIX
Papier aus verantwortungsvollen Quellen
Paper from responsible sources
FSC® C105338

If you have any concerns about our products,
you can contact us on
**ProductSafety@springernature.com**

In case Publisher is established outside the EU,
the EU authorized representative is:
**Springer Nature Customer Service Center GmbH**
**Europaplatz 3, 69115 Heidelberg, Germany**

Printed by Libri Plureos GmbH
in Hamburg, Germany